WATER AND BIOLOGICAL MACROMOLECULES

TOPICS IN MOLECULAR AND STRUCTURAL BIOLOGY

Series Editors

Stephen Neidle
Institute of Cancer Research
Sutton, Surrey, UK

Watson Fuller
Department of Physics
University of Keele, UK

Recent titles

Protein–Nucleic Acid Interaction
Edited by Wolfram Saenger and Udo Heinemann (1989)

Calcified Tissue
Edited by David W. L. Hukins (1989)

Oligodeoxynucleotides: Antisense Inhibitors of Gene Expression
Edited by Jack S. Cohen (1989)

Molecular Mechanisms in Muscular Contraction
Edited by John M. Squire (1990)

Connective Tissue Matrix, Part 2
Edited by David W. L. Hukins (1990)

New Techniques of Optical Microscopy and Microspectroscopy
Edited by Richard J. Cherry (1990)

Molecular Dynamics: Applications in Molecular Biology
Edited by Julia M. Goodfellow (1990)

Topics in Molecular and Structural Biology
Volume 17

Water and Biological Macromolecules

Edited by

ERIC WESTHOF

Institut de Biologie Moléculaire et Cellulaire
Centre National de la Recherche Scientifique
Strasbourg, France

First published 1993 by
THE MACMILLAN PRESS LTD
Houndmills, Basingstoke, Hampshire RG21 2XS
and London
Companies and representatives
throughout the world

ISBN 0–333–55116–8

A catalogue record for this book is available
from the British Library

Printed and bound in Great Britain by
Mackays of Chatham PLC, Chatham, Kent

Contents

Preface ix

The Contributors xii

Part 1 Water

1 Water structure
H. F. J. Savage 3
1 Introduction 3
2 Structure determination: Diffraction studies 5
3 Liquid water 11
4 Ice polymorphs 11
5 Small crystal hydrates 19
6 Structural regularities between water molecules 21
7 Water structure around biological molecules 30
8 Summary 39

2 Thermodynamic and dynamic properties of water
H.-D. Ludemann 45
1 Introduction 45
2 Density and isothermal compressibility of liquid water 46
3 Dynamic properties 48
4 Rotational correlation times and the lifetime of hydrogen bonds in water 50
5 Aqueous solutions of simple hydrophobic solutes 55
6 Concluding remarks 56

Part 2 Proteins

3 Hydration of amino acids in protein crystals
J. M. Goodfellow, N. Thanki and J. M. Thornton 63
1 Introduction 63

2 Methods 64
3 Results 67
4 Discussion 90

4 Water structure of crystallized proteins: High-resolution studies
M. Frey 98
1 Introduction 98
2 Crystallographic methods 101
3 Other methods 129
4 Perspectives 135

5 Hydration of protein secondary structures — the role in protein folding
C. Y. Sekharudu and M. Sundaralingam 148
1 Introduction 148
2 Hydration of alpha-helices 149
3 Hydration of 3_{10} helices and reverse turns 154
4 Hydration of β-sheets 155
5 Conclusions 159

Part 3 Nucleic Acids

6 Molecular dynamics simulations on the hydration, structure and motions of DNA oligomers
D. L. Beveridge, S. Swaminathan, G. Ravishanker, J. M. Withka, J. Srinivasan, C. Prevost, S. Louise-May, D. R. Langley, F. M. DiCapua and P. H. Bolton 165
1 Introduction 165
2 MD simulation 171
3 MD studies on prototype systems 173
4 MD studies on DNA *in vacuo* 183
5 MD studies on DNA with solvent 189
6 Restrained MD and refinement 200
7 MD studies of DNA variants and complexes 203
8 Brownian dynamics simulations 209

7 Structural water bridges in nucleic acids
E. Westhof 226
1 Introduction 226
2 Hydration around phosphate groups 229
3 5′-Phosphate–water–base bridges 231
4 3′-Phosphate–water–sugar bridges 231
5 3′-Phosphate–water–base bridges 231

6 Other phosphate–water–base bridges 233
7 Other phosphate–water–sugar bridges 235
8 Sugar–water–base bridges 235
9 Base–water–base bridges 238
10 Conclusions 239

8 Hydration sites and hydration bridges around DNA helices
F. Vovelle and J. M. Goodfellow 244
1 Introduction 244
2 Hydration site approach 245
3 Monte Carlo calculations 253
4 Discussion 260

9 Light scattering spectroscopy studies of the water molecules in DNA
N. J. Tao 266
1 Introduction 266
2 Raman and Brillouin spectroscopies 268
3 Structure of DNA hydration shell studied by Raman spectroscopy 269
4 Dynamics of DNA hydration shell studied by Brillouin spectroscopy 277
5 Conclusions 289

Part 4 Polysaccharides and Lipids

10 Polysaccharide interactions with water
S. Pérez 295
1 Introduction 295
2 Analysis of the structural methods 296
3 Water location in crystalline polysaccharides 301
4 Water in the crystalline polymorphism of polysaccharides 302
5 Difficulties and limits of water location in crystalline polysaccharides 313
6 Discussion and conclusion 315

11 The role of structural water molecules in protein–saccharide complexes
Y. Bourne and C. Cambillau 321
1 Introduction 321
2 Periplasmic binding proteins: Complexes with monosaccharides and disaccharides 322

3 Enzymes 324
4 Lectins 326
5 Discussion and Conclusion 334

12 Lipid hydration
G. Cevc 338
1 Lipids 338
2 Sources of lipid hydration 340
3 Temperature effects 348
4 Ion and solute effects 349
5 Modelling lipid–water interactions 351
6 Consequences of lipid hydration 358
7 Lipid aggregation in water 373
8 Hydration force 374
9 Relevance of lipid hydration 376

Part 5 Thermodynamics

13 Hydration forces
C. J. van Oss 393
1 Introduction 393
2 Theory 394
3 Negative interfacial tensions and polar repulsion 401
4 Quantitative expression of hydrophilicity and 'hydrophobicity' 403
5 Hydration orientation 407
6 Consequences and examples of hydration forces 413
7 Solvation forces in non-aqueous media 419
8 How specific interactions overcome hydration forces 422

14 Solvation thermodynamics of biopolymers
A. Ben-Naim 430
1 Introduction 430
2 Definitions 431
3 The ingredients of ΔG^*_α 436
4 Some illustrative numerical examples 441
5 Critique of the group additivity assumption for the solvation Gibbs energy 446
6 Examples of biochemical processes affected by solvation 450
7 Solvation and structural changes in the solvent 455
8 Conclusion 458

Index 461

Preface

The Makonde people of Mozambica, well-known for their wood sculptures, tell the following legendary story of their origins. A long time ago came to the Rovuma valley a living being, not yet quite human. He did not eat, drank little, never went into the river and did not cut his hair. Once out of boredom, he took a piece of wood and carved something looking like him. In the evening, he left the wooden sculpture standing on the earth. The next morning, the statue was alive and it was a woman. They both swam in the river and settled next to the river. Later, the woman gave birth, but the child was still-born. They moved to another spot close to the water and, again, they had a still-born baby. They moved several times to different sites along the river without any more luck. Finally, they decided to run away from water, took refuge on a dry table-land, and the woman gave birth to a large descendance. From then on, Makonde built their villages one-hour's walking distance from a water spot.

This legend renders the constant love–hate story of life and water. Indeed, although life, as we know it, is inconceivable without water, life could only emerge and develop together with systems for controlling and confining water. Whatever the nature of the prebiotic soup (warm little ponds or clays), alternation of wet and dry periods is required. From lipid bilayers to the amniotic egg, life evolved around the containment of water. This subtle and parsimonious use of water constitutes a recurrent theme in the present book.

The book is organized in five sections. The first describes the peculiar and narcissistic properties of water in the solid and liquid states. Analysis of the hydrogen bonded water networks in the ice polymorphs indicate that they are characterized by a maximization of the number of hydrogen bonds and a minimization of the short-range repulsive restrictions. Although the formation and breaking of the hydrogen bonds occur on the picosecond time scale, the three-dimensional hydrogen bond network has a retarding influence on the rotational and translational molecular mobility, as measured by nuclear magnetic resonance techniques. A vision of water in constant conflict between order and chaos illustrates the links between structure and movement and stresses how molecular dynamics is constrained by molecular geometry.

The three central sections treat the main types of biological macromolecules: proteins, nucleic acids, polysaccharides, and lipids. They all exploit rather extensively the data obtained by X-ray diffraction analysis of crystals. The fundamental driving force for the folding of a macromolecule into a three-dimensional architecture with potential biological functions is the hydrophobic effect, arising much less from the macromolecule's phobia of water than from the self-loving properties of water molecules. Depending on the chemical entities constituting the macromolecule, there is a constant interplay between the hydrophobic effect and the specific interactions of hydrophilic sites on the macromolecule with water molecules. The wealth of structural information gathered by crystallographers on protein crystals allows extensive comparative studies on the arrangement of water molecules around amino acids and elements of secondary structures. Furthermore, careful analysis of those static tertiary structures led to understanding of the role of water molecules in the formation and denaturation of the main repetitive secondary structures of proteins. In contrast to the case of proteins, the folding and stability of nucleic acids are not principally governed by the hydrophobic effect, as indicated by the thermodynamic characteristics of their denaturation. Since the discovery of the A to B transition in DNA made by Franklin and Gosling and which was so instrumental for the establishment of the double-stranded DNA helix by Watson and Crick, the role of water in stabilizing or inducing DNA conformations has been repeatedly stressed. The chapters describe the beautiful and striking arrangements of water molecules around nucleic acids both from the point of view of their observation, stressing the geometry and periodicity of nucleic acid helices, and of their theoretical intepretation, emphasizing movements and water exchanges. Polysaccharides and lipids delineate extreme cases. Stereochemistry imposes the precise spacings and arrangements of hydroxyl groups in polysaccharides so that hydrophilic interactions with water molecules are facilitated. Bilayers formed by polar lipids, although suspectible to specific interactions, display striking electrostatic properties partly governed by geometry. In the last section, theoretical approaches for understanding the macroscopic observations and integrating the microscopic descriptions of the previous sections are developed. Thus, the nature and roles of hydration forces in macromolecular complexation and cell–cell interactions are explained, as well as phenomena such as entropy-enthalpy compensation and the thermodynamic treatment of water bridging.

Most of the chapters accentuate the structural aspects of water molecules around biological macromolecules. Water is so pervasive that attempting to cover all of its features becomes an impossible task. Each contributor has taken special care to refer to other review articles or books, either more general or specialized, so that the interested reader can complement his knowledge. Finally, with respectful awe, mention should be made to the books and reviews edited by F. Franks, *Water: A Comprehensive Treatise*, published by Plenum Press, and *Water Science Reviews*, published by Cambridge University Press.

The power of iconography is well appreciated outside the realm of science. Indeed, scientists consider pictures just as mirrors of nature, more or less accurate, and besides, there is always the danger of equating ideas with descriptions. Nevertheless, the insertion of a large number of figures was actively pursued in each contribution because descriptions of molelar patterns, of the way water molecules repeatedly arrange themselves around macromolecules, while certainly not helping to organize our thought, constitute a rich soil on which our insight and intuition feed and grow. At the onset of each section, an artistic drawing tries to convey the tension between geometry and dynamics so strongly inherent to the nature of water. I thank Marc Westhof (Hoboken, N. J.) for letting me publish these drawings in black-and-white and not with their original colours.

I thank all the authors for their contributions and I hope they learnt as much from writing as I did from editing.

Strasbourg, 1992 E. W.

The Contributors

A. Ben-Naim
Department of Physical Chemistry
The Hebrew University of Jerusalem
Jerusalem 91904
Israel

D. L. Beveridge with S. Swaminathan, G. Ravishanker, J. M. Withka, J. Srinivasan, C. Prevost, S. Louise-May, D. R. Langley, F. M. DiCapua and P. H. Bolton
Wesleyan University
Department of Chemistry
Middletown, CONN 06457
USA

C. Cambillau with Y. Bourne
Laboratoire CCMB Faculté Nord
Boulevard P. Dramard
13326 Marseille CDX 15
France

G. Cevc
Medizinische Biophysik
Klinikum r.d.I., TU München
Ismaningerstr. 22
8000 München 80
Germany

M. Frey
Laboratoire de Cristallographie et de Cristallogénèse des Protéines
DSV/DIEP/LCCP, CENG 85X
38401 Grenoble CDX 01
France

J. M. Goodfellow with N. Thanki and J. M. Thornton
Department of Crystallography
Birkbeck College
University of London
Malet Street
London WC1E 7HX
UK

H. D. Lüdemann
Institut für Biophysik und Physikalische Biochemie
Universität Regensburg
Postfach 397
8400 Regensburg
Germany

S. Pérez
Institut National de la Recherche Agronomique
Rue de la Géraudière, BP 527
44026 Nantes CDX
France

H. F. J. Savage
Department of Chemistry
University of York
Heslington
York YO1 5DD
UK

M. Sundaralingam with C. Sekkarudu
Department of Chemistry
The Ohio State University
120 West 18th Avenue
Columbus
OH 43210-1173
USA

N. Tao
Department of Physics
Florida International University
Miami
FL 33199
USA

C. J. van Oss
Department of Microbiology
University at Buffalo
School of Medicine and Biomedical Sciences
207 Sherman Hall
Buffalo
New York 14214-3078
USA

F. Vovelle
Centre de Biophysique Moléculaire
Centre National de la Recherche Scientifique
1A Avenue de la Recherche Scientifique
45701 Orléans CDX 2
France

E. Westhof
Institut de Biologie Moléculaire et Cellulaire
Centre National de la Recherche Scientifique
15 Rue René Descartes
67084 Strasbourg CDX
France

Part 1
Water

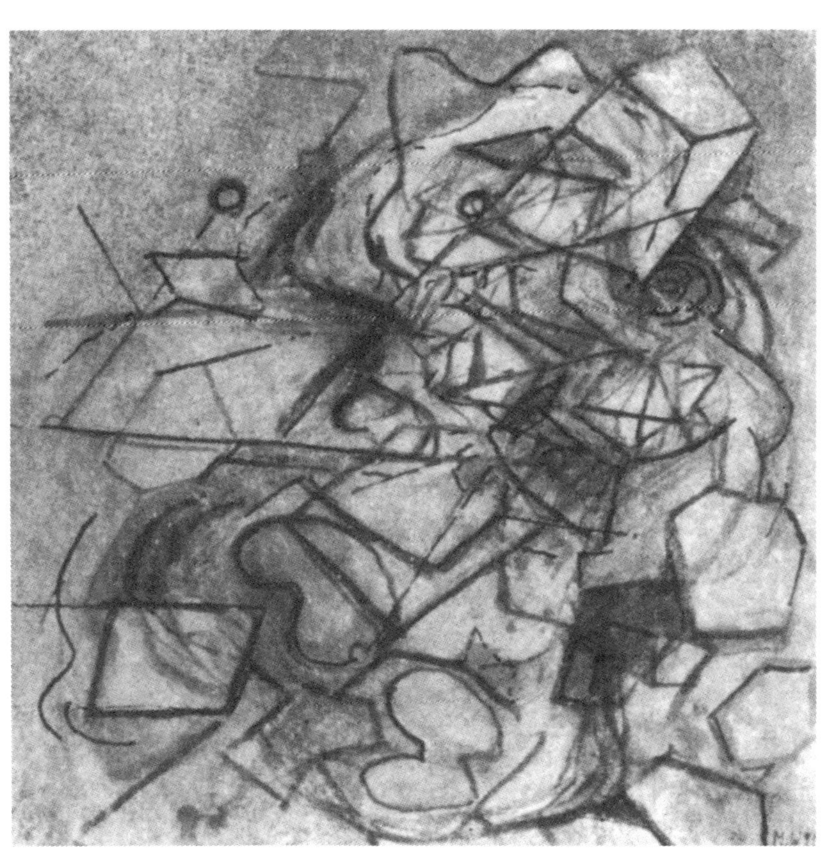

1
Water Structure

H. F. J. Savage

1 Introduction

Water is the most abundant naturally occurring inorganic liquid, covering approximately two-thirds of the earth's surface. As a liquid it occupies a position between the highly mobile structureless gases of the air and the highly structured immobile aggregates of the rocks and sediments of the earth's surface: possessing properties of both motion and structure. Organic life has evolved from this aqueous environment and we need no more direct evidence of this than our dependence on water to carry out the basic molecular processes of life.

Within the surrounding environment and under laboratory conditions, water molecules can exist in a wide variety of states: (a) crystalline ice (ordinary and ice polymorphs); (b) amorphous ice (non-crystalline); (c) crystalline hydrates (organic and inorganic); (d) liquid water (ordinary, supercooled and vapour); (e) aqueous solutions (ionic and non-ionic); and (f) gaseous state (monomers and clusters). In this chapter the structural properties of water are concentrated on; however, for the interested reader, the general properties and characteristics of the various forms and states of water have been extensively reviewed in several published books and journals. These include: *Structure and Properties of Water*, by Eisenberg and Kauzmann (1969); *Water: a Comprehensive Treatise* (Franks, 1972–1982); *Ice Physics*, by Hobbs (1974); and *Water Science Reviews* (1985–).

The structure of water in both its solid and liquid forms is usually perceived as a tetrahedrally coordinated lattice of H-bonded water molecules. This is certainly the case for the naturally occurring solid form of ice Ih (Figure 1.1), which contains an almost perfect tetrahedral lattice. However, within the more disordered forms of the liquid, amorphous ice and aqueous solutions, there are no underlying repeating crystalline lat-

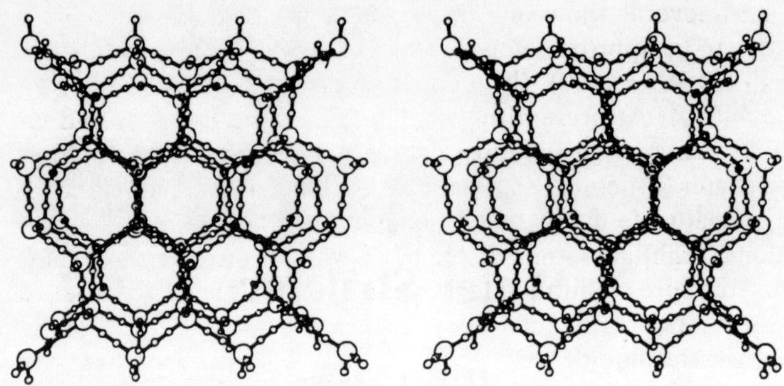

Figure 1.1 Structure of ice Ih, viewed in stereo down the hexagonal axis, as deduced by neutron diffraction. Large spheres denote oxygen positions and the small spheres denote half-hydrogen (deuterium) positions in the disordered model. After Hamilton (1968)

tices. The water molecules occupy a wide range of conformational space.

Within the known structures of the ice polymorphs and crystal hydrates, very large deviations from expected tetrahedrality and normal H-bond geometries are observed. For example, the O· · ·O· · ·O H-bond angles range from 70° to 150° (i.e. 109 ± 40°), and the O· · ·O H-bond distances range between 2.6 Å and 3.2 Å (cf. 2.76 ± 0.01 Å in ice Ih). Although some very weak correlations between distances and angles may be found (e.g. between H· · ·O distance and O–H· · ·O angle), these H-bond characteristics are generally too soft to be used as reliable stereochemical restraints (cf. covalent bonds) in describing water structure in its various phases. The overall structure of water appears to be somewhat more complicated and flexible than the standard tetrahedral ice Ih model.

When details of the short-range non-bonded interactions between water molecules (i.e. O· · ·O, O· · ·H, H· · ·H contacts) are taken into consideration, a tighter set of restraints than those associated with H-bond geometries is found to occur. These non-bonded restraints appear to effectively control the local orientational structure and can be used to rationalize the large structural deviations that commonly exist within the different water phases.

Most of the experimental information concerning the structure of water is derived from X-ray and neutron diffraction studies of the condensed water phases. Diffraction from the liquid and amorphous states yield information only about the average distances between atoms, whereas from the crystalline state individual relative atomic positions are obtainable.

Many different kinds of molecules crystallize together with water. Such hydrates range from small organic (e.g. amino acids) or inorganic systems containing a few waters per asymmetric unit to larger macromolecular systems (e.g. proteins), where a significant fraction of the crystal volume is water and several thousand water molecules may be present. Crystal hydrates can be approximately divided into three categories based on size: (a) small (M.W. <1000), (b) medium (M.W. = 1000–4000) and (c) large (M.W. >4000). At present the small and medium hydrates and ice polymorph crystals furnish the most detailed experimental information available on water structure, especially when studied by neutron diffraction, which is invaluable in locating reliable hydrogen positions.

Section 2 highlights some of the problems that arise in the examination of water structure by diffraction methods — in particular, disorder, scarcity of high-resolution data, and use of neutron diffraction. Details of the water structure in the liquid, ice, amorphous ice and hydrate phases are presented in Sections 3, 4 and 5. Subsequently, in Section 6, the structural features of the water–water interactions found in accurate neutron ice and hydrate structures are described, mainly focusing on regularities of short-range non-bonded contacts between water molecules.

Biological macromolecules are inherently complex, with many different atomic groups present, giving rise to structures which are very irregular and amorphous in character. Details of the water structure at the biomolecular–solvent interface are not well understood, and in Section 7 some of the recent work on the analysis of water structure around biomolecules is outlined. Emphasis is placed on the use of high-resolution neutron diffraction, possible existence of clathrate water cages around non-polar groups and the use of structural restraints in analysing the solvent structure around protein molecules. The final section summarizes the current state of our knowledge on water structure with reference to its main structural features.

2 Structure Determination: Diffraction Studies

For the crystalline state, the main experimental methods used in examining molecular structures are neutron and X-ray diffraction. These techniques provide a reliable direct method of analysing the details of the atomic positions in ice and hydrate crystals: the results depend negligibly on interpretative methods, and thus give us relatively unbiased data on interatomic geometries between molecules. Details of the theory and application of diffraction techniques can be found in Stout and Jensen (1968), Woolfson (1970) and Bacon (1975).

Problems Encountered with Larger Hydrate Crystals

Several problems arise in the analysis of hydrate crystals, mainly relating to the size of the molecules present in the crystals — for example, proteins and DNA. These difficulties are usually manifested in (a) the limited resolution of diffraction data for larger systems and (b) inherent disorder present in larger hydrates.

As outlined in the introduction, hydrate crystals range from small structures containing only one or a few water molecules to macromolecules such as proteins, containing several hundred or thousand water molecules. In analyses of small and medium-sized hydrates, the resolution is usually sufficiently high (1.0 Å or better) to reveal individual peaks for all atoms present, of both hydrate molecule and solvent. However, as the size of the system increases, the attainable resolution of the data is usually lower: between 1.5 Å and 2.5 Å for large proteins. This is mainly because of increased lattice and molecular disorder: for example, in protein crystals, as the water content increases, the protein–protein interactions tend to decrease, allowing increased mobility of whole protein molecules. In addition, the noise level at lower resolutions in electron (X-ray case) or neutron density maps is often quite high and it becomes more difficult to differentiate unambiguously between noise and signal when assigning solvent positions.

In larger hydrates one or more layers of hydration may be present and the first shell usually appears as fairly well-ordered peaks, with the water sites making many H-bond contacts to the molecule. In the layers further away from the surface of the hydrate molecule (e.g. protein or DNA), the solvent density becomes weaker and more diffuse, eventually merging with the featureless background continuum of the bulk solvent regions. Interpretation of the outer layers is usually almost impossible.

Neutron Diffraction

In X-ray diffraction the scattering power of hydrogen is low (ratio between oxygen and hydrogen is O:H = 8:1), and except for high resolution data on very small structures, only the non-hydrogen atoms are observable at reasonable levels of accuracy. Thus, information on exact H-bonding orientations of waters will be missing for a majority of hydrate structures. For neutron diffraction, on the other hand, hydrogen is a relatively strong scatterer (ratio O:H = 5.80: 3.74), comparable in magnitude to C, N and O atoms, and giving accuracies of the order of 0.01–0.02 Å for bond lengths involving hydrogens in small structures (cf. >0.1 Å in X-ray structures). However, as hydrogen is also a very strong incoherent scatterer, neutron scattering patterns from hydrogenous samples tend to have a much higher background. To avoid this, neutron diffraction measurements are often made from samples in which the solvent and easily exchangeable protons on the molecule have been replaced by deuterium by crystallizing from a D_2O-based solvent.

Far fewer hydrate structures have been studied by neutrons than by X-rays. Nevertheless, neutron data on such structures provide a much more reliable source from which detailed information about interatomic interactions can be gained about both the oxygen and hydrogen atoms. To date, the following numbers of neutron hydrate structures have been studied: (a) ~300 small hydrates at resolutions <1.0 Å, (b) ~10 medium hydrates at resolutions <1.0 Å and (c) ~10 large hydrates at resolutions of 1.2–2.2 Å.

Although neutron diffraction appears to be ideal for investigating solvent organization, additional problems also arise. For example, it may not be clear in D_2O solvent regions whether a given peak should be assigned as an oxygen or a deuterium of a water molecule, as the scattering powers of O and D are similar (O:D = 5.80:6.67). Both appear as positive peaks in neutron maps. Such difficulties easily lead to problems in interpreting partly ordered solvent regions. In these cases (a) H_2O instead of D_2O solvent can be used, since hydrogens appear as negative peaks in neutron maps, and/or (b) X-ray data for the same crystal hydrate can be used to identify the oxygen positions — assuming that the neutron and X-ray structures both possess the same solvent structure.

Modelling Disorder

The interpretation and modelling of solvent density within disordered regions is a significant problem. The handling of static disorder is fairly straightforward, but where alternative sites tend to overlap (especially in the case of neutron maps) care must be taken when interpreting the local structure. An example of this is shown in Figure 1.2 for one of the solvent regions in the medium-sized crystal hydrate of coenzyme B_{12} (Savage, 1986b). The X-ray oxygen positions (Figure 1.2b) are useful guidelines in the interpretation of the neutron diffraction map (Figure 1.2a). Also, other diffraction data on the same system are very helpful.

Modelling dynamic disorder is more of a problem, especially when the solvent density does not have a relatively simple shape that can be represented by spheres (isotropic) or ellipsoids (anisotropic). One way to interpret the solvent is as follows: first assign main sites to the more ordered regions of solvent density and then assign 'continuous' sites to the more diffuse and elongated regions between the main sites (an extension of this is to assign a grid of solvent sites). An example of this is shown in Figure 1.3.

From the main sites, alternative water networks can be formulated, using the standard criteria of H-bonding (O· · ·O = 2.6–3.2 Å, H· · ·O = 1.5–2.4 Å) and non-bonded restrictions. Figures 1.17 and 1.18 show the main solvent networks in coenzyme B_{12} hydrate crystals (Savage, 1986b; Bouquiere *et al.*, 1992).

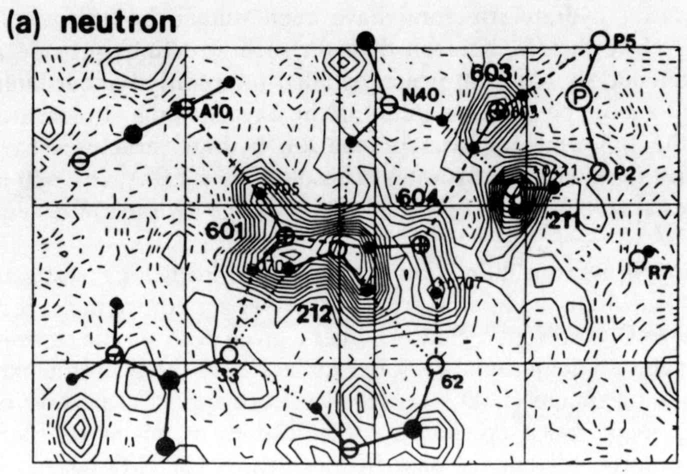

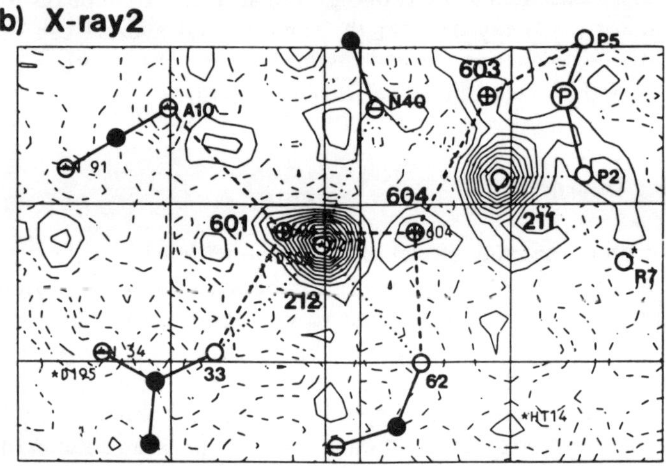

Figure 1.2 Solvent density around disordered side-chain N40 of the coenzyme B_{12} molecule: (*a*) neutron, (*b*) X-ray $F_o - F_c$ difference Fourier maps (the alternative position for the side-chain N640 is not shown, but lies ~1.8 Å behind N40. Two partially occupied solvent networks are present—
network A: waters 211 and 212; occupancies, 0.9(X-ray), 0.6(neutron);
network B: waters 601, 603, 604; occupancies, 0.1(X-ray), 0.4(neutron)

Liquid Diffraction

Diffraction from liquids and amorphous substances yields only structural information about correlations between pairs of atoms — the pair correlation function, PCF — which gives the probability of finding an atom within a certain distance of another. For example, the X-ray diffraction pattern

(a)

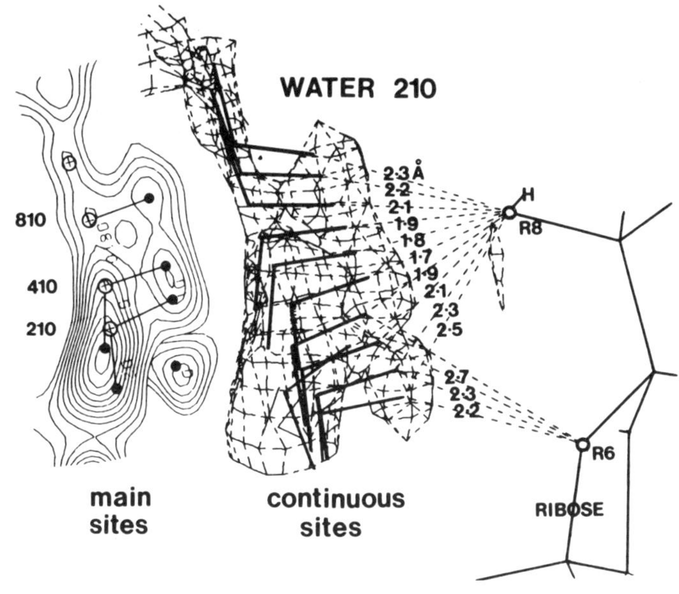

(b)

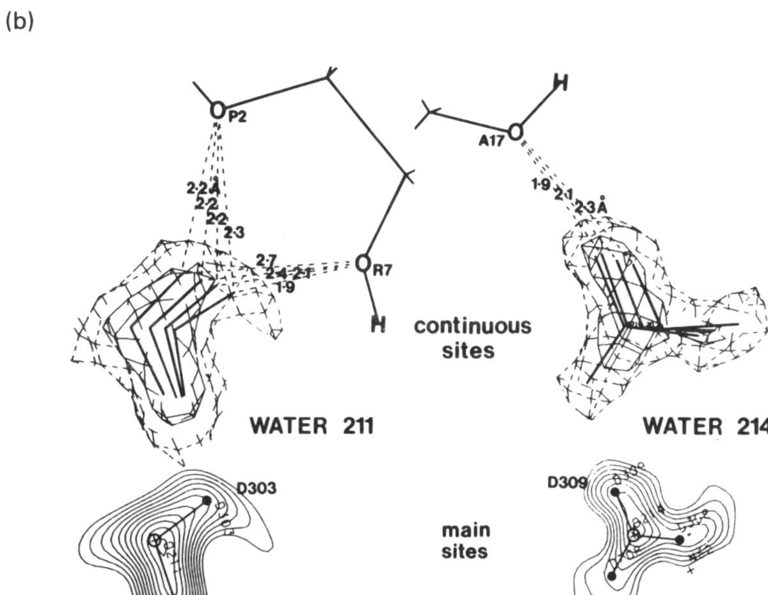

Figure 1.3 Interpretation of solvent regions in coenzyme B_{12} structural analysis: (*a*) over the 210 solvent region, (*b*) over the 211 and 214 solvent regions. 'Main' sites initially assigned to well-defined solvent density. 'Continuous' sites assigned to the elongated and diffuse regions of solvent density

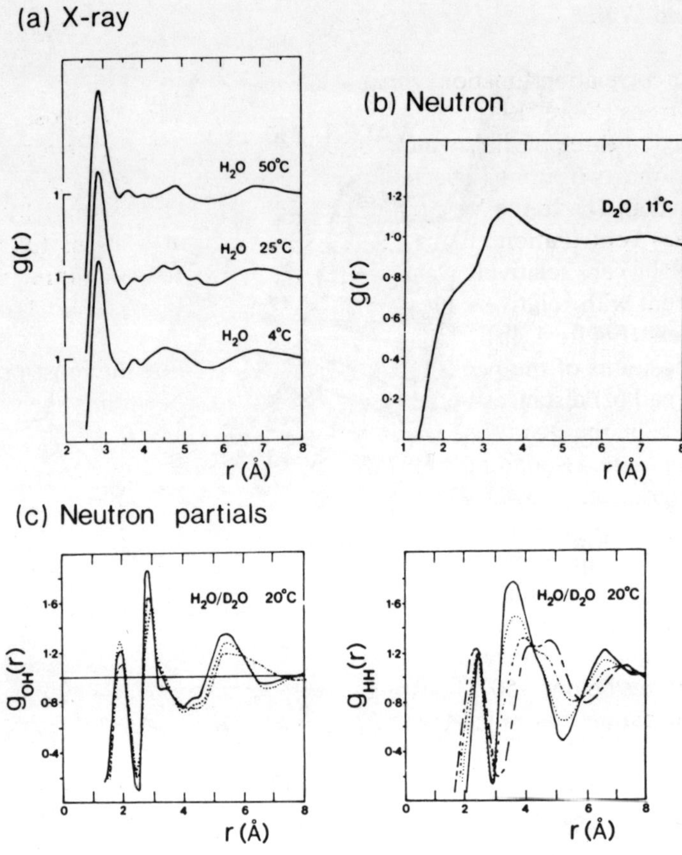

Figure 1.4 Pair correlation functions for liquid water: (*a*) composite X-ray PCF at three different temperatures (after Narten *et al.*, 1967); (*b*) composite neutron PCFs (after Dore, 1985); (*c*) partial neutron PCFs: $g_{OH}(r)$ and $g_{HH}(r)$ functions calculated from different combinations of datasets (after Dore, 1985)

for liquid water is shown in Figure 1.4. Some angular information is obtainable, but only from the most commonly observed distances (i.e. peaks in PCFs). In addition to these difficulties, the molecules in liquids and solutions have mobility that cannot be directly observed in diffraction experiments, since both time and space are sampled, giving only a time-averaged picture of the structure. Thus, there is apparently no direct access to the details of the atomic arrangements at any instance within liquid water. The next-best approach is to analyse the 'water of crystallization' in crystalline hydrate systems.

3 Liquid Water

The pair correlation functions for water, using X-rays (Narten *et al.*, 1967) and neutrons (Dore, 1985), are shown in Figure 1.4. The composite PCF is composed of three partials which correspond to the cross-pair correlations between the two atomic species present, oxygen and hydrogen: $g_{OO}(r)$, $g_{OH}(r)$ and $g_{HH}(r)$. In the X-ray case the composite PCF is essentially that of the $g_{OO}(r)$ contribution (Figure 1.4a), since hydrogen atoms (compared with oxygen) are relatively weak scatterers. For neutrons all three partials are present with relatively high weights, and the composite PCF is quite featureless (Figure 1.4b).

The positions of the peaks in the partial PCFs show the most probable (and spread of) distances between a particular pair of atomic species. The first peak in the X-ray $g_{OO}(r)$ is found at ~2.85 Å (Figure 1.4a) and corresponds to H-bonding between water molecules. The width of this peak ranges from a lower limit of ~2.5 Å to an upper limit of ~3.3 Å. The second $g_{OO}(r)$ peak is much broader than the first and is centred around 4.5 Å. This position corresponds mainly to next-nearest-neighbour non-bonded contacts, similar to those found in ices Ih and Ic. The presence of this second peak suggests that some degree of tetrahedrality is present in the liquid stucture. However, the second peak seems to be present only for temperatures below ~50 °C. At higher temperatures the second peak dies away, indicating that much less or almost no tetrahedral-type structure is present.

Neutron partials, $g_{OH}(r)$ and $g_{HH}(r)$ (Figure 1.4c), appear to be somewhat difficult to separate out (Dore, 1985) and are generally not as accurate as the X-ray $g_{OO}(r)$. In spite of this problem, analyses of neutron scattering data carried out by different groups (Palinkas *et al.*, 1977; Thiessen and Narten, 1982; Dore, 1985) have shown that the first peaks of the $g_{OH}(r)$ and $g_{HH}(r)$ partials occur at relatively similar positions (up to ~0.1 Å differences): $g_{OH}(r)$ at ~2.0 Å, corresponding to H-bonding, and $g_{HH}(r)$ at ~2.4 Å, corresponding to H· · ·H repulsions. However, the agreement between different analyses for the second peaks is not as good, with discrepancies of the order of 0.5 Å occurring.

4 Ice Polymorphs

At present eleven different forms of crystalline ice are known: ice Ih (hexagonal), ice Ic (cubic) and ices II–X. Figure 1.5 shows the pressure–temperature phase diagram for water over the solid and liquid regions. Of the known ice phases, only ordinary ice Ih has been found naturally. Ice Ic exists at ambient pressures and temperatures of less than ~110 K, but transforms irreversibly to ice Ih above this temperature. There is evidence

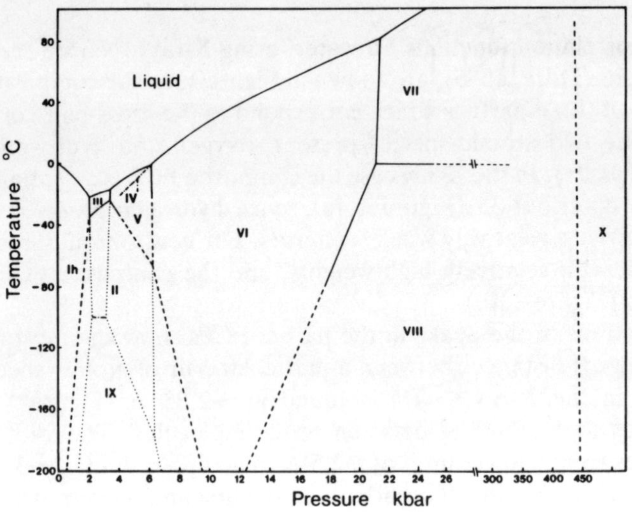

Figure 1.5 Pressure–temperature phase diagram of ice and liquid water: solid lines represent measured stable boundaries; dashed–dotted lines represent measured metastable boundaries; dotted lines are metastable continuations of a phase into an adjacent region. Ice Ic (stable below ~110 K) is not shown, but occurs in the lower part of the ice Ih region

that ice Ic occurs at high altitudes in the atmosphere (Whalley, 1983). Ices II–IX are formed at pressures of between 2 kbar and 22 kbar, while ice X exists at substantially higher pressures of 440 kbar and above (Polian and Grimsditch, 1984). The structure of ice X is not known, but is postulated to be of ionic character, containing symmetric hydrogen bonds (Schweizer and Stillinger, 1984).

The hydrogens in ices II and VIII are fully ordered, with only one fully occupied hydrogen position between each pair of H-bonded oxygen atoms. In the remainder (apart from ice X), the hydrogens are either partially or fully disordered: that is, two (half or partially occupied) hydrogen positions are found between H-bonded oxygens. In several cases there is controversy as to whether the hydrogens are partially ordered or fully disordered, especially at lower temperatures (e.g. ice VI: Kamb, 1973; Johari and Whalley, 1979; Kuhs *et al.*, 1984).

Figures 1.6–1.9 show the basic crystalline structures for the ice polymorphs I–IX. A summary of the ranges of the H-bond distances, O· · ·O· · ·O angles, ring sizes and other properties are listed in Table 1.1.

(a) Ice Ih. Each water forms four H-bonds with O· · ·O distances of 2.76 Å to nearest-neighbour waters. The O· · ·O· · ·O H-bond angles are

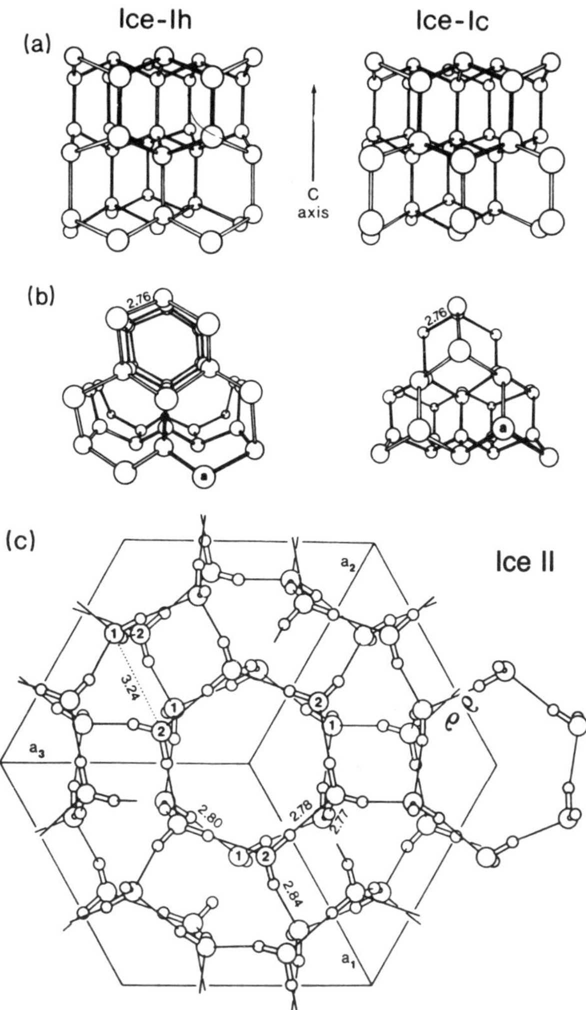

Figure 1.6 (*a*) Structure of ices Ih and Ic (oxygen positions only) viewed perpendicular to *c*-axis of the ice Ih lattice. The filled bonds show the difference in the hexagonal ring conformations along the *c*-axis (of ice Ih): boat form in ice Ih and chair form in ice Ic. (*b*) Structure of ice Ih and Ic viewed along the *c*-axis. The hexagonal rings are arranged in columns in ice Ih, but are staggered in ice Ic. H-bond distances are in angstroms. (*c*) Ordered structure of ice II viewed along hexagonal *c*-axis. The H-bond O· · ·O lengths and a close non-bonded O· · ·O contacts (dotted line) are given in angstroms After Kamb *et al.* (1971)

all tetrahedral to within ~0.2°. The closest next-nearest-neighbour O· · ·O contacts are 4.5 Å, and the resulting overall structure is very open (Figure 1.6a, b). The covalent O–D bond lengths, as deduced by neutron diffraction (Peterson and Levy, 1957; Kuhs and Lehmann, 1983), appear to be significantly longer (1.01 Å) than generally observed in the other ice

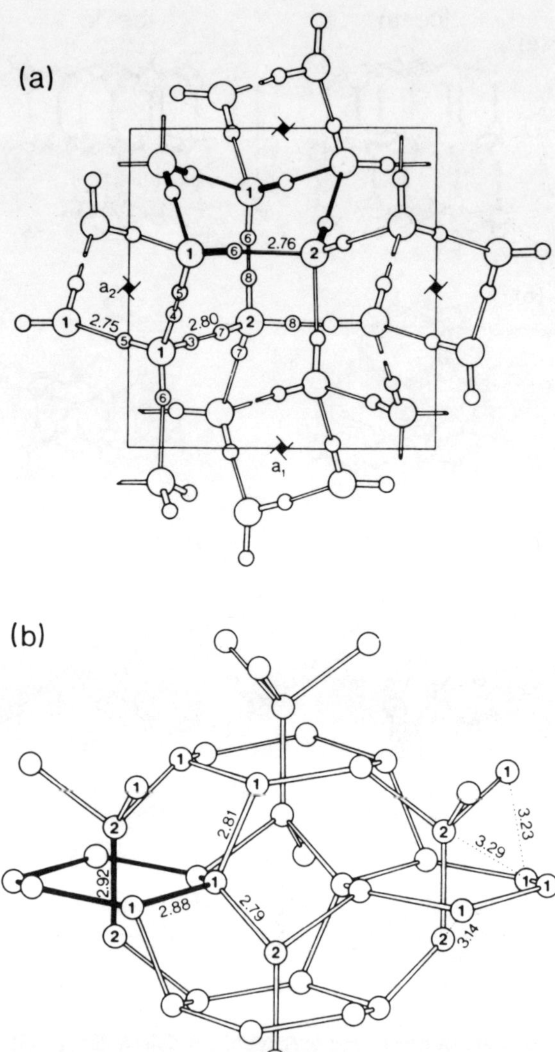

Figure 1.7 (*a*) Structure of ice IX(III), viewed in projection along the *c*-axis. The filled bonds show one of the five-membered H-bonded rings in the unit cell. Two networks of protons are present with occupancies of 0.96 (5, 6 and 7) and 0.04 (3, 4 and 8). O· · ·O H-bond lengths are given in angstroms. After Laplaca *et al.* (1973). (*b*) Oxygen positions in the X-ray structure of ice IV. The filled bonds represent a structural unit comprising a hexagonal ring with an H-bond passing through it. The H-bond lengths and close non-bonded O· · ·O contacts (dotted lines) are given in angstroms. After Engelhardt and Kamb (1981)

structures, the average length being 0.98 Å. However, when possible disorder of the oxygen positions is taken into consideration along with spectroscopic data, the O–D bond lengths fall in line with the average value (Kuhs and Lehmann, 1986).

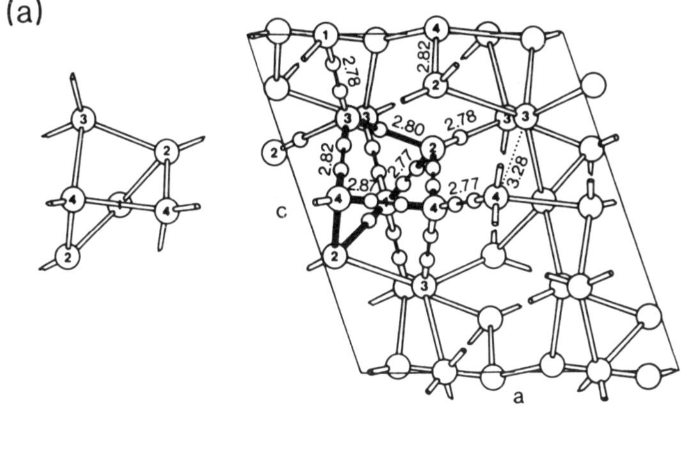

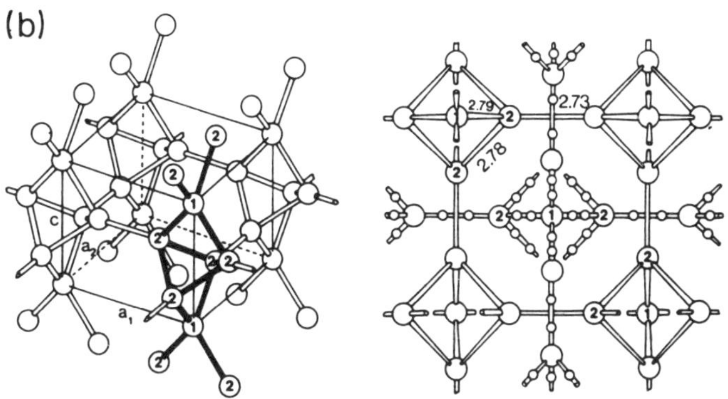

Figure 1.8 (*a*) Structure of ice V viewed in projection along the *b*-axis (right). Some deuteron sites (partially occupied) are also included. The basic structural unit (left) comprises a four- and five-membered ring. The H-bond lengths and close non-bonded O· · ·O contacts (dotted lines) are given in angstroms. After Hamilton *et al.* (1969). (*b*) Structure of ice VI. One of the two interpenetrating lattices is shown on the left (filled bonds show a structural unit). On the right, two lattices are shown in projection along the *c*-axis of the unit cell. The H-bond lengths are given in angstroms. After Kamb (1965)

(b) Ice Ic. The structure of this ice phase (Shimaoka, 1960) is essentially similar to that of ice Ih. Both ices contain layers of hexagonal rings in the chair configuration (Figure 1.6a). The main difference lies in the connections between the layers of 'chair' rings (see the hexagonal rings with filled bonds in Figure 1.6a, b). In ice Ih the hexagonal rings along the vertical *c*-axis are of boat form (layers are eclipsed with respect to vertical bonds; Figure 1.6b), while those in ice Ic are of chair form (staggered with respect to vertical bonds: Figure 1.6b), and the overall structure is analogous to the cubic form found in diamond.

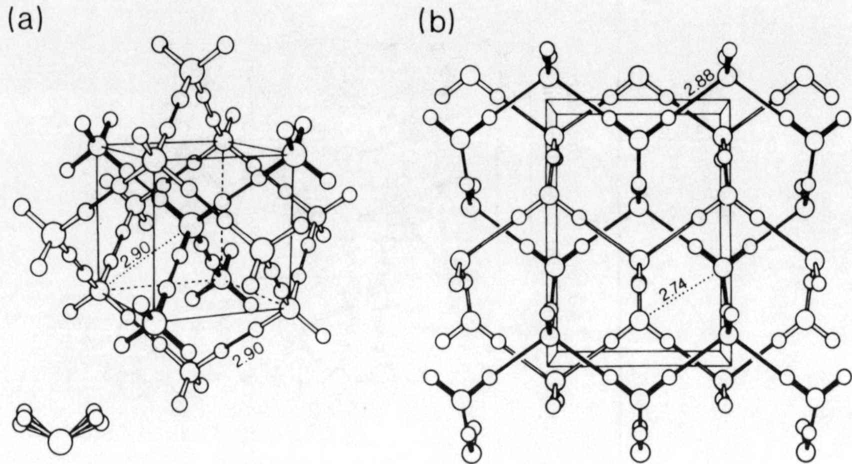

Figure 1.9 (*a*) Average positions in the structure of ice VII: a model for the disorder of the deuterons is shown at bottom left (after Jorgensen and Worlton, 1985). (*b*) Structure of ice VIII (after Kuhs *et al.*, 1984): the two interpenetrating lattices in each of these phases are shown by open and filled bonds. The H-bond lengths and short non-bonded O· · ·O contacts (dotted lines) in both structures are given in angstroms

(c) Ice II. This is a fully ordered phase of ice (Kamb, 1964; Kamb *et al.*, 1971). The structure (Figure 1.6c) is composed of columns of almost flat hexagonal rings linked by H-bonds that are somewhat distorted from tetrahedral coordination. The O· · ·O H-bonds range from 2.77 Å to 2.84 Å, the O–H· · ·O angles from 166° to 178° and the O· · ·O· · ·O H-bond angles are greatly distorted from tetrahedrality, lying in the range 81–130°. One of the main structural features of this ice phase is a close intrusive next-nearest-neighbour contact of 3.24 Å which is not an H-bond.

(d) Ices III and IX. With respect to the oxygen positions, these two ices have basically the same structures. Ice III is the high-temperature form, in which the hydrogen positions are fully disordered, while ice IX is a lower-temperature form (see the metastable region of this ice in Figure 1.5), in which the hydrogens are almost fully orientationally disordered (Figure 1.7a). These ice structures (Kamb and Prakash, 1968; LaPlaca *et al.*, 1973) are composed mainly of 5-membered rings (also some 7-membered). No hexagonal rings are present. The O· · ·O H-bonds lie between 2.75 Å and 2.80 Å, while the O· · ·O· · ·O H-bond angles are more distorted at larger values than in ice II, ranging from 91° to 144°.

(e) Ice IV. This is a metastable form that exists in the ice V region of the ice *P*–*T* phase diagram (Figure 1.5). The structure is interesting in that an H-bond is formed through the centre of a hexagonal ring (Figure 1.7b). Only the X-ray structure is known (Engelhardt and Kamb, 1981), in which the O· · ·O H-bonds are 2.79–2.92 Å and the O· · ·O· · ·O angles are between 88° and 128°. There are also a large number of close non-bonded

Table 1.1 Structural properties of ice polymorphs

Ice polymorph	*No. of nearest neighbours and distances (Å) of H-bonds*	*No. of next-nearest neighbours and distances <3.6Å*	*Hydrogen order*	*O–O–O angles(°)*	*Density ($g.cm^{-3}$)*	*No. of waters per unit cell*	*Ring sizes present*				
							4	*5*	*6*	*7*	*8*
Ice Ih	4 2.76	–	disordered	109	0.931	4	–	–	√	–	–
Ice Ic	4 2.75	–	disordered	109.5	0.93	2	–	–	√	–	–
Ice II	4 2.77–2.84	9 3.24–3.60	ordered	81–128	1.18	12	–	–	√	–	√
Ice III	4 2.75–2.80	1 3.45	disordered[a]	87–144	1.16	12	–	√	–	√	√
Ice IX(III)	4 2.75–2.80	1 3.45	partial	87–144	1.16	12	–	√	–	√	√
Ice IV	4 2.79–2.92	9 3.14–3.29	disordered[a]	88–128	1.27	16	–	–	√	–	√
Ice V	4 2.76–2.80	7 3.28–3.49	disordered[a]	84–128	1.23	10	√	√	√	√	√
Ice VI	4 2.80–2.82	17 3.44–3.46	disordered[a]	76–128	1.31	12	√	–	–	–	√
Ice VII	4 2.90	4 2.90	disordered	109.5	1.50	4	–	–	√	–	–
Ice VIII	4 2.88	5 2.74–3.14	ordered	109.5	1.50	8	–	–	√	–	–

[a] May be some partial ordering at low temperatures.

O· · ·O contacts between 3.14 Å and 3.29 Å around each water molecule.

(f) Ice V. The structure (Kamb *et al.*, 1967) is composed of 4-, 5- and 6-membered rings. Figure 1.8(a) shows the overall structure and the basic unit which is made up of one 4- and two 5-membered rings. The O· · ·O H-bonds range from 2.76 Å to 2.87 Å and the O· · ·O· · ·O H-bond angles range from 84° to 128°. The protons in this structure were seen to be partially ordered in a neutron structural analysis (Hamilton *et al.*, 1969).

(g) Ice VI. As the ice phases become more densely packed, this is the first phase to form two independently H-bonded interpenetrating lattices. Figure 1.8(b) shows the X-ray structure, deduced by Kamb (1965). The neutron structure has been deduced by Kuhs *et al.* (1984), and their results suggest that the protons are fully orientationally disordered in long range and are not partially ordered at low temperatures, as indicated by earlier static permittivity experiments (Johari and Whalley, 1979). The O· · ·O H-bonds are 2.73–2.79 Å and the O· · ·O· · ·O angles lie between 77° and 128°. The next-nearest-neighbour O· · ·O contacts are surprisingly long, with an average of 3.4 Å, compared with the short non-bonded contacts present in the above ices.

(h) Ices VII and VIII. These structures are the most densely packed forms of ice known that maintain the basic character of asymmetric hydrogen bonding (the denser form of ice X is thought to contain symmetric hydrogen bonds with ionic character). They each comprise two interpenetrating (ice Ic) lattices in which all the O· · ·O· · ·O H-bond angles are very close to the tetrahedral value. The neutron structure of ice VIII has been examined by two groups (Jorgensen *et al.*, 1984; Kuhs *et al.*, 1984) and was shown to be fully ordered (both oxygens and deuteriums). Each water in this structure (Figure 1.9b) is surrounded by eight nearest neighbours: four forming H-bonds and four making non-bonded contacts. The O· · ·O H-bonds distances are 2.88 Å, longer than the two shortest O· · ·O contacts of 2.74 Å. Three other short O· · ·O contacts also occur: 3.03 Å, 3.03 Å and 3.14 Å. In the X-ray structure of ice VII (Kamb and Davis, 1964) all of the eight nearest neighbours have the same distance of 2.90 Å. However, in neutron analyses of this phase (Kuhs *et al.*, 1984; Jorgenson and Worlton, 1985) the results and interpretations indicate that both the oxygens and deuteriums (Figure 1.9a) are probably disordered: ~0.1 Å for the oxygen sites and >0.1 Å for the deuterium sites. Ice VII appears to be the disordered phase of fully ordered ice VIII.

Within the known ice polymorphs, the O· · ·O H-bond distances range from 2.73 Å to 2.92 Å and all the water molecules form four-coordinated H-bonded networks. However, there are very large deviations of the O· · ·O· · ·O H-bond angles (77–144°) from the 'expected' tetrahedral value of 109.5°. In addition to this, some very short non-bonded O· · ·O contacts between 2.74 Å and 3.29 Å also occur.

(i) Amorphous Ice. Amorphous phases of ice can be prepared in several ways: vapour condensation at low temperature (Narten *et al.*, 1976), quenching the liquid, squeezing ice under high pressures (Mishima *et al.*, 1984) and heating the amorphous ice phases produced by the previous methods. The only structural information that can be directly obtained for the amorphous and liquid phases relates to the pair correlation functions obtained from scattering experiments.

Figure 1.10(a) shows the oxygen–oxygen PCFs for amorphous ice derived from X-ray diffraction by Narten *et al.* (1976). Three phases are shown: one condensed at 77 K which is a low-density form (0.94 g/cm); one at 10 K which is a high-density form (1.1 g/cm); and one at 10 K which is then warmed to 77 K (high-density form). The first peak is very well defined and is located at the same position as for polycrystalline ice Ih at 2.76 Å (Figure 1.10b). The second peak for ice Ih falls (as expected) very sharply at 4.5 Å, but in the amorphous phases the corresponding peak is very broad, indicating that the local orientations of the water molecules (O· · ·O· · ·O angles, etc.) vary considerably from tetrahedrality. In the high-density forms an extra peak appears at ~3.3 Å and is probably due to the intrusion of other water molecules in between the nearest (2.76 Å) and next-nearest-neighbour (4.5 Å) shells, similar to the high-pressure ice structures.

5 Small Crystal Hydrates

Several surveys of available small hydrate structures analysed by neutron diffraction have been carried out (Ferraris and Franchini-Angela, 1972; Pedersen, 1974; Chiari and Ferraris, 1982; Jeffrey and Maluszynska, 1990). The covalent O−H bond lengths and H−O−H angles for the water molecule have values of 0.96(2) Å and 107.0(2.0)°, respectively, which are close to the monomer gas phase values (0.96 Å and 104.5°). The number of H-bonds formed by individual water molecules within these structures is between 2 and 6, most forming 3 or 4. Water molecules involved in 5–6 H-bonds are usually involved in bifurcated H-bond arrangements (three-centred interactions).

The O· · ·O H-bond distances range from 2.6 Å to ~3.2 Å, which is a significantly larger range than that found in the ice polymorphs: 2.73–2.92 Å. The wider spread of the hydrate O· · ·O H-bonds corresponds well with the range found in the first peak of the liquid X-ray PCF : ~2.5 Å to ~3.3 Å. The O−H· · ·O H-bond angles in hydrates vary between ~120° and 180°, again greater than the distribution in ice structures : 150–180°.

While the variations in the H-bond geometries appear to be larger in hydrates than in ice polymorphs, the ranges of non-bonded geometries appear to be very similar. For example, the O· · ·O· · ·O H-bond angles vary from 70° to 150° in hydrates, which is similar to the range in ices :

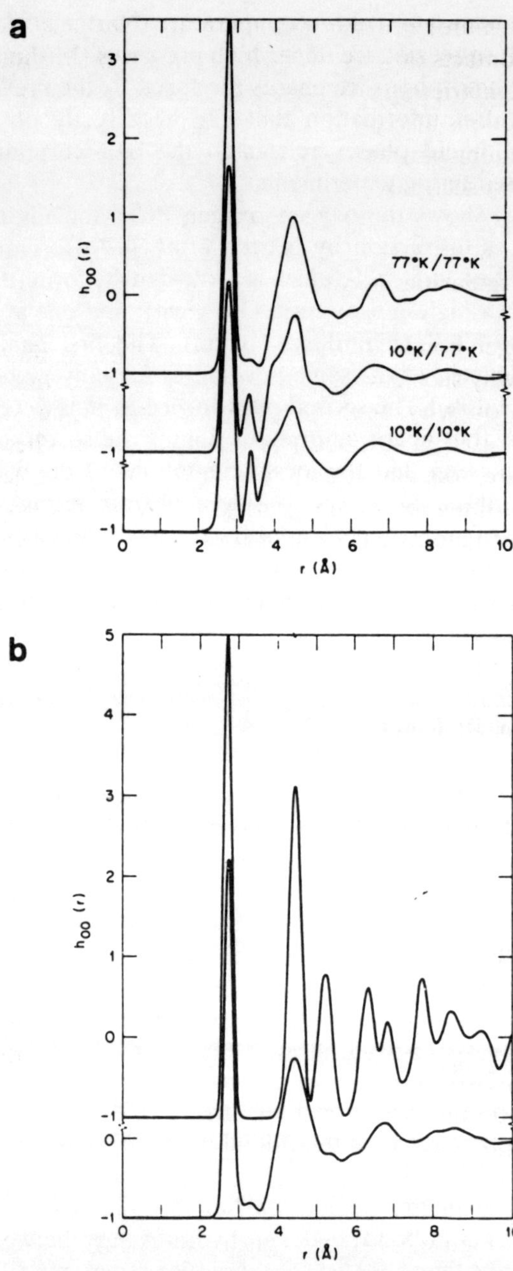

Figure 1.10 Oxygen–oxygen pair correlation functions for amorphous ice and polycrystalline ice Ih obtained from X-ray diffraction (after Narten *et al.*, 1976). (*a*) Amorphous ice: 77 K deposit measured at 77 K (top); 10 K deposit measured at 77 K and at 10 K (two lower curves). (*b*) Polycrystalline ice Ih (upper) and amorphous ice at 77 K/77 K (lower).

76–144°. Furthermore, in both hydrate and ice structures O· · ·O non-bonded water contacts down to 3.1–3.2 Å occur. These similarities in the non-bonded geometries strongly indicate that, in addition to H-bonding, repulsive interactions also play a significant role in determining the non-covalent interactions between water molecules, and these are further discussed in the next section.

6 Structural Regularities between Water Molecules

Although hydrogen bonds are directional in character, they do not appear to fully explain why such a wide range of H-bond geometries (e.g. O· · ·O = 2.6–3.2 Å) or why such large deviations from tetrahedrality (O· · ·O· · ·O = 109° ± 40°) occur in water structure. These geometrical characteristics are too variable to be used as effective stereochemical restraints in describing the structural details of H-bonded water networks.

A better understanding can be gained when the local short range contacts are considered. Details of the O· · · ·O, H· · · ·O and H· · · ·H atomic geometries have been examined in ice and small hydrate structures (Savage and Finney, 1986, Savage, 1986a) and some interesting regularities in terms of minimum contact distances are apparent. Four different short-range regularities have been identified (illustrated schematically in Figure 1.11a) and are listed as follows.

RR1: O· · · ·O repulsion of the H-bonds: Standard plots of the O· · · ·O and H· · · ·O H-bond distances against the O–H· · · ·O H-bond angle involving water molecules are shown in Figure 1.12. Usually a regression line is drawn on the plots representing an overall decrease in the angle as the distance increases. The scatter of the geometries is too much outside the accuracy of these data points (~0.02 Å for distance) for a correlation to be significant. It appears that the H-bonds can bend, when required, but there is a limit on the amount of bending and this is represented by the full lines in Figure 1.12 — H-bond bending limit curve. The actual angles and distances a water H-bond may have appear to depend on the local packing arrangements determined by the surrounding short-range contacts of RR2, RR3 and RR4.

RR2: O· · · ·O non-bonded contacts: This regularity refers to the significant anisotropy of the water oxygen–oxygen contacts, which range from 3.1 Å to 3.6 Å. Figure 1.13 shows the minimum O· · · ·O non-bonded contacts in ices and hydrates with respect to two orientational angles defined in the inset. Three main regions are apparent and can be assigned the following van der Waals radii: region A ~1.8 Å over the lone-pair region; region B ~1.7 Å between the hydrogens; and region C ~1.6 Å between the hydrogens and the lone-pair region.

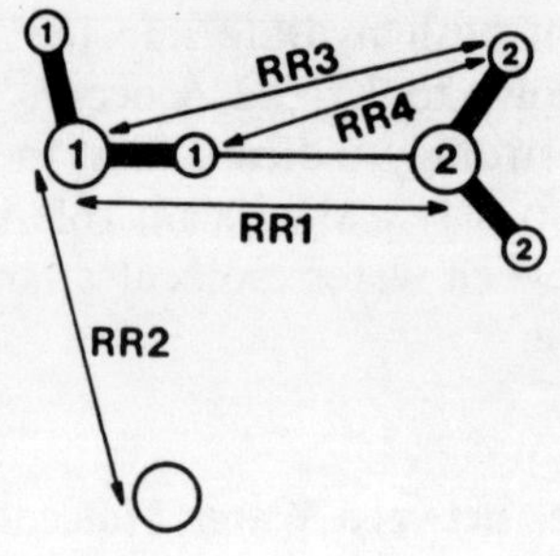

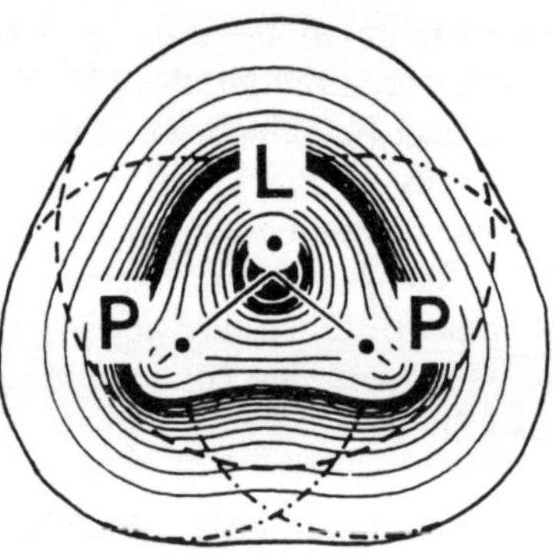

Figure 1.11 (*a*) Schematic figure of the four repulsive restraints: RR1, O· · ·O repulsion of H-bond; RR2, O· · ·O non-H-bonded repulsions; RR3, H2· · ·O1 oxygen–remote-neighbour hydrogen non-bonded interactions; RR4, H· · ·H non-bonded repulsion. (*b*) Displaced van der Waals model for the RR non-bonded short-range interactions around a water molecule. Contours represent the water electron density (from quantum mechanical calculations), dashed circles represent displaced van der Waals spheres: centred at L of the oxygen (radius = 1.50 Å) and P for hydrogen (radius = 1.30 Å). The L and P centres are displaced by 0.2 Å from the nuclear positions of O and H. Note: two L sites are present in approximately the tetrahedral lone pair positions. Figure adapted from Hermannson (1984)

RR3: H· · · ·O interactions: In the configuration O1–H1· · · ·O2–H2 there are two main types of interactions. The first, H1· · · ·O2, are the normal H-bonds with distances of 1.5–2.4 Å. The second, H2· · · ·O1, are the longer remote non-bonded interactions and these are seen to have a minimum contact distance of around 3.0 Å in the ice and hydrate structures at ambient pressures. Such H2· · · ·O1 contacts for water in α cyclodextrin (Klar *et al.*, 1980) and coenzyme B_{12} (Savage, 1986b) crystals are shown in Figure 1.14. Figure 1.15(a) shows a plot of H2–O2· · · ·O1 angle of the O1–H1· · · ·O2–H2 configuration against the O1· · · ·O2 H-bond distance. A limiting line (full line) can again be drawn representing the limits

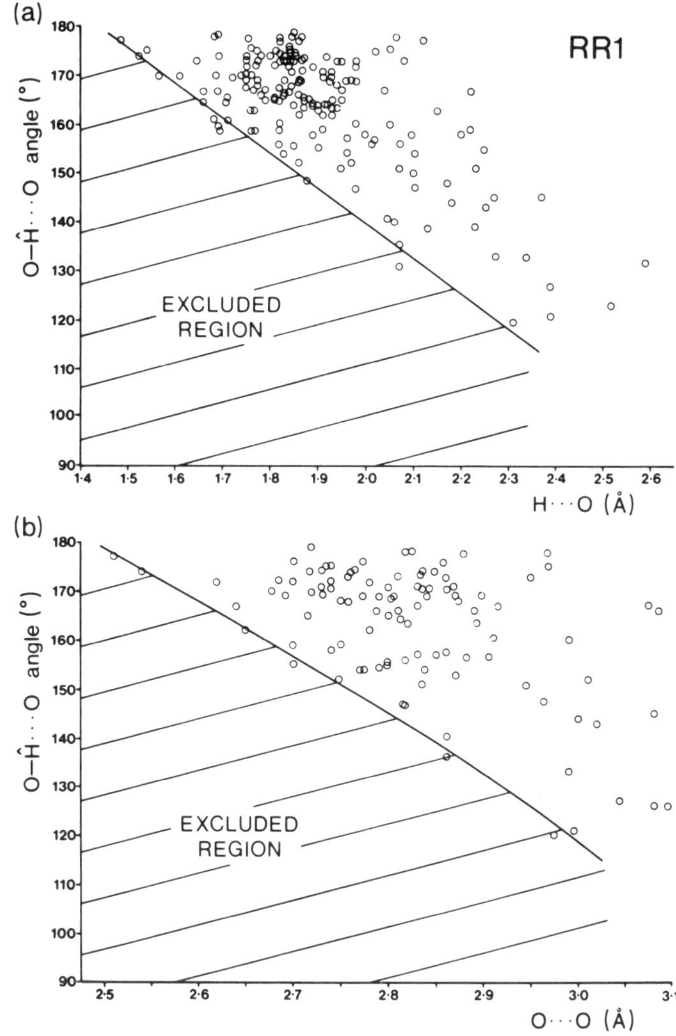

Figure 1.12 Plot of O–H· · · ·O angles versus H· · ·O distance for: (*a*) H-bonds in small neutron structures (after Olovsson and Jonsson, 1976), (*b*) water H-bond geometries found in high-resolution neutron hydrate and ice structures. The solid curves approximately represent minimum allowed values

for H-bond geometries. As the O1· · · ·O2 H-bond distance decreases, the H2–O2· · · ·O1 angle increases in order to maintain the minimum RR3 H2· · · ·O1 contact distance of ~3.0 Å.

RR4: H· · · ·H interactions These refer to the H· · · ·H repulsions involving water hydrogens. Contacts down to 2.1 Å occur, with a spread of 2.1–2.6 Å. Figure 1.15(b) shows a plot of the H· · · ·H contacts against the sum of the angles (χ) subtended at the hydrogens (see inset). As the angle χ

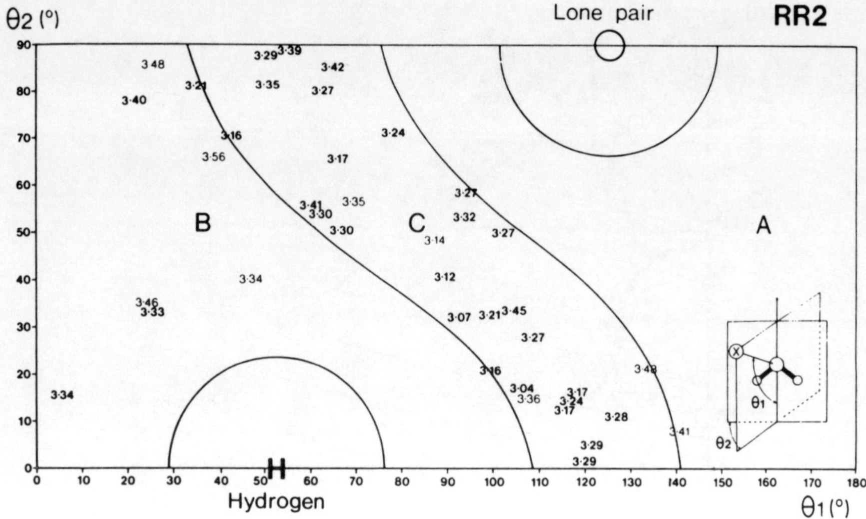

Figure 1.13 Θ 1/2 orientational angle (defined in inset) plot for non-bonded O· · · ·O,C contacts (A) around 4-coordinated water molecules. Bold numbers are O(W)· · · ·O contacts and lighter numbers are O(W)· · · ·C contacts

increases, the H· · · ·H contacts tend to decrease and again a limiting line can be drawn.

To a first approximation, the angular dependence/anisotropy of the RR minimum short-range contacts can be rationalized in terms of four van der Waals spheres that are displaced with respect to the inherent asphericity of the electron density (due to polarization) over a water molecule. Figure 1.11(b) shows the total molecular electron density for a water molecule obtained from quantum mechanical calculations (Hermansson, 1984). The surface of the van der Waals sphere for an atom is usually considered to correspond to the outer electron density levels of the atom. On this basis, four spheres can be used to explain the anisotropic character of the RR contacts. The centres of the four spheres (two for oxygen and two for hydrogen) reside at locations that are displaced as follows:

Oxygen: 0.2 Å from the oxygen nucleus along the lone-pair direction roughly (tetrahedral positions assumed) corresponding to the lone-pair positions: centres at L.

Hydrogen: 0.2 Å from the hydrogen nucleus towards the oxygen nucleus along the O−H bond (centres at P), corresponding to the hydrogen positions in high resolution X-ray studies: centres at P.

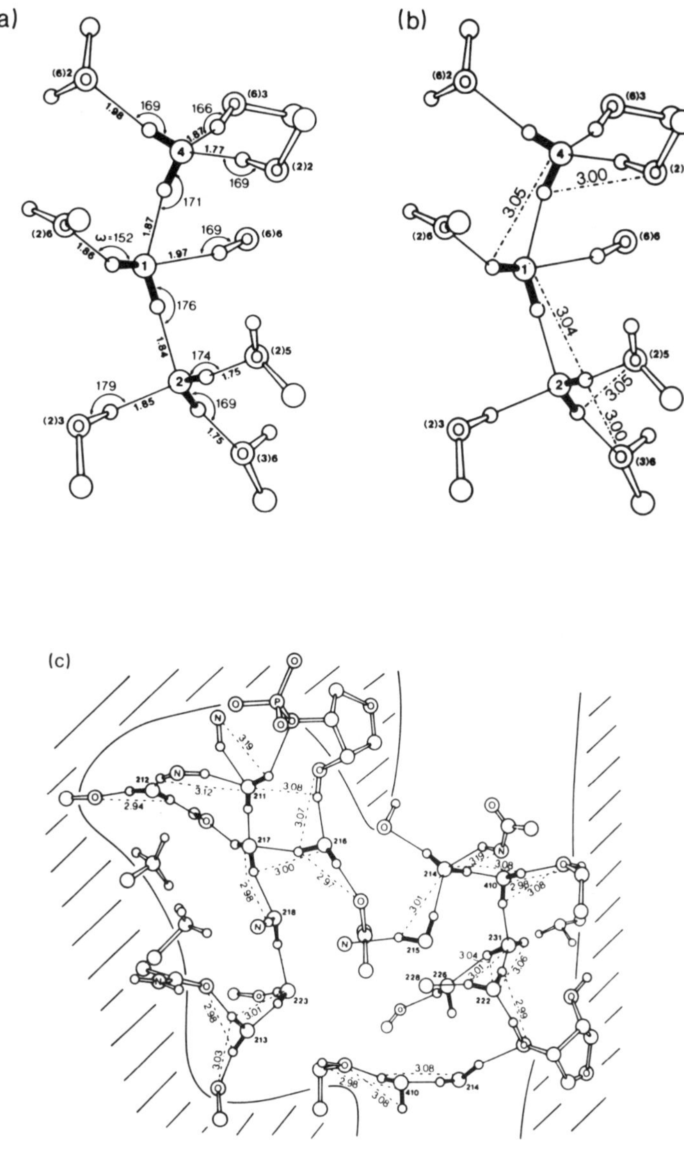

Figure 1.14 H-Bonded water network structures in: (*a*) α-cyclodextrin, showing the range of H-bond distances (in angstroms) and particularly H-bond angles (in degrees); (*b*) α-cyclodextrin, showing the minimum oxygen remote hydrogen H2· · ·O1 contacts (RR3, in angstroms) of ~3.0 Å; (*c*) coenzyme B_{12} hydrate, showing the H2· · ·O1 contacts (RR3) around water molecules

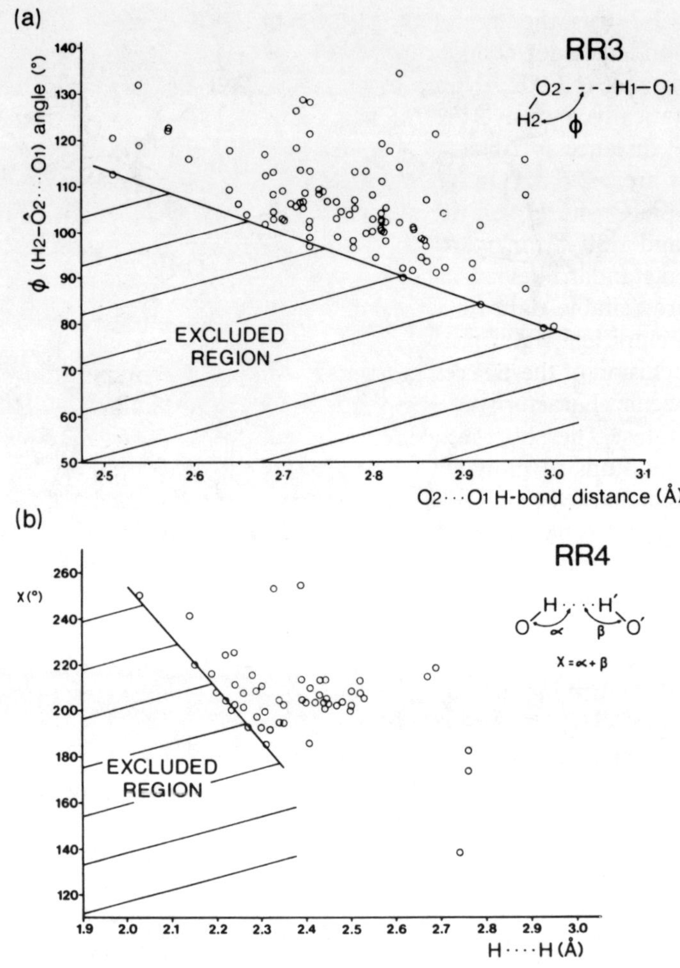

Figure 1.15 (*a*) Plot of H2−O2· · · ·O1 angles (ø) against O2· · · ·O1 H-bond distances for waters involved in H-bonds. The solid line represents the minimum allowed values. (*b*) Plot of water–water H· · · ·H contacts against χ, the sum of the angles alpha and beta subtended at the hydrogen atoms (see inset)

The displaced spheres are drawn as dashed circles in Figure 1.11(b). With the centre positions established, it remains to assign the radii of the van der Waals spheres. This was done in an analysis of the distances between the L and P centres (i.e. the interwater LL, LP and PP distances) in ~16 different structures of ordered ices II, IX and VIII and cyclodextrin hydrates determined by neutron diffraction. Two L and two P centres were computed for each water in the experimental structures. The distances between these centres were only included in the analysis where the overall electrostatic energy of the water dimer configuration was repulsive or very close to it.

Table 1.2 lists the minimum LL, LP and PP contacts for different non-H-bonded dimer configurations. The average minimum intermolecular LL distance is 3.002 Å (number of data = 18). All others are greater for electrostatically repulsive dimer configurations. The average of the minimum PP distance is 2.600 Å (number of data = 26), and all other PP distances are >2.6 Å. On halving the minimum LL and PP distances, the radii of the O and H van der Waals spheres can be assigned: 1.50 Å for oxygen and 1.30 Å for hydrogen to an accuracy of 0.02 Å — i.e. average estimated standard deviations (e.s.d.) of distances. These values will provide more reliable data on which to model and fit anisotropic repulsive terms in empirical water potentials.

The inclusion of the RR restraints in addition to H-bonding allows many of the specific characteristics of water structure to be explained in both ices and hydrates. The short-range restrictions appear to largely control the final orientations of individual water molecules. For example, the basic tetrahedral characteristics of water structure can be related to a minimization of the RR repulsive interactions (especially RR3: remote H2· · · ·O1 contacts) and a maximization of the number (four) and strengths (2.7–2.8 Å) of H-bonds formed.

There are two ways of obtaining a more compact water structure than that of ice Ih: (1) by decreasing some of the O· · ·O· · ·O H-bond angles, and/or (2) by forming interpenetrating H-bonded lattices (as found in ices VI, VII and VIII). In the former some of the O· · ·O H-bond distances have to increase to maintain the RR3 minimum H2· · ·O1 remote contacts (Figure 1.16). These particular contacts appear to play a prominent role in determining the orientational structure between waters. For instance, acceptor waters which form very short H-bonds of <2.7 Å tend towards a trigonal planar coordination due to the necessity of maintaining minimum H2· · ·O1 contacts of ~3.0 Å. For longer H-bonds of 2.9–3.2 Å, significant distortions from expected tetrahedrality are allowed, which are often required to maximize H-bonding around a water.

The H-bonded coordinations around waters in small hydrates range from 2 to 6, with values of 3 or 4 being the most frequent. In the ice polymorph structures 4 is the minimum number. In general, there is a distinct lack of directionality of the H-bonds over the lone pair regions — an observation well established in all small-molecule crystals (Olovsson and Jonsson, 1976). The local H-bonded coordination values depend on the number and presence of local polar groups and maintaining the minimum RR values. In regions where a significant number of apolar groups are present, waters that are <3 coordinated sometimes occur.

The RR restraints segregate the geometries of a water molecule into an excluded (region below the limit line in the RR plots of Figures 1.12 and 1.15) and an allowed accessible space (above each curve). The RR restraints can be applied to larger protein systems in which only the water

Table 1.2a Minimum LL distance present for dimer configurations where the electrostatic energy is generally repulsive (> 0.0 kcal/mol)

	T	P	*O· · ·O*	*(e.s.d.)*	*L· · ·L*	*Multiplicity*	*L· · ·P*	*P· · ·P*	*Energy (kcal/mol)*	
									TIPS2	*MCY*
Ice II :(N)	80 K	1 bar	3.223	(0.030)	3.005	(1)	2.847	2.658	+0.04	+0.68
Ice II :(N)	110 K	1 bar	3.240	(0.020)	3.029	(1)	2.861	2.660	+0.33	+0.99
Ice II :(N)	110 K	1 bar	3.234	(0.012)	3.011	(1)	–	–	–	–
Ice II :(N)	220 K	2.75 kbar[b]	3.213	(0.020)	3.002	(1)	2.880	2.652	−0.06	+0.61
Ice II :(N)	197 K	2.91 kbar[b]	3.197	(0.020)	2.977	(1)	2.856	2.656	+0.15	+0.83
Ice II :(N)	195 K	3.60 kbar[b]	3.179	(0.020)	2.967	(1)	2.850	2.664	+0.16	+0.85
Ice II :(N)	249 K	4.10 kbar[b]	3.209	(0.020)	2.986	(1)	2.840	2.662	−0.13	+0.48
Ice II :(N)	220 K	4.76 kbar[b]	3.190	(0.020)	2.975	(1)	2.851	2.642	−0.06	+0.61
Ice II :(N)	195 K	4.77 kbar[b]	3.204	(0.020)	2.991	(1)	2.855	2.630	+0.10	+0.82
Ice II :(N)	262 K	4.80 kbar[b]	3.200	(0.020)	2.976	(1)	2.866	2.676	+0.09	+0.76
Ice VIII:(N)	10 K	1 bar[a]	3.245	(0.010)	3.023	(4)	3.644	4.308	+3.29	+3.25
Ice VIII:(N)	269 K	1 bar[a]	3.214	(0.011)	2.992	(4)	3.623	4.295	+3.51	+3.52
Ice IX :(N)	110 K	1 bar	3.450	(0.003)	3.224	(1)	3.046	2.880	+1.07	+0.61
Ice IX :(N)	110 K	1 bar	3.455	(0.015)	3.218	(1)	3.039	2.877	+0.91	+0.40
Ice IX :(N)	110 K	2.80 kbar[b]	3.328	(0.015)	3.100	(1)	2.929	2.779	+1.07	+0.54
Ice IX :(N)	164 K	2.80 kbar[b]	3.322	(0.015)	3.088	(1)	2.932	2.811	+1.03	+0.48

[a] Ice VIII data collected under pressure (≈26 kbar) and extrapolated to ambient pressure (1 bar).
[b] From Londono, J. D. (1989). Thesis, University of London.

Table 1.2b Minimum PP distance present for dimer configurations where the electrostatic energy is repulsive (> 0.0 kcal/mol)

	T	P	*O···O*	*L···L*	*L···P*	*P···P*	*Multiplicity*	*Unregularized*	*(e.s.d.)*	*Energy (kcal/mol)*	
										TIPS2	*MCY*
Water–water:											
Ice II :(N)	80 K	1 bar	3.572	3.458	3.105	2.573	(1)	2.351	(0.030)	+1.50	+2.21
Ice II :(N)	110 K	1 bar	3.605	3.512	3.106	2.581	(1)	2.381	(0.020)	+1.16	+1.93
Ice II :(N)	220 K	2.75 kbar[b]	3.579	3.463	3.115	2.603	(1)	2.403	(0.025)	+1.42	+2.02
Ice II :(N)	197 K	2.91 kbar[b]	3.596	3.489	3.137	2.602	(1)	2.392	(0.025)	+1.39	+1.97
Ice II :(N)	195 K	3.60 kbar[b]	3.583	3.490	3.077	2.546	(1)	2.331	(0.025)	+1.39	+2.10
Ice II :(N)	249 K	4.10 kbar[b]	3.530	3.407	3.093	2.552	(1)	2.365	(0.025)	+1.41	+2.03
Ice II :(N)	220 K	4.76 kbar[b]	3.575	3.459	3.106	2.583	(1)	2.393	(0.025)	+1.47	+2.13
Ice II :(N)	195 K	4.77 kbar[b]	3.567	3.450	3.105	2.575	(1)	2.371	(0.025)	+1.51	+2.11
Ice II :(N)	262 K	4.80 kbar[b]	3.619	3.510	3.154	2.613	(1)	2.388	(0.025)	+1.54	+2.21
Ice VIII:(N)	10 K	1 bar[a]	3.403	3.199	2.887	2.594	(4)	2.463	(0.010)	+0.98	–
Ice VIII:(N)	10 K	1 bar[a]	3.123	3.044	2.840	2.612	(4)	2.527	(0.010)	+3.43	–
Ice VIII:(N)	269 K	1 bar[a]	3.406	3.204	2.890	2.602	(4)	2.474	(0.011)	+0.98	–
Ice VIII:(N)	269 K	1 bar[a]	3.142	3.066	2.850	2.616	(4)	2.528	(0.011)	+3.43	–
β-cyclodextrin:(N)	120 K	1 bar[a]	3.247	3.127	2.905	2.554	(1)	2.400	(0.020)	+1.86	–
Water–hydroxyl:											
β-cycdtn:(N)	120 K	1 bar				2.612	(1)	2.391	(0.010)		
Hydroxyl–hydroxyl:											
α-cyclodextrin:(N)	296 K	1 bar				2.604	(1)	2.390	(0.010)		
β-cyclodextrin:(N)	120 K	1 bar				2.572	(1)	2.373	(0.010)		
β-cyclodextrin:(N)	120 K	1 bar				2.610	(1)	2.375	(0.010)		

[a] Ice VIII data collected under pressure (≈ 26 kbar) and extrapolated to ambient pressure (1 bar).
[b] From Londono, J. D. (1989). Thesis, University of London.

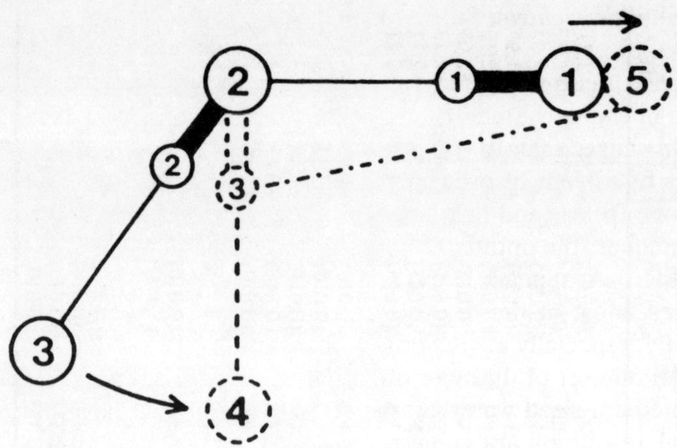

Figure 1.16 Schematic diagram of how the operation of RR3 relates an increase of H-bond distance to a decrease in O· · ·O· · ·O H-bond angle. Starting from a tetrahedral conformation of O1· · ·O2· · ·O3 with H-bond lengths of 2.76 Å, then if O3 moves to O4 (assuming H-bond linearity), the H3· · ·O1 (RR3) distance becomes less than the acceptable minimum contact of ~3.0 Å. To relieve this strain, the O2· · ·O1 H-bond distance should increase (O1 moves to O5), to retain a minimum H3· · ·O1 contact of 3.0 Å

oxygen positions are known, but no hydrogen sites (Savage, 1986a). We can use the RR restraints along with standard H-bond criteria (e.g. O· · · ·O H-bond distance = 2.5–3.2 Å, etc.) to assign the missing water hydrogens and formulate water networks around the protein molecules: an example is shown below (p. 37 and Figure 1.20).

7 Water Structure around Biological Molecules

Water is the main solvent environment for a large majority of biomolecules: these include proteins, nucleic acids, carbohydrates and many other smaller molecules, such as ATP. Within a water medium, the stabilization and operations of biomolecules are not well understood. The free energy balance for most protein molecules is quite marginal (~10–20 kcal/mol) and most of the contributions involve solvent interactions. Hence, details at the atomic level of the water structure will lead to a better understanding of the solvent contributions involved in biological processes. Extensive reviews of the analysis and implications of water structure around biomolecules have been given by Kuntz and Kauzmann (1973), Cooke and Kuntz (1974), Finney (1979), Finney *et al.* (1982), Edsall and McKenzie (1978, 1983), Savage (1986a), Saenger (1987), Westhof (1987), Finney and Savage (1988), Thanki *et al.* (1988) and Savage and Finney (1992).

High-resolution Neutron Studies

In principle, neutron diffraction appears to be ideal for analysing water structure in crystal systems the size of a small protein, typically in molecular weight range 3000–10 000. However, even for small proteins there are problems of solvent disorder and lack of atomic resolution data, as outlined above (p. 5) (and below, p. 37). Using present-day neutron diffraction techniques, the optimal largest size for obtaining good high-resolution data of <1.0 Å appears to be around molecular weight 2000–4000. For these sizes, most of the disorder of the solvent can be interpreted quite confidently, especially at very low temperatures of below 120 K, and also if more than one set of diffraction data is available for a particular system.

Two medium-sized water–biomolecular systems that have been studied to <1.0 Å resolution are vitamin B_{12} coenzyme (Lenhert, 1968; Savage *et al.*, 1987; Bouquiere, 1990) and cyclodextrins (Klar *et al.*, 1980; Betzel *et al.*, 1984; Zabel *et al.*, 1986, 1988; Steiner *et al.*, 1990). Some of the details of these analyses are described below.

Vitamin B_{12} Coenzyme

This biomolecule has a MW of ~1800 and crystallizes with 16–18 water molecules per asymmetric unit (~70 per unit cell). It contains a wide variety of biological groups (hydroxyls, amides, carbonyls, methyls, benzene ring, adenine, sugar rings) that interact with solvent, and thus is a good candidate for a water–biomolecular analysis. Two sets of neutron data have been collected: one at 279 K to 0.95 Å resolution (Savage *et al.*, 1987) and a second at 15 K to 0.90 Å resolution (Bouquiere, 1990). Both sets of data used crystals grown in D_2O, which presents difficulties in analysing solvent densities (p. 7). Oxygen and deuterium have similar positive scattering lengths; hence, the chemical identity of positive solvent peaks can easily be confused, especially in disordered regions.

For the 279 K data, a majority of the solvent peaks were interpretable with the aid of two X-ray data sets that were also collected (for example, see Figure 1.2). The latter data gave reliable oxygen positions which were then used in the interpretation of solvent regions in the neutron structure. The solvent structure of the 15 K data was found to be very much more ordered and interpretable than for the 279 K data. Although the crystal for the 15 K data was grown in D_2O, all the 'expected water deuterium positions' showed up as hydrogen atoms (that is, negative peaks) with high occupancies, indicating that almost full H/D exchange of the water deuteriums occurred (Bouquiere *et al.*, 1992). Although unexpected, this was a fortunate result, as it made the solvent interpretation very much easier than in the 279 K case, since the negative hydrogen peaks were instantly distinguishable from the positive oxygen peaks.

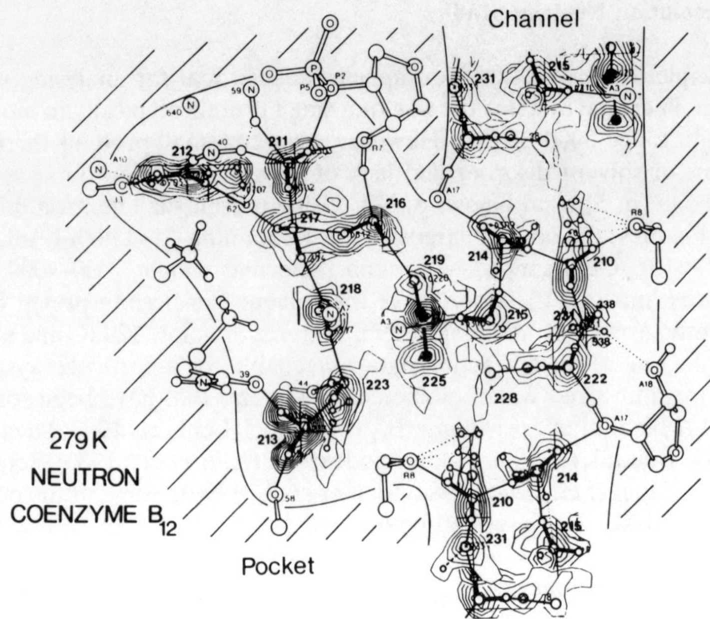

Figure 1.17 Summary of the main solvent density in the 279 K and neutron structure of coenzyme B_{12}. Solvent density is obtained from a $F_o - F_c$ difference Fourier synthesis in which the solvent was omitted from the phases. The main water density and positions are projected onto the *bc*-plane of the unit cell. An acetone molecule is present, separating the two main solvent regions, which are labelled pocket and channel. The main H-bonded water networks are included as full lines connecting the water and polar atom of the coenzyme B_{12}. The waters in the pocket form an ordered network, but in the channel the waters appear to be quite disordered

The solvent densities, the main solvent positions and H-bonding within the 279 K and 15 K structures are shown in Figures 1.17 and 1.18, respectively. The solvent peaks in the 15 K structure appear more ordered than in the 279 K structure. In particular, this can be seen in the channel region. The solvent density at 279 K is mostly continuous and water positions are difficult to locate. However, at 15 K, the density is much more discrete. See, for example, around waters 210, 228 and 222: the positive oxygen peaks lie at intervals of 2.8 Å apart and two distinct water networks can be easily formulated. In the 279 K structure a substantial number of water deuterium positions could not be located for several of the partially occupied water networks, while at 15 K over 95% of the water hydrogens were visible (Bouquiere *et al.*, 1992).

The main solvent H-bonded networks in each structure are also shown in Figures 1.17 and 1.18. The solvent is distributed over two main regions — a pocket region and a channel region. At 279 K an acetone molecule is present at ~0.5 occupancy separating the pocket and channel. However, at

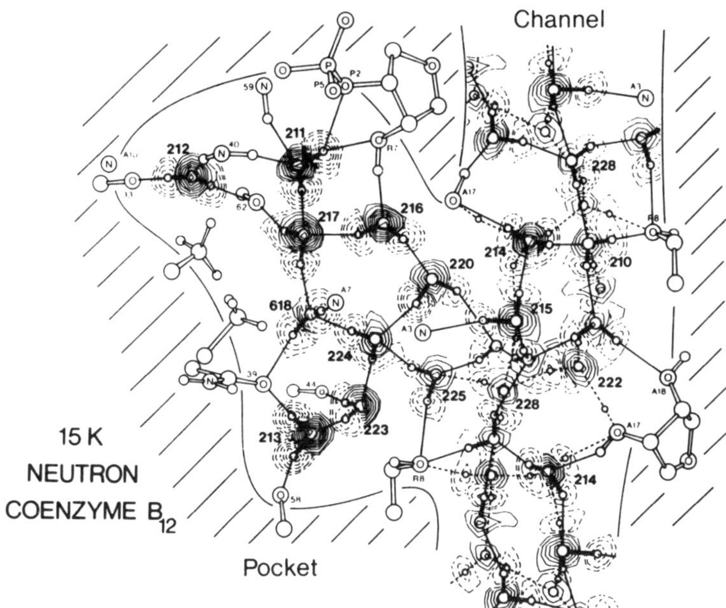

Figure 1.18 Summary of the main solvent density in the 15 K and neutron structure of coenzyme B_{12}. Solvent density is obtained from a $F_o - F_c$ difference Fourier synthesis in which the solvent was omitted from the phases. The main water density and positions are projected onto the *bc*-plane of the unit cell. The acetone has a very low occupancy and is not shown, for clarity. A continuous water network structure is present between the pocket and channel regions. In the pocket region the water is well-ordered, forming a main H-bonded water network (full lines). In the channel the water is also ordered, forming two distinct alternative networks: one shown by full-line H-bonds and the other by dashed-line H-bonds

15 K the acetone has a very low occupancy of ~0.1 and is not present in the main networks. Instead, the pocket and channel are completely filled with water molecules forming continuous water networks consisting of many 4-, 5- and 6-membered water rings (Bouquiere *et al.*, 1992) — similar to the ring structures in the ice polymorphs (Section 4). Furthermore, several of the incomplete partial water networks present at 279 K were present as complete networks in the 15 K structure, with better precision.

The estimated standard deviations of the water distances in the 15 K structure were substantially better than those of the 279 K structure: 0.013–0.025 Å at 15 K, compared with 0.06–0.15 Å at 279 K. Overall, the water structure at 15 K appears to have been 'frozen' in with very little apparent dynamic disorder present — as judged by the substantially reduced number of alternative water oxygen positions present. Most of the remaining solvent disorder is probably of a static nature, with different networks being occupied in different asymmetric units.

Cyclodextrins

Cyclodextrins are essentially carbohydrate complexes of molecular weight 1000–1300 which are somewhat smaller than coenzyme B_{12}. They are composed of 6–8 glucose rings which are bonded together through 1–4 oxygen linkages forming ring structures. Three forms have been studied by neutron diffraction:

(1) α-Form: contains 6 sugars in the ring with 6 waters in the crystals. Neutron data collected at 295 K to 1.0 Å resolution (Klar *et al.*, 1980).

(2) β-Form: contains 7 sugars with up to 11 waters in the crystals. Three sets of neutron data collected: at 295 K to 0.6 Å (Betzel *et al.*, 1984), containing 11 waters; at 120 K to 0.8 Å (Zabel *et al.*, 1986), containing 11 waters; and at 15 K to 0.93 Å (Steiner *et al.*, 1990), containing 8 waters and an ethanol molecule.

(3) γ-Form: contains 8 sugars and 12–14 waters in the crystals. Neutron data at 110 K to ~1.0 Å (Zabel *et al.*, 1988).

Of the three above forms, the solvent structure in β-cyclodextrin has been studied in detail the most and is discussed further here. At 295 K the 11 waters appear to be statistically disordered over ~16 sites. At 120 K only 1 of the 11 waters is disordered, partially occupying 2 sites. In the 15 K structure, 2 of the 8 waters occupy alternative sites. Figure 1.19 shows a stereoview of the H-bond chains between the water and hydroxyl oxygens in the 120 K structure.

Dynamic and static disorder of the solvent molecules is difficult to separate out from diffraction data, since both time and space are sampled, giving only an averaged picture of the structure. However, at very low temperatures, such as 110 K, it appears to be possible to distinguish to some extent between the two. This has been shown to be apparent in the β-cyclodextrin study.

Calorimetry has indicated that the β-form goes through an exothermic phase transition at ~227 K (Fujiwara *et al.*, 1983). An ordering of the hydrogens has been observed in the neutron structures below this transition temperature. At 295 K the polar hydrogens between most H-bonded oxygens occupy 2 half H-sites (O−H· · ·H−O), while at 15 K and 120 K most of the hydrogens fully occupy only 1 hydrogen site between pairs of oxygens (O−H· · ·O). The reduction in the occupancy of H-sites suggests that a majority of the hydrogens are dynamically disordered above 227 K (occupying 2 sites), and become ordered below the transition temperature. If static disorder were present, then most of the 2 half H-sites between the oxygen pairs would be occupied below 227 K. The dynamic disorder has been termed flip-flop H-bonds, whereby sequential O−H· · · O−H· · ·O−H switch dynamically to the alternative arrangement

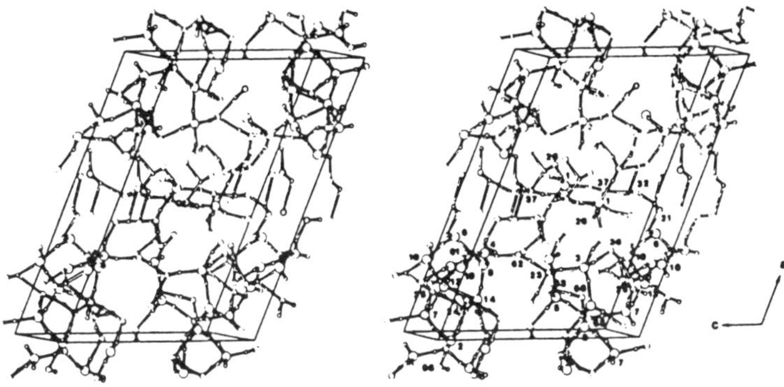

Figure 1.19 Stereoview of unit cell of β-cyclodextrin structure at 120 K. All the carbon and C–H hydrogen atoms omitted for clarity. Covalent O–D bonds shown by solid lines, H-bonds by open lines. Circles in increasing size represent D, O (hydroxyl) and O (water) atoms. Individual atoms denoted by two numbers are cyclodextrin hydroxyls

H–O· · · ·H–O· · · ·H–O with the hydrogen atoms rotating about each oxygen (Saenger *et al.*, 1982).

Static disorder of whole water molecules also occurs in the ß-structures below 227 K: 2 of the 8 waters in the 15 K structure occupy alternate sites; 1 of the 11 waters in the 120 K structure occupies 2 sites. In the 15 K coenzyme B_{12} structure 12 of the 18 waters can occupy alternative sites.

Clathrate Water Structures

One of the key areas of interest in water–macromolecular interactions is the structure of water around apolar groups and the involvement of such interactions in the hydrophobic effect and protein stabilization (Kauzmann, 1959; Franks, 1975; Tanford, 1980). The classical picture derived from small hydrate crystals containing apolar groups (Jeffrey, 1969) is one of water molecules forming regular cage structures enclosing the apolar groups. The water molecules make H-bonds only with other water molecules or nearby polar atoms and clearly avoid H-bonded-type interactions with apolar groups, thus preventing the loss of H-bonds. Clathrate-like cages of water molecules are observed around certain apolar molecules and, because of the small size of these hydrate systems, the waters in these cages are well ordered and well characterized in structural studies. Partial clathrate cages are observed in the coenzyme B_{12} hydrate structure shown in Figure 1.18: on the left-hand side of the pocket region two methyl groups point into the solvent region and the local waters (217 and 618) form a bridge-like structure over them, forming H-bonds to local polar atoms. In addition, a ring of 5 waters (217, 216, 220, 224 and 618) spans over a bed of apolar groups (not shown in Figure 1.18) that are located below the ring.

In larger biomolecular hydrates, such as protein crystals, the solvent close to non-polar groups appears to be relatively disordered. For most protein data, it is not possible to locate clathrate around apolar groups to the extent found in small hydrates (Jeffrey, 1969). In the case of crambin, several distorted five-membered rings are present (Teeter, 1984), but there are no extensive partial cages. It appears that in most crystal hydrates with water–apolar contacts present, the waters tend to maximize the number of H-bonds they make — which does not necessarily mean they form regular water cages, but instead make the best of the environment they find themselves in. This underlines the versatility and flexibility of water molecules, attributes which appear central to the role of water in biomolecular stability and interactions.

Protein and Nucleic Acid Crystals

The majority of protein and nucleic acid hydrate systems are relatively large and present problems in obtaining high-resolution data (Section 2). By volume, such hydrates contain anywhere from 25% to 90% water, generally filling the unoccupied spaces between the biomolecules. The nature of the water structure in the first solvent shell usually appears to be fairly well ordered, with a majority of these waters making H-bond contacts to the biomolecule. In the layers further away from the surface the solvent usually becomes significantly disordered and bulk-liquid-like.

Over 600 protein and more than 30 large nucleic acid structures have been solved using X-ray diffraction. However, the data for most of these structures lie at resolutions of 1.5–3.0 Å — that is, substantially above atomic resolutions of ~1.0 Å. Only a small proportion of proteins and nucleic acids have been studied to better than 1.5 Å resolution, mainly because the crystals involved do not diffract well at higher resolutions, owing to molecular and lattice disorder inherent in large biomolecular crystals. Several of the high-resolution analyses in which the solvent regions have been well characterized are crambin (Hendrickson and Teeter, 1981), rubredoxin (Watenpaugh *et al.*, 1978) and insulin (Baker *et al.*, 1988). Crambin has been studied to ~0.9 Å resolution, with ~80% of the solvent observed to be well ordered: 73 solvent sites were assigned, 4 of them as ethanol molecules. Around some of the apolar groups, several distorted pentagonal water rings were present (Teeter, 1984), forming hydrogen-bonded bridge structures between the polar groups of the protein. In 2Zn insulin crystals (1.5 Å resolution) ~70% of the solvent content was modelled using ~340 solvent sites, while in rubredoxin (1.2 Å resolution) 127 sites were assigned, accounting for 50% of the solvent.

From X-ray analyses of most proteins, only the water oxygen positions are obtained with any reliability, and the geometries of the hydration of polar and apolar protein groups are thus restricted to the non-hydrogen atoms (unless calculated positions are included). Overall, there is reasonable agreement of the hydrogen bond geometries with the more accurate geometries obtained from small-molecule data. The O(W)· · · ·O(P) distances vary from 2.2 Å to 3.5 Å, with a main peak around 2.9 Å; O(W)· · · ·N(P) distances range from 2.4 Å to 3.5 Å, with a peak around 3.0 Å. Distances below ~2.5 Å are mainly due to alternative partially occupied solvent sites. Y· · · ·O(W)· · · ·Y angles (where Y = O or N) range between ~70° and ~150° (cf. 109 ± 40° for ices and small hydrates) with the majority around 100–120°. To obtain more detailed and accurate geometries especially involving the hydrogen positions, neutron diffraction (close to 1.0 Å resolution) is required using relatively small biomacromolecular systems. Several neutron analyses have been carried out on small proteins at resolutions below 1.5 Å — crambin at 1.2 Å (46 amino acids: Teeter and Kossiakoff, 1984) and hen lysozyme at ~1.4 Å (129 amino acids: Mason *et al.*, 1984). The details of these analyses are much less precise than their equivalent high-resolution X-ray studies and have as yet to be fully reported.

As mentioned above, the solvent regions in 2Zn insulin crystals (Baker *et al.*, 1988) have been extensively studied and oxygens have been assigned. Figure 1.20(a) shows the electron density over one of the solvent regions along with water sites. In Figure 1.20(b) one of several possible water networks formulated using the RR structural restraints given in Section 6 is shown. Several alternative sites of <2.5 Å are present and alternative H-bonded networks can also be formulated.

Characteristics of Water Structure around Biomolecules

In the well-determined biomolecular structures (examined by neutron diffraction to <1.0 Å resolution), the water structure is similar to that found in the ices and small hydrates: wide ranges of H-bond geometries occur and the RR restrictions are not violated.

One regular feature present in the biomolecular hydrates of pp. 31–35, the smaller hydrates and the ice polymorphs, is the maximization of H-bonding of water hydrogen atoms. Over 95% of these atoms participate in H-bonds (100% for the ice polymorphs). This tendency to maximize the number of polar hydrogens making H-bonds also appears to occur in proteins. A recent survey of high-resolution protein (X-ray) structures suggests that ~95% of all polar hydrogen atoms participate in H-bonds.

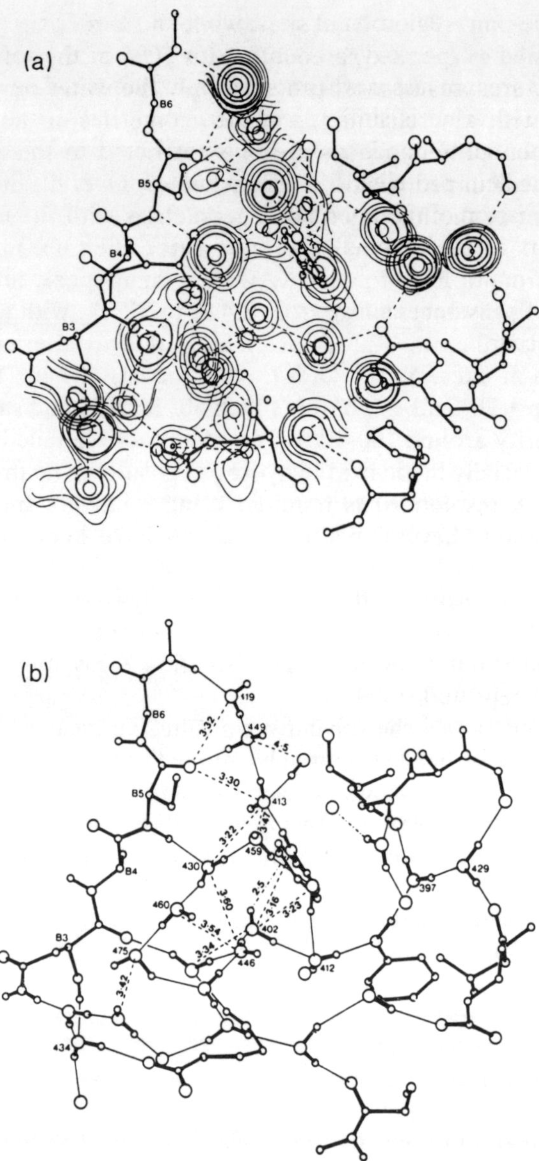

Figure 1.20 Analysis of the water structure over one region of solvent density adjacent to residues B3–B7 in porcine 2Zn insulin crystals. (*a*) Solvent density at 1.5 Å resolution from X-ray refinement. (*b*) One of several possible water networks constructed in which all the non-bonded minimum constraints are satisfied (broken lines represent non-bonded O· · · ·O contacts)

Over the lone-pair region of water oxygens, the fraction of potential H-bonds formed (2 per oxygen assumed as potential maximum) varies between structures: in ices it is 100%, while in hydrates it varies from 75 to 85%. This latter value is similar to a value of ~80%, estimated for the fraction of potential H-bonds made by carbonyl oxygens (potential of 2 per oxygen assumed) in proteins.

The maximization of the H-bonding of polar hydrogens probably plays a significant part in the stabilization of protein structures within a water medium. Any major loss of potential H-bonding would tend to destabilize a protein–solvent system and would thus have to be compensated for when such events occur.

8 Summary

The basic characteristics of water structure can be related to a maximization of the number of H-bonds and the minimization of the short-range RR repulsive restrictions. The attractive H-bonds are crucial in holding the overall structure together, while the RR short-range restrictions play a prominent role in determining the local orientational structure and the actual H-bond geometries between water molecules, accounting for the large deviations from tetrahedrality.

The RR restrictions form an excluded volume constraint which excludes significant regions of the configurational space of water geometries. Furthermore, the RR restrictions appear to be anisotropic with respect to their angular geometries. Using a QM electron density model, the anisotropy can be modelled using four displaced van der Waals spheres, significantly simplifying the description of the non-bonded interactions.

Inclusion of anisotropic repulsive interactions in current water potential functions may improve the present level of agreement between simulation and experimental results for liquid water. Very few simulations are able to reproduce the correct height and width for both the first (2.76 Å) and second (4.5 Å) peaks in the experimental PCF (Figure 1.4a). The majority of simulations overemphasize the first peak in their attempts to reproduce the 'water-like' second peak, making the overall water structure too tetrahedral. At present, one of the best potentials that gives reasonable agreement with the experimental PCF is the MCY potential of Matsuoka *et al.* (1976). The first peak has the correct shape, while the second peak at 4.25 Å is fairly close to the experimental value of 4.5 Å. This potential is one of the few water force fields that include specific O· · ·H and H· · ·H repulsions. Most potentials focus on the electrostatics to handle the orientational characteristics of water structure, only using a single-centred isotropic core sited on or near the oxygen to account for short-range repulsions. Clearly, an anisotropic repulsive model is required, as evidenced from the MCY potential and recognition of aspherical RR restrictions (Section 6).

Many of the features of water structure found within ices and hydrates can be explained using the RR restrictions. For example, in the ice polymorphs the RR2 O· · ·O and RR3 remote H2· · ·O1 contacts become fully strained in the various phases as higher pressures are encountered. Decreases in volume can be achieved by adopting structures that have large deviations from tetrahedrality that do not violate the RR excluded volume restraints, and/or form interpenetrating H-bonded lattices (self-clathrates, as in ices VI, VII and VIII). Details of the dynamics of water molecules from diffraction experiments are limited, as only the space–time-averaged positions are obtained. Such details are relevant to entropic contributions of water to biomolecular-solvent stability.

In addition to rationalizing water itself, the RR regularities can be used directly in the interpretation of solvent in macromolecular crystals. The regularities can be used as restraints in the same way as protein restraints are used in fitting a protein molecule to electron density in macromolecular structure analyses. This enables more consistent hydration structures to be obtained around biomacromolecules from crystallographic studies, where the available resolution is too low (1.2–2.5 Å) and the solvent too disordered to allow direct and unambiguous determination of the water structure. Undoubtedly, diffraction data at very high resolution (<1.0 Å), at very low temperatures (<120 K), and preferably using neutron scattering on intermediate-sized hydrate systems, provide the most pertinent information on the water structure around biological macromolecules.

References

Bacon, G. E. (1975) *Neutron Diffraction*, 3rd edn. Clarendon Press, Oxford

Baker, E. N., Blundell, T. L., Cutfield, J. F., Cutfield, S. M., Dodson, E. J., Dodson, G. G., Crowfoot Hodgkin, D. M., Hubbard, R. E., Isaacs, N. W., Reynolds, C. D., Sakabe, K., Sakabe, N. and Vijayan, N. M. (1988). The structure of 2Zn pig insulin crystals at 1.5A resolution. *Phil. Trans. Roy. Soc. London*, **319**, 369–456

Betzel, C., Saenger, W., Hingerty, B. E. and Brown, G. M. (1984). Circular and flip-flop hydrogen bonding in β-cyclodextrin undecahydrate. *J. Am. Chem. Soc.*, **106**, 7545–7557

Bouquiere, J. (1990). High resolution neutron studies of solvent networks in vitamin B_{12} coenzyme at 15 Kelvin. *Acta Cryst. IUCr. Abstracts*, **A46S**, C106

Bouquiere, J. P., Finney, J. L., Lehmann, M. S., Lindley, P. F. and Savage, H. F. J. (1992). High resolution neutron study of vitamin B_{12} coenzyme at 15 kelvin: structure analysis and comparison with the structure at 279 kelvin. *Acta Cryst. B.* (submitted)

Chiari, G. and Ferraris, G. (1982). The water molecule in crystalline hydrates studied by neutron diffraction. *Acta Cryst.*, **B38**, 2331–2341

Cooke, R. and Kuntz, I. D. (1974). Properties of water in biological systems. *Ann. Rev. Biophys. Bioeng.*, **3**, 95–126

Dore, J. C. (1985). Structural studies of water by neutron diffraction. In *Water Science Reviews*, Vol. 1. Cambridge University Press, Cambridge, pp. 3–92

Edsall, J. T. and Mackenzie, H. A. (1978). Water and proteins. I. The significance and structure of water; its interaction with electrolytes and non-electrolytes. *Adv. Biophys.*, **10**, 137–207

Edsall, J. T. and McKenzie, H. A. (1983). Water and proteins. II. The location and dynamics of water in protein systems and its relation to their stability and properties. *Adv. Biophys.*, **16**, 53–183

Eisenberg, D. and Kauzmann, W. (1969). *Structure and Properties of Water*. Oxford University Press, London

Engelhardt, H. and Kamb, B. (1981). Structure of ice IV, a metastable high pressure phase. *J. Chem. Phys.*, **75**, 5887–5899

Ferraris, G. and Franchini-Angela, M. (1972). Survey of the geometry and environment of water molecules in crystalline hydrates studied by neutron diffraction. *Acta Cryst.*, **B28**, 3572–3583

Finney, J. L. (1979). The organization and function of water in protein crystals. In Franks, F. (Ed.), *Water: A Comprehensive Treatise*, Vol. 6. Plenum Press, New York, pp. 47–122

Finney, J. L., Goodfellow, J. M. and Poole, P. L. (1982). The structure and dynamics of water in globular proteins. In Davies, D. B., Danyluk, S. and Saenger, W. (Eds), *Structural Molecular Biology*. Plenum Press, New York, p. 387

Finney, J. L. and Savage, H. F. J. (1988). Solvation of proteins. In Dogonadze, R. R., Kalman, E., Kornyshev, A. A. and Ulstrup, J. (Eds), *The Chemical Physics of Solvation*. Elsevier, pp. 603–663

Franks, F. (Ed.) (1972–1982). *Water: A Comprehensive Treatise*, Vols 1–7. Plenum Press, New York

Franks, F. (1975). The hydrophobic interaction. In Franks, F. (Ed.), *Water: A Comprehensive Treatise*, Vol. 4. Plenum Press, New York, pp. 1–94

Fujiwara, T., Yamazaki, M., Tomizu, Y., Tokuoka, R., Tomita, K. I. and Saenger, W. (1983). *J. Chem. Soc. Japan*, **2**, 181–187

Hamilton, W. C. (1968). On hydrogen bonding in inorganic crystals: some generalizations, some recent results, and some new techniques. In Rich, A. and Davidson, N. (Eds), *Structural Chemistry and Molecular Biology*. Freeman, San Francisco, pp. 466–483

Hamilton, W. C., Kamb, B., Laplaca, S. J. and Prakash, A. (1969). Deuteron arrangements in high pressure forms of ice. In Riehl, N., Bullemer, B. and Engelhardt, H. (Eds), *Physics of Ice*. Plenum Press, New York, pp. 44–58

Hendrickson, W. A. and Teeter, M. M. (1981). Structure of the hydrophobic protein crambin determined directly from the anomalous scattering of sulphur. *Nature*, **290**, 107–113

Hermansson, K. (1984). In *The Electron Distribution in the Bound Water Molecule*. PhD Thesis, Acta Universitatis Upsaliensis, p. 9

Hobbs, P. V. (1974). *Ice Physics*. Clarendon Press, Oxford

Jeffrey, G. A. (1969). Water structure in organic hydrates. *Accounts Chem.*, **2**, 344–352

Jeffrey, G. A. and Maluszynska, H. (1990). The stereochemistry of the water molecules in the hydrates of small biological molecules. *Acta Crystallogr.*, **B46**, 546–549

Johari, G. P. and Whalley, E. (1979). Evidence for a very slow transformation in ice VI at low temperatures. *J. Chem. Phys.*, **70**, 2094–2097

Jorgensen, J. D., Beyerlein, R. A., Watanabe, N. and Worlton, T. G. (1984).

Structure of D_2O ice VIII from *in situ* powder neutron diffraction. *J. Chem. Phys.*, **81**, 3211–3214

Jorgensen, J. D. and Worlton, T. G. (1985). Disordered structure of ice D_2O ice VII from *in situ* neutron powder diffraction. *J. Chem. Phys.*, **83**, 329–333

Kamb, B. (1964). A proton-ordered form of ice. *Acta Cryst.*, **17**, 1437–1449

Kamb, B. (1965). Structure of ice VI. *Science*, **150**, 205–209

Kamb, B. (1973). Crystallography of ice. In Whalley, E., Jones, S. J. and Gold, L. W. (Eds), *Physics and Chemistry of Ice*. Royal Society of Canada, Ottawa, Ontario, pp. 28–41

Kamb, B. and Davis, B. L. (1964). Ice VII, the densest form of ice. *Proc. Natl Acad. Sci. USA*, **52**, 1433–1439

Kamb, B., Hamilton, W. C., LaPlaca, S. J. and Prakash, A. (1971). Ordered proton configuration in ice II, from single crystal neutron diffraction. *J. Chem. Phys.*, **55**, 1934–1945

Kamb, B. and Prakash, A. (1968). Structure of ice III. *Acta Cryst.*, **B24**, 1317–1327

Kamb, B., Prakash, A. and Knobler, C. (1967). Structure of ice V. *Acta Cryst.*, **22**, 706–715

Kauzmann, W. (1959). Some factors in the interpretation of protein denaturation. *Adv. Protein Chem.*, **14**, 1–63

Klar, B., Hingerty, B. and Saenger, W. (1980). Topography of cyclodextrin inclusion complexes. XII. Hydrogen bonding in the crystal structure of α-cyclodextrin hexahydrate: the use of a multicounter detector in neutron diffraction. *Acta Cryst.*, **B36**, 1154–1165

Kuhs, W. F., Finney, J. L., Vettier, C. and Bliss, D. V. (1984). Structure and hydrogen bond ordering in ices VI, VII and VIII by neutron powder diffraction. *J. Chem. Phys.*, **81**, 3612–3623

Kuhs, W. F. and Lehmann, M. S. (1983). The structure of ice lh by neutron diffraction. *J. Phys. Chem.*, **87**, 4312–4313

Kuhs, W. F. and Lehmann, M. S. (1986). The structure of ice lh. In *Water Science Reviews*, Vol. 2. Cambridge University Press, Cambridge, pp. 1–65

Kuntz, I. D. and Kauzmann, W. (1973). Hydration of proteins and polypeptides. *Adv. Protein Chem.*, **28**, 239–345

LaPlaca, S. J., Hamilton, W. C., Kamb, B. and Prakash, A. (1973). On a nearly proton ordered structure for ice IX. *J. Chem. Phys.*, **58**, 567–580

Lenhert, P. G. (1968). The structure of vitamin B12. VII. The X-ray analysis of the vitamin B12 coenzyme. *Proc. Roy. Soc. A*, **303**, 45–84

Mason, S. A., Bentley, G. A. and McIntyre, G. J. (1984). Deuterium exchange in lysozyme at 1.4 Å resolution. In Schoenborn, B. P. (Ed.), *Neutrons in Biology*. Plenum Press, New York, pp. 323–334

Matsuoka, O., Clementi, E. and Yoshimine, M. (1976). CI study of water dimer potential surface. *J. Chem. Phys.*, **64**, 1351–1361

Mishima, O., Calvert, L. D. and Whalley, E. (1984). 'Melting ice' I at 77 K and 10 kbar: A new method of making amorphous solids. *Nature*, **310**, 393–395

Narten, A. H., Danford, M. D. and Levy, H. A. (1967). X-ray diffraction study of liquid water in the temperature range 4-200 °C. *Discuss. Faraday Soc.*, **43**, 97–107

Narten, A. H., Venkatesh, C. G. and Rice, S. A. (1976). Diffraction pattern and structure of amorphous solid water at 10 and 77 K. *J. Chem. Phys.*, **64**, 1106–1121

Olovsson, I. and Jonsson, P.-G. (1976). X-ray and neutron diffraction studies of hydrogen bonded systems. In Schuster, P., Zundel, G. and Sandorfy, C. (Eds.), *The Hydrogen Bond*, Vol. 2. North-Holland, Amsterdam, p. 393

Palinkas, G., Kalman E. and Kovacs, P. (1977). Liquid water II: Experimental

atom pair correlation functions of liquid D_2O. *Mol. Phys.*, **34**, 525–537
Pedersen, B. (1974). The geometry of hydrogen bonds from donor water molecules. *Acta Crystallogr.*, **B30**, 289–291
Peterson, S. W. and Levy, H. A. (1957). A single-crystal neutron diffraction study of heavy ice. *Acta Cryst.*, **10**, 70–76
Polian, A. and Grimsditch, M. (1984). New high-pressure phase of H_2O: ice X. *Phys. Rev. Lett.*, **52**, 1312–1314
Saenger, W. (1987). Structure and dynamics of water surrounding biomolecules. *Ann. Rev. Biophys. Biophys. Chem.*, **16**, 93–114
Saenger, W., Betzel, Ch., Hingerty, B. and Brown, G. M. (1982). Flip-flop hydrogen bonding in a partially disordered system. *Nature*, **296**, 581–583
Savage, H. F. J. (1986a). Water structure in crystalline solids: ices to proteins. *Water Science Reviews*, Vol. 2. Cambridge University Press, Cambridge, pp. 67–148
Savage, H. F. J. (1986b). Water structure in vitamin B_{12} coenzyme crystals. *Biophys. J.*, **50**, 947–80
Savage, H. F. J. and Finney, J. L. (1986). Repulsive regularities of water structure in ices and crystalline hydrates, *Nature*, **322**, 717–720
Savage, H. F. J. and Finney, J. L. (1992). New approaches in studying biomolecule-water interactions. In Gomez-Puyou, A. (Ed.), *Biomolecules in Organic Solvents*. CRC Press, Boca Raton, Florida, pp. 1–32
Savage, H. F. J., Lindley, P. F., Finney, J. L. and Timmins, P. A. (1987). High-resolution neutron and x-ray refinement of vitamin B12 coenzyme, $C_{72}H_{100}CoN_{18}O17P.17H_2O$. *Acta. Cryst.*, **B43**, 280–295
Schweizer, K. S. and Stillinger, F. H. (1984). High pressure phase transitions and hydrogen-bond symmetry in ice polymorphs. *J. Chem. Phys.*, **80**, 1230–1240
Shimaoka, K. (1960). Electron diffraction study of ice. *J. Phys. Soc. Japan*, **15**, 106–119
Steiner, T., Mason, S. A. and Saenger, W. (1990). Cooperative O- H· · ·O hydrogen bonds in β-cyclodextrin-ethanol-octahydrate at 15 K: a neutron diffraction study. *J. Am. Chem. Soc.*, **112**, 6184–6190
Stout, G. H. and Jensen, L. H. (1968). *X-ray Structure Determination*. Macmillan, New York
Tanford, C. (1980). *The Hydrophobic Effect*. Wiley, New York
Teeter, M. M., (1984). Water structure of a hydrophobic protein at atomic resolution: pentagonal rings of water molecules in crystals of crambin. *Proc. Natl Acad. Sci. USA*, **81**, 6014–6018
Teeter, M. M. and Kossiakoff, A. A. (1984). The neutron structure of the hydrophobic plant protein crambin. In Schoenborn, B. P. (Ed.), *Neutrons in Biology*. Plenum Press, New York, pp. 335–348
Thanki, N., Thornton, J. M. and Goodfellow, J. M. (1988). Distributions of water around amino acid residues in proteins. *J. Mol. Biol.*, **202**, 637–657
Thiessen, W. E. and Narten, A. H. (1982). Neutron diffraction study of light and heavy water mixtures at 25°C. *J. Chem. Phys.*, **77**, 2625–2662
Watenpaugh, K. D., Margulis, T. N., Sieker, L. C. and Jensen, L. H. (1978). Water structure in a protein crystal: rubredoxin at 1.2 Å resolution. *J. Mol. Biol.*, **122**, 175–190
Water Science Reviews (1985–). Vols 1–5. Cambridge University Press, Cambridge
Westhof, E. (1987). Hydration of oligonucleotides in crystals. *Int. J. Biol. Macromol.*, **9**, 186–192
Whalley, E. (1983). Cubic ice in nature. *J. Phys. Chem.*, **87**, 4174–4179
Woolfson, M. M. (1970). *An Introduction to X-ray Crystallography*. Cambridge University Press, Cambridge

Zabel, V., Hingerty, B., Mason, S. A. and Saenger, W. (1988). Neutron diffraction study of γ-cyclodextrin. $14D_2O$ at 110K. *Am. Cryst. Soc. Ann. Meeting*, **16**, 64
Zabel, V., Saenger, W. and Mason, S. A. (1986). Neutron diffraction study of the hydrogen bonding in β-cyclodextrin undecahydrate at 120 K: from dynamic flip-flops to static homodromic chains. *J. Am. Chem. Soc.*, **108**, 3664–3673

2
Thermodynamic and Dynamic Properties of Water

H.-D. Lüdemann

1 Introduction

Liquid water is the essential biological solvent for all life processes occurring in our planetary system. Human existence and most terrestrial life is confined to pressures around 0.1 MPa (≡ 1 bar). However, a large fraction of the biological production occurs in the oceans and thus at high pressures. The known biosphere on earth extends down into the abyssal depth of the oceanic trenches, where pressures up to 110 MPa at temperatures around 274 K are found.[1, 2]

The structural properties of liquid water at biological temperatures (300 ± 50 K) are determined by the random and distorted hydrogen bond network formed by the almost tetrahedral arrangement of the two lone pairs and the two protons of the individual water molecules. This tetrahedral local order with its low inherent space-filling ability predominant also in the many polymorphs of ice confers to the neat liquid a unique set of thermodynamic and dynamic properties and is the cause of the many irregularities observed in the aqueous solutions of ionic, polar and nonpolar compounds. Liquid water has until now defied all attempts at a comprehensive quantitative description of the temperature and density, ρ,T, dependence of its physicochemical properties by theoreticians and experimentalists as well as computer simulators.

Structural proposals offered in the literature can be classified into two categories: continuum models and mixture models. During recent decades the fashion has several times swung from one class of model to the other. It is this author's prejudice that only continuum descriptions, as, for instance, provided by the random network models of Bernal and Fowler[3] and Sceats and Rice[4,5] and their later modifications[6] are in accord with all experimental evidence and with the results of computer simulations.

A very lucid concept for the discussion of water structure has been offered by Eisenberg and Kauzmann.[7] They consider three different time domains. The structure observed over time intervals of 10^{-14}–10^{-16} s, i.e. in periods short compared with the intramolecular vibrations and librations, is named instantaneous structure or I-structure. The range 10^{-12}–10^{-14} s, which is comparable to or longer than the inverse IR and Raman frequencies, but short on the time-scale of translational diffusion over distances of the molecular diameter, is named vibrationally averaged structure or V-structure. On the longer time-scale we find the diffusionally averaged or D-structure. In the following it will be attempted to present a few physical properties of water and to discuss them qualitatively.

2 Density and Isothermal Compressibility of Liquid Water

Figure 2.1 gives the particle number densities at saturation vapour pressure for light and heavy water. Both liquids exhibit a temperature of maximum density, T_{MD} (see insert, Figure 2.1); on further cooling the liquids expand and this expansion becomes most pronounced in the supercooled metastable region. (The physics of supercooled liquid water will not be discussed in this context. The interested reader is referred to some recent reviews.[8–10]) The particle number density at saturation vapour pressure for D_2O is larger than for light water at temperatures $T > 300$ K, while for lower temperature the reverse is true. This is a consequence of the formation of the open tetrahedral structure with almost linear hydrogen bonds. Because of the smaller amplitude zero point vibrations of the deuterons, the hydrogen bonds are stronger in heavy water and thus force the liquid already at higher temperatures into the formation of the open network.

Figure 2.2 gives the isothermal compressibility κ_T for water and for a normal distorted network liquid ethylene glycol. In ethylene glycol, as in all normal liquids, κ_T decreases with falling temperature, while for water it passes through a pronounced minimum around 320 K and increases significantly upon further cooling. The fluctuating open tetrahedral network structure that develops upon cooling is obviously compressed more easily than the thermally distorted local arrangements that dominate at high temperatures.

Also, the molar heat capacities, C_p, of light and heavy water both show a peculiar temperature dependence (cf. Figure 2.3). For all normal liquids, including hydrogen-bonded liquids such as the lower alcohols, C_p de-

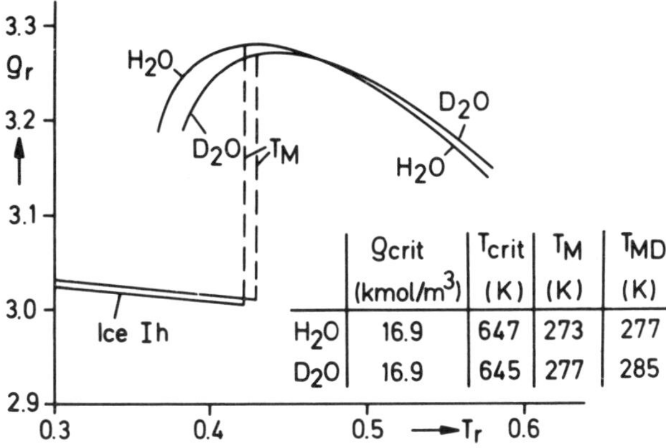

Figure 2.1 Reduced particle number density as a function of reduced temperature.

$\rho_r = \left(\frac{\rho(p, T)}{\rho_{crit}}\right)$ for light and heavy water at saturation vapour pressure

$T_r = \frac{T}{T_{crit}}$

$T_{MD} \equiv$ temperature of maximum density

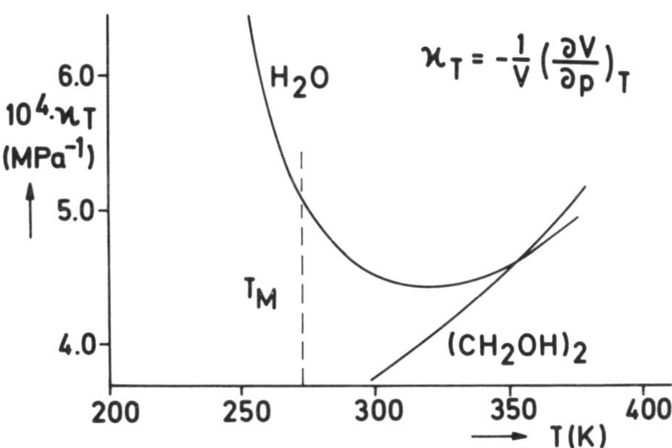

Figure 2.2 Isothermal compressibility of water and ethanediol-1, 2

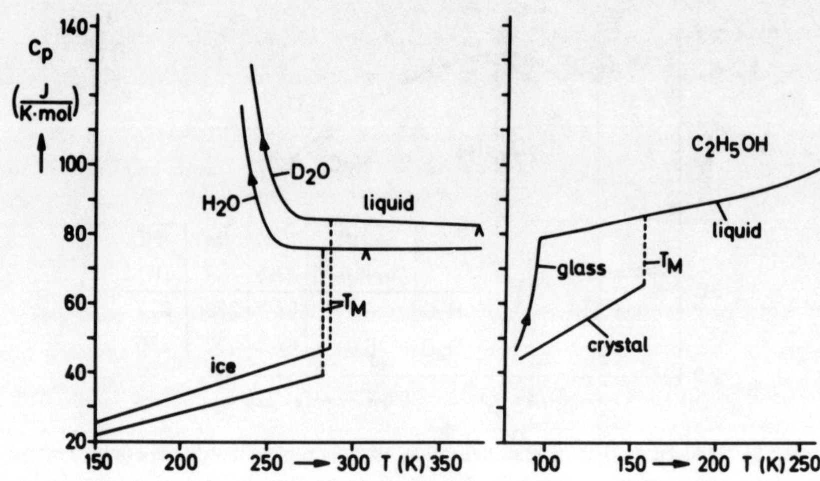

Figure 2.3 Temperature dependence of the molar heat capacity at constant pressure C_p of H_2O and C_2H_5OH (Λ = temperature of the C_p minimum (H_2O = 308 K; D_2O = 373 K))

creases with falling temperature. C_p for liquid water passes through a shallow minimum and, in the metastable supercooled range, the increase with falling temperature becomes very steep. This is generally explained by the transient formation of the local tetrahedral structures with four hydrogen bonds per water molecule. The peculiar concentration and temperature dependence of C_p in many aqueous solutions of non-polar compounds[11] is thus already manifest in neat liquid water.

3 Dynamic Properties

In Figure 2.4 the temperature dependence of the viscosity, η, of water is compared with that of methanol, a liquid in which the hydroxyl group can participate in hydrogen bonding with its two lone pairs and one proton leading to two-dimensional associates (chains, branched chains and rings) only. The significant difference with the three-dimensional network forming liquid water is seen in the stronger temperature dependence of η and in the pronounced curvature of the Arrhenius plot, which is most pronounced in the metastable supercooled region.

In Figure 2.5 the self-diffusion coefficient, D, of water[12, 13] is given together with results for methylfluoride[14] and two lower alcohols[15] (CH_3OH, C_2H_5OH). Again the main difference between the alcohols and water is the steeper slope at low temperatures. Methylfluoride has the same gas phase dipole moment (1.85 Debye) as water, and is very similar

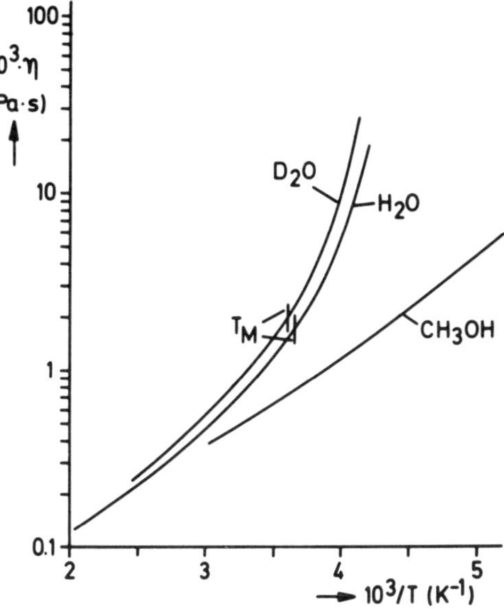

Figure 2.4 Arrhenius plot of the shear viscosities of light and heavy water and methanol

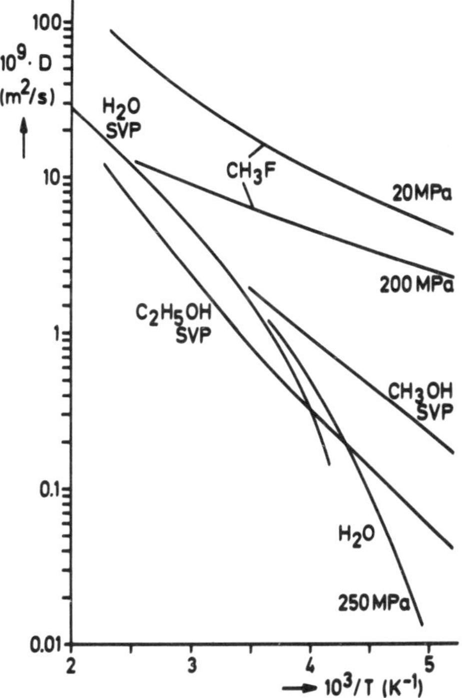

Figure 2.5 Temperature and pressure dependence of the self-diffusion coefficient of water, methanol, ethanol and methylfluoride (SVP = saturation vapour pressure)

in mass, molar volume and gas phase dipole moment (1.70 Debye) to methanol. However, it lacks the ability to form hydrogen bonds, and therefore this liquid is ideally suited to reveal the retarding influence of hydrogen bond formation upon translational mobility.

Because of the low boiling point of this substance, high pressure has to be applied in order to keep the sample at liquid-like densities. The two isobars obtained at 20 MPa and 200 MPa show the influence of density variation upon D. At the lowest temperatures reached at ambient pressure, the self-diffusion coefficient in methanol and water is slowed down by more than an order of magnitude compared with CH_3F, and this shows the contribution of hydrogen bond formation to the reduction of the translational mobility. At elevated pressures where water can be supercooled to temperatures around 200 K, the 250 MPa isobar of D given in Figure 2.5 shows a typical set of results. In this region of the p,T space the difference between water and the monoalcohols also becomes very pronounced.[13] Compared with methanol, D is reduced by an order of magnitude, while the self-diffusion in CH_3F is approximately a factor of 100 faster. Obviously, the hydrogen bond network in water is well developed under these conditions and thus slows translational mobility.

4 Rotational Correlation Times and the Lifetime of Hydrogen Bonds in Water

In order to quantify the influence of the hydrogen bond network, the rotational correlation times as determined by dielectric relaxation,[16] NMR relaxation[17–19] and depolarized Rayleigh scattering[20–22] will be discussed in the following.

Figure 2.6 compiles the results. The depolarized Rayleigh spectrum of liquid water between 360 K and 253 K could be described by two Lorentzians. The width of the broader line $l_B(w)$ is assigned to the formation and breaking of the hydrogen bonds and characterized by τ_{HB}, the lifetime of a hydrogen bond:

$$l_B\,(w) = [1 + \omega^2\tau_{HB}^2]^{-1} \tag{2.1}$$

It reveals an essentially Arrhenian temperature dependence in the temperature range studied. From the slope of this line an activation energy of 1.85 kcal/mol is derived. The lifetime of a hydrogen bond in liquid water at biological temperatures around 300 K is thus $\tau_{HB} < 1$ ps. Formation of solvation shells around apolar surfaces of molecules will at most increase τ_{HB} by one or two orders of magnitude.

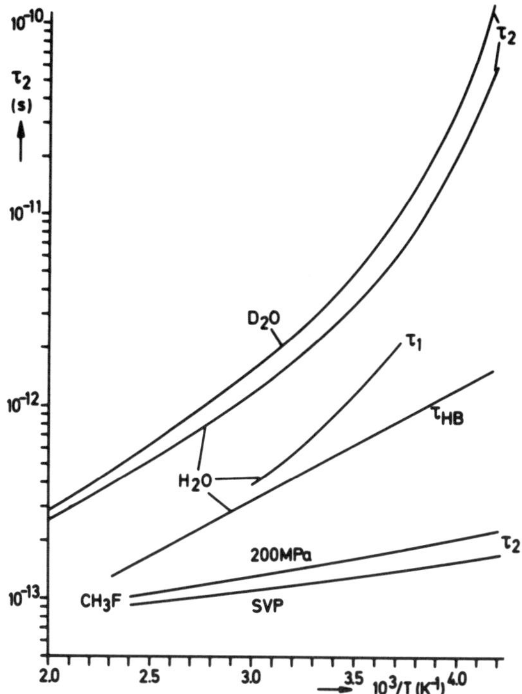

Figure 2.6 Hydrogen bond lifetime τ_{HB} (from Rayleigh scattering[18, 20]) and the two rotational correlation times τ_1 (from dielectric relaxation[14]) in light water and the rotational correlation time τ_2 (from NMR relaxation[15, 16, 23]) in methylfluoride; light and heavy water. For the definition of τ_1 and τ_2 see text

Kaatze[16] studied the complex permittivity of water at frequencies up to 57 GHz and could describe his data and older results from the literature by a Debye-type relaxation function[23] which implies a simple rotational diffusion model characterized by a single dipole correlation time, τ_1:

$$\varepsilon'(\omega) = \varepsilon_\infty + \frac{\varepsilon_0 - \varepsilon_\infty}{1 + \omega^2\tau_1^2}\,;\ \varepsilon''(\omega) = \frac{(\varepsilon_0 - \varepsilon_\infty)\omega\tau_1}{1 + \omega^2\tau_1^2} \tag{2.2}$$

NMR relaxation measurements can provide information about single-particle mobility. In water the nuclei ${}^1_1H$, ${}^2_1H$, ${}^{17}_8O$ have nuclear spins $I \neq O$. For protons with $I = 1/2$ relaxation at temperatures below ~350 K occurs exclusively by dipole–dipole interaction. The deuterons ($I = 1$) and the oxygen-17 nuclei ($I = 5/2$) relax by electric quadrupole relaxation.[24, 25] In the following, the results obtained from the analysis of the quadrupolar nuclei will be discussed. Detailed descriptions of the proton relaxation[19, 25] and a rigorous treatment of the quadrupolar relaxation can be found in the literature.[25] The quadrupole relaxation is dominated by the interaction of the nuclear quadrupole moment with the electric field gradient (e.f.g.) at the nucleus. To a good approximation the e.f.g. is generated in liquid water exclusively by intramolecular charges. It is oriented along the OD bond for the deuteron and perpendicular to the plane spanned by the three atoms of the water molecule for the oxygen-17. The interaction fluctuates in time because of the reorientation of the diffusing molecule. The spin–lattice relaxation rate, R_1, of the nuclei can then be expressed as:

$$R_1 = K\chi^2_{\text{eff}}[2g(\omega\tau_2) + 8g(2\omega\tau_2)] \tag{2.3}$$

The constant K depends on the nuclear spin, I, and χ_{eff} is the effective nuclear quadrupole coupling constant with the e.f.g. averaged over the fast librations of the water molecules. The slower rotational diffusion of the molecules provides the main contribution to the spectral intensity, $g(\omega)$, at the Larmor frequency, ω. If the tumbling of the molecules is isotropic and characterized by a single correlation time τ_2, the spectral intensity is given by

$$g(\omega) = \tau_2/(1 + (\omega\tau_2)^2) \tag{2.4}$$

For low-molecular-weight liquids of normal viscosity the correlation time τ_2 is much shorter than ω^{-1} and, in this region, R_1 is thus independent of frequency. It is only at extreme undercooling that the time scale of the molecular motions of water becomes comparable to the currently available NMR frequencies ($\omega < 10^9$ s^{-1}), and it is this exotic range of the p,T space that proved unambiguously that relaxation of water can be described by the isotropic Debye model with a single correlation time τ_2. It is worth mentioning that the treatment of the nuclear relaxation data as given by Equation (2.3) applies the concept of the V- and D-structures, as described in Section 1.

Figure 2.6 includes τ_2 obtained for light and heavy water at ambient pressure and compares it with the results obtained for the dipolar CH_3F molecule. The formation and continuous restructuring of the random hydrogen bond network thus slows down rotational reorientation of the water molecules by more than an order of magnitude. In the temperature range where both τ_1 (H_2O) and τ_2 (H_2O) were obtained,

$$\tau_1 = \frac{\tau_2}{3} \tag{2.5}$$

This further corroborates the description of the spectral intensity by an isotropic rotational diffusion model (since a rotational jump model would give τ_2/τ_1 ratios less than 3) and thus supports the analysis of the NMR and dielectric relaxation data.[26] It must, however, be stated, that the results of the analysis of the quasielastic and inelastic neutron scattering[27] are at

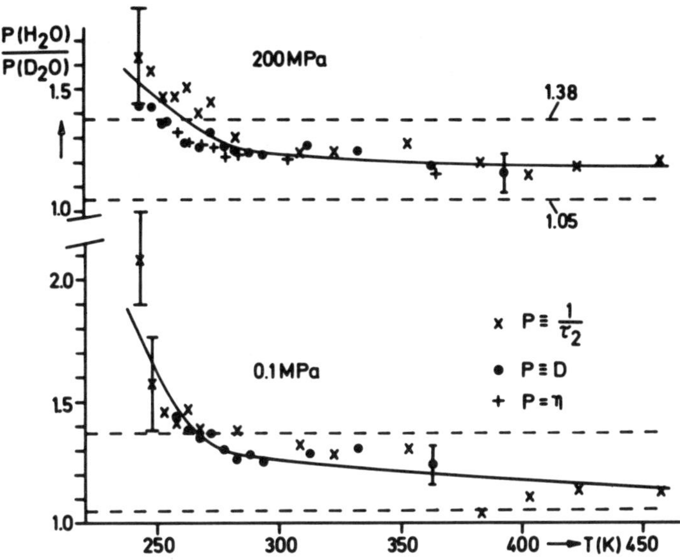

Figure 2.7 Dynamic isotope effect $p(H_2O)/p(D_2O)$ for the self-diffusion coefficient D, the shear viscosity η and the rotational correlation time τ_2 in H_2O and D_2O. The solid lines drawn through the points are a guide to the eye only. The broken horizontal lines at 1.38 and 1.05 correspond to the square root of the moments of inertia and the square root of the masses of the two molecules, respectively

variance with the description given above. The ratios between the self-diffusion coefficients obtained for light and heavy water and the shear viscosities are given in Figure 2.7 together with the ratios between the rotational correlation times τ_2 obtained from oxygen-17 spin–lattice relaxation time measurements. These ratios are called the dynamic isotope effect of the respective property. These two ratios are identical within the limits of experimental error. From simple liquid dynamics, however, one would expect D to depend on the square root of the mass ratio (= 1.05), while the τ_2 ratios should correspond to the square root of the ratio of the moments of inertia (= 1.38). Especially in the supercooled range, both ratios reveal a pronounced temperature dependence. The p,T dependence of the mobility of the water molecules is obviously not determined by the mechanics of the single free water molecule but is a collective property reflecting the dynamics of the hydrogen bond network.

Figure 2.8 illustrates another peculiarity of liquid water, resulting from its network structure. Here the pressure dependence of self-diffusion and rotational diffusion as characterized by $1/\tau_2$ is plotted at two temperatures. The 363 K isotherm for H_2O and the 200 K isotherm for CH_3F exhibit the usual relative density dependence of these two properties. Increasing pressure reduces the translational mobility significantly more than the rotational diffusion. Rotational motions are less sensitive to changes in density than translational displacements of molecules, because in the latter

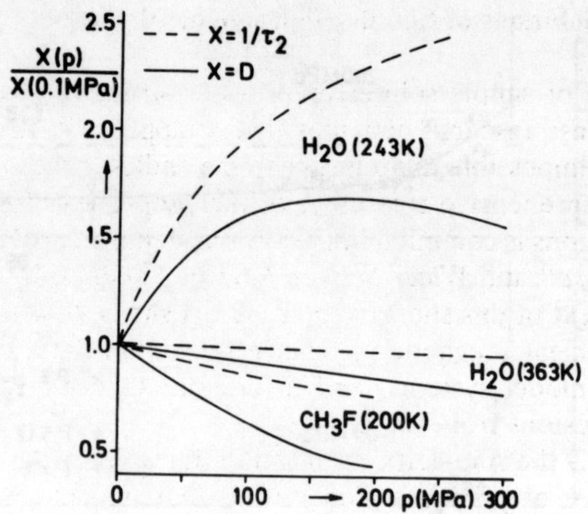

Figure 2.8 Reduced isotherms of the pressure dependence of the translational and rotational mobility in light water. For comparison a normal dipolar liquid, methylfluoride, is included

type of motion 'free volume' has to be provided for the moving particle. This normal behaviour is strikingly reversed in the supercooled region, where the translational mobility is enhanced along the 243 K isotherm of H_2O by a factor of 1.6, while the inverse rotational correlation time increases by a factor of approximately 2.5. Also, the pressure where the maximal mobility is observed along each isotherm appears to be significantly lower for D ($\approx$150 MPa). This anomaly has already been observed in the stability range of liquid water,[30] but becomes much more pronounced in the supercooled liquid. The pressure dependence of η,[28, 29] the shear viscosity, and $1/\tau_1$, the dielectric relaxation time,[30] reveal quantitatively the same trends as the properties given in Figure 2.8.

The anomalies of water originate from the open network structure of this liquid. The ability of the water molecule to participate with its two lone pairs and its two hydrogens in four hydrogen bonds enforces an approximately tetrahedral local arrangement with rather poor packing efficiency. Removal of thermal energy forces the molecules into low-energy configurations which resemble local structures observed in low-pressure polymorphs of ice and clathrates. The resulting quasitetrahedral force field presented by the surrounding network constrains a molecule to a few well-defined orientational states which are separated by potential barriers. The lifetime of these fully H-bonded local structures must increase upon cooling, with a consequent slowing down of diffusive modes. Compression of the sample brings next-nearest neighbours closer to a central molecule with nearest-neighbour correlations being largely unchanged. The potential energy surface for rotation is then less sharply defined and thus rotation is facilitated. The increasing number of next-nearest neighbours, however, starts to impede translational mobility already at pressures where rotation is still enhanced by further compression.

5 Aqueous Solutions of Simple Hydrophobic Solutes

The influence of simple solutes upon water structure continues to be an area of immense research activities, and it appears in the context of this short chapter impossible to do justice to the endless stream of publications. A rather comprehensive treatment of the properties of ionic and apolar aqueous solutions is compiled in the various volumes of *Water: A Comprehensive Treatise*[31] and *Water Science Reviews*,[32] edited by Felix Franks.

In the context of this short chapter it must suffice to summarize some of the recent nuclear magnetic resonance relaxation data, obtained on two hydrophobic model systems, and to attempt to draw some preliminary general conclusions from these data.

In Figure 2.9 the spin–lattice relaxation times T_1 for the deuterons in the binary mixture of perdeuterated tertiary butanol and the ^{17}O nuclei in enriched heavy water are given over a wide temperature range together with the self-diffusion coefficients of the alcohol in the mixture.[33] The results obtained for this system are typical of water/monoalcohol mixtures, which are generally assumed to be good model systems for hydrophobic hydration and hydrophobic interaction. In order to characterize the rotational mobility of the water molecules, the oxygen-17 nucleus had to be studied, since the deuterons of the water and the alcohol hydroxyl group yield only one unresolved signal. The analysis of the relaxation of this signal could furthermore be complicated by the chemical exchange between two deuteron sites. Oxygen-17 is a quadrupolar nucleus and its relaxation is under the conditions of these experiments described by Equation (2.3), after insertion of the appropriate nuclear properties. The T_1 thus monitor directly rotatory diffusion.

Below and around room temperature small additions of t-butanol reduce significantly the rotatory and also the translatory mobility (not shown) of the water molecules. The concentration dependence of the dynamics of the alcohol molecules is much smaller. This is generally explained by the formation of clathrate-like hydrate cages around the non-polar moieties of the alcohol molecules, the so-called 'icebergs'. In using this concept it is very important to give the 'lifetime' of these cages or rather the reduction in correlation time that water molecules experience at the hydrophobic surface. The correlation time of neat water at 300 K is around 2 ps; it does increase by at most a factor of 5 to a value around 10 ps in the alcohol solutions. The clathrate-like cages are thus very flickery icebergs indeed.

Including into the analysis of the molecular dynamics of this system the study of the dipolar proton permits conclusions about the temperature dependence of the self-association of the alcohol molecules[33,34] and thus can yield insight into the temperature and pressure dependence of the hydrophobic association. The analysis can be summarized by the following qualitative statements[35]:

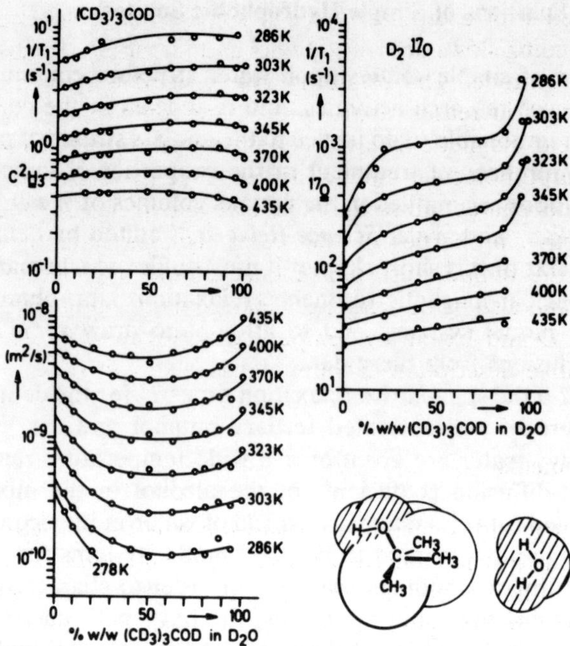

Figure 2.9 Deuteron spin–lattice relaxation times T_1 of the methyl deuterons and the oxygen-17 nucleus of enriched heavy water in the binary system $(CD_3)_3COD/D_2O$ and self-diffusion coefficient of butanol in a wide temperature and composition range. The molecules in the lower right of the figure are given with their van der Waals radii. Polar surfaces are shaded. The non-polar surfaces of the methyl groups make up approximately 80% of the total surface of the alcohol

(1) The anomalous concentration dependence of many thermodynamic properties of water–alcohol mixtures for temperatures around and below 300 K is predominantly caused by the influence of this non-polar compound upon the water structure. Self-association or hydrophobic interaction does at most yield a minor contribution.

(2) Hydrophobic association becomes very pronounced for higher temperatures. It passes through a maximum around 350 K.

(3) For the water–alcohol systems increasing pressure slightly reduces the microheterogeneities, and thus suppresses hydrophobic association.

6 Concluding Remarks

Qualitatively the local structure and dynamics of water can be well described by the random hydrogen bond network concept. The formation and breaking of the hydrogen bonds occur on the picosecond time-scale, and compared with a dipolar liquid of similar molar volume and mass, hydrogen bonding slows down rotational and translational mobility by 1–2 orders of magnitude. The retarding influence of the three-dimensional H-bond network increases rapidly with falling temperatures. At the time-scale of nuclear magnetic dielectric relaxation, i.e. in the time domain of the D-structure, molecular mobility is described by an isotropic rotational diffusion model characterized by single Debye correlation times τ_1 and τ_2.

The collective properties of the water dynamics are most strikingly revealed by the temperature dependence of the dynamic isotope effect (Figure 2.7) and also by the very peculiar anomalous pressure dependence observed for D and $1/\tau_2$ at low temperatures (Figure 2.8). A comprehensive quantitative description of these effects is still awaited.

Acknowledgements

Our own contributions have been sponsored generously by the Deutsche Forschungsgemeinschaft and the Fonds der Chemischen Industrie. Most of our work listed here was done by dedicated students and technicians who worked in the laboratory during the last decade.

References

1. Jannosch, H. W., Marquis, R. E. and Zimmermann, A. M. (eds) (1987). *Current Perspectives in High Pressure Biology*. Academic Press, London
2. Gage, J. D. and Tyler, P. A. (1991). *Deep-Sea Biology*. Cambridge University Press, Cambridge
3. Bernal, J. D. and Fowler, R. H. (1933). A theory of water and ionic solutions, with particular reference to hydrogen and hydroxyl ions. *J. Chem. Phys.*, **1**, 515
4. Rice, S. A. and Sceats, M. G. (1981). A random network model for water. *J. Phys. Chem.*, **85**, 1108
5. Sceats, M. G., Stavola, M. and Rice, S. A. (1980). A random network model calculation of the free energy of liquid water. *J. Chem. Phys.*, **72**, 6183
6. Henn, A. R. and Kauzmann, W. (1989). Equation of state of a random network, continuum model of liquid water. *J. Phys. Chem.*, **93**, 3770
7. Eisenberg, D. and Kauzmann, W. (1969). *The Structure and Properties of Water*. Oxford University Press, Oxford
8. Angell, C. A. (1983). Supercooled water. *Ann. Rev. Phys. Chem.*, **34**, 593
9. Angell, C. A. (1982). In Franks, F. (Ed.), *Water: A Comprehensive Treatise*, Vol. 7. Plenum Press, New York, pp. 1
10. Lang, E. W. and Lüdemann, H.-D. (1982). Anomalies of liquid water. *Angew. Chem. Int. Ed. Engl.*, **21**, 315
11. Franks, F. (1975). The hydrophobic interaction. In Franks, F. (Ed.), *Water: A Comprehensive Treatise*, Vol. 4. Plenum Press, New York, pp. 1
12. Weingärtner, M. (1982). Self diffusion in liquid water: A reassessment. *Zeitschr. Phys. Chem.*, **132**, 129
13. Prielmeier, F. X., Lang, E. W., Speedy, R. J. and Lüdemann, H.-D. (1988).

The pressure dependence of self diffusion in supercooled light and heavy water. *Ber. Bunsenges. Phys. Chem.*, **92**, 1111

14. Lang, E. W., Prielmeier, F. X., Radkowitsch, H. and Lüdemann, H.-D. (1987). High pressure NMR study of the molecular dynamics of liquid methylfluoride and deutero-methylfluoride. *Ber. Bunsenges. Phys. Chem.*, **91**, 1017
15. Karger, N., Vardag, T. and Lüdemann, H.-D. (1990). Temperature dependence of self diffusion in compressed monohydric alcohols. *J. Chem. Phys.*, **93**, 3437
16. Kaatze, U. (1989). Complex permittivity of water as function of frequency and temperature. *J. Chem. Eng. Data*, **34**, 371
17. Lang, E. W. and Lüdemann, H.-D. (1981). High pressure O-17 longitudinal relaxation time studies in supercooled H_2O and D_2O. *Ber. Bunsenges. Phys. Chem.*, **85**, 603
18. Lang, E. W., Lüdemann, H.-D. and Piculell, L. (1984). Nuclear-magnetic relaxation rate dispersion in supercooled heavy water under high pressure. *J. Chem. Phys.*, **81**, 3820
19. Lang, E. W., Girlich, D., Lüdemann, H.-D., Piculell, L. and Müller, D. (1990). Proton spin-lattice relaxation rate in supercooled H_2O and $H_2^{17}O$ under high pressure. *J. Chem. Phys.*, **93**, 4796
20. Montrose, G. J., Bacaro, J. A., Marshall-Coakley, J. and Litowitz, T. A. (1974). Depolarized Rayleigh scattering and hydrogen bonding in liquid water. *J. Chem. Phys.*, **60**, 5025
21. Danninger, W. and Zundel, G. (1981). Intense depolarized Rayleigh scattering in Raman spectra caused by large proton polarisibility of hydrogen bonds. *J. Chem. Phys.*, **74**, 2769
22. Conde, G. and Teixeira, J. (1983). Hydrogen bond dynamics studied by depolarized Rayleigh scattering. *J. Phys. (Paris)*, **44**, 525
23. Debye, P. (1929). *Polar Molecules*. Chemical Catalogue, New York
24. Spiess, H. W. (1978). Rotation of molecules and nuclear spin relaxation. In Diehl, P., Fluck, E. and Kosfeld, R. (Eds), *NMR — Basic Principles and Progress*, Vol. 15. Springer, Berlin, Heidelberg, New York, pp. 55 ff
25. Lang, E. W. and Lüdemann, H.-D. (1991). High pressure NMR studies on water and aqueous solutions. In Diehl, P., Fluck, E. and Kosfeld, R. (Eds), *NMR — Basic Principles and Progress*, Vol. 24. Springer, Berlin, Heidelberg, New York, pp. 129 ff
26. Kraus, P. (1977). *Liquids and Solutions*. Marcel Dekker, New York, Basel, p. 131
27. Chen, S. H. (1991). Quasi-elastic and inelastic neutron scattering and molecular dynamics of water at supercooled temperatures. In Dore, J. C. and Teixera, J. (Eds), *Hydrogen Bonded Liquids*. Kluwer, Dordrecht, Boston, London, pp. 289 ff
28. Harlow, A. (1967). *Further Investigation into the Effect of High Pressure on the Viscosity of Liquids*. PhD Thesis, University of London
29. De Fries, T. and Jonas, J. (1977). Molecular motions in compressed liquid heavy water at low temperatures. *J. Chem. Phys.*, **66**, 5393
30. Pottel, R., Asselborn, E., Eck, R. and Tresp, V. (1989). Dielectric relaxation rate and static dielectric permittivity of water and aqueous solutions at high pressures. *Ber. Bunsenges. Phys. Chem.*, **93**, 676
31. Franks, F., (Ed.) (1972–1982). *Water: A Comprehensive Treatise*, Vols 1–7. Plenum Press, New York
32. Franks, F., (Ed.) (1985–1991). *Water Science Reviews*, Vols 1–5. Cambridge University Press, Cambridge
33. Woznyj, M. (1985). *Hochdruck Kernresonanzuntersuchungen zur molekularen*

Dynamik und Struktur des binären Systems t-Butanol/D_2O. Dissertation, Universität Regensburg

34. Has, M. (1991). *Kernresonanzuntersuchungen an einem Modellsystem zur Untersuchung der hydrophoben Wechselwirkung bei hohen Drücken*. Dissertation, Universität Regensburg. S. Roderer Verlag, Regensburg
35. Lüdemann, H.-D. (1987). Water and aqueous solutions under pressure. In Jannosch, H. W., Marquis, R. E. and Zimmermann, A. M. (Eds). *Current Perspectives in High Pressure Biology*. Academic Press, London, pp. 273 ff

Part 2
Proteins

3
Hydration of Amino Acids in Protein Crystals

Julia M. Goodfellow, Narmada Thanki and Janet M. Thornton

1 Introduction

The interactions of water molecules with proteins have been of interest for a long time. This stems from the role of water in protein folding, during which apolar side-chains tend to be buried in the centre of the protein (thus reducing their contact with aqueous solvent). On folding, intramolecular hydrogen bonds are formed at the expense of protein–solvent hydrogen bonds. The fully folded protein retains many interactions with its aqueous media, since not all polar main chain or side-chain groups are involved in intramolecular hydrogen bonds, leaving unfilled potential hydrogen bond donor and acceptor atoms. Many apolar groups are also found on the surface in contact with water molecules.

Analysis of protein hydration is complicated by the fact that the protein surface may be highly mobile and that solvent–protein interactions are generally rather weak compared with the intramolecular forces which stabilize protein structure. In crystals the mobility of proteins can be estimated from temperature factors (B-values) and side-chains on the surface have relatively high B-values compared with those for other residues in the centre of the protein. Recent NMR studies on proteins in solution indicate that many of the surface side-chains are very flexible. Molecular dynamics simulations have also led to the view of proteins as highly mobile structures (Levitt, 1983; van Gunsteren and Berendsen, 1984). The temperature factors of water molecules in protein structures tend to be higher than the average temperature factor for the protein. Moreover, determination of solvent molecule positions is complicated by a number of factors, including the mobility of the water molecules, the relative weak stereochemical constraints which appear to define a hydrogen bond and the errors in the experimental data. Molecular dynamics simulations of proteins surrounded by solvent have also confirmed the mobility of the interfacial solvent molecules (Ahlstrom *et al.*, 1988).

Although protein–solvent interactions have been studied by many techniques, we are going to concentrate on structural aspects of these interactions. Thus, the data are taken from X-ray crystallography of protein crystals and mainly concern those water molecules which are hydrogen bonded to polar or charged main-chain and side-chain groups. Although previous studies have focussed on main-chain hydration (Finney, 1979; Baker and Hubbard, 1984; Sundaralingam and Sekhurudu, 1989), high-resolution data and modern refinement techniques have allowed us to extract an overview of protein hydration not just based on a limited number of hydration sites from one protein structure but using an accumulation of data from a number of the most accurate reliable structures (Thanki *et al.*, 1988, 1990, 1992; Goodfellow *et al.*, 1989).

2 Methods

The data were taken from the Brookhaven Protein Databank (Bernstein *et al.*, 1977). Originally, only 16 protein structures were found to meet with our criteria of 1.7 Å resolution or better and a *R*-factor of less than 0.26. As our study progressed, another 8 structures became available which met these restrictions and are included in later analyses. These structures are listed in Table 3.1. The Brookhaven datafiles contain information on water molecules in one asymmetric unit. In order not to omit any interactions, we generated all symmetry-related water molecule sites around the protein (from adjacent unit cells in the crystal) and included all symmetry-related sites which lay within 5.0 Å of the protein surface. The numbers of water sites in each protein are given in Table 3.2.

These data were checked for errors in the following way. Any residues (such as those at the amino or carboxyl terminus) which were poorly resolved in the electron-density maps (as stated by the authors) were removed from our data files. If two alternative sites were given for a residue, only the highest occupancy site was included. If the occupancies were equal (i.e. 50% each), then the data were treated as if there were two distinct residues. Further residues were excluded when the local stereochemistry was poor as measured by fitting individual side-chains to a reference side-chain and calculating the root mean square deviation in the fitting; if this was larger than 0.1 Å (see below), the residue was excluded. Water molecules which were found to be in very close contact with protein atoms usually had either very low occupancies or very high temperature factors and were removed from the file.

Table 3.1 Protein structures used in analysis of hydration

Protein	*File*[a]	R[b]	*Reference*
Actinidin	2ACT	0.171	Baker (1980)
Cytochrome (P450)	2CPP	0.190	Poulos *et al.* (1987)
Cytochrome c (peroxidase)	2CYP	0.202	Finzel *et al.* (1984)
Cytochrome c (rice)	1CCR	0.190	Ochi *et al.* (1983)
Cytochrome c (tuna)	4CYT	0.173	Takano and Dickerson (1980)
Carboxypeptidase A	1CPA	0.190	Rees *et al.* (1983)
DHFR (*L. casei*)	3DFR	0.152	Bolin *et al.* (1982)
DHFR (*E. coli*)	4DFR	0.155	Bolin *et al.* (1982)
Endothiapepsin	–	0.140	Blundell *et al.* (1990)
Erabutoxin B	3EBX	0.140	Smith *et al.* (1988)
Erythrocruorin	1ECD	0.190	Steigemann and Weber (1979)
γ II Crystallin	–	0.210	Najmudin and Slingsby (1989)
Insulin dimer	1INS	0.179	Baker *et al.* (1985)
$\lambda V_L I_g$ (Human Bence-Jones)	2RHE	0.149	Furey *et al.* (1983)
Human lysozyme	1LZ1	0.187	Artymuik and Blake (1981)
Myglobin (deoxy)	1MBD	0.159	Phillips (1980)
Phospholipase A2	1BP2	0.171	Dijkstra *et al.* (1981)
Plastocyanin (Cu^{2+})	1PCY	0.160	Guss and Freeman (1983)
Proteinase K	2PRK	0.167	Betzel *et al.* (1988)
BPTI	5PTI	0.200	Wlodawer *et al.* (1984)
Ribonuclease A	1RN3	0.260	Borkakoti *et al.* (1982)
Thermolysin	3TLN	0.213	Holmes and Matthews (1982)
β-Trypsin	1TPP	0.191	Walter *et al.* (1982)
Rubredoxin	5RXN	0.137	Watenpaugh *et al.* (1980)

[a] File = file name in Brookhaven Protein Data Bank.
[b] *R* = *R*-factor.

An interaction between a water molecule and a polar or charged protein atom has been defined solely in terms of a structural criterion. This criterion was chosen as the distance between the non-hydrogen atoms being less than or equal to 3.5 Å. No further criterion based on hydrogen atom positions is possible, because these atoms are not usually included in the refinement of a protein structure. This cut-off of 3.5 Å is intended to include all hydrogen bond interactions (typically less than 3.2 Å) but to exclude most van der Waals contacts. In the analysis of apolar side-chains, the distance cut-off is extended to 5.0 Å in order to include the non-bonded contacts. We focus on interactions of solvent molecules with specific polar or charged atoms in each type of side-chain, as indicated in Table 3.3. Some water molecules are found to interact with more than one protein atom. Thus, the number of interactions is often larger than the total number of water molecules. The total number of interacting water molecules is calculated for all atoms as well as for the main chain and side-chain atoms separately.

The distribution of water molecules around each type of side-chain is studied by combining the data for all occurrences of the side-chain of interest within all the proteins. A reference side-chain geometry is defined

Table 3.2 Water sites associated with protein structures

Protein	*SS*[a]	*NRES*[b]	*NHOH*[c]	*AVB*[d]
2ACT	α+β	218	272	36.04
2CPP	all α	414	204	27.13
2CYP	all α	294	262	41.06
1CCR	all α	111	460	–
4CYT	all α	103	54	26.13
1CPA	α β	307	192	25.28
3DFR	α β	162	264	?8.74
4DFR	α β	159	428	49.97
–	all β	326	322	35.19
3EBX	all β	62	111	25.08
1ECD	all α	136	94	20.94
–	all β	174	118	39.72
1INS	α+β	21, 30	87	36.25
2RHE	all β	114	186	34.15
1LZ1	α+β	130	143	52.84
1MBD	all α	153	338	45.76
1BP2	all α	123	106	26.77
1PCY	all β	99	44	27.00
2PRK	αβ	279	178	27.25
5PTI	α+β	58	63	32.73
1RN3	α+β	124	78	34.84
3TLN	α+β	316	173	32.92
1TPP	α+β	223	93	19.20
5RXN	all β	54	102	40.51

[a] SS = type of secondary structure. α = helical; β = sheet.
[b] NRES = total number of residues in protein.
[c] NHOH = original number of waters in protein data.
[d] AVB = average temperature factor of water molecules.

3.3 Hydrogen bonding capacity

Residue	*GRP*[a]	*HYB*[b]	*NLP*[c]	*NH*[d]
ASP	COO^-	sp^2	4	–
GLU	COO^-	sp^2	4	–
ASN	NH_2	sp^2	0	2
	C=O	sp^2	2	–
GLN	NH_2	sp^2	0	2
	C=O	sp^2	2	–
SER	OH	sp^3	2	1
THR	OH	sp^3	2	1
TYR	OH	sp^2	1	1
ARG	NH	sp^3	–	1
	$2NH_2$	sp^2	–	4
LYS	NH_3	sp^3	–	3
TRP	NH	sp^2	–	1
HIS	NH^+	sp^2	–	1
	NH	sp^2	–	1
MC	NH	sp^2	–	1
	CO	sp^2	2	–

[a] GRP = charged or polar atomic group.
[b] HYB = hybridization.
[c] NLP = number of lone pairs.
[d] NH = number of polar hydrogen atoms.

Table 3.4 Reference atoms and atom centres

Residue	*Fixed atoms*	*Centres*	*NPOL*[a]
ARG	NE, CZ, NH1, NH2	NE, NH1, NH2	5
ASN	CG, OD1, ND2	OD1, ND2	4
ASP	CG, OD1, OD2	OD1, OD2	4
GLN	CD, OE1, NE2	OE1, NE2	4
GLU	CD, OE1, OE2	OE1, OE2	4
HIS	CG, ND1, CD2, CE1, NE2	ND1, NE2	2
LYS	CD, CE, NZ	NZ	3
SER	CA, CB, OG	OG	3
THR	CB, OG1, CG2	OG1	3
TYR	CG, CD1, CD2, CE1, CE2, CZ, OH	OH	2
MC–O	CA_i, C_i', O_i	O	2
TRP	CG, CD1, NE1	NE1	1
MC–N	C_{i-1}', O_{i-1}, N_i, CA_i	NH	1

[a] NPOL refers to the total number of possible hydrogen bonding sites associated with each residue.

for each side-chain, using the data of Momany *et al.* (1975). The transformation matrix was then calculated between each occurrence of the side-chain of interest and the reference side-chain, using a number of reference atom sites, as defined in Table 3.4. The same matrix was used to transform the coordinate system of any water molecules interacting with each side-chain under analysis. This procedure led to a picture of side-chain hydration in which all occurrences of a side-chain were superimposed on the reference side-chain together with the surrounding solvent molecules.

These three-dimensional distributions of solvent molecules around reference geometry side-chains have been visualized, using FRODO molecular graphics program (Jones, 1978) on an Evans and Sutherland PS390. We have also analysed the data numerically in terms of spherical polar coordinates centred on the polar atom of interest, as shown diagramatically in Figure 3.1. This analysis was interpreted in terms of the potential hydrogen bonding capacity of the polar or charged side-chain atoms, as shown in Table 3.3.

3 Results

Overview of Protein Hydration

To determine an overall picture of protein hydration, we calculated the percentage of residues of each type interacting with solvent molecules for main-chain and side-chain atoms (Figure 3.2). We can clearly see that the percentage of main-chain atoms interacting with water molecules varies from 31% for methionines up to 67% for glycines. As expected, the main

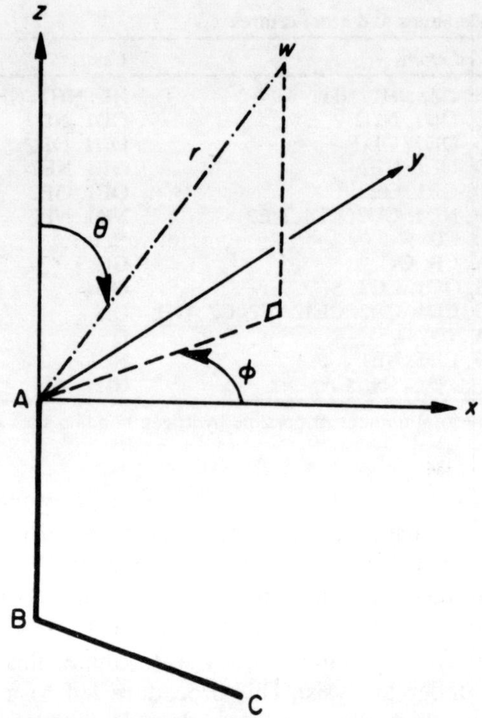

Figure 3.1 Diagrammatic representation of the spherical polar coordinates, r, θ and φ. The 'reference' main-chain atoms A, B and C lie in the xz-plane, where A is either the carbonyl O or amino N atom around which the water distribution is calculated. The parameters for the water position (W) are as follows:

$$x = r \sin \theta \cos \varphi;\ y = r \sin \theta \sin \varphi;\ z = r \cos \theta$$

Thus, the limits are $0° < \theta < 180°$ with $\theta = 0°$ along z-axis and $\theta = 180°$ along $-z$-axis. $-180° < \varphi < 180°$ with $\varphi = 0°$ along x-axis, $\varphi = 180°$ along $-x$-axis, $\varphi = 90°$ along y-axis and $\varphi = 270°$ along $-y$-axis

chains of large hydrophobic residue types are poorly hydrated, while the small hydrophilic side-chains allow the water molecules to approach the polar main chain. Glycine, with no side-chain, is the most highly hydrated. A more extreme picture emerges from the analysis of overall side-chain hydration. In this case we can clearly see that apolar or hydrophobic side-chains have very few interactions less than 3.5 Å with solvent molecules, whereas the polar and charged side-chains have a high percentage of residues interacting with water molecules. Thus, the hydrophilic side-chains tend to occur above the line representing main-chain hydration, while the hydrophobic side-chains tend to fall below the percentage hydration of main-chain atoms.

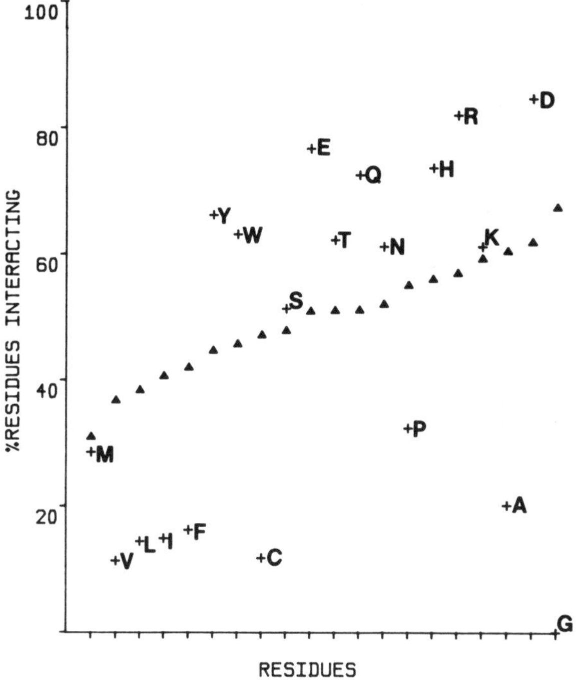

Figure 3.2 Plot of the percentage side-chain atoms (+) and percentage main-chain chain atoms (▲) interacting with water molecules within 3.5 Å against residues sorted in increasing order of interactions with main-chain atoms

Main-chain Hydration

Our initial study using the first 16 high-resolution structures showed that there were indeed many water molecules interacting with both main-chain N and main-chain O atoms (Thanki *et al.*, 1988). As found by others (e.g. Baker and Hubbard, 1984), there were many more interactions with the carbonyl groups compared with the NH group. Both the distributions from the computer graphics and the numerical analysis in terms of spherical polar coordinates indicated that that there was clustering on either side of the carbonyl oxygen, indicating hydrogen bonding with the oxygen lone pairs but only a single cluster directly above the N in line with the N–H bond.

This analysis included all main-chain N and O atoms irrespective of their position within secondary structural elements in each of the proteins. We decided that it would be of interest to look at the effect of intramolecular hydrogen bonds (due to the presence of secondary structure) on the

Table 3.5 Main chain hydration

		TOT[a]	*NWINT*[b]	*%RES*[c]
α-Helix:				
bonded	CO	935	216	21.5
free	CO	315	200	45.1
bonded	NH	897	20	2.1
free	NH	360	154	37.8
β-Sheet:				
bonded	CO	698	127	17.3
free	CO	213	189	62.0
bonded	NH	718	46	6.1
free	NH	193	112	53.9

[a] TOT = total number of residues.
[b] NWINT = number of interactions with water molecules < 3.5 Å.
[c] %RES = percentage of residues interacting with solvent.

distribution of water molecules. Thus, the initial analysis was repeated, using this time the full 24 high-resolution structures which were then available and defining secondary structure according to Kabsch and Sander (1983).

α-Helices

The main-chain CO and NH groups within α-helical secondary structure are divided into two groups, depending on whether they take part in intramolecular hydrogen bonding ('bonded') or lie at the ends of helices ('free'). We found that 45% of free CO groups but only 21.5% of 'bonded' CO groups are hydrated (Table 3.5). In comparison, 38% of 'free' NH groups are hydrated. Although it is unlikely that a 'bonded' NH can take part in another hydrogen bond interaction, we did look for such interactions but found that only 2% of such 'bonded' NH groups are hydrated. We would have found even fewer interactions if we had reduced the distance criterion below 3.5 Å.

The stereo plots and histograms of the spherical polar coordinates describing the solvent molecule distributions around 'bonded' and 'free' CO groups show distinct differences in these distributions (Figure 3.3). The θ plots show little or no interactions with $30° < \theta < 45°$ for 'bonded' CO, whereas a large number of such interactions are seen in the plots for the 'free' CO groups. The φ plots are also different, with very few interactions in the range $-180° < \varphi < 0°$ for solvent molecules interacting with those CO groups already involved in intramolecular hydrogen bonds. In fact, the φ distribution for water molecules interacting with 'bonded' CO shows distinct clustering into one region, viz. $0° < \varphi < 60°$. The other regions are prohibited by steric effects due to the local helical structure.

The distributions around the 'free' NH groups in alpha helices show that

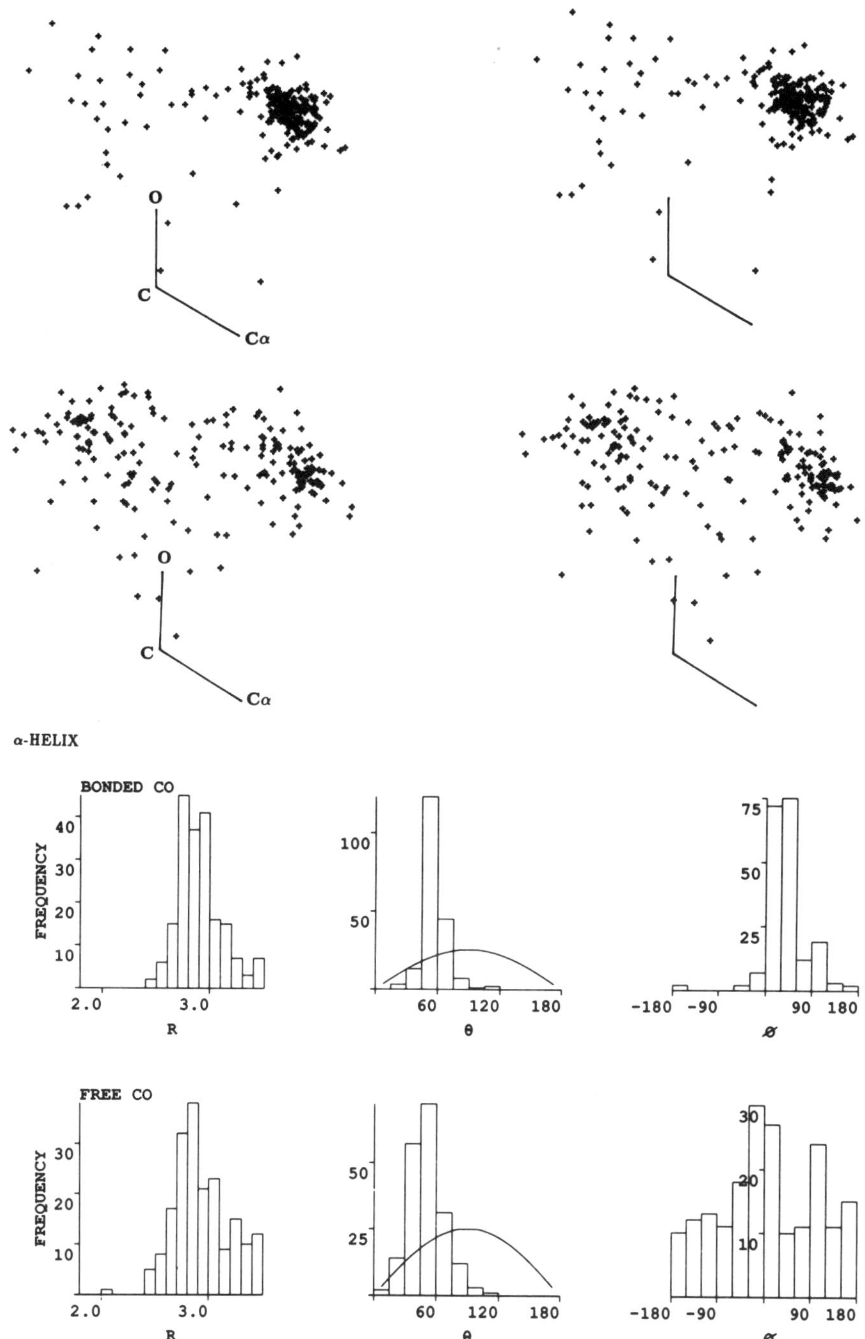

Figure 3.3 Stereo plots and histograms of the distribution of water molecules around 'bonded' and 'free' main-chain carbonyl groups within α-helical conformations

most water molecules interact with reasonable hydrogen bonding distance and around $\theta = 0°$ (Figure 3.4). The peaks in the phi distributions show that the water molecules are mainly clustered around +90° and −90°, i.e. directly below and directly above the plane containing the C, N and CA atoms. The very few solvent interactions with the 'bonded' NH groups are at relatively long distances and for this reason may be considered not to be hydrogen bonded.

Some of the water molecules which interact with a main-chain CO or NH group are also found to form bridging interactions with other main-chain atoms. These tend to occur at the carboxy and amino termini of helices in order to 'cap' the ends, as can be seen in Figures 3.5 and 3.6.

β-Sheets

In β-sheet regions of secondary structure, it is also necessary to distinguish between those CO and NH groups which form intramolecular hydrogen bonds ('bonded') and those which do not ('free'). Sixty-two per cent of 'free' CO groups are found to be hydrated and 17% of 'bonded' CO groups. Similarly, a large percentage (54%) of 'free' NH groups are hydrated and the expected small number (6%) of 'bonded' NH groups. Thus, we see that the 'free' CO and NH groups within sheets are more hydrated on average than those residues in α-helices.

Again, we see distinct differences in the distribution of solvent molecules around 'bonded' and 'free' CO groups (Figure 3.7). These differences are mainly due to steric hindrance and can be seen in both the θ and φ plots. Few interactions occur in the range $30° < \theta < 45°$ and around $\varphi = 0°$ for 'bonded' CO groups. If we compare the distributions for solvent molecules around 'bonded' CO in β-sheets (Figure 3.7) with those in α-helices (Figure 3.3) we see that, although the r and θ distributions are very similar, the φ distributions are not the same. In α-helices the main peak in the φ distribution occurs between 0° and 60° (almost in the plane of O, C and CA atoms), whereas two peaks occur at +90° and −90° in sheets.

Relatively few interactions occur between solvent molecules and 'bonded' NH and many of these are at distance > 3.2 Å (Figure 3.8). The interactions with 'free' NH groups within sheets occur at reasonable hydrogen bond distances and around $\theta = 0°$. Peaks in the φ distributions are seen at +90° and −90°.

Our analysis of solvent bridging sites in β-sheets shows that water molecules occur on the 'edge' of strands, in the 'middle' of strands and on the 'ends'. These interactions are shown schematically for an antiparallel β-sheet in Figure 3.9. In general, they appear to extend the strand hydrogen bonding networks or in some cases — for example, from Human Bence Jones protein — they can be seen to 'extend' the sheets by appearing to form another 'strand' (Figure 3.10).

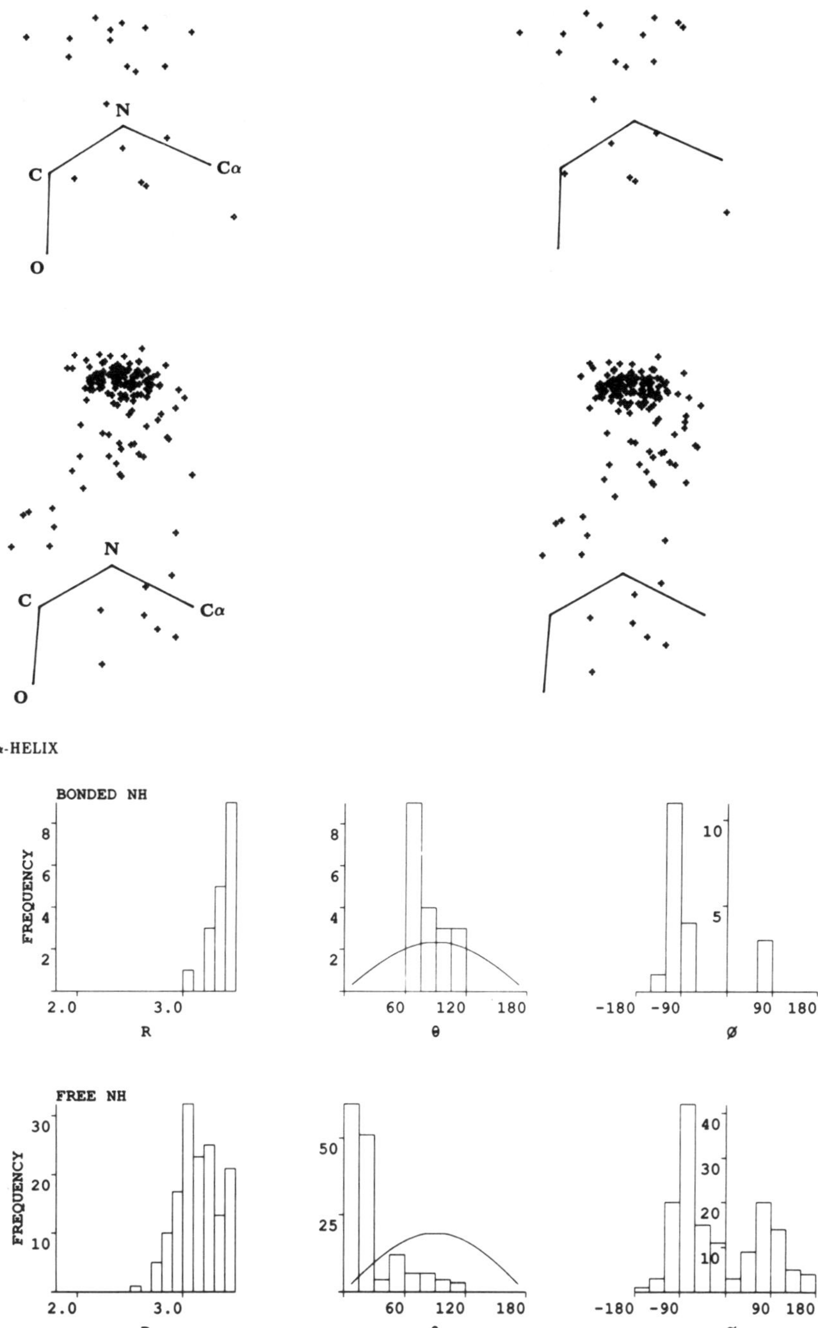

Figure 3.4 Stereo plots and histograms of the distribution of water molecules around 'bonded' and 'free' main-chain amide NH groups within α-helical conformations

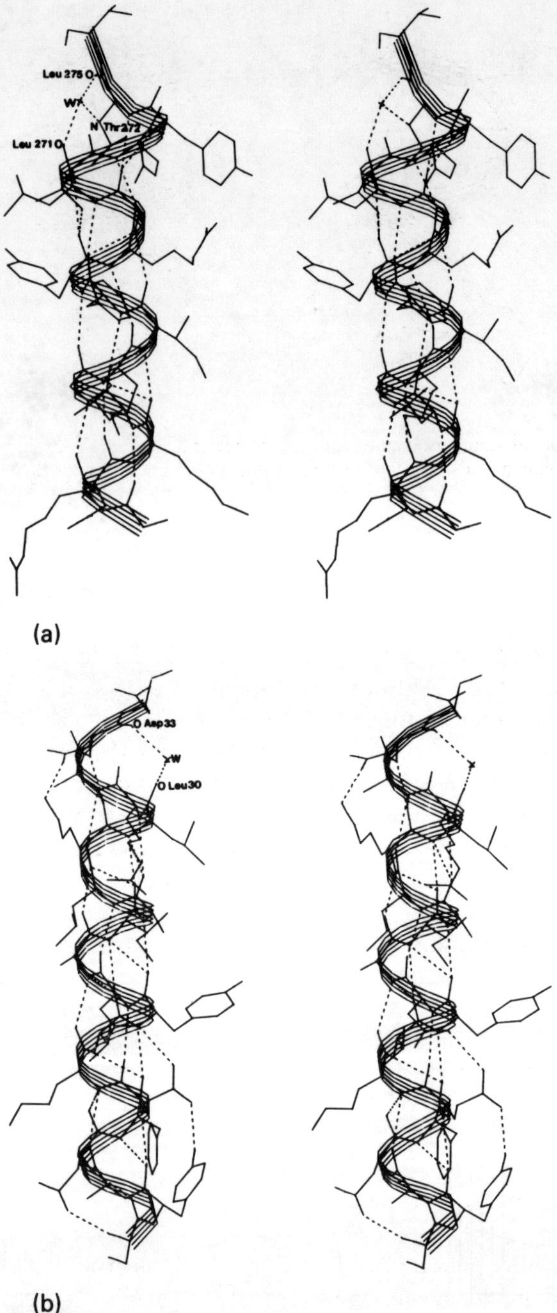

Figure 3.5 Plots of water molecules forming bridges between main-chain atoms of (*a*) O_i, N_{i+1} and O_{i+4} and (*b*) O_i and O_{i+3} at the carboxyl termini of helical segments in thermolysin and cytochrome C

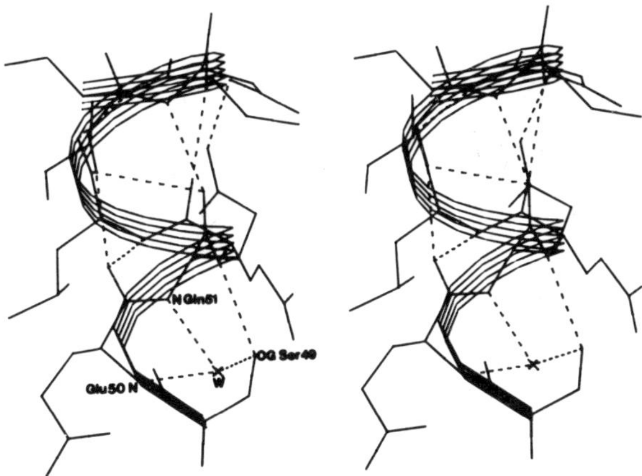

Figure 3.6 Plots of water molecules forming bridges between main-chain atoms of N_i, N_{i-1} and the side-chain atom OG_{i-2} of serine at the amino terminus of a helical segment in actinidin (Baker, 1980)

β-Turns

Our analysis of β-turns in 24 proteins listed in Table 3.1 has focused on Type I ($\alpha_R\alpha_R$), Type II ($\beta\alpha_L$) and Type VIII ($\alpha_R\beta$), as these occur most frequently. The four residues numbered *i* to *i*+3, of which the turns are comprised, must be considered separately and the percentage of each main-chain CO or NH group within the turn which interacts with solvent molecules is shown in Table 3.6. The CO groups tend to be more hydrated than the NH groups. A number of solvent sites are found which bridge two or more residues within the turn. The most commonly occurring are listed in Table 3.7 and two of the most frequently occurring types of solvent bridging sites are displayed schematically in Figure 3.11.

In general, both CO and NH groups did not appear to be as hydrated as one might expect, given that turns tend to occur on the surface of the protein. Further consideration of these β-turns led to the realization that they are very tight and that there is not always enough space for solvent molecules to bridge the protein atoms in a turn.

Hydration of Amino Acid Side-chains

Although we have generated solvent distributions around all 20 amino acid side-chains (Thanki *et al.*, 1988), we now focus on some specific side-chains.

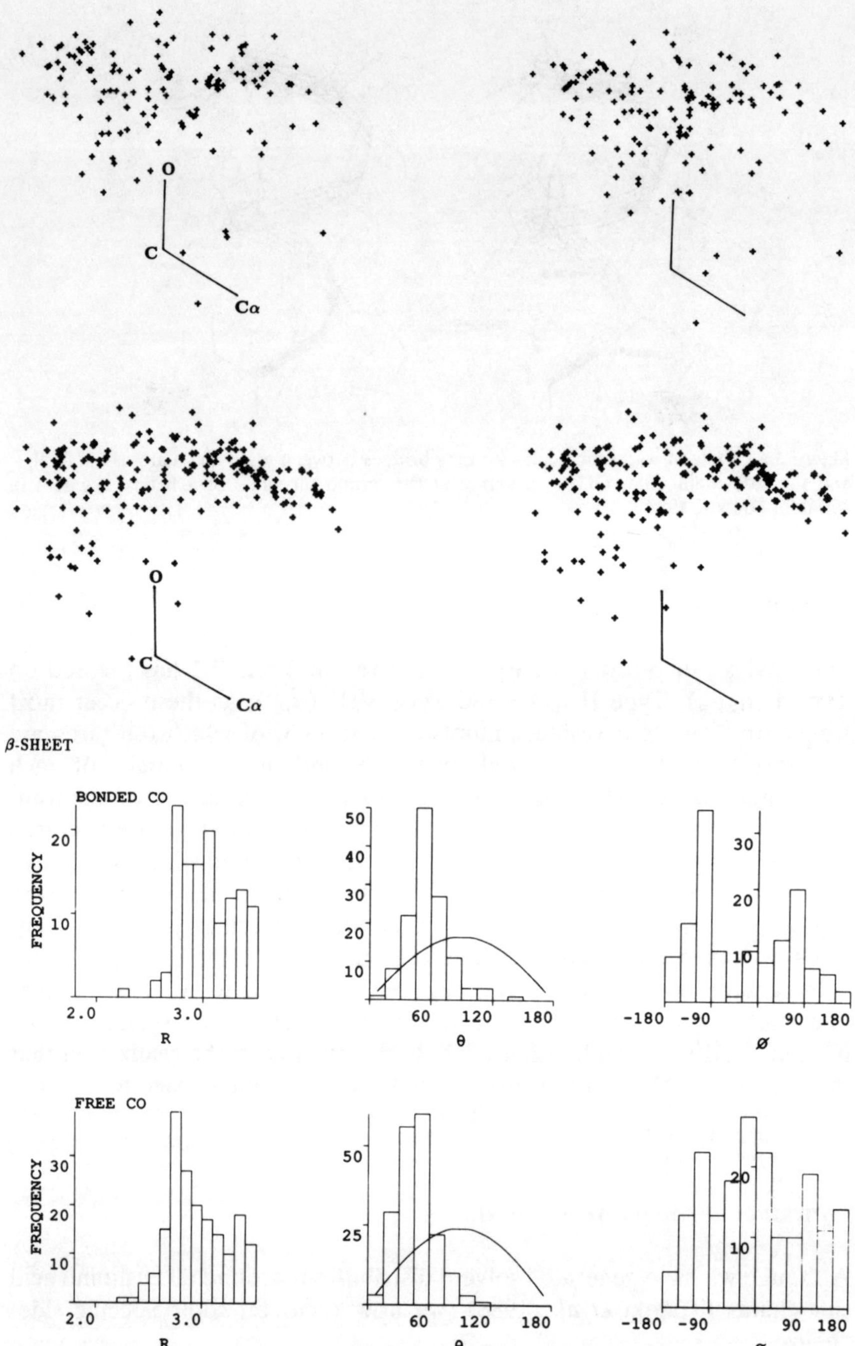

Figure 3.7 Stereo plots and histograms of the distribution of water molecules around 'bonded' and 'free' main-chain carbonyl groups within β-sheet conformations

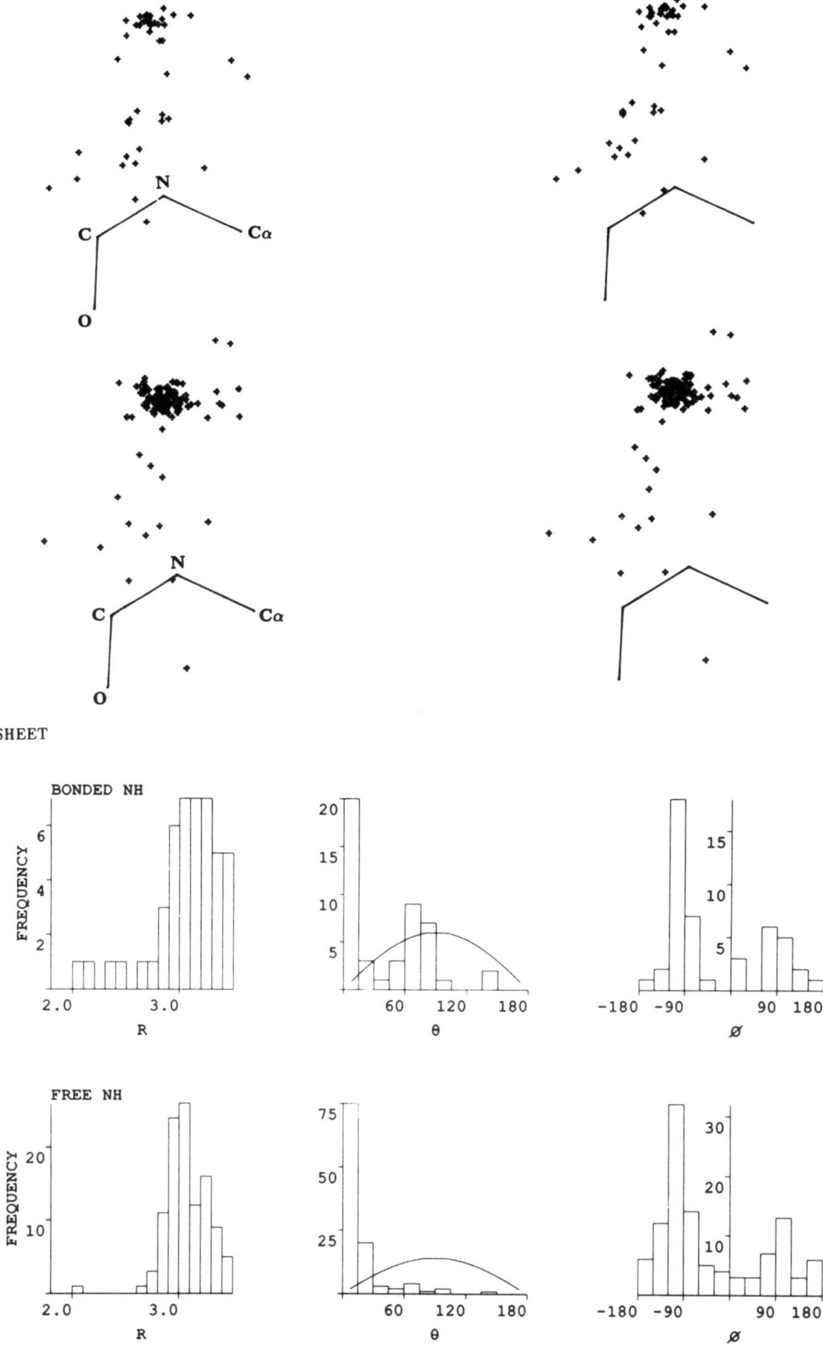

Figure 3.8 Stereo plots and histograms of the distribution of water molecules around 'bonded' and 'free' main-chain amide NH groups within β-sheet conformations

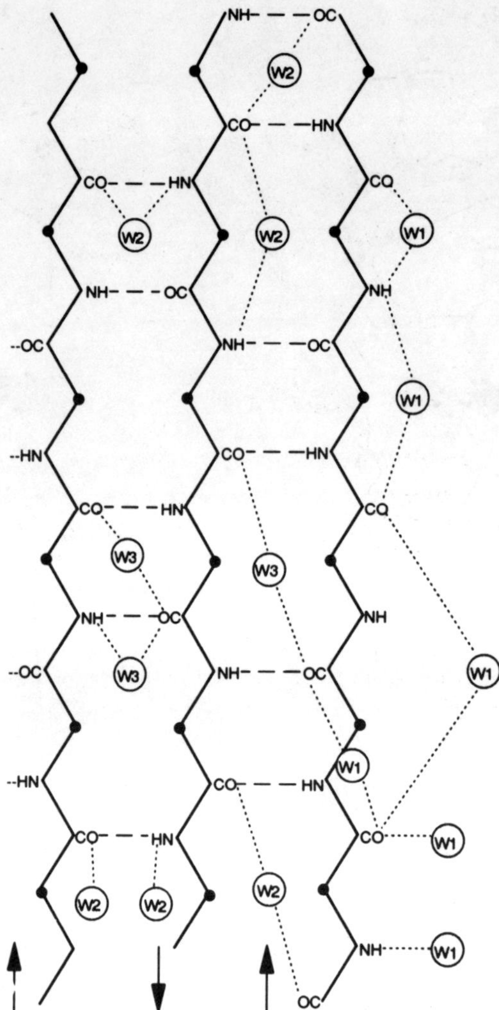

Figure 3.9 Schematic diagram of 'edge' (W1), 'end' (W2) and 'middle' (W3) categories of interactions of water molecules with main-chain atoms in antiparallel β-sheets

Charged Side-chains

The residues aspartic and glutamic acid have quite clear hydration patterns (Figure 3.12). The distributions of water molecules show clustering such that a water molecule can interact with the two lone pairs on each carboxylic acid oxygen. The numerical analysis shows that the water molecules are found within hydrogen bonding distance (with a distance peak at around 2.8 Å) with θ angles around 60°, consistent with sp^2 hybridization of the

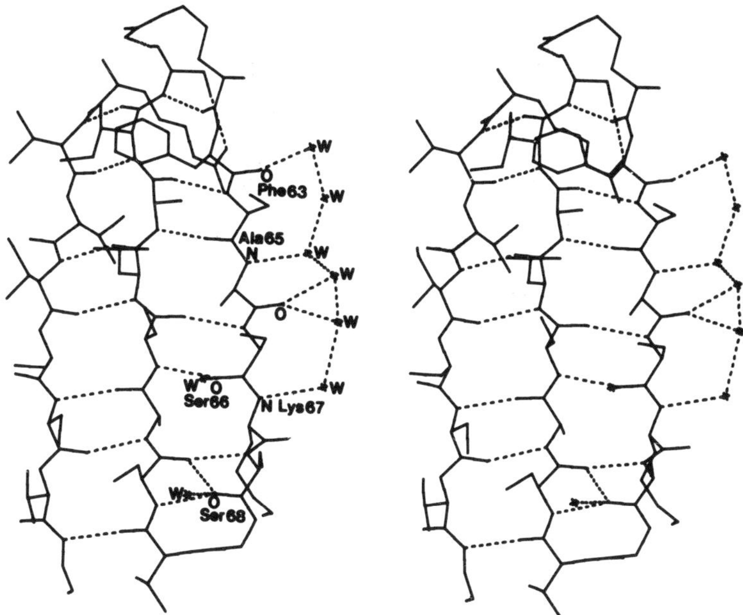

Figure 3.10 Stereo plot showing part of an antiparallel β-sheet structure in a protein Human Bence-Jones protein (Furey *et al.*, 1983), illustrating an 'extension' of the sheet structure by the water molecule network

Table 3.6 Hydration of β-turns

	Type I ($\alpha_R\alpha_R$)	*Type II* ($\beta\alpha_L$)	*Type III* ($\alpha_R\beta$)
NTURN[a]	164	30	41
BOND/FREE[b]	72:92	23:7	0:41
NW[c]-CO_i	34%	35%	59%
bond/free[b]	22:34	8:3	0:24
NW–NH_{i+1}	32%	42%	29%
NW–CO_{i+1}	48%	39%	44%
NW–NH_{i+2}	17%	35%	27%
NW–CO_{i+2}	43%	55%	46%
NW–NH_{i+3}	11%	13%	17%
bond/free[b]	8:10	2:2	0:7

[a] NTURN = number of turns of given type.
[b] BOND/FREE = number turns with $CO_i\cdots NH_{i+3}$ hydrogen bonds: number of turns without this hydrogen bond.
[c] NW = number of interactions with water molecules.

oxygen atoms. The φ plots are also well defined, with distinct peaks at 0° and 180°, i.e. in the plane of the acidic group (Figure 3.13).

The solvent distributions around NZ of lysine and the NE, NH1 and NH2 atoms of arginine are broader (Figure 3.14). Most but not all the distance plots show peaks at reasonable hydrogen bonding distance (Figure

Table 3.7 Bridging interactions in turns

	Interaction	*No.*
Type I		
NH_i	$-W-NH_{i+1}$	4
CO_i	$-W-CO_{i+1}$	4
CO_i	$-W-CO_{i+3}$	14
CO_i	$-W-NH_{i+3}$	5
NH_{i+1}	$-W-NH_{i+2}$	11
CO_{i+1}	$-W-CO_{i+3}$	12
CO_{i+1}	$-W-NH_{i+3}$	5
NH_{i+2}	$-W-NH_{i+3}$	4
NH_{i+3}	$-W-CO_{i+3}$	5
TYPE II		
CO_i	$-W-CO_{i+3}$	7
TYPE VIII		
NH_{i+1}	$-W-NH_{i+2}$	8
NH_{i+1}	$-W-CO_{i+2}$	4
NH_{i+2}	$-W-CO_{i+2}$	7

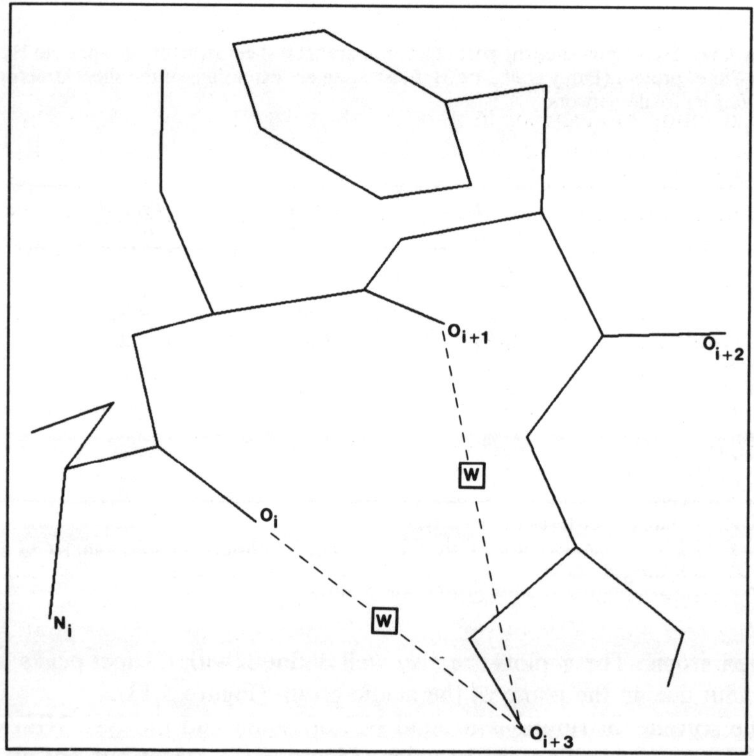

Figure 3.11 Schematic diagram of solvent-bridging interactions in Type I ($\alpha_R\alpha_R$) β-turns

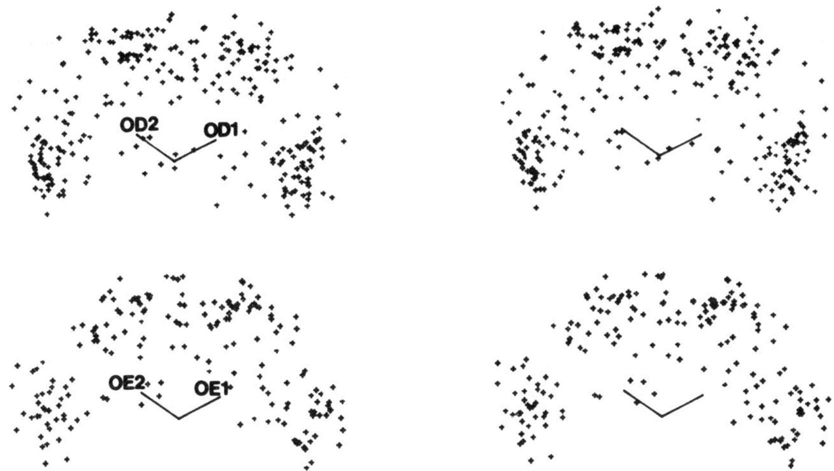

Figure 3.12 Stereo plots of the distribution of water molecules around (above) Asp OD1 and OD2, (below) Glu OE1 and OE2

3.15). The peaks in the θ plots tend to occur at higher values than for the acidic groups. This would be expected for the NZ but the NE of arginine would be expected to be sp^3 hybridized. This discrepancy may be due to delocalization of electrons in the guanidyl group leading to mixed sp^2–sp^3 character. Some water molecules are found which can interact with either both NH1 and NH2 or both NH1 and NE atoms. These bridging interactions may also lead to distortion of the hydrogen bond geometry. The φ distribution for NE of arg shows one peak at 180°, as there is only one hydrogen atom with which the water molecules may hydrogen bond. The NH1 and NH2 arg atoms show φ peaks at 0° and 180°, indicating that the water molecules are found predominantly in the plane of the guanidyl group. In contrast, the φ distribution for NZ of lysine shows little or no preferred orientation. This probably reflects the high mobility of lysine residues in proteins and rotation around the CE–NZ bond.

Polar Side-chains

The hydration patterns around the hydroxyl groups of serine, threonine and tyrosine have been studied in some detail (Thanki *et al.*, 1990). Initially one might expect them to be similar, but in fact chemical considerations show that the OH group of tyrosine is likely to have some sp^2 character due to the delocation of electrons in the ring system. In contrast, the OG group of serine and OG1 of threonine are thought to be sp^3 hybridized.

Our initial analysis of all serines, threonines and tyrosines showed

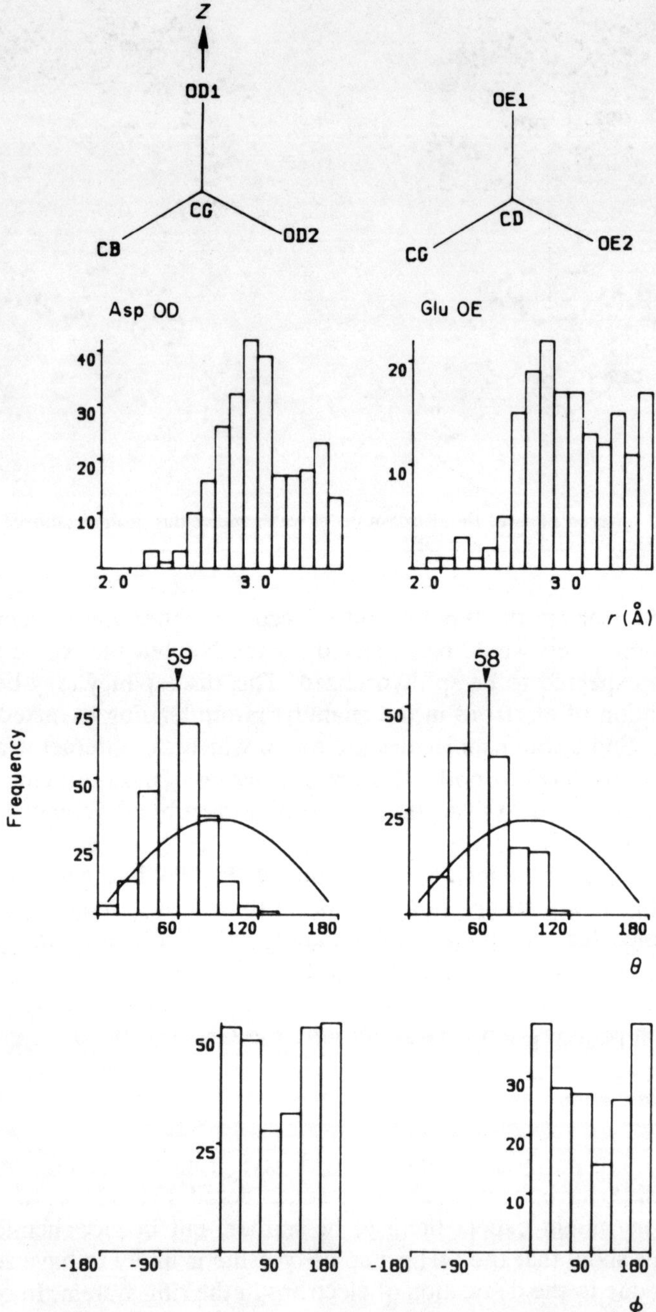

Figure 3.13 Histograms of the spherical polar coordinate (r, θ, φ) distributions of water molecules around (left) Asp OD (i.e. OD1 and OD2 combined) and (right) Glue OE (i.e. OE1 and OE2 combined)

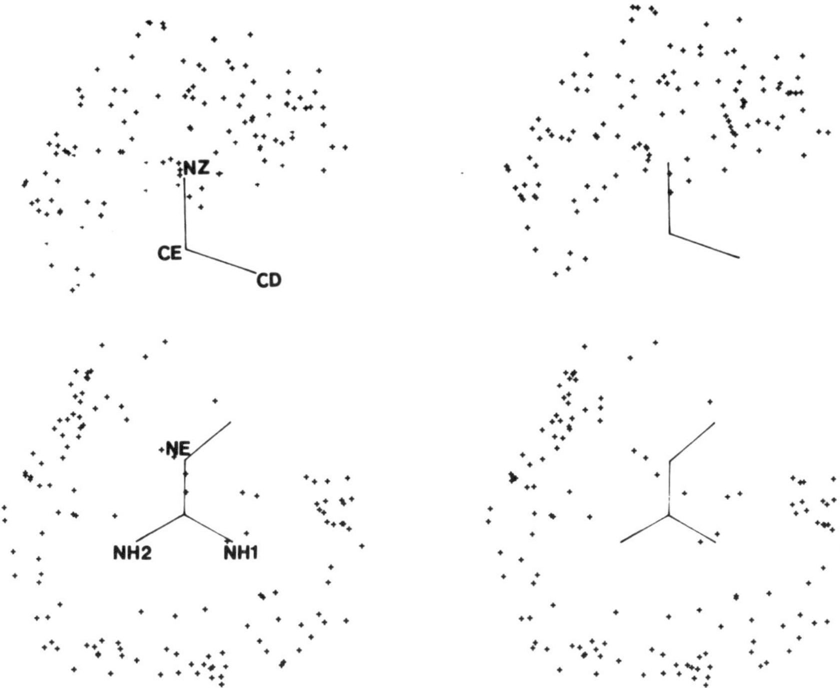

Figure 3.14 Stereo plots of the distribution of water molecules around (above) Lys NZ and (below) Arg NH1, NH2 and NE

clearly that there are significant differences in the hydration patterns reflecting the different stereochemistry (Figures 3.16, 3.17). It is seen clearly that water molecules cluster either side of the OH of tyrosine in the plane of the ring (i.e. peaks at $\varphi = 0°$ and $\varphi = 180°$). The distributions of water molecules around the hydroxyl groups of serine and threonine are very broad, with little hint of any clustering.

Further analysis of these distributions (using 24 protein structures) has indicated that there are different hydration patterns, depending on the nature of the secondary structure of the main-chain atoms. Thus, the hydration patterns for serine and threonine depend on whether these residues occur within α-helical or β-sheet regions (Figures 3.18, 3.19). This is not the case for tyrosine.

For serine residues in β-sheets, the φ region between 0° and 90° is hardly populated but is allowed if this residue occurs within an α-helical region. In the solvent distributions for threonine, the values of φ in the β-sheet region peak at $\varphi = 180°$, whereas they peak at $-90°$ for these residues in α-helical regions. Inspection of space filling models shows that these 'forbidden' regions are due to steric exclusion by other atoms preventing solvent molecules interacting with the side-chain hydroxyl group.

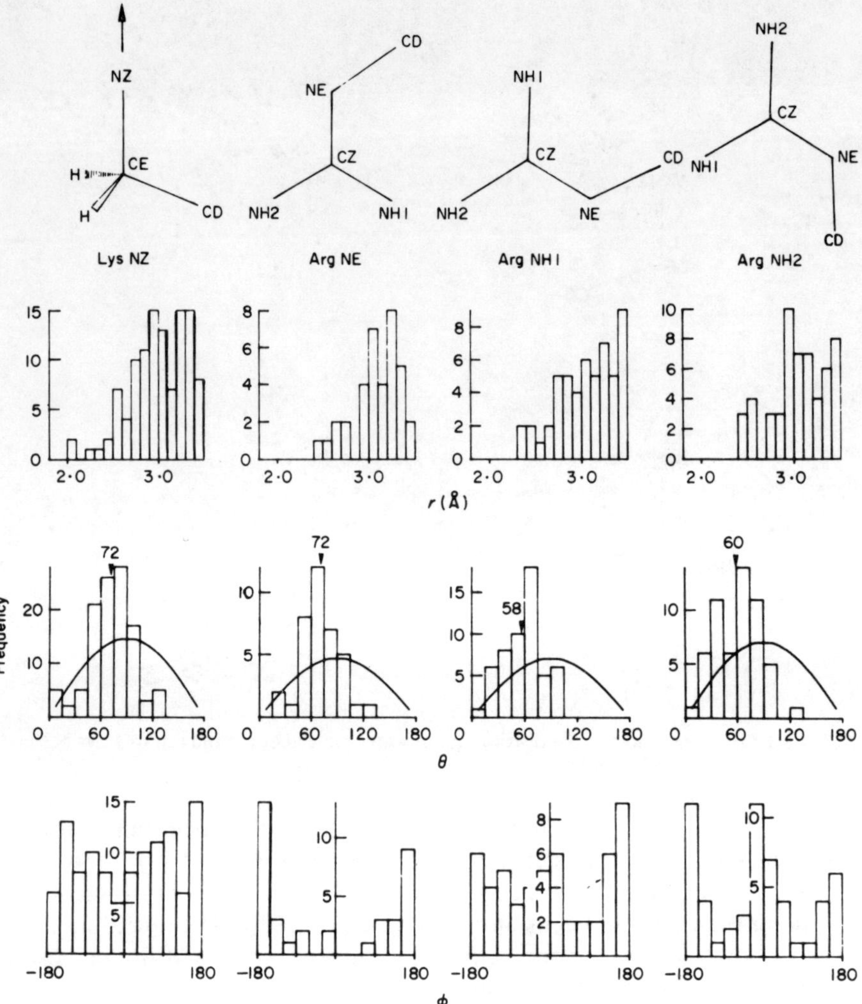

Figure 3.15 Histograms of spherical polar coordinate (r, θ, φ) distributions of water molecules around Lys NZ, Arg NE, Arg NH1 and Arg NH2

When we consider the θ angles, we find that they are very similar irrespective of whether the solvent distribution is around hydroxyl groups in α-helices or β-sheets. The r distributions (corresponding to the hydrogen bond distance between the non-hydrogen atoms) are still very broad in contrast to those of tyrosine.

We have also undertaken energy minimization calculations in order to find minimum energy sites which could correspond to probable hydration sites. These studies show that two of the three hydration sites around serine and threonine OG atoms are bridging sites in which the water

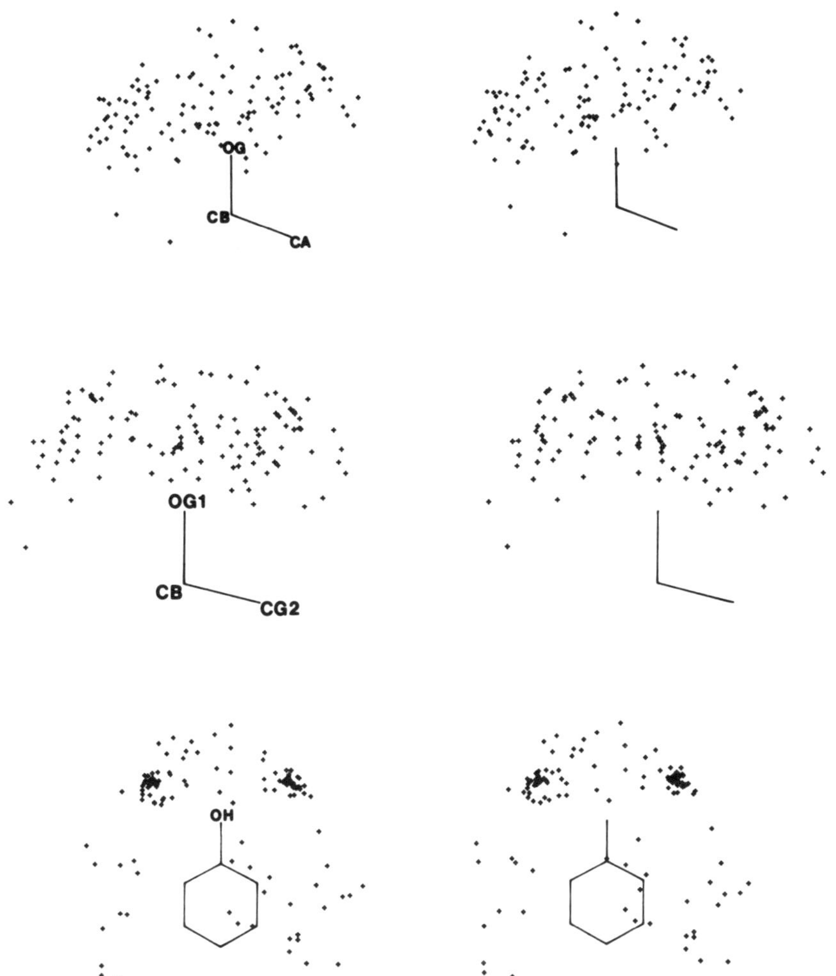

Figure 3.16 Stereo plots of the distribution of water molecules around (top) serine, (middle) threonine and (bottom) tyrosine

molecule interacts not only with the side-chain polar atom, but also with main-chain carbonyl residues. The location of these latter polar groups is clearly dependent on the secondary structure conformation. In α-helices these second interactions occur with CO_{i-4} and in β-sheets they occur with CO_{i-1}. These interactions are only possible because of the main-chain conformation and the specific χ_1 torsion angles for these residues, which themselves depend on the secondary structure (McGregor *et al.*, 1987).

It is well known that serine and threonine hydroxyl groups can form hydrogen bonds directly with carbonyl oxygens of residues $i-3$ and $i-4$ within an α-helix (Gray and Matthews, 1984), whereas we are finding

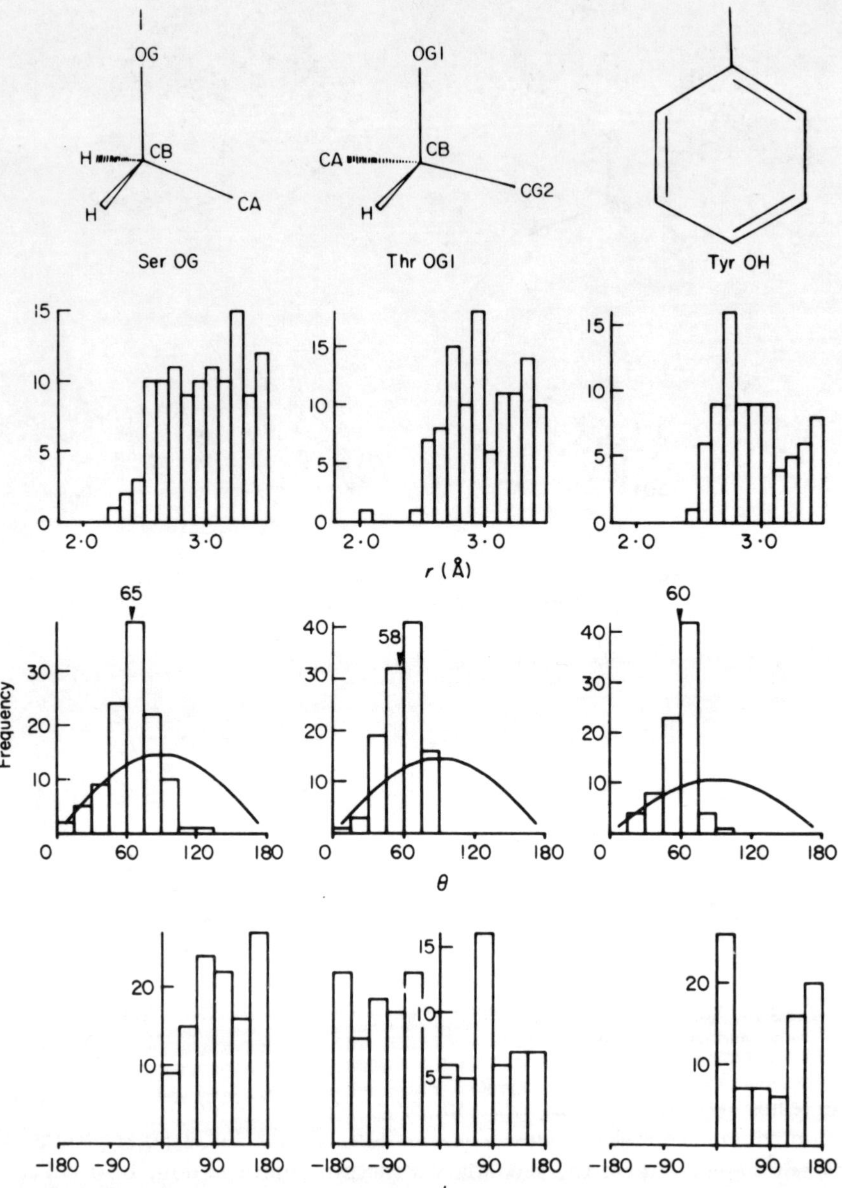

Figure 3.17 Histograms of the spherical polar coordinate (r, θ, φ) distributions of water molecules around serine OG, threonine OG1 and tyrosine OH

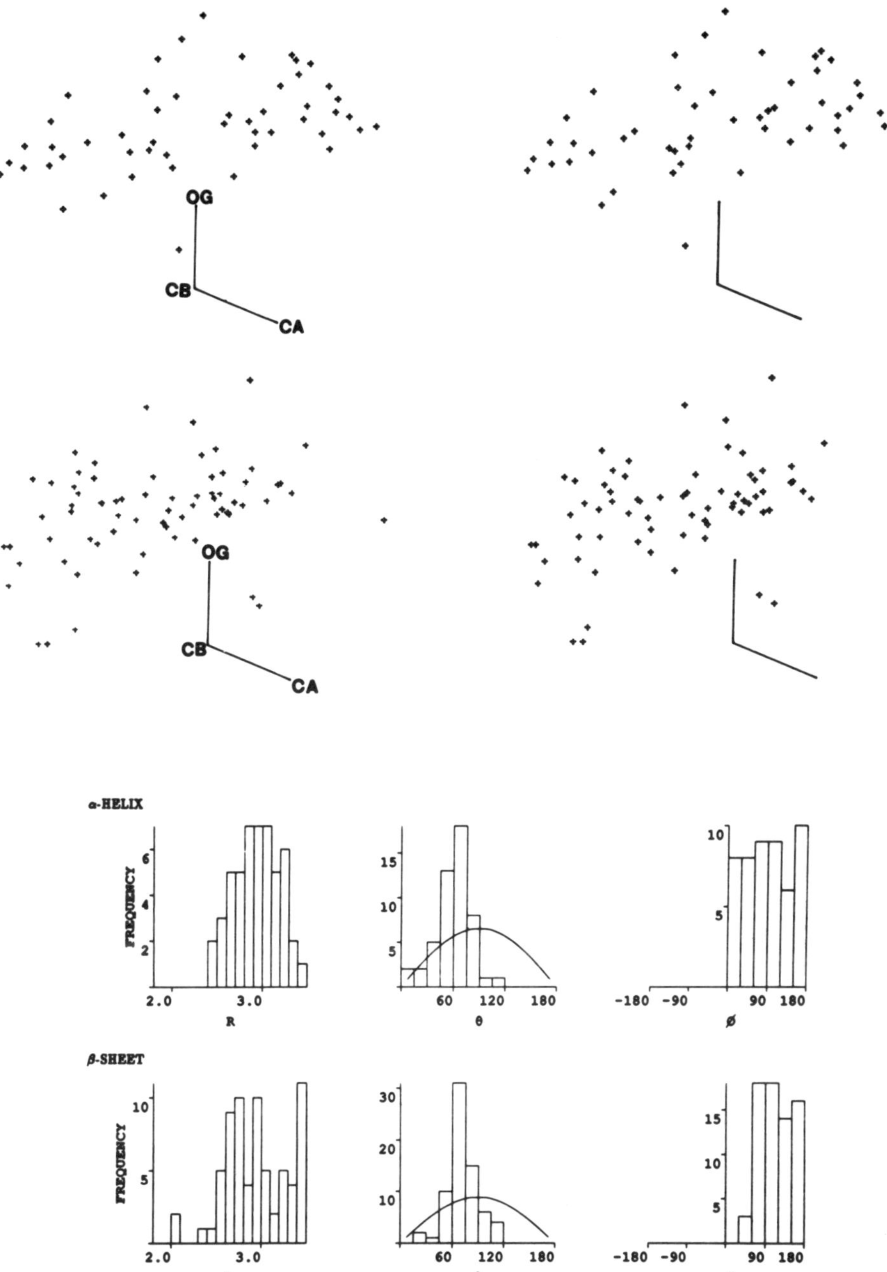

Figure 3.18 Stereo pictures of the distribution of water molecules interacting with OG atom of serine within (first row) an α-helical and (second row) β-sheet conformation. Quantitative analysis of these distributions based on polar coordinates centred on the OG atoms is presented for (third row) α-helical and (fourth row) β-sheet conformations

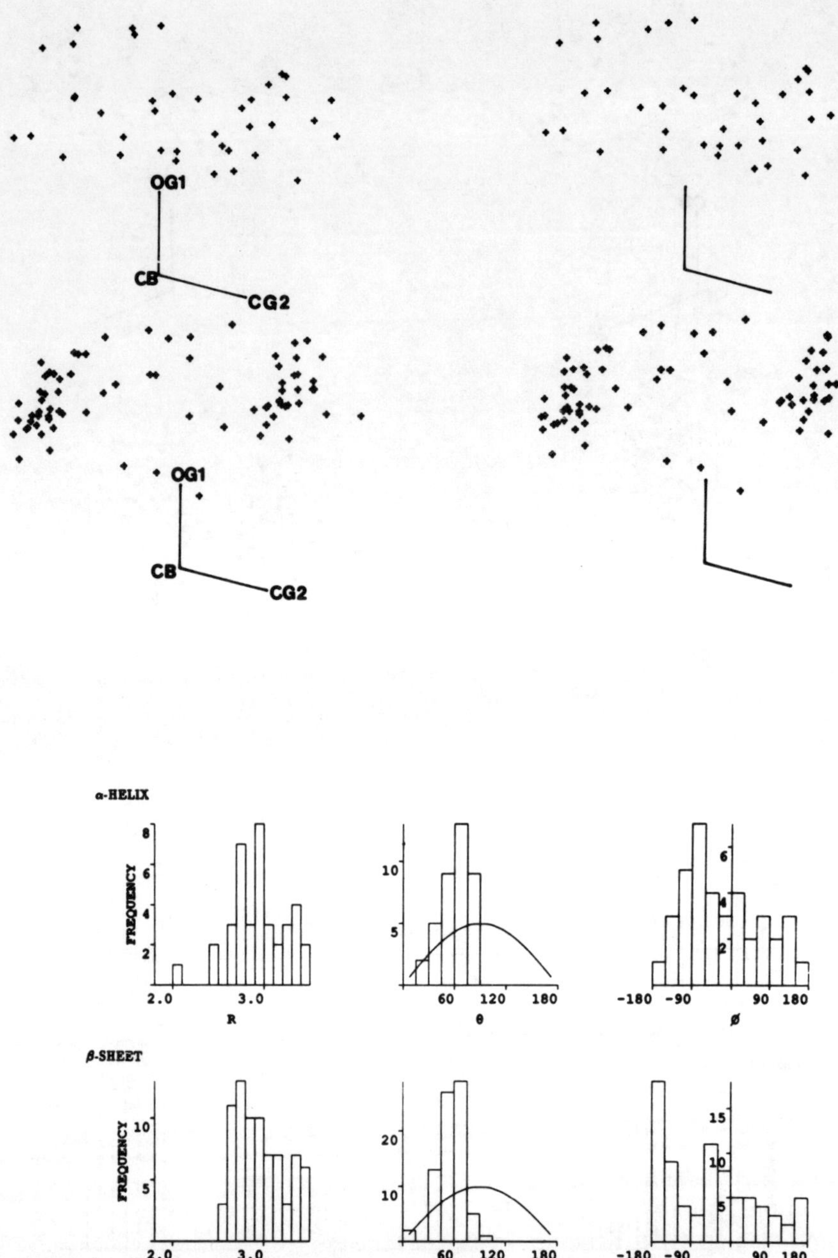

Figure 3.19 Stereo pictures of the distribution of water molecules interacting with OG1 atom of threonine within (first row) an α-helical and (second row) β-sheet conformation. Quantitative analysis of these distributions based on polar coordinates centred on the OG atoms is presented for (third row) α-helical and (fourth row) β-sheet conformations

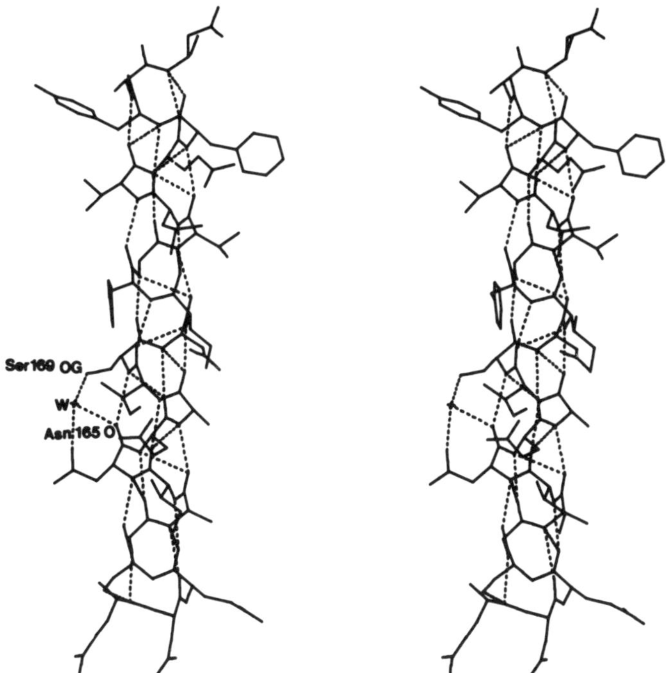

Figure 3.20 Stereo picture showing water molecules bridging between OG of serine (residue *i*) and main-chain carbonyl oxygen (residue *i*−4) within an α-helix from the thermolysin structure (Holmes and Matthews, 1985)

water-mediated hydrogen bonding between the side-chains and main-chain atoms in our energy calculations. Further analysis of the experimental data shows that these water-mediated bridging interactions between serine or threonine side-chains and main-chain carbonyls can also be seen in protein crystal structures (Figures 3.20, 3.21).

Apolar Side-chains

From a full analysis of the hydration of apolar residues (Walshaw and Goodfellow, 1993), it is clear that there are many water molecules which occur within 5.0 Å of these side-chains (see Figure 3.22a for distribution around phenylalanine ring), although at 3.5 Å (our cutoff for polar group interactions) there are many fewer (see Figure 3.22b for distribution around alanine CB). Distance plots show peaks in the number of water molecule interactions at around 4.0 Å, indicative of van der Waals interactions (Figure 3.22c). Our current analysis of these interactions is showing distinct preferences for solvent molecules around phenylalanine to lie in the plane of the ring and opposite carbon· · ·carbon bonds. This is consistent with the analysis of side-chain· · ·side-chain interactions of Singh and

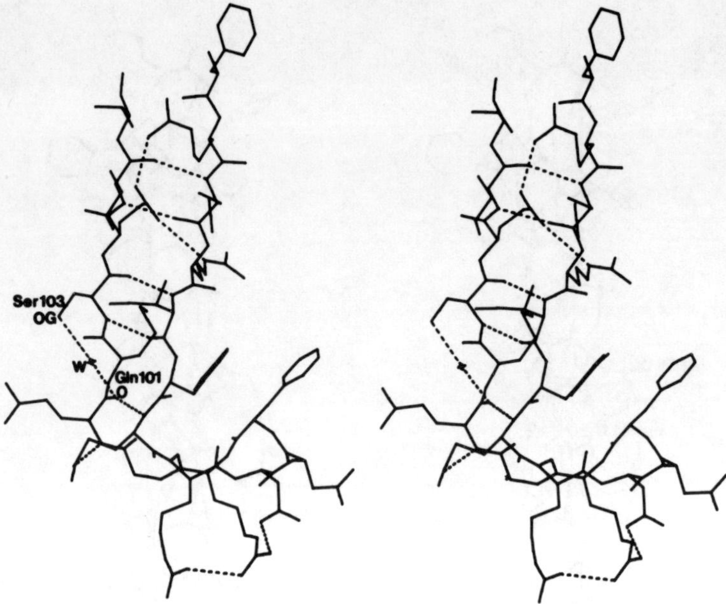

Figure 3.21 Stereo picture showing water molecules bridging between the OG atom of serine (residue *i*) and a main-chain atom along the same strand from residue $i-2$ taken from the γ-crystallin structure (Najmudin and Slingsby, 1989)

Thornton (1990) and the quantum mechanical calculations of Thomas *et al.* (1982).

Aromatic Side-chains

Solvent molecules within <3.5 Å of either histidine or tryptophan side-chains interact primarily with the ring nitrogen atoms as seen in Figure 3.23. Numerical analysis showns that some peaks in the r distributions occur at 2.7 Å and others at higher values more typical of hydrogen bonds involving nitrogen atoms (Figure 3.24). This is thought to be due to the fact that one of the nitrogens in His will have a lone pair, while the other is protonated. The greater electronegativity of the lone-pair nitrogen is thought to account for the short hydrogen bond distance (Dixon and Lipscomb, 1986). Although the θ plots are rather broad, the φ plots are very clear with solvent molecules occurring only in the plane of the ring at $\varphi = 180°$. Steric clashes prevent interactions at $\varphi = 0°$.

4 Discussion

Our interest in analysing the interaction of solvent molecules around the surface of proteins was twofold. First, we wished to learn more about the

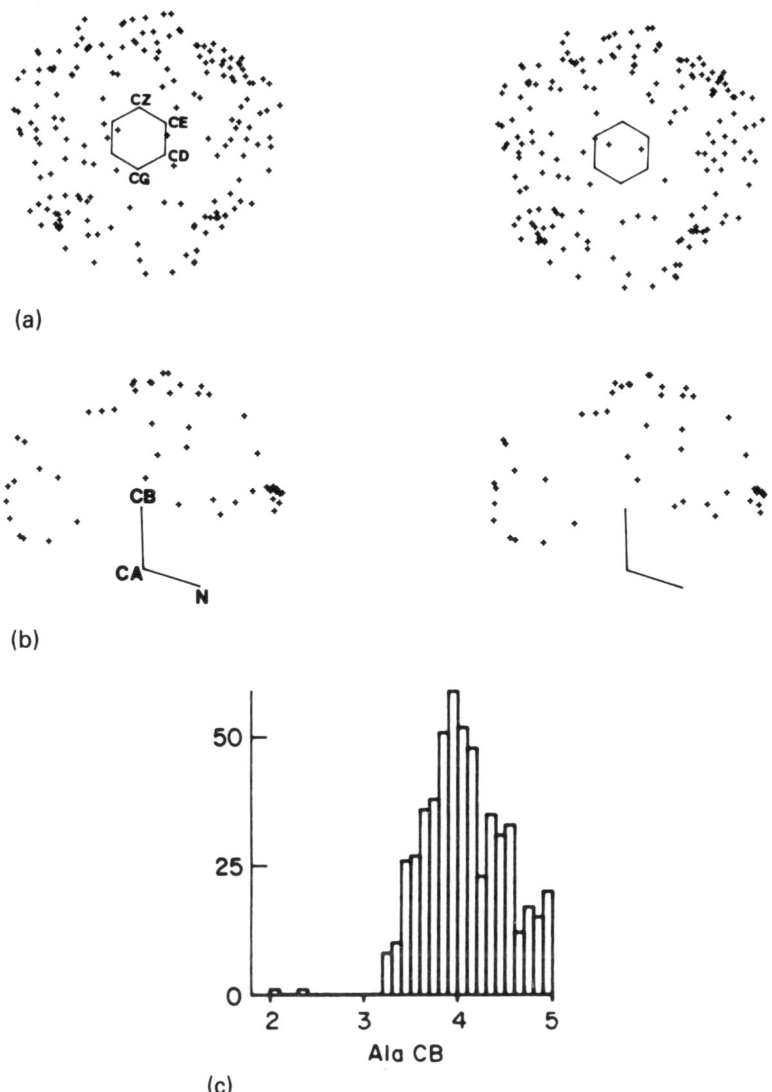

Figure 3.22 Distribution of water molecules (a) within 5.0 Å of phenylalanine ring, (b) within 3.5 Å of alanine CB and (c) the histogram of distances between solvent molecules and CB of alanine

role of water molecules in stabilizing protein structures, and second, we wished to use them both for knowledge-based prediction and for testing potential energy functions used in energy calculations.

Perhaps the most important information to come from these studies is that specific distributions of solvent molecules can be seen around all amino acid side-chains. Those around charged and polar side-chains

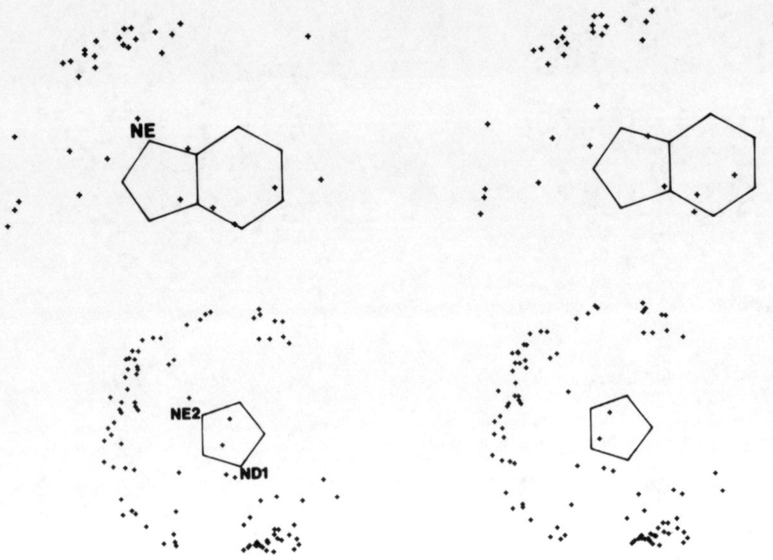

Figure 3.23 Stereo plot of the distribution of water molecules (above) within 3.5 Å of His ring, (below) within 3.5 Å of Trp ring

appear to be generally consistent with the known stereochemistry of the protein atoms. Thus, X-ray crystallography of high-resolution well-refined protein structures can provide accurate information of solvent interactions, albeit in hydrated crystals.

Analysis of those distributions which are very broad (such as serine and threonine), with little discernible clustering, has been shown to be due to superposition of several distributions, as the hydration sites depend on the nature of the local secondary structure and the χ_1 torsion angle. Within α-helices and β-sheets, solvent bridging sites (between side-chain polar groups and main-chain atoms) can occur which are specific to the local secondary structure.

Hydration of main-chain CO and NH groups is also dependent on the secondary structure in which they occur, as specific regions of space may be sterically inaccessible to solvent molecules, because of the presence of the helical or extended backbone. Again, bridging interactions are seen to occur which have the effect of appearing to 'extend' the hydrogen-bonding interactions in the local secondary structure, especially at the end of helices, on the edge of sheets and stabilizing β-turns.

We (Pitt and Goodfellow, 1991) have already begun to use these data on hydration in order to model solvent sites around proteins not in our initial dataset. The most simplistic knowledge-based modelling of solvent sites (based on peaks in the r, θ, φ distributions) have led to prediction rates of

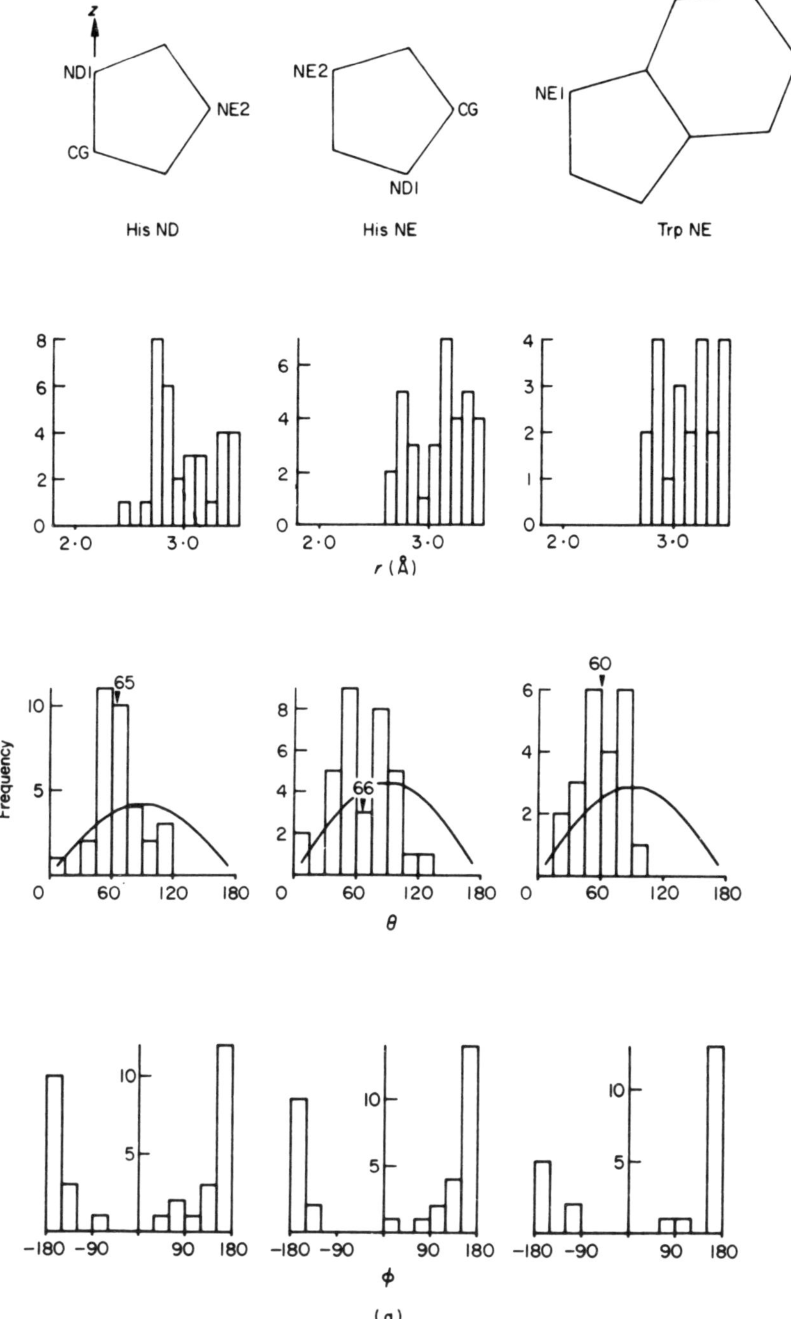

Figure 3.24 Histograms of the spherical polar coordinate distributions of water molecules around His ND, His NE and Trp NE.

60% (Pitt and Goodfellow, 1991). Our success rate is better than this for solvent molecules with low crystallographic *B*-values and worst for solvent molecules whose experimentally determined positions are highly mobile. A more sophisticated method based on contouring the solvent molecule distributions has proved more successful.

These experimental data also provide a stringent test for energy calculations. Energy minimization methods have been used to predict solvent sites around some side-chains (Goodfellow and Vovelle, 1989; Goodfellow *et al.*, 1990). The calculations for serine and threonine residues within polypeptides predict the secondary structure-specific bridging sites which are found experimentally. Similar calculations for tyrosine residues indicate that the solvent distributions around the OH atom should not depend on secondary structure. Again, no such dependence has been observed so far in our analysis of the experimental data.

However, detailed analysis of the energy minima due to solvent interactions around amino acid side-chains show that although traditional potential energy functions are reasonable at predicting distances (r) and the θ angle of the interactions (see Figure 3.1), they are not so good at predicting the φ distributions, especially when the interaction is with lone pairs. This is not surprising, given the fact that the dominating electrostatic terms are based on partial atomic charges alone. Currently, we are using more complex, distributed multipoles on each atom (Faerman and Price, 1990) to represent the electrostatic interactions to test this hypothesis.

Acknowledgements

We acknowledge support from the Science and Engineering Research Council under grants GR/F/81071 and GR/F/21183 (Molecular Recognition Initiative) and GR/E/57550 (Computational Science Initiative). JMG would like to thank The Wellcome Trust for a Research Leave Fellowship. Figures 3.1, 3.2, 3.12–3.17 and 3.22–3.24 have previously appeared in *J. Mol. Biol.*, **202** (1988) and are reproduced by permission of Academic Press Inc. (London) Ltd. Figures 3.18–3.21 have been published in *Protein Engineering*, **3** (1990) and reproduced by permission of Oxford University Press.

References

Ahlstrom, P., Teleman, O. and Jonsson, B. (1988). Molecular dynamics simulations of interfacial water structure and dynamics in a parvalbumin solution. *J. Am. Chem. Soc.*, **110**, 4198–4203

Artymuik, P. J. and Blake, C. C. F. (1981). Refinement of human lysozyme at 1.5 Å resolution. *J. Mol. Biol.*, **152**, 737–762

Baker, E. N. (1980). Structure of Actinidin, after refinement at 1.7 Å resolution. *J. Mol. Biol.*, **141**, 441–484

Baker, E. N., Dodson, E., Dodson, G., Hodgkin, D. and Hubbard, R. E. (1985). The water structure in 2 Zn insulin crystals. In Moras, D., Drenth, J., Strandberg, B., Suck, D. and Wilson, K. (Eds), *Crystallography of Molecular Biology*. Plenum Press, New York, pp. 179–192

Baker, E. N. and Hubbard, R. E. (1984). Hydrogen bonding in globular proteins. *Prog. Biophys., Mol. Biol.*, **44**, 97–179

Bernstein, F. C., Koetzle, T. F., Williams, G. J. B., Meyer, E. F., Brice, M. D., Rodgers, J. R., Kennard, O., Shimanouchi, T. and Tasumi, M. (1977). The protein data bank: a computer-based archival file for macromolecular structures. *J. Mol. Biol.*, **112**, 535–542

Betzel, C., Pal, G. P. and Saenger, W. (1988). Synchrotron X-ray data collection and restrained least squares refinement of the crystal structure of proteinase K at 1.5 Å resolution. *Acta Cryst.*, **B44**, 163–172

Blundell, T. L., Jenkins, J. A., Sewell, B. T., Pearl, L. H., Cooper, J. B., Tickle, I. J., Veerapandian, B. and Wood, S. P. (1990). X-ray analyses of aspartic proteinases. *J. Mol. Biol.*, **211**, 919–941

Bolin, J. T., Filman, D. J., Matthews, D. A., Hamlin, R. C. and Kraut, J. (1982). Crystal structures of *E. coli* and *L. casei* dihydrofolate reductase at 1.7 Å resolution. *J. Biol. Chem.*, **257**, 13650–13662

Borkakoti, N., Moss, D. S. and Palmer, R. A. (1982). Ribonuclease A: Least squares refinement of the structure at 1.45 Å resolution. *Acta Cryst.*, **B38**, 2210–2217

Dijkstra, B. W., Kalk, K. H., Hol, W. G. J. and Drenth, J. (1981). Structure of bovine pancreatic phospholipase A2 at 1.7 Å resolution. *J. Mol. Biol.*, **147**, 97–123

Dixon, A. D. and Lipscomb, W. N. (1986). Electronic structure and bonding of amino acids containing first row atoms. *J. Biol. Chem.*, **251**, 5992–6000

Faerman, C. H. and Price, S. A. (1990). A transferable distributed multipole model for the electrostatic interactions of peptides and amides. *J. Am. Chem. Soc.*, **112**, 4915–4926

Finney, J. L. (1979). The organization and function of water in protein crystals. In Franks, F. (Ed.), *Water: A Comprehensive Treatise*, Vol. 6. Plenum Press, New York, pp. 47–122

Finzel, B. C., Poulos, T. L. and Kraut, J. (1984). Crystal structure of yeast cytochrom C peroxidase refined at 1.7 Å resolution. *J. Biol. Chem.*, **259**, 13027–13036

Furey, W., Wang, B. C., Yoo, C. S. and Sax, M. (1983). Structure of a novel Bence-Jones protein (Rhe) fragment at 1.6 Å resolution. *J. Mol. Biol.*, **167**, 661–692

Goodfellow, J. M., Jones, D. M., Laskowski, R. A., Moss, D. S., Saqi, M., Thanki, N. and Westlake, R. (1990). Use of parallel processing in the study of protein–ligand binding. *J. Comp. Chem.*, **11**, 314–325

Goodfellow, J. M., Thanki, N. and Thornton, J. M. (1989). Preliminary analysis of water molecule distributions in proteins. *Mol. Simulation*, **3**, 167–182

Goodfellow, J. M. and Vovelle, F. (1989). Biomolecular energy calculations using transputer technology. *Eur. Biophys. J.*, **17**, 167–172

Gray, T. M. and Matthews, B. W. (1984). Intrahelical hydrogen bonding of serine, threonine and cysteine residues within α-helices and its relevance to membrane bound proteins. *J. Mol. Biol.*, **175**, 75–81

Guss, J. M. and Freeman, H. C. (1983). Structure of oxidized poplar plastocyanin at 1.6 Å resolution. *J. Mol. Biol.*, **169**, 521–563

Holmes, M. A. and Matthews, B. W. (1982). Structure of thermolysin refined at 1.6 Å resolution. *J. Mol. Biol.*, **160**, 623–639

Jones, T. A. (1978). A graphics model building and refinement system for macromolecules. *J. Appl. Crystallogr.*, **11**, 268–272

Kabsch, W. and Sander, C. (1983). Dictionary of protein secondary structure pattern recognition of hydrogen bonded and geometrical features. *Biopolymers*, **22**, 2577–2637

Levitt, M. (1983). Molecular dynamics of native protein. I. Computer simulation of trajectories. *J. Mol. Biol.*, **168**, 595–620

McGregor, M. J., Islam, S. A. and Sternberg, M. (1987). Analysis of the relationship between side-chain conformation and secondary structure in globular proteins. *J. Mol. Biol.*, **198**, 295–310

Momany, F. A., McGuire, R. F., Burgess, A. W. and Scheraga, H. A. (1975). Energy parameters in polypeptides VII. *J. Phys. Chem.*, **79**, 2361–2381

Najmudin, S. and Slingsby, C. (1989). Personal communication

Ochi, H., Hata, J. H., Tanaka, N., Kakuda, M., Sakurai, T., Aihara, S. and Morita, Y. (1983). Structure of rice ferricytochrome C at 2.0 Å resolution. *J. Mol. Biol.*, **166**, 407–418

Phillips, S. E. V. (1980). Structure and refinement of oxymyoglobin at 1.6 Å resolution. *J. Mol. Biol.*, **142**, 531–554

Pitt, W. and Goodfellow, J. M. (1991). Modelling of solvent positions around polar side-chains in proteins. *Protein Engng*, **4**, 531–537

Poulos, T. L., Finzel, B. C. and Howard, A. J. (1987). High resolution crystal structure of cytochrome P450cam. *J. Mol. Biol.*, **195**, 687–700

Rees, D. C., Lewis, M. and Lipscomb, W. N. (1983). Refined crystal structure of carboxypeptidase A at 1.54 Å resolution. *J. Mol. Biol.*, **168**, 367–387

Singh, J. and Thornton, J. M. (1990). SIRIUS. An automated method for the analysis of the preferred packing arrangements between protein groups. *J. Mol. Biol.*, **211**, 595–615

Smith, J. L., Corfield, P. W. R., Hendrickson, W. A. and Low, B. W. (1988). Refinement at 1.4 Å resolution of a model of erabutoxin: treatment of ordered solvent and discrete disorder. *Acta Cryst.*, **A44**, 357–368

Steigemann, W. and Weber, E. (1979). Structure of erythrocruorin in different ligand states at 1.4 Å resolution. *J. Mol. Biol.*, **127**, 309–338

Sundaralingam, M. and Sekhurudu, Y. C. (1989). Water inserted in α-helical segments implicate reverse turns as folding intermediates. *Science*, **244**, 1333–1337

Takano, T. and Dickerson, R. E. (1980). Redox conformation changes in refined tuna cytochrome C. *Proc. Natl Acad. Sci. USA*, **77**, 6371–6375

Thanki, N., Thornton, J. M. and Goodfellow, J. M. (1988). Distribution of water around amino acids in proteins. *J. Mol. Biol.*, **202**, 637–657

Thanki, N., Thornton, J. M. and Goodfellow, J. M. (1990). Influence of secondary structure on the hydration of serine, threonine and tyrosine residues in proteins. *Protein Engng*, **3**, 495–508

Thanki, N., Umrania, Y., Thornton, J. M. and Goodfellow, J. M. (1992). Analysis of protein main chain hydration. *J. Mol. Biol.*, **221**, 669–691

Thomas, K. A., Smith, G. M., Thomas, T. B. and Feldmann, R. J. (1982). Electronic distributions within protein phenylalanine aromatic rings are reflected by the three dimensional oxygen atom environments. *Proc. Natl Acad. Sci. USA*, **79**, 4843–4847

Van Gunsteren, W. F. and Berendsen, H. (1984). Computer simulation as a tool for tracing the conformational differences between proteins in solution and in the crystalline state. *J. Mol. Biol.*, **176**, 559–564

Walshaw, J. and Goodfellow, J. (1993). Distribution of solvent molecules around apolar side-chains in protein crystals. *J. Mol. Biol.*, submitted

Walter, J., Steigemann, W., Singh, T. P., Bartunik, H., Bode, W. and Huber, R. (1982). On the disordered activation domain in trypsinogen. *Acta Cryst.*, **B38**, 1462–1472

Watenpaugh, K. D., Sieker, L. C. and Jensen, L. H. (1980). Crystallographic refinement of rubredoxin at 1.2 Å resolution. *J. Mol. Biol.*, **138**, 615–633

Wlodawer, A., Walter, J., Huber, R. and Sjolin, L. (1984). Structure of bovine pancreatic trypsin inhibitor. *J. Mol. Biol.*, **180**, 301–329

4

Water Structure of Crystallized Proteins: High-Resolution Studies

Michel Frey

1 Introduction

Most of the current knowledge concerning the water structure in and around biological macromolecules has been derived at near-atomic resolution from high-resolution X-ray and neutron diffraction studies of crystals (for review, see Finney, 1979; Edsall and McKenzie, 1983; Baker and Hubbard; 1984; Savage and Wlodawer, 1986; Saenger, 1987; Westhof, 1988; Westhof and Beveridge, 1990; Rupley and Caveri, 1991). Protein crystals, like nucleic acid crystals, are made up of a lattice of macromolecules which delimit channels of variable dimensions filled with solvent molecules. They constitute, thus, a suitable system to study the interactions between protein and water molecules at the atomic level, under conditions which are, to some extent, close to those found in solution.

The early determinations of the three-dimensional structures of proteins by X-ray diffraction, together with thermodynamical and dynamical studies, yielded a picture of some important structural and functional properties of water associated with proteins (Finney, 1979; Edsall and McKenzie, 1983; Baker and Hubbard, 1984; Teeter, 1991): waters are seen as mobile molecules which preferentially occupy sites either in the interior of the protein or in crevices distributed over the molecular surface. Since many of such water molecules may establish strong hydrogen bonds with proteins (e.g. Watenpaugh *et al.*, 1978; Edsall and McKenzie, 1983), they are frequently called 'bound' waters.

Analysis of crystal structures has shown that water molecules constitute the sole hydrogen bond partners of *c.* 30% of the main-chain CO and NH bonds and *c.* 60% of the polar side-chains groups. It has also been observed that waters have a strong preference to be bound to protein oxygen atoms rather than to nitrogen atoms and, thus, that they use their protons rather than their lone pairs of electrons in hydrogen bonding. This structural feature reflects: (i) the relatively greater number of solvent exposed oxygen atoms found in proteins; (ii) the ability of oxygen atoms to be involved in more than one hydrogen bond; and (iii) the greater geometrical flexibility of the hydrogen bonding on oxygen atoms, as compared with that for nitrogen atoms (Baker and Hubbard, 1984).

Internal waters —i.e. those shielded from bulk solvent— are observed to bridge polar or charged groups, or both, the bonding potentials of which would have been otherwise unsatisfied. These waters are a constitutive part of the protein, as illustrated by the fact that they are often found in similar locations in various crystal forms of the same molecule or homologous proteins (e.g. Blake *et al.*, 1983). In a similar manner, many waters located on the molecular surface appear as constitutive of the protein, since they help to stabilize the three-dimensional structure by connecting parts of the polypeptidic chain through hydrogen bonds or water-mediated salt bridges.

Waters have also been found in the catalytic cavities, where they may be distributed over groups of sites which mimic the atomic arrangement of the substrate (described as 'substrate mimicry' (e.g. James *et al.*, 1980, Figure 4.1; Blake *et al.*, 1983)). Here, waters help to stabilize the local structure by 'spreading' the buried electrostatic charge over a large volume and maintaining a local optimal stereochemistry for catalytic activity (Finney, 1979). Moreover, the groups of waters located in active sites contribute to the binding energy between the protein and its substrate, since they are —wholly or partially— easily displaced on substrate binding.

In parallel, a wealth of data on the protein–solvent interactions had been derived from structural, thermodynamical and dynamical studies on the hydration process of globular proteins (for review, Finney, 1979; Edsall and McKenzie, 1983; Rupley *et al.*, 1983). For example, it has been demonstrated that the hydration of lysozyme is a stepwise process which involves (1) the interactions of water with charged groups; (ii) the formation and growth of clusters of waters over most of the protein surface; and (iii) the hydration of the remaining exposed portions of the surface. The enzymatic activity becomes detectable at a hydration level of 0.20 gram of water per gram of protein (g/g), and then increases in parallel with the hydration level and the mobility of bound ligand. The fact that changes in the activity still occur at high hydration levels (i.e. above 0.38 g/g) suggests that the most loosely 'bound' waters are also involved in the catalytic properties of the protein. Furthermore, coupling between the motions of water and proteins have been observed. Thus, the protein–solvent interactions are dynamically associated with the interdomain motions of the protein (at time-scale *c.* 10^{-8} s), as well as with the local fluctuations of the protein surface (time-scale *c.* 10^{-11} s).

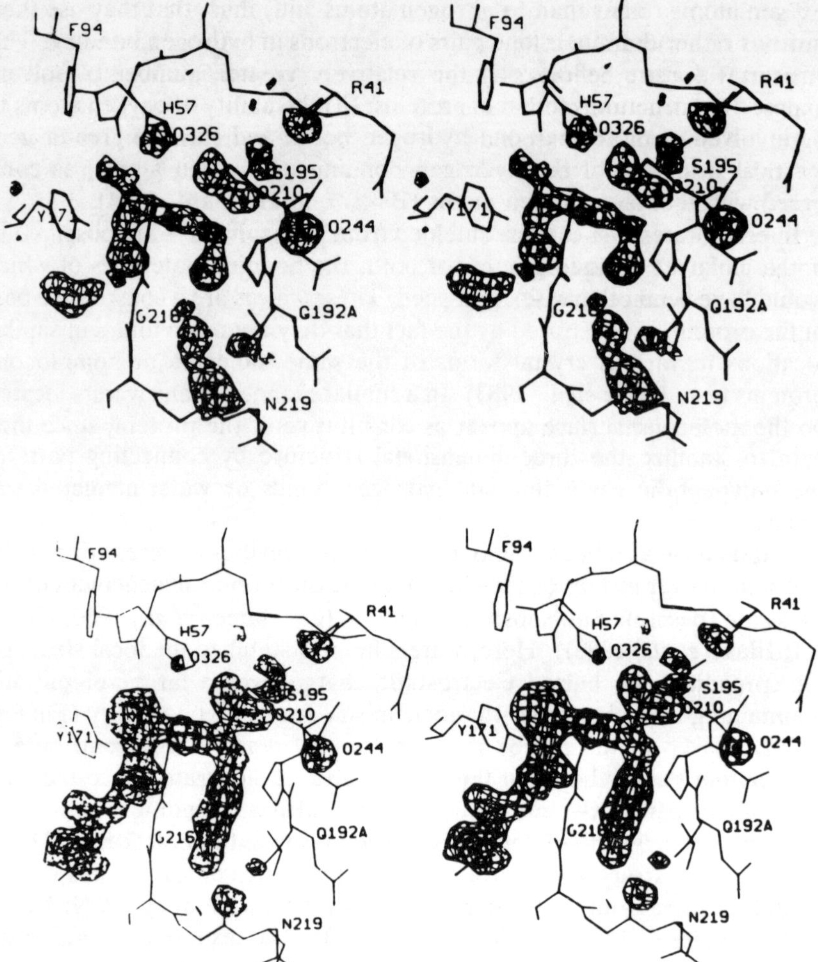

Figures 4.1 *Streptomyces griseus* protease A: stereoscopic view of the electron density (e.d) distribution in the active site (up) for the free enzyme; and (down) with the tetrapeptide Ac–Pro–Ala–Pro–Tyr–OH bound. O326 and O244 are internally bound waters. In the free enzyme the density corresponds to sites of waters which do not make hydrogen bonds with the protein and thus are easily displaced upon binding of the substrate. Note the remarkable similarity between both densities. Maximum resolution 1.8 Å; R = 0.13 and 0.122. From James *et al.* (1980). Copyright © 1980 Academic Press

Protein–solvent interactions have also have been analysed by Monte Carlo methods. For example, simulations of the structure and energetics of solvent in the triclinic egg-white lysozyme crystal have confirmed that the water structure adjacent to the protein and that of bulk water are different, since many of the waters 'bound' to the protein have a considerably lower energy than bulk waters (Hagler and Moult, 1978).

Tremendous progress has occurred for the last few years in all the domains of structural biology. In this context, the first part of the present chapter will describe the strategy currently followed to obtain and to analyse the structure of the solvent associated with proteins by crystallographic methods. The second part will attempt, with the help of selected examples, to give an overview of the structural and functional properties of protein–water interactions at the atomic level. Finally, some of the future prospects for protein–solvent interactions studies will be outlined, in view of the current methodological developments in crystallography, nuclear magnetic resonance (NMR), molecular dynamics (MD) and site-directed mutagenesis.

2 Crystallographic Methods

Protein Crystals

The quality of a crystallographic study leans above all on the growth of well-ordered crystals diffracting at high resolution (i.e. 1.8 Å or better). Moreover, since the fraction of the crystal volume occupied by solvent may range from *c.* 27% to *c.* 65% (Matthews, 1968), the solvent structure in and around the crystallized protein may depend critically on factors such as volume and composition of the solvent within the crystal (e.g. Salunke *et al.*, 1985), pH, ionic strength . . . (e.g. Suguna *et al.*, 1987; Figure 4.2).

It follows that the most suitable crystalline environments in which to examine protein–water interactions involve low salt concentrations, sufficiently large solvent volumes and nearly physiological pHs. Along these lines, a ferricytochrome, whose molecular and solvent structures had previously been determined with crystals grown from solutions at very high ionic strengths (9.5 M of ammonium sulphate), has recently been recrystallized in a medium whose ionic strength was closer to physiological conditions (45 mM; Walter *et al.*, 1990). The resulting X-ray model is greatly awaited, since a recent comparison of the crystal structures of bacteriophage T4 lysozyme, determined at different ionic strengths, suggests that the presence of high salt has actually a limited effect on the hydrogen bonding pattern of solvent (Bell *et al.*, 1991).

Measurements of crystal density combined with crystallographic data analysis may also provide information on the respective fractions of 'bound' and bulk solvent within the crystals, and, in some cases, on the precise composition of the 'bound' solvent (see, for example, Scanlon and Eisenberg, 1975; Hagler and Moult, 1978). Selenate-exchanged (SO_4^{2-}, SeO_4^{2-}) mother-liquor is also an efficient way to identify the sulphate ions, although, to the author's knowledge, this method has only been reported for α-chymotrypsin crystals (Tulinsky and Wright, 1973; Blevins and

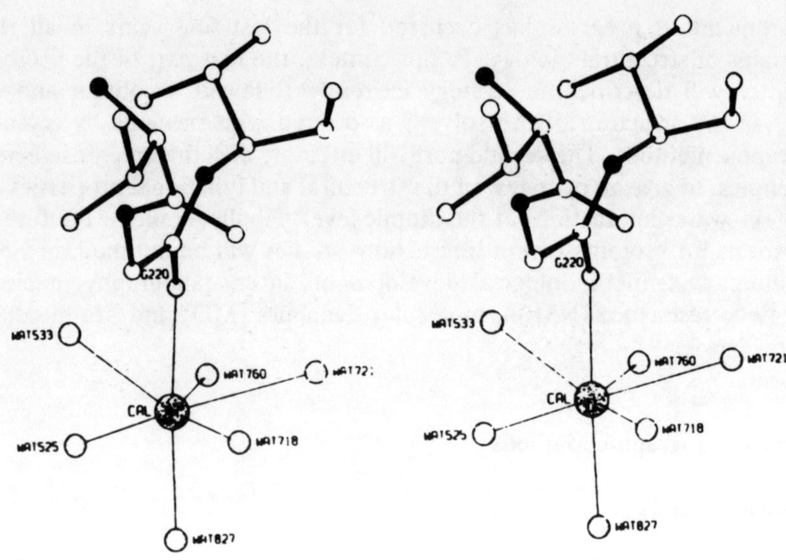

Figure 4.2 *Rhizopus chinensis* aspartic proteinase: a calcium atom from the crystallization medium which contained 20 mM calcium acetate, binds to the crystallized protein. Stereoscopic view. Maximum resolution 1.8 Å; $R = 0.143$. From Suguna *et al.* (1987). Copyright © 1987 Academic Press

Tulinsky, 1985). In any event, all these experiments could be very helpful and, in some instances, critical (Blevins and Tulinsky, 1985) for a proper chemical assignment of solvent peaks observed in electron density maps.

X-Ray Data Collection

Accurate measurements of the full X-ray diffraction pattern, i.e. including very-low- and very-high-resolution data, are required to determine an overall picture of the solvent structure within the crystal. This can now be achieved for large proteins, owing to the availability of: (i) powerful and well-monochromatized X-ray sources; (ii) two-dimensional detectors; and (iii) efficient data reduction software, including absorption corrections (e.g. Kopfmann and Huber, 1968; North *et al.*, 1968; Helliwell *et al.*, 1984; Sheriff and Hendrickson, 1987; Messerschmidt *et al.*, 1990).

These new experimental approaches reduce the effects of radiation damage on crystals, and thus allow the collection of higher-resolution data. Moreover, the use of the intense, short-wavelength (i.e. 1.0 Å or less) radiation available on synchrotrons reduces absorption effects, which are an important source of systematic errors. Thus, the collection of a 1.4 Å data set with the 1.0 Å wavelength radiation of the EMBL beam line at the DESY synchrotron (Hamburg, Germany) and an image plate detector led

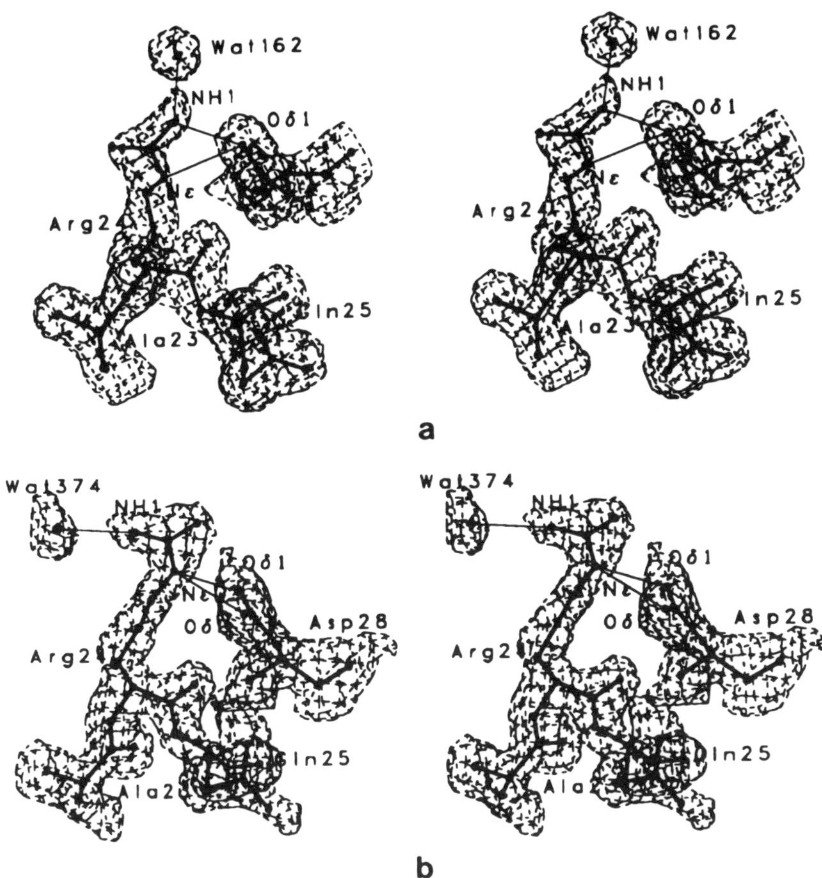

Figure 4.3 *Desulphovibrio vulgaris* flavodoxin: stereoscopic view of the electron density distribution around Arg24 in (*a*) at room temperature; in (*b*) at − 150°C. Maximum resolution ~ 2.0 Å; R = 0.20–0.17. From Watt *et al.* (1991). Copyright © 1991 Academic Press

to the determination of a precise atomic model of a thermitase (M_r 28 380) and of its associated solvent, despite a strongly anisotropic crystal habit (0.8 mm × 0.4 mm × 0.1 mm) Teplyakov *et al.*, 1990).

To reduce further the effects of radiation damage and of dynamic disorder, X-ray data collection may be performed at low temperatures (e.g. Watt *et al.*, 1991; Figure 4.3; Frauenfelder *et al.*, 1979; and for a recent review, Hope, 1990). Under these conditions, absorption corrections can be made more accurately, since the samples do not have to be mounted inside glass capillaries in the presence of mother-liquor. However, the use of cryotechniques requires that neither drastic increase in mosaicity nor significant change in the crystal structure is induced during and after the cooling procedures.

Molecular Modelling and Refinement

The determination of the molecular model of a protein through crystallographic methods is performed with various computer programs and graphics techniques, by interleaving Fourier series calculations, model building and refinement cycles. The refinement process uses either energy or restrained least squares minimization techniques (EREF, Jack and Levitt, 1978; PROLSQ, Hendrickson and Konnert, 1980; CORELS, Sussmann *et al.*, 1977; TNT, Tronrud *et al.*, 1987) or molecular dynamic calculations (MD), coupled with crystallographic and energy refinement (XPLOR, Brünger *et al.*, 1987, Brünger, 1988; MDXREF, Fujinaga *et al.*, 1989). The molecular dynamics approach allows one to reach more efficiently a global minimum, with less manual rebuilding of intermediate models into electron density maps. MD software also includes programs which predict hydrogen atom positions with an energy minimization procedure which is limited to their immediate environment (Brünger and Karplus, 1988; Figure 4.4).

Ordered Solvent

All strategies currently followed to model and refine the protein–solvent structure handle molecules according to their degree of ordering within the crystal (see, for example, Blake *et al.*, 1983). Peaks above a predetermined threshold are automatically searched for in the asymmetric unit of difference F_o–F_c Fourier maps. The threshold (e.g. 0.22 *e*/peak in the final stage of the refinement of α-lytic protease: Fujinaga *et al.*, 1985) is chosen as a function of the current map noise level. These selected peaks are modelled as water molecules: (i) if they have a convex shape; (ii) if they appear also in $2F_o$–F_c or, preferably, $3F_o$–$2F_c$ maps (Luzzatti, 1953); (iii) if there are no short contacts either with the protein or with neighbouring solvent molecules; and (iv) if the resulting surrounding hydrogen bonding geometry is stereochemically plausible (Chiari and Ferraris, 1982; Baker and Hubbard, 1984; Jeffrey and Maluszynska, 1990).

The sites of the most highly ordered solvent molecules can be readily identified in difference Fourier electron density maps at resolutions between 3 Å and 2 Å (e.g. Bartunik *et al.*, 1989; Figure 4.5a). At this stage of the study, unusually high or large peaks have to be carefully checked, since they may reflect the presence within the crystal of other, expected or unexpected, electron-dense components of the mother-liquor such as ions (Figure 4.5b) or organic compounds.

In very accurate structure determinations, such as that of *Desulphovibrio desulphuricans* rubredoxin at 1.5 Å resolution (Stenkamp *et al.*, 1990), the electron density of many hydrogen atoms appears clearly in the final electron density maps. Therefore, it is advisable in such cases to take into

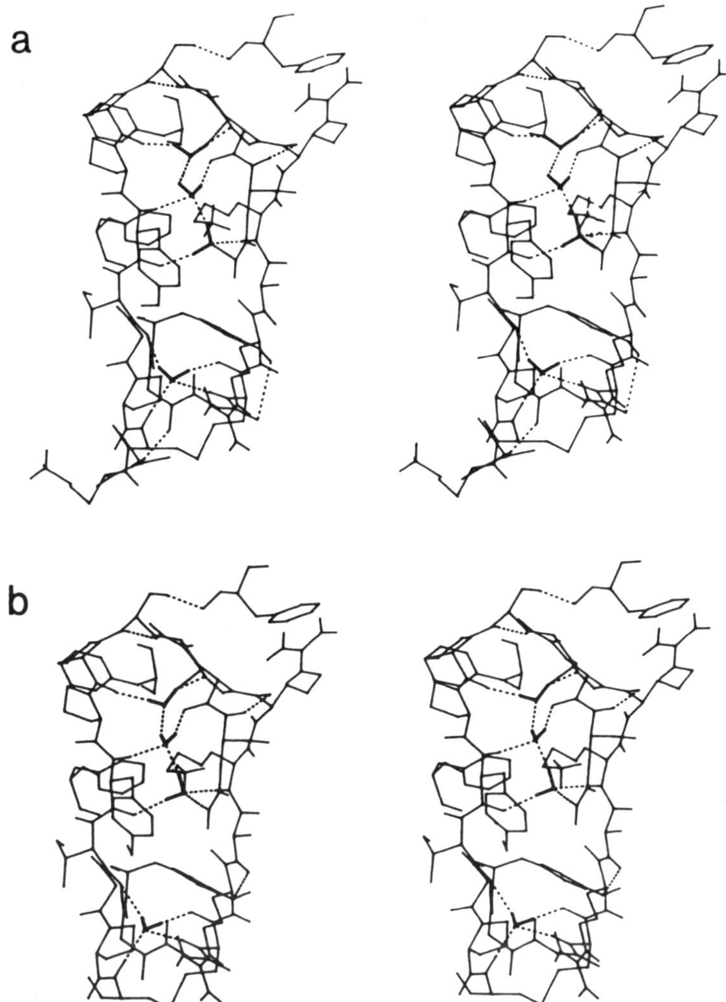

Figure 4.4 Bovine pancreatic trypsin inhibitor (BPTI): stereoscopic view of the 5 Å environment of the four ordered waters; comparison of the hydrogen atoms positions of the waters in (*a*) calculated by empirical energy placement; in (*b*) observed by neutron diffraction. Hydrogen bonds (dotted lines) are indicated if (i) the heavy donor–acceptor distance is less than 4 Å; and (ii) the donor–hydrogen–acceptor angle is less than 60° (angles are 0° for a linear hydrogen bond). From Brünger and Karplus (1988). Copyright © 1988 John Wiley and Sons

account the scattering contribution of the hydrogen atoms in the refinement process by using, for example, a modified scattering factor for water oxygen atoms (Jensen, 1990).

It is very important to include solvent molecules (waters, ions or small organic molecules) in the calculations as soon as they have been localized and identified with certainty, since neglecting solvent may lead to

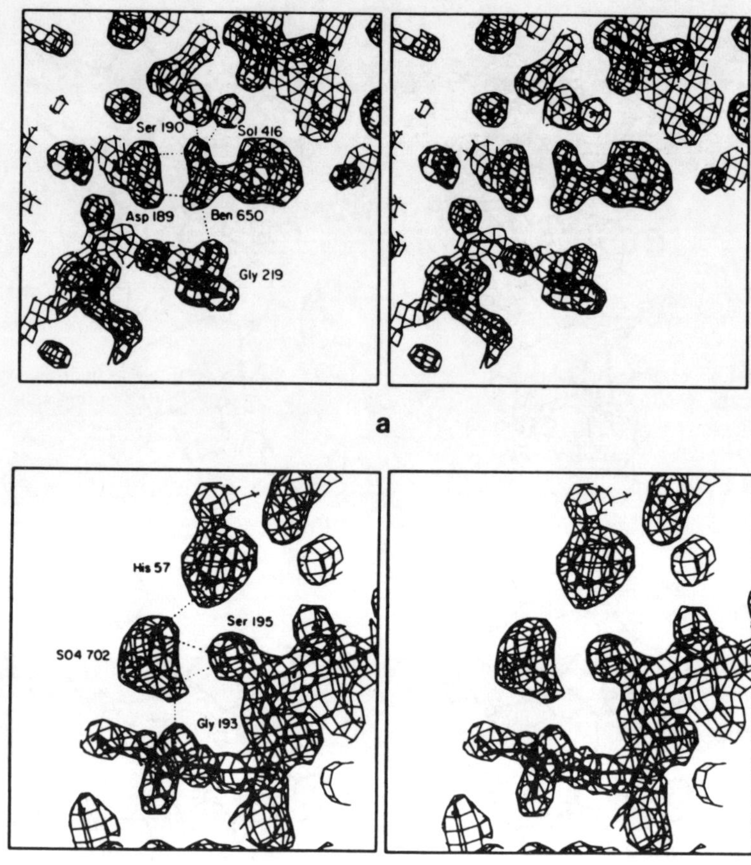

Figure 4.5 Bovine pancreatic β-trypsin: stereoscopic view of the electron density of the crystal structure with low molecular packing density. In (*a*) benzamidine and part of the waters bound in the specificity pocket; in (*b*) sulphate molecule at the active site with its hydrogen bonding to His57 Nε2, Ser195 Oγ and Gly193 N. Maximum resolution 1.5. Å; $R \sim 0.168$. From Bartunik *et al.* (1989). Copyright © 1989 Academic Press

erroneous positioning of hydrophilic side-chains of proteins (Sielecki *et al.*, 1979) or to distorted geometries of nucleic acids (Kennard *et al.*, 1986). As an alternative, one may include high-resolution data progressively in the refinement (e.g. Suguna *et al.*, 1987). In any case, it should be emphasized that the overall scale factor has to be recalculated throughout the refinement process, prior to further refinement cycles, whenever a substantial number of solvent molecules have been included or modified in the model.

Biases in the refinement process may arise from the requirement that a solvent atom should be within a predetermined distance of a polar protein atom. It has therefore been suggested to redetermine several times the

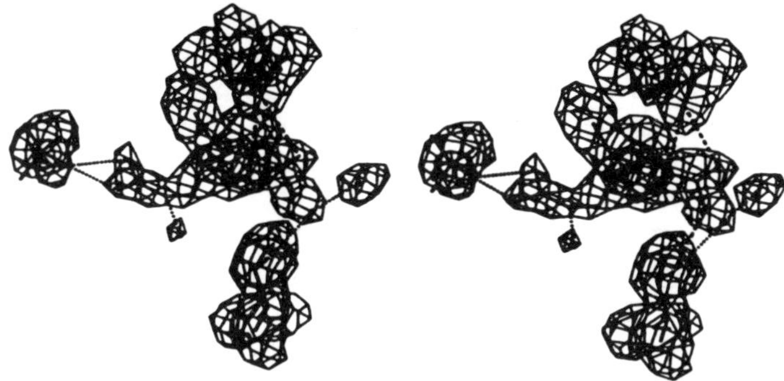

Figure 4.6 Myohemerythrin: stereoscopic view of the electron density and hydrogen bonding of the two possible conformers of Arg 37. In the conformer I (pointing right) the guanidinium salt bridges (thick dashed lines) to Asp34 (*top*) of the same molecule and to Thr52 (*bottom*) of a neighbouring molecule. Conformer II (pointing left) salt bridges (thin dotted lines) to a sulphate ion (*left*), while a water molecule (*right*) occupies a site vacated by conformer I and hydrogen bonds to Thr52. Maximum resolution 1.7–1.3 Å; $R = 0.159$. From Smith *et al.* (1986). Copyright © 1986 American Chemical Society

solvent structure during the process (e.g. Fujinaga *et al.*, 1985): subsets, or the totality, of solvent atoms are removed from the calculations, and the remaining atoms are refined for a few cycles. The solvent atoms are then again localized by an iterative procedure. Biases are also generated by imposition of non-crystallographic symmetry. In such cases, the most productive approach appears to restrain non-crystallographic symmetry at low resolution and to relax— or to remove— the non-crystallographic restraints at high resolution in the final stages of the refinement (Tulinsky and Blevins, 1986).

Multisite Configuration

Very frequently, peaks appear too close to either protein atoms or other solvent atoms. These short contacts usually indicate the presence of multisite conformation, i.e. the affected atoms cannot occupy the same position in the cell at the same time. The modelling and the refinement of the mutually exclusive side-chain conformers and solvent sites has been extensively discussed and described in several papers (e.g. Smith *et al.*, 1986, 1988; Figure 4.6; Jensen and Watenpaugh, 1986; Teeter and Whitlow, 1986; Kuriyan *et al.*, 1986, 1991).

Temperature and Occupancy Factors

When localized and identified, atoms are generally assigned a temperature factor (B in Å^2) and an occupancy factor (Q), representing the probability of the presence of a solvent molecule at the site under consideration. Initial estimates of the values of B are generally chosen as a function of the number of potential hydrogen bonds, and those of Q according to peak heights on F_o–F_c Fourier maps. For disordered atoms, the sum of the Q-factors of atoms belonging to different conformers of the same side-chain, or to mutually exclusive solvent molecules, should not exceed unity (Smith *et al.*, 1988).

For refinement, owing to the high correlation between the B- and Q-factors, damped shifts are applied on alternate cycles (Hendrickson, 1980). The sites of solvent molecules whose temperature or occupancy factors fall outside reasonable limits (typically Bs > 30 Å^2 or < 5 Å^2 and Qs < 0.3) are discarded from the calculations, but the corresponding peaks have to be re-examined carefully on subsequent difference Fourier maps. The peaks could, then, be reinserted in the calculations if they meet the criteria described above. The same procedure is used for physically meaningless combinations of B- and Q-values, i.e. high B/high Q or low B/low Q, keeping in mind that high B/high Q may reflect multiple unresolved discrete water molecules (Smith *et al.*, 1988).

The use of occupancy factors is still discussed by many authors (Kundrot and Richards, 1987; Bhat, 1989; Jensen, 1990; van der Sluis and Spek, 1990), which explains why in most cases only the temperature factor is refined. However, our opinion is that the Q-factor remains a simple and efficient tool to model multisite conformation when identified.

Bulk Solvent

The most widely used method to model the solvent molecules, apart from those in immediate contact with the protein, has been summarized by Moews and Kretzinger (1975). It consists of the addition of a constant solvent density to the protein model beyond a preset distance from any surface protein atom, with a smoothing function used for the interface area between protein and solvent (Phillips, 1980; Blake *et al.*, 1983). The contribution of the solvent to the structure factors is then deduced by Fourier inversion and assignment of a temperature factor.

A result of this approach has been the recent computation of a 1.7 Å resolution electron density map of the water structure within insulin crystals (67 % v/v of solvent: Badger and Caspar, 1991). Briefly, a 'flat' solvent map was calculated outside the density of the protein. Afterwards this map was added to a difference Fourier map calculated from the current protein and discrete solvent models. Finally, a new set of calculated amplitudes and phases was extracted from this combined map by Fourier inversion and was included in additional refinement cycles. The final map showed a set of peaks in the vicinity of the protein corresponding to discrete water molecules. Beyond this zone, characteristic fluctuations appeared in the solvent density above the noise level, and were assigned to superimposed structural arrangements of water molecules. This model has been supported by the low value of the final crystallographic *R*-factor ($R = 0.06$ for all the data), several tests on the reliability of the density map and stereochemical considerations. However, it has recently been suggested that the structural features which appeared in the solvent continuum might result either from experimental errors or from protein model deficiencies, due mainly to uncertainties as to the atomic temperature factors (Jensen, 1991).

Completion of the Refinement

The refinement of a molecular model is generally considered as completed when, in the final $F_o - F_c$ Fourier map, no peak is found which could be explained in terms of additional solvent molecule, and when no significant peak (e.g. higher than 0.2 $e/Å^3$) is seen in the protein region. The quality of the final map can be judged by the values of peak heights, for those peaks which have not been accounted for (typically, $\pm$ 0.4 $e/Å^3$). Finally, water molecules are classified according either to Q^2/B (James and Sielecki, 1983) or to their peak heights on the final difference Fourier maps.

Such an exhaustive inspection of difference Fourier maps has led recently to the nearly complete modelling of the solvent structure of *Desulphovibrio vulgaris* rubredoxin crystals (Adman *et al.*, 1991): the final solvent model includes one sulphate group and 100 water molecules distributed among 180 sites. The remarkably low value of the corresponding crystallographic *R*-factor ($R = 0.098$ for all diffraction data to 1.5 Å resolution) is close to the value one would expect from the accuracy of the X-ray data. This indicates that, in favourable circumstances (in this case the solvent occupies *c.* 27% of the crystal volume), the discrete solvent structure can be nearly comprehensively described with a great degree of confidence. In this study, it is also of interest to note that the contribution of the disordered solvent molecules to the X-ray data can still be detected at 2.5 Å resolution (see Figure 11 in Adman *et al.*, 1991).

Neutron Diffraction

For the determination of protein structures, neutron diffraction presents several well-known advantages over X-ray diffraction: (i) a more precise localization of hydrogen atoms (e.g. Wlodawer *et al.*, 1989; Figure 4.7); (ii) a nearly unbiased determination of the solvent structure, including weakly scattering waters, on H_2O/D_2O solvent difference maps; and (iii) no radiation damage to the crystals (see, for review, Schoenborn, 1984; Kossiakoff, 1985; Savage and Wlodawer, 1986). Moreover, joint neutron and X-ray data refinements have been shown to provide more information than the use of either type of data alone (Wlodawer and Hendrickson, 1982).

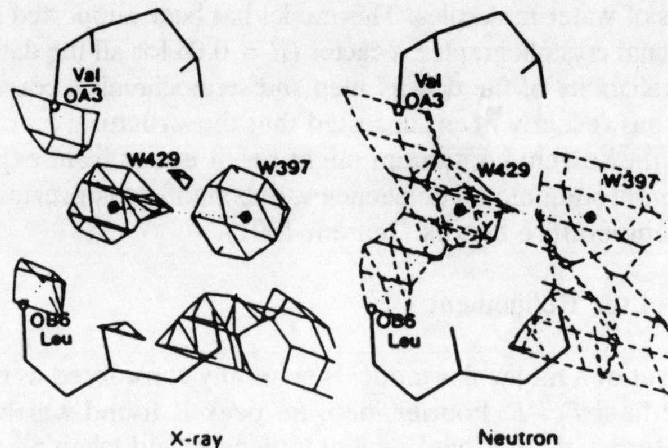

Figure 4.7 Insulin: X-ray (solid lines) and positive neutron (dashed lines) density around two water sites located at the surface of the protein. The continuous density of the neutron map is due to unresolved deuterium atoms and indicates the direction of hydrogen bonds. Maximum resolution 1.5 Å (X-ray), 2.2 Å (neutron); R = 0.182 (X-ray), 0.191 (neutron). From Wlodawer *et al.* (1989). Copyright © 1989 International Union of Crystallography

However, neutron diffraction requires rather large crystals and long data collection time, owing to the limitation of reactor fluxes.

Several neutron diffraction studies of small protein crystals have been reported (e.g. Mason *et al.*, 1984; Teeter and Kossiakoff, 1984; Wlodawer *et al.*, 1984). The strategies which were used to model and refine protein and solvent structures were similar to those described above for X-rays.

Two recent examples illustrate well the power of neutrons to elucidate the structure of solvent. In the first case, high-resolution studies (*c.* 2–1.8 Å) of triclinic lysozyme crystals, soaked in either ethanol or dimethyl sulphoxide (DMSO), have shown that most of the ethanol or DMSO molecules are in hydrophobic contacts with the protein. Since these contacts do not perturb the protein structure, it has been suggested that the denaturation of lysozyme is a consequence of dehydration and cannot be due to an infiltration of the solvent into the protein (Lehmann *et al.*, 1985, 1989). In the second study, analysis of the low-resolution terms of the neutron diffraction data of myoglobin crystals combined with careful density measurements led to the identification of four (over a total of seven) previously undetected sulphate ions in the structure (Schoenborn, 1988; Cheng and Schoenborn, 1990).

Quality of the Model

A first method to estimate the plausibility of the protein solvent structure is to calculate (i) the mean surface area of the protein covered by one water molecule (*c.* 20 Å^2), taking into account the molecular packing contacts,

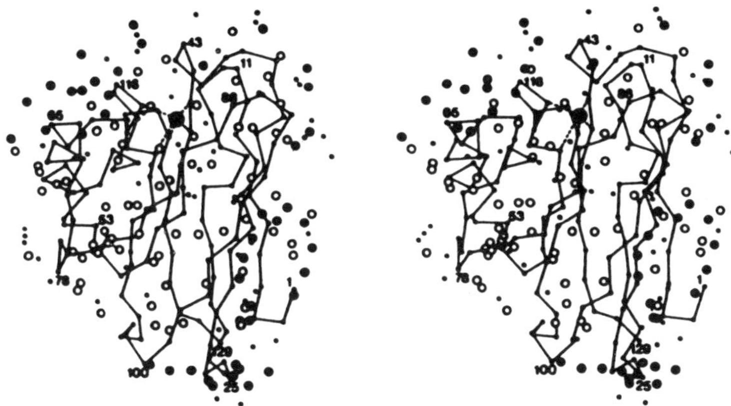

Figure 4.8 *Alcaligenes denitrificans* azurin: stereoscopic view of the water sites around one of the two crystallographically independent molecules. Waters found in equivalent positions in both protein molecules are represented by large open circles. The large filled circles represent waters involved in intermolecular contacts. Maximum resolution 1.8 Å; $R = 0.157$. From Baker (1988). Copyright © 1988 Academic Press

and (ii) the percentage of discrete waters relative to the total number of waters per protein within the crystal (e.g. Adman *et al.*, 1991).

Furthermore, a mean value of the errors in the atomic positions of protein and solvent may be obtained from Luzzatti plots, calculated from the differences between the observed and calculated data (Luzzatti, 1952). The characteristic increase of the crystallographic R-factor which is frequently observed on these plots at low resolution is a clear indication of incomplete solvent modelling. A more detailed estimate of the error can be deduced according to the atom-type and B-factors by Cruickshank's method (1949), as developed by Chambers and Stroud (1979) and by Read *et al.* (1983).

Comparisons of molecular models determined independently—i.e. without symmetry restraints (Tulinsky and Blevins, 1986)—or in different crystal environments, or, in some instances, derived from X-ray and neutron data, have probably provided the best indicators about the quality and reproducibility of protein and solvent structures. Thus, in crystals of the blue copper protein azurin from *Alcaligenes denitrificans*, 182 *independently* refined waters are associated with the two molecules present in the asymmetric unit. These waters are related by the non-crystallographic twofold axis, with r.m.s. deviations less than 1 Å for 104 of them (Baker, 1988; Figure 4.8). A similar comparison between a 1.45 Å resolution X-ray model and a 2.0 Å resolution joint neutron and X-ray model of ribonuclease A has shown a mean difference of 0.374 Å between the location of the protein atoms in the two structures, and of less than 1.0 Å for 60 of the 120 solvent molecules (Wlodawer *et al.*, 1986).

Similarly, half of the solvent molecules, mainly those with low temperature factors, present in two distinct crystal forms of bovine pancreatic trypsin inhibitor (BPTI), occupy the same sites, within distances less than 1.0 Å, and show similar hydrogen bonding patterns (Wlodawer *et al.*, 1987). This conservation of the 'bound' solvent structure has also been observed in homologous proteins such as human and tortoise egg-white lysozyme (Blake *et al.*, 1983) and rubredoxins (Watenpaugh *et al.*, 1978; Frey *et al.*, 1987; Adman *et al.*, 1991).

All these results confirm that crystal packing or the presence of salt ions do not have a dominating influence on the interaction between proteins and 'bound' water molecules. As expected, the differences in the water structures are due either to the crystallization medium or to charged amino acids substitutions.

Results

Structural Properties

The over-all picture of the structural properties of waters associated with proteins which was described in Section 1 remains essentially valid: (i) a substantial number of water molecules establish hydrogen bonds with the protein (Watenpaugh *et al.*, 1978; Figure 4.9); (ii) many of these 'bound' waters are found in similar locations in different crystal forms of the same protein or of homologous proteins (e.g. Bell *et al.*, 1991); (iii) waters are constitutive elements of proteins, where they contribute to the stabilization of the three-dimensional structures (e.g. Baker and Hubbard, 1984; Frey *et al.*, 1987; Read and James, 1988; Figures 4.10, 4.11, 4.12), including stabilization of catalytic sites (e.g. Suguna *et al.*, 1987; Kamphuis *et al.* 1985; Figures 4.13, 4.14); and (iv) waters may play a key role in catalytic activity.

A wealth of systematic studies on high-resolution X-ray structures deposited with the Protein Data Bank (Bernstein *et al.*, 1977) has added new insights on the structural properties of waters associated with proteins. Thus, it has been shown that waters are usually found in protein cavities which may be relatively large (e.g. 180 Å^3: Rashin *et al.*, 1986). In the few cavities where no ordered water molecules were localized crystallographically, water is most probably present in a disordered form. Likewise, it has been established that the hydration pattern around amino acid residues in proteins shows distinct non-random distributions of waters around all side-chains (Thanki *et al.*, 1988; Pitt and Goodfellow, 1991; Goodfellow *et al.*, this book).

It has also been shown that water networks contribute to the stabilization of the protein secondary structures (e.g. Richardson, 1981; Figure

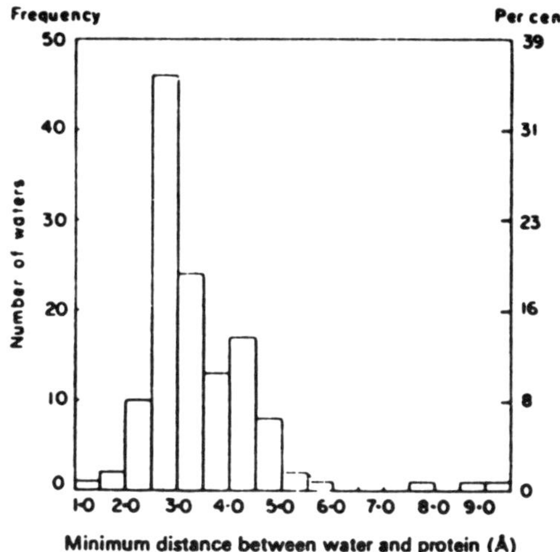

Figure 4.9 *Clostridium pasteurianum* rubredoxin: distribution of waters as a function of their minimum distance from protein. Maximum resolution 1.2 Å; $R = 0.127$. From Watenpaugh *et al.* (1978). Copyright © 1978 Academic Press

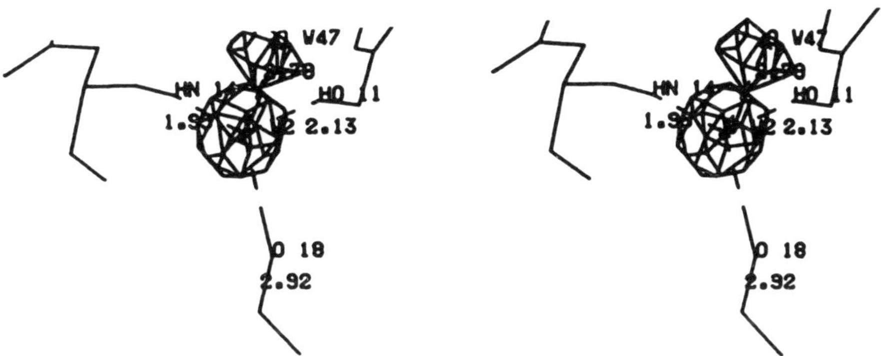

Figure 4.10 *Desulphovibrio gigas* rubredoxin: stereoscopic view of the electron density of the water molecules located at W2 and W47. W2 stabilizes the loop Pro15–Gly37 through hydrogen bonds (dashed lines) with Tyr 11Oη, Asp 14N, Gly 18O and W47. A water is always found in W2 location in all homologous rubredoxins. Distances are in Å. Maximum resolution 1.4 Å; $R = 0.136$. From Frey *et al.* (1987) Copyright © 1987 Academic Press

4.15; Thanki *et al.*, 1991), including that of helix distortions, which are thought to optimize the internal molecular packing and to accommodate sequence changes (Blundell *et al.*, 1983; Barlow and Thornton, 1988). Moreover, examination of high-resolution (1.8 Å or better) X-ray models deposited with the Protein Data Bank (Bernstein *et al.*, 1979) has led to the proposal that the hydration of alpha-helical segments (Sundaralingam and Sekharudu, 1989; Figure 4.16) implicate reverse turns as folding intermedi-

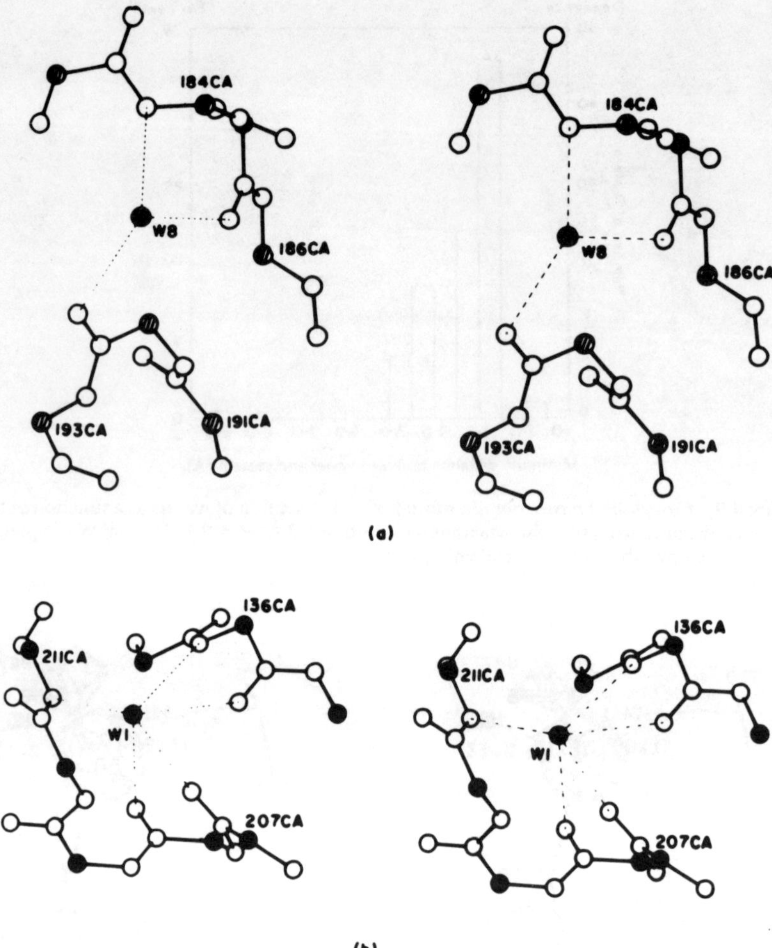

Figure 4.11 Actinidin: in (*a*) W8 has 3 ligands in an approximately trigonal arrangement, 2 proton acceptors (185O 2.88 Å; 192O 2.91 Å) and 1 proton donor (184N 2.83 Å); in (*b*) W1 has 4 close neighbours, 2 proton acceptors (208O 2.74 Å; 136O 2.87 Å) and 2 proton donors (211N 2.87 Å; 136N 3.14 Å), as well as a fifth more distant neighbour (207O 3.33 Å). Three of these ligands (211N, 208O, 136 N) are tetrahedrally disposed and the fourth tetrahedral position lies between 136O and 207O, suggesting a three-centre hydrogen bond (the proton directed between 136O and 207O). Stereoscopic views. Maximum resolution 1.7 Å; R = 0.146. From Baker (1980). Copyright © 1980 Academic Press

ates, and that, consequently, they could play a role in the protein chain folding.

The frequent occurrence of cavities filled with waters in interdomain regions (e.g. Furey *et al.*, 1983; Pflugrath and Quiocho, 1988; Figures 4.17, 4.18) suggests that waters play a role in molecular motions. Thus, several

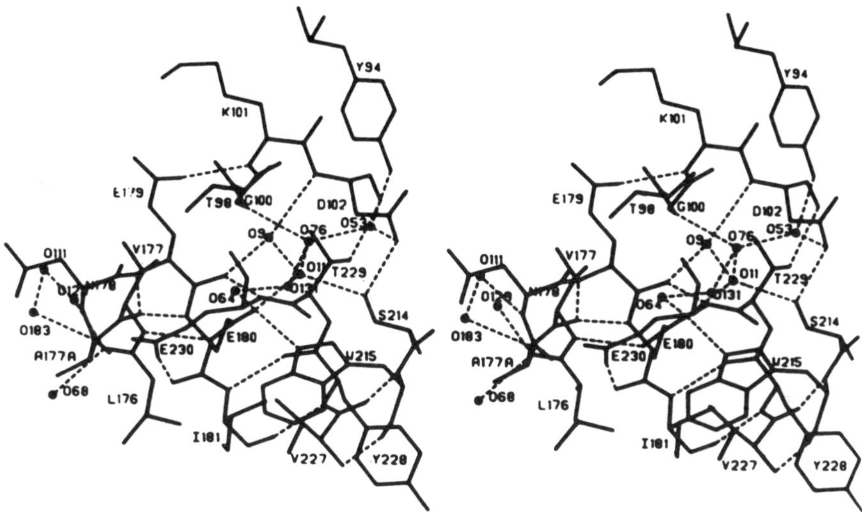

Figure 4.12 *Streptomyces griseus* trypsin: the negative charge of Glu180 is not balanced by a positively charged group but by the peptide dipole Leu176–Val177, by the hydrogen bond from Trp215 and by several solvent molecules. The water molecule O53 is hydrogen bonded to Asp102. Stereoscopic views. Maximum resolution 1.7 Å; $R = 0.161$. From Read and James (1988). Copyright © 1988 Academic Press

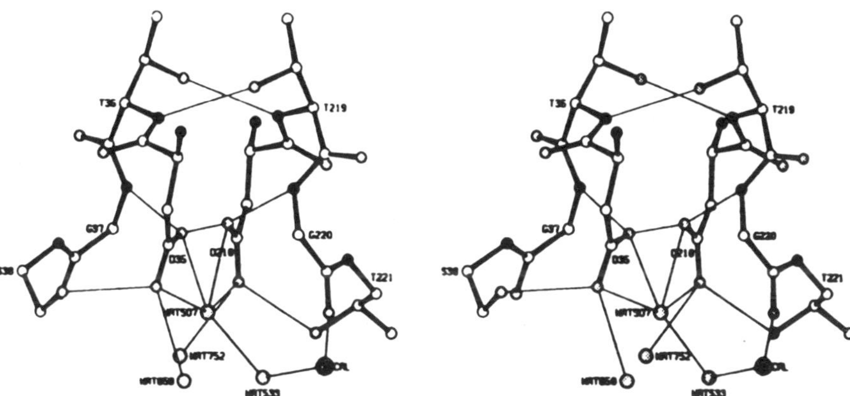

Figure 4.13 *Rhizopus chinensis* aspartic proteinase: hydrogen bonds at the active site. Note that the pseudo dyad axis that relates the two domains passes close to W507 and lies in the plane of the diagram. Maximum resolution 1.8 Å; $R = 0.143$. From Suguna *et al.* (1987). Copyright © 1987 Academic Press

waters might contribute to the respective motions of the C- and N-terminal domains of phage T4 lysozyme (Weaver and Matthews, 1987; Figure 4.19). Along the same lines, insertion or removal of water molecules between dimers in phosphofructokinase seems to be coupled to the allosteric regulation of the enzyme's activity (Schirmer and Evans, 1990; Figure 4.20).

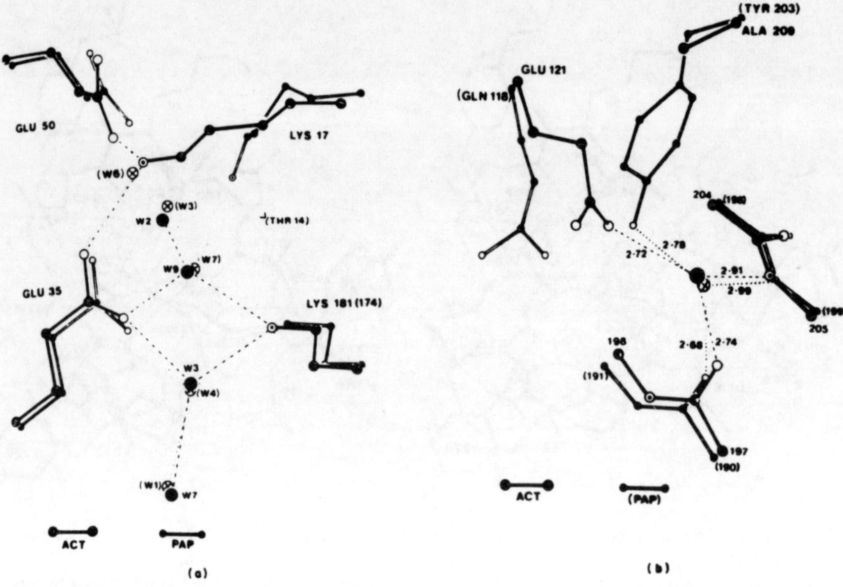

Figure 4.14 Thiol proteases: effect of sequences changes on the structure of actinidin and papain (papain residue numbers are indicated in parentheses). In (*a*) papain Lys 17Nζ is hydrogen bonded to Thr 14Oγ1 (Val in actinidine). However, W6 occupies the location of the Nζ of Lys17 in actinidin, so that the hydrogen bonding network is conserved. Note the conserved water molecule. The effect of substituting Tyr203 by Ala is seen in (*b*). Thr 203Oη in papain corresponds to Glu 121Oδ2 in actinidin (Gln118 in papain) such that hydrogen bonding (dotted lines) to an internal water molecule remains unchanged. (Maximum resolution 1.7 Å; R = 0.146; actinidin, Baker, 1980; Maximum resolution 1.65 Å; R = 0.161; papain, Kamphuis *et al.*, 1984). From Kamphuis *et al.* (1985). Copyright © 1985 Academic Press

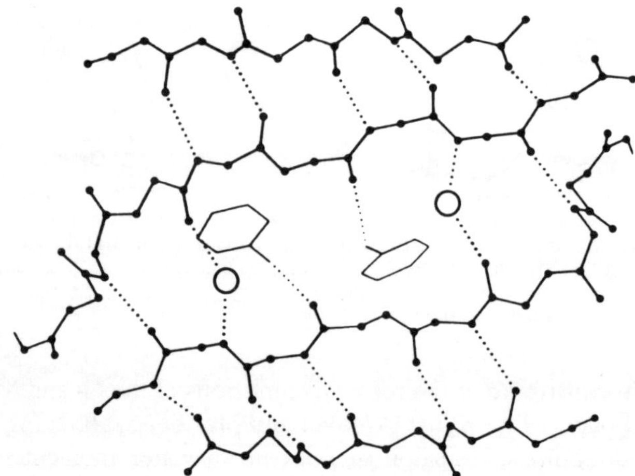

Figure 4.15 Prealbumin: waters (open circles), bridging main chain groups that are too far apart to continue β-sheets hydrogen bonding. A hydrogen bond involving tyrosine side-chains is also visualized. After Richardson (1981). Copyright © 1981 Academic Press

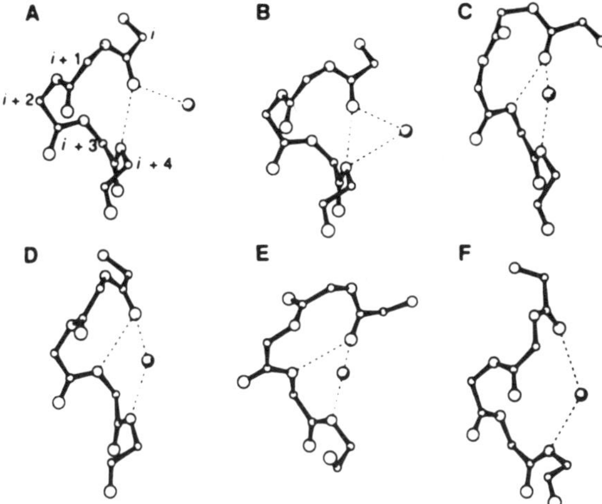

Figure 4.16 Helix hydrated segments in protein crystal structure. In (*A*) bound water (on the left side) to peptide carbonyl CO uniquely. In (*B*) bound water to peptide CO and NH. Note that H is involved in a three-centre hydrogen bond. In (*C, D, E*) disruption of the helix with the formation of a type III, type I and type II turn, respectively. In (*F*) insertion of a water in the helix resulting in open turn. These hydrated segments may be regarded as 'snapshots' of the intermediates involved in folding. From Sundaralingam and Sekharudu (1989). Copyright © 1989 AAAS

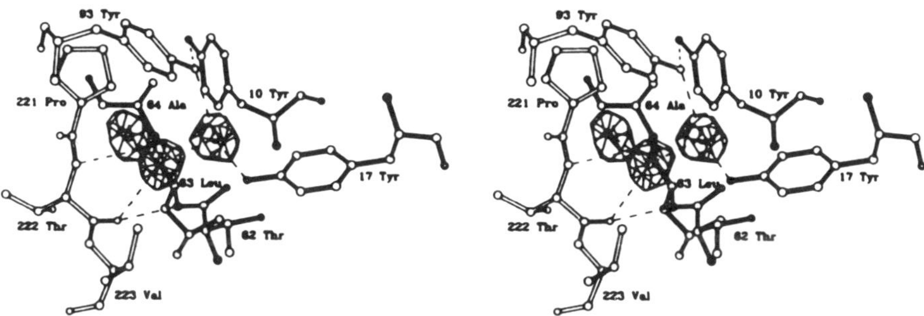

Figure 4.17 *Salmonella typhimurium* sulphate-binding protein: stereoscopic view of the tyrosine residues and electron density of buried waters in centre of the N-domain between the two major units (filled circles) and the minor unit (open bonds). Hydrogen bonds are indicated (dotted lines). Maximum resolution 2.0 Å; $R = 0.14$. From Pflugrath and Quiocho (1988). Copyright © 1988 Academic Press

Water Networks

One striking feature of the solvent structure, brought to light by several high-resolution studies, concerns the existence of water networks either at the interior (e.g. James and Sielecki, 1983; Baker and Hubbard, 1984; Figures 4.21, 4.22) or at the surface of proteins, or at the interface between

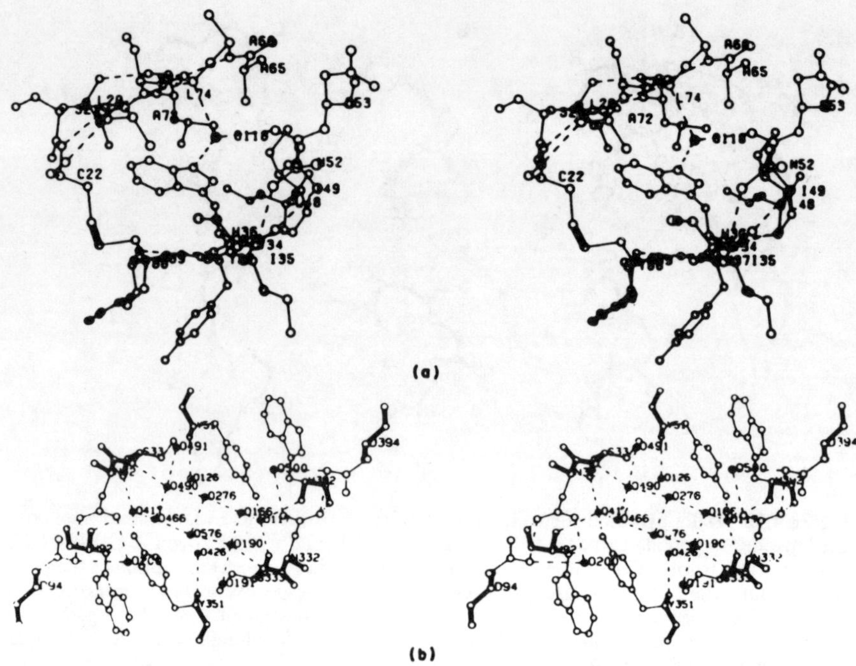

Figure 4.18 A λ-type Bence-Jones protein (Rhe) fragment: main chain (thick lines) and side-chains (thin lines). Stereoscopic views: in (*a*) of a water O 118 hydrogen bonded to Tyr 36 located in the core of the protein among hydrophobic residues; in (*b*) of waters located in the cavity between variable domains in the dimer — O 417 stabilize four hydrogen bonds with three hypervariable regions residues. Maximum resolution 1.6 Å; $R = 0.149$. From Furey *et al.* (1983). Copyright © 1983 Academic Press

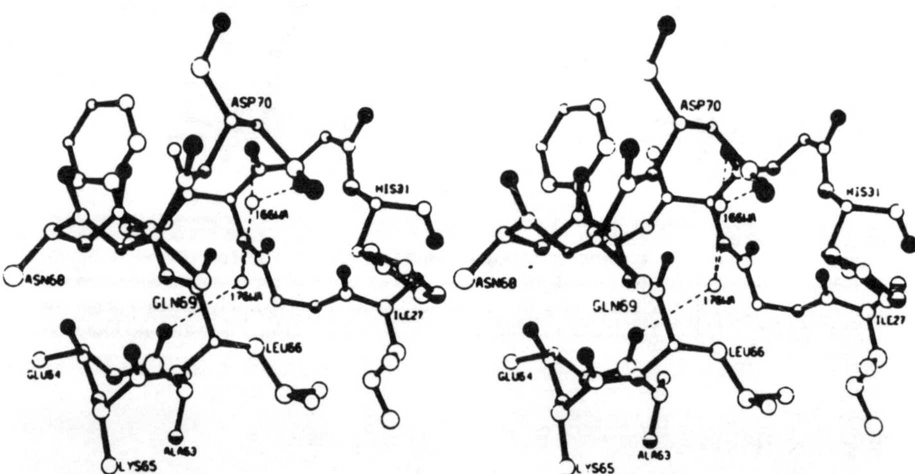

Figure 4.19 Phage T4 lysozyme: stereoscopic view of the two solvent molecules which form a channel behind the long α-helix (residues 60 to 79) that connects the N-terminal and C-terminal domains. Oxygen (filled), nitrogen (half-filled), and carbon atoms (open) have a diameter proportional to their temperature factor *B*. Maximum resolution 1.7 Å; $R = 0.193$. From Weaver and Matthews (1987). Copyright © 1987 Academic Press

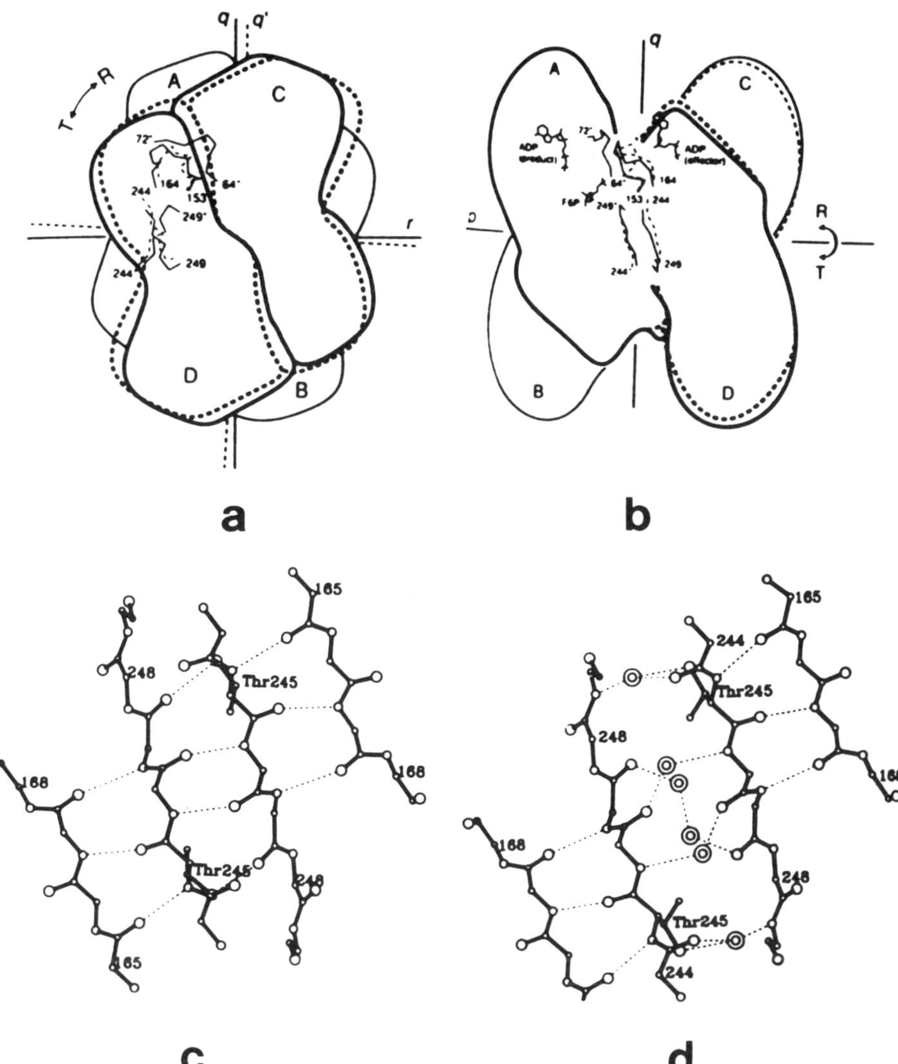

Figure 4.20 Phosphofructokinase (PFK): in (*a*) and (*b*) schematic sketch of the superimposed R-state (dashed lines) and T-state (solid lines). Part of the Cα tracing in the interface is represented for A (labels with *) and D subunits of the tetramer. *p*, *q* and *r* are the dyad axes. The views are drawn in (*a*) along the *p*-axis and in (*b*) along the *r*-axis: dimer CD performs a relative rotation of ~ 7° approximately around the *p*-axis. In (*c*) and (*d*) interactions between dimers across the *p*-axis (i.e. A and D; B and C): in the T-state (*c*) the interactions are direct; in the R-state (*d*) the interactions are mediated by a layer of waters. Maximum resolution 2.5 Å; $R = 0.183$. From Schirmer and Evans (1990). Copyright © 1990 Macmillan Magazines Ltd

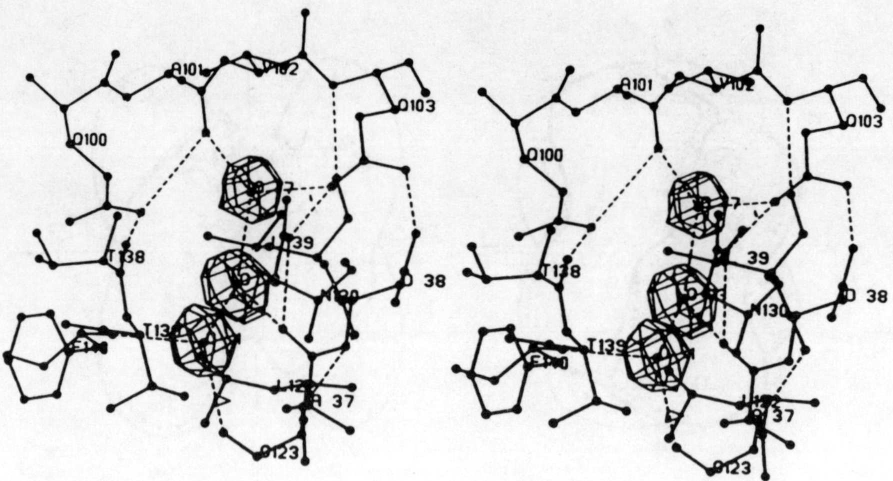

Figure 4.21 Penicillopepsin: stereoscopic view of the electron density of a cluster of three internal waters O 1, O 3, O 7. Each water donates two hydrogen bonds and receives one (dotted lines). Note the hydrophobic 'wall', due to the two leucines L39 and L122, on one side of the channel. Maximum resolution 1.8 Å; R = 0.136. From James and Sielecki (1983). Copyright © 1983 Academic Press

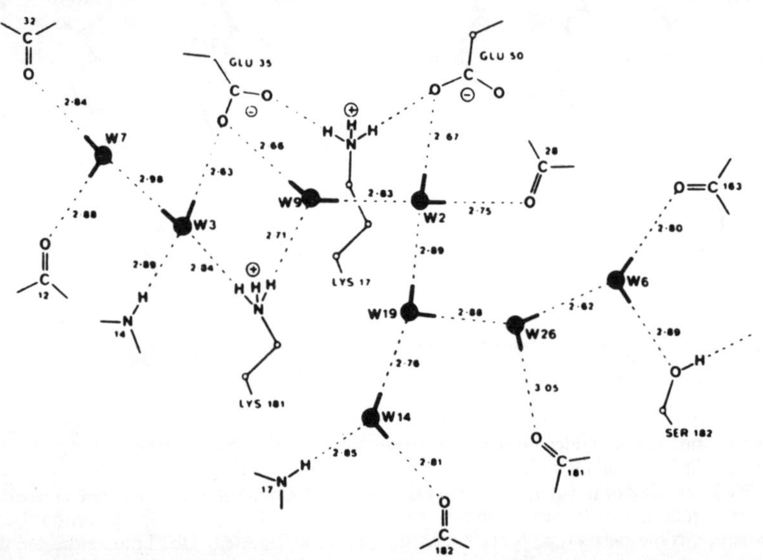

Figure 4.22 Actinidin: hydrogen bond network in the interior of the molecule (between its two domains). Presumed proton orientations are shown. Note the bridging of the charged Lys 181 and Glu 35 by the two waters W 3 and W 9. Distances are in Å. Maximum resolution 1.7 Å; R = 0.146, Baker (1980). Figure from Baker and Hubbard (1984). Copyright © 1984 Pergamon Press

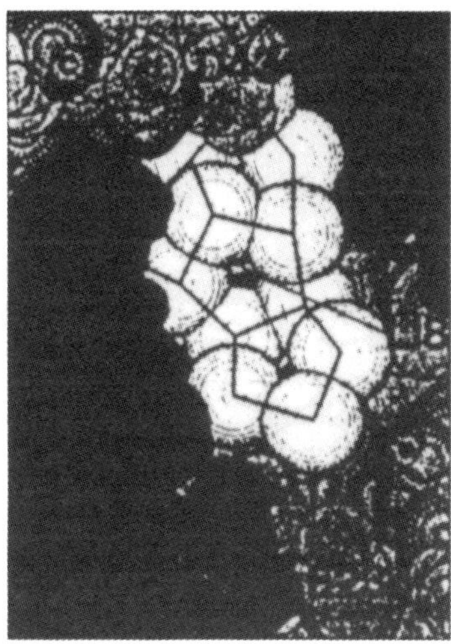

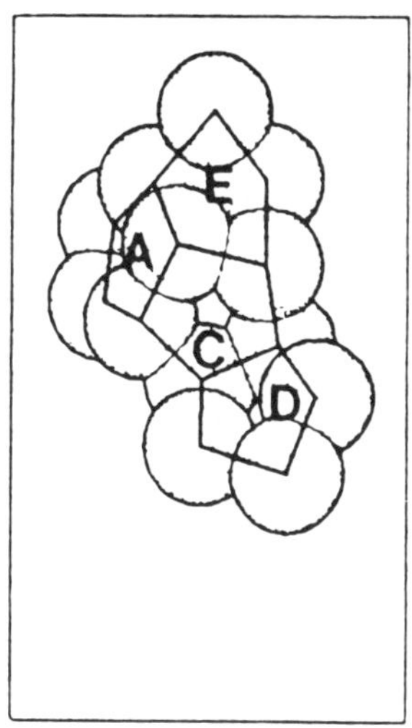

Figure 4.23 Crambin: the pentagonal rings, A, C, D, E of waters (oxygen as light circles) cap leucine 18 (L18), of which the methyl group is seen through the hole of the C ring. Adjacent protein molecules are shaded. Maximum resolution 0.945 Å; R = 0.129 (X-ray). Maximum resolution 1.2–1.5 Å; R = 0.144 (neutrons). From Teeter (1984). Copyright © 1984 National Academy of Sciences USA

monomers (e.g. Teeter, 1984; Jensen and Watenpaugh, 1986; Karplus and Schulz, 1987; Wright, 1987; Smith *et al.*, 1988; Louie and Brayer, 1990). These networks are generally defined following criteria which have been described in detail by Savage for the water structure in vitamin B_{12} coenzyme crystals (Savage, 1986a, b).

A typical example of water networks is found in *Clostridium pasteurianum* rubredoxin crystals (M_r *c.* 6000; crystal solvent content 43% b/v). In the 1.2 Å resolution electron density map, discrete waters are observed in a layer around the molecular surface which extend to *c.* 6 Å around the molecular surface (Watenpaugh *et al.*, 1978; Figure 4.9). Nearly all these waters are included within networks, which contain 2–19 water sites: 8 networks are involved in intermolecular contacts within the crystal packing (linking networks) and 7 others link donor and acceptors of the same molecule (Jensen and Watenpaugh, 1986). Not surprisingly, linking networks are larger than non-linking networks.

By contrast, nearly all the waters found in crambin crystals belong to a single network (Teeter, 1984; Teeter and Whitlow, 1986), with clusters of pentagonal water rings located in a zone delimited by hydrophobic intermolecular contacts between three neighbouring crambin molecules (Figure 4.23). Several mutually incompatible networks may also exist within a crystal (e.g. Smith *et al.*, 1986; Figure 4.24). This, again, reflects the essentially dynamic structure of the solvent undergoing endless rearrangements, with bonds being strained, broken and formed again.

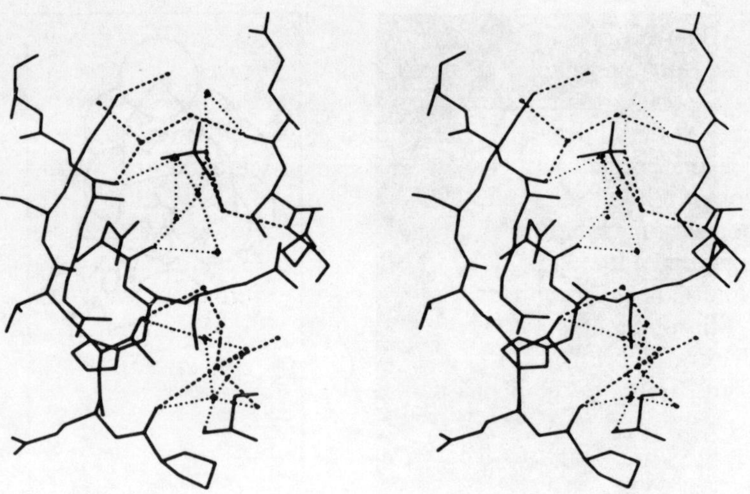

Figure 4.24 Erabutoxin: stereoscopic view of the two mutually exclusive sets of hydrogen bonds (dotted and dashed lines) which determine two solvent networks containing six molecules each. Maximum resolution 1.4 Å; $R = 0.141$. From Smith *et al.* (1986). Copyright © 1986 American Chemical Society

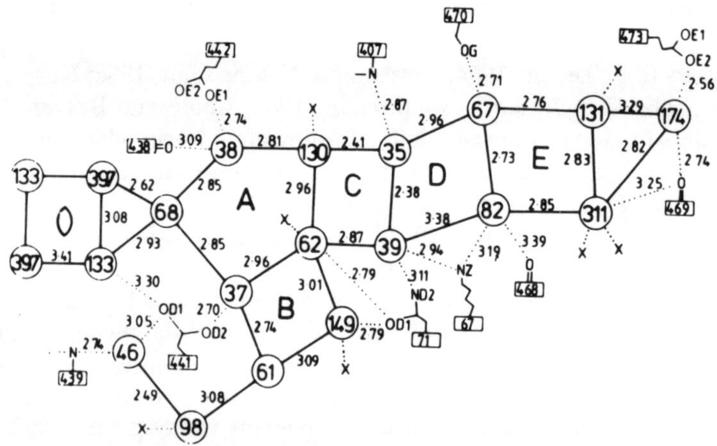

Figure 4.25 Glutathione reductase: core of the solvent network (large circles) in the inter-subunit cavity. Amino acid residues numbers are denoted in rectangles. The interatomic distances between neighbouring atoms are indicated in Å. An (X) indicates a solvent molecule closer than 3.5 Å to a water molecule of the network. The pentagonal and tetragonal arrangements (A to E) are nearly planar (if 38 is neglected). The dyad axis is indicated (on the left). Maximum resolution 1.54 Å; $R = 0.186$. From Karplus and Schulz (1987). Copyright © 1987 Academic Press

It is of interest to note that the characteristic pentagonal arrangement of water molecules found in crambin crystals has also been seen in several other proteins, such as glutathione reductase (Karplus and Schulz, 1987; Figure 4.25). However, such pentagons have been observed infrequently.

This could be due to the fact that the ordering of water molecules in contact with hydrophobic groups is mainly influenced by networks bridging surface polar groups, rather than by the interactions with hydrophobic patches at the molecular surface (Blake *et al.*, 1983).

Not unexpectedly, water plays an essential role in crystal packing stabilization (e.g. James *et al.*, 1980). A detailed analysis of the probable influence of packing forces on the morphology of several protein crystals has confirmed that individual water molecules or water networks reinforce the intermolecular contacts within the crystal through hydrogen bonding. For thin-plate crystals, water molecules are often the sole possible mediators to bridge successive layers of protein molecules, which are stacked parallel to the plane of the plate (Frey *et al.*, 1988).

Bulk Solvent

The nature of the solvent structure beyond the vicinity of the protein surface (the so-called 'first hydration shell') has been, and is still, discussed (see, for example, Finney, 1979; Rupley *et al.*, 1983). Solvent plumes extending from many exposed charged side-chains into the bulk solvent were observed in a 1.2 Å resolution electron density map of *Clostridium pasteurianum* rubredoxin (Jensen and Watenpaugh, 1986). Moreover, it has been proposed very recently that non-random arrangements of waters in cubic insulin crystals could extend over longer distances (*c.* 30 Å) between molecular surfaces (Badger and Caspar, 1991). This feature, which is still strongly questioned (Jensen, 1991), could reflect short-range repulsive hydration forces opposing the close approach of a hydrophilic surface, or other long-range interactions. This solvent structure, if confirmed, would be of major interest, since it compared with the arrangement of water molecules near molecular surfaces as found in physiological conditions (i.e. within living cells: Badger and Caspar, 1991).

Table 4.1 *Clostridium pasteurianum* rubredoxin: number of waters according to the number of hydrogen bonds to protein atoms (vertical) and total number of hydrogen bonds (horizontal). From Watenpaugh *et al.* (1978). Copyright © 1978 Academic Press

Number of hydrogen bonds to protein	*Number of hydrogen bonds*									*Total*
	0	*1*	*2*	*3*	*4*	*5*	*6*	*7*	*8*	
0	13	10	18	12	5				1	59
1		4	14	16	5	3				42
2			1	5	7	3	1			17
3				3	2	3				8
4					1					1
Total	13	14	33	36	20	9	1		1	127

Catalytic Activity

The importance of water molecules in catalytic activity is now a well-established fact, whatever reaction mechanism applies (see, for example, Warshel *et al.*, 1989b). This is the case for the well-studied serine proteases: it has been proposed that the large rates and specificity of these molecules could be related either to the simple elimination of bound water (Dewar and Storch, 1985) or to the replacement of water by another specific polar environment, which may 'solvate' the ionic transition state more than water does (Warshel *et al.*, 1989a).

More recently, several structures of phospholipase A_2 have shown that three of the four water molecules consistently bound to the catalytic surface stabilize both the active site as a substrate and a calcium atom, prior to the binding and stabilizing of the substrate and transition state (Scott *et al.*, 1990). By contrast, the same calcium binding site of a human phospholipase A_2 site lacks two of the above-mentioned water molecules, while the interaction of the calcium with the transition state analogue is identical with that previously reported (Scott *et al.*, 1991). This difference in the solvent cage of the calcium ion might be due to changes in bulk solvent structure or reflect partial occupancy of the primary calcium ion binding site by competing sodium ions.

Water molecules have also been found to play a role in electron carrier mechanisms. Thus, in oxidized cytochromes c, a conserved buried water molecule is shifted by 0.9 Å towards the haem iron atom from the position in the reduced form (Takano and Dickerson, 1981a, b; Bushnell *et al.*, 1990; Figure 4.26). The close proximity of the haem iron (6.3 Å) and the evolutionary conservation of the surrounding residues to which this water molecule is hydrogen bonded suggests that it plays an important role in the function of these cytochromes: the reduction of the distance between the water molecule and the iron observed for the oxidized state may be related to the increased positive charge centred at the haem iron. Likewise, for the His35Leu azurin mutant, it has been proposed that water molecules might directly mediate electron transfer by a 'through hydrogen bond' mechanism (Nar *et al.*, 1991; Figure 4.27).

Intermolecular Recognition

Not surprisingly, water molecules have been found to be involved in the stabilization of intermolecular complexes: ligand–protein, protein–protein (for review, see Janin and Chothia, 1990) and nucleic acid–protein complexes (for review, see Harrison and Aggarwal, 1990; Steitz, 1990; Westhof and Beveridge, 1990; Harrison, 1991).

For example, in the complex of L-arabinose binding protein with L-arabinose, two hydrogen-bonded waters, located in the binding site,

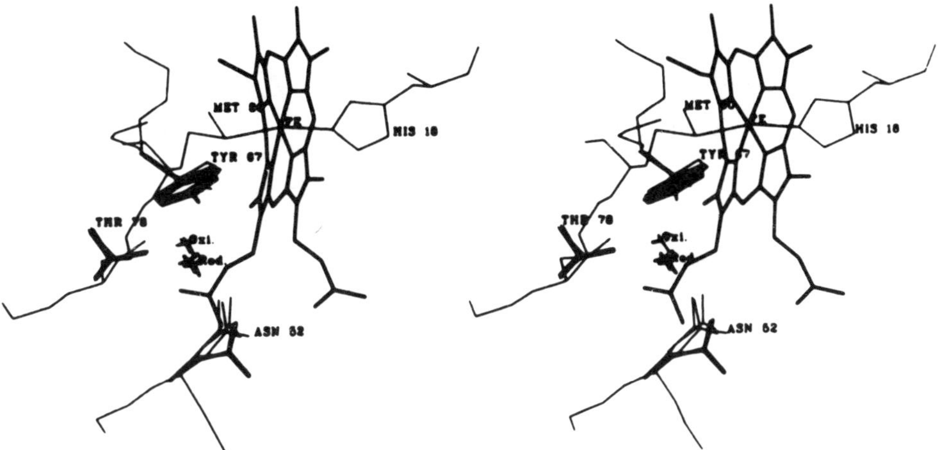

Figure 4.26 Eukaryotic cytochromes: stereoscopic view of the local environment of an internally bound water molecule adjacent to the central haem group and found in common in all eukaryotic cytochromes. The superimposed waters at the sites Oxi. and Red. are from structures in the oxidized state (horse, tuna and rice cytochromes) and in the reduced state (yeast iso-1 and tuna), respectively. The protein residues are represented by thin lines for the oxidized forms and by thick lines for the reduced forms. The water molecule forms hydrogen bonds with Asn 52, Tyr 67 and Thr 78. The water shift (0.9 Å) toward the haem iron atom upon oxidation is associated with displacements of Asn 52 and Tyr 67. Maximum resolution 1.94 Å; $R = 0.17$. From Bushnell *et al.* (1990); see also Takano and Dickerson (1981a, b). Copyright © 1990 Academic Press

reinforce the binding of L-arabinose (Quiocho *et al.*, 1989; Figure 4.28). By contrast, the same waters create an unfavourable interaction in the case of D-fucose binding, accounting for the lower K_d of that substrate. Moreover, D-galactose, which is as good a substrate as D-arabinose, also binds tightly to the protein through the displacement of one of these hydrogen-bonded waters, together with a correlated shift and rearrangement of the hydrogen bonding pattern of the second water molecule.

The high-resolution X-ray structures of proteases liganded to various inhibitors have yielded a wealth of accurate data on the structural role of waters at protein–ligand interfaces. In the representative structures of two complexes of a turkey ovomucoid inhibitor with either α-chymotrypsin or *Streptomyces griseus* serine protease (SGPB), several internal water molecules are equivalent structurally, while the solvent structures at the interface are different, as a result of sequential and conformational changes in the contact region (Fujinaga *et al.*, 1987; Figure 4.29). Thus, a water-mediated ionic interaction between α-chymotrypsin and its inhibitor was not observed for SGPB, which accounts for the enhanced specificity of α-chymotrypsin for a specific arginine residue of the inhibitor. In these two cases, as in many others, the waters which are 'buried' in the surface crevices are not displaced upon inhibitor binding. This confirms that they have to be considered as constitutive elements of the protein.

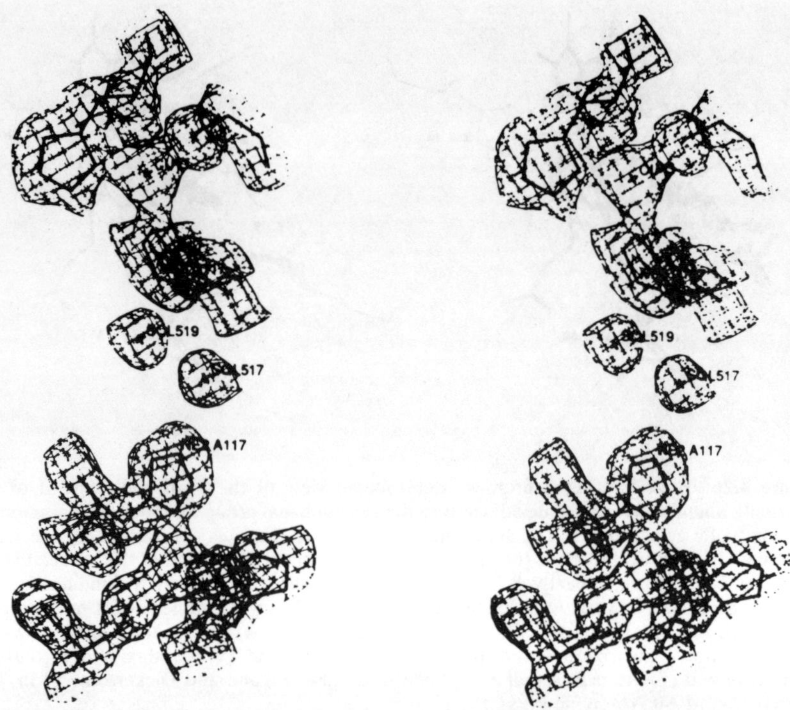

Figure 4.27 Azurin His35Leu mutant: stereoscopic view of the electron density around the copper redox centres of the subunits A and C of the tetramer. The two copper atoms (x Cu) are distant by 14.7 Å. The cavity between the two imidazole rings of His 117 (N_ε to N_ε = 6.6 Å) formed by the hydrophobic patch residues is filled by the two solvent molecules sol 517 and sol 519. These molecules might mediate electron transfer between the two copper atoms by a 'through hydrogen-bond' mechanism. Maximum resolution 1.9 Å; R = 0.171. From Nar *et al.* (1991). Copyright © 1991 Academic Press

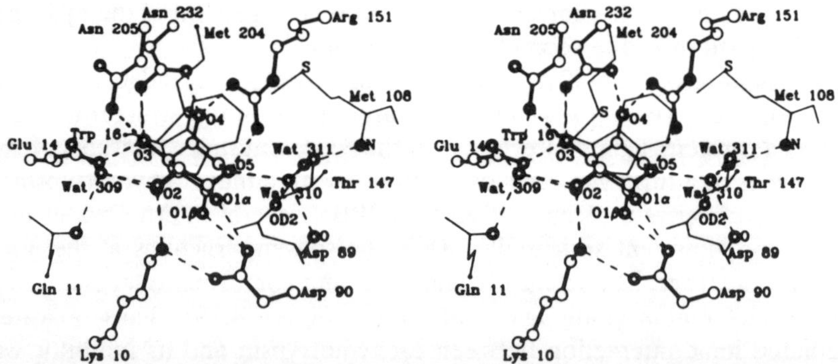

Figure 4.28 L-Arabinose binding protein: stereoscopic view of the two waters at the sites Wat 310 and Wat 311 which contribute further to the tight binding of L-arabinose and D-galactose to the protein, including, in the latter case, the replacement of a water by $-CH_2OH$ in Wat 311. By contrast, these two waters together with Asp 89 create an unfavourable interaction with the methyl group of D-fucose. Maximum resolution 1.7–1.9 Å; $R \sim$ 0.135. From Quiocho *et al.* (1989). Copyright © 1989 Macmillan Magazines Ltd

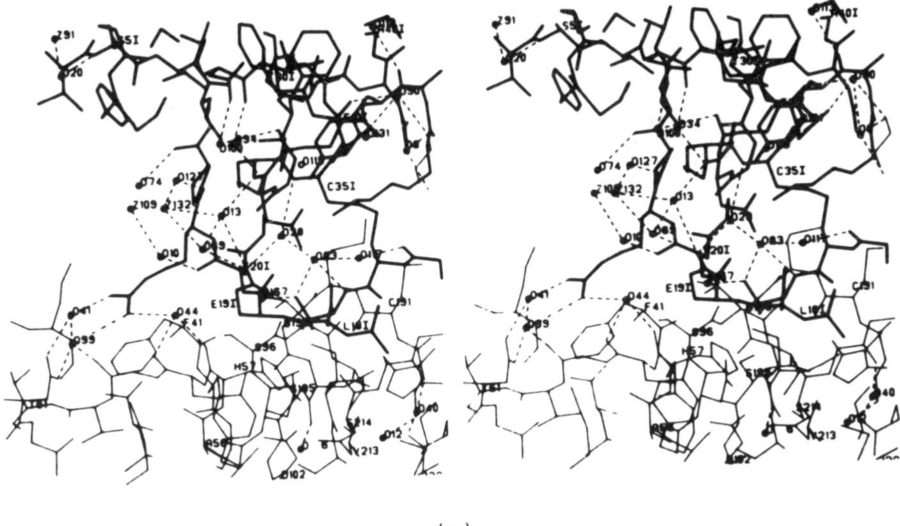

(a)

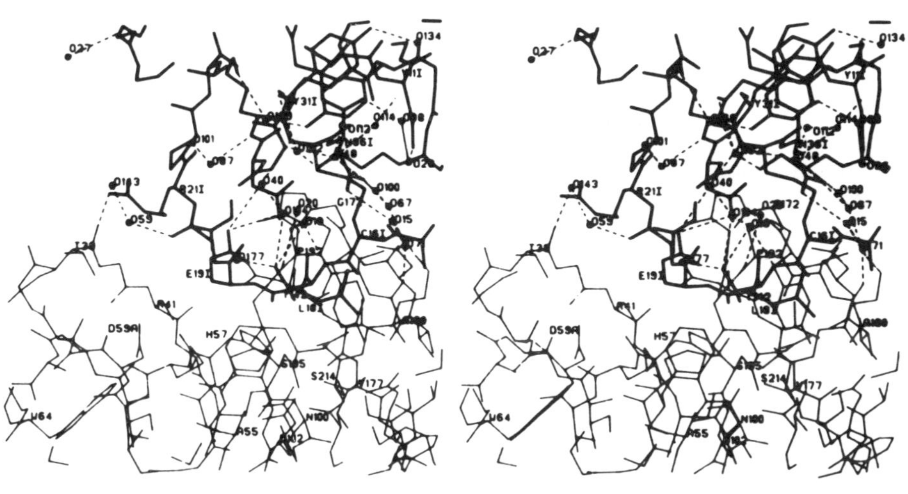

(b)

Figure 4.29 Complex of α-chymotrypsin with its inhibitor turkey ovomucoid third domain (OMTKY3): stereoscopic view of the ordered solvent molecules in the region P_2 Thr 171, P'_1 Glu 191 and P′3 Arg 211 on OMTKY3 (thick lines on both figures). In (a) complex with α-chymotrypsin (thin lines); in (b) complex with *Streptomyces griseus* serine proteinase (SGPB, thin lines). The solvent structure in the vicinity of Thr 171 Oγ and Glu 191 Oε1 and Glu 191 Oε2, which appears to be conserved in both complexes, is also found in the free form of the highly homologous inhibitor ovomucoid third domain from the silver pheasant (OMSVP3, Bode *et al.*, 1985). Maximum resolution 1.8 Å; $R = 0.168$. From Fujinaga *et al.* (1987). Copyright © 1987 Academic Press

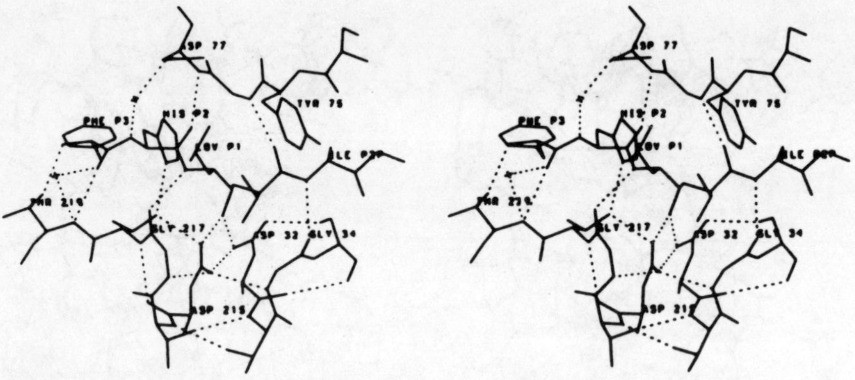

Figure 4.30 Endothiapepsin and H-261 complex: stereoscopic view of the hydrogen bonds (broken lines) and van der Waals contacts between the enzyme and the inhibitor in the P3–P′1 region. The stars indicate waters or ions. Maximum resolution 1.6 Å; $R = 0.14$. From Veerapandian *et al.* (1990). Copyright © 1990 Academic Press

Similarly, the complex between aspartic proteinase endothiapepsin and a highly potent renin inhibitor (Veerapandian *et al.*, 1990; Figure 4.30) is stabilized by the free energy decrease due to the loss of the solvent contact area of both the free enzyme and the inhibitor in the complexed structure, and the replacement of 15 of the 20 water molecules occupying the active site of the uncomplexed enzyme. Here, again, extended water networks form bridges which help to stabilize the enzyme–inhibitor complex.

As mentioned above, water molecules also play a role in the DNA or RNA sequence recognition. They complete the unsatisfied bonds of the protein surface to which the nucleic acid surface must be complementary for optimal binding (Aggarwal *et al.*, 1988; Otwinowski *et al.*, 1988; Figures 4.31, 4.32). By analogy, this may be also the case for the toxin II from the scorpion *Androctonus hector*. In this protein a network of ordered waters occupy a cavity close to the C-terminus, which constitutes a particular structural feature of the surface of this potent class of toxins, and may, at least partially, be responsible for their specific mode of action (Fontecilla-Camps *et al.*, 1988; Figure 4.33).

In some cases crystallographic studies may reveal unexpected properties of the solvent. For example, the determination of the X-ray structure of a complex between *p*-hydroxybenzoate hydroxylase—an NAPDH-dependent enzyme—and the coenzyme analogue adenosine-5-diphosphoribose (ADPR) in the reduced state has shown (i) that the FAD was removed by ADPR (van der Laan *et al.*, 1989) and (ii) that the binding pocket leaved by the flavin ring is filled with waters mimicking the ring, and, possibly, with ions which were present in high concentrations in the crystallization medium. Otherwise, the architecture of the active site and the binding of the substrate remains essentially unaffected (van der Laan *et al.*, 1989; Figure 4.34).

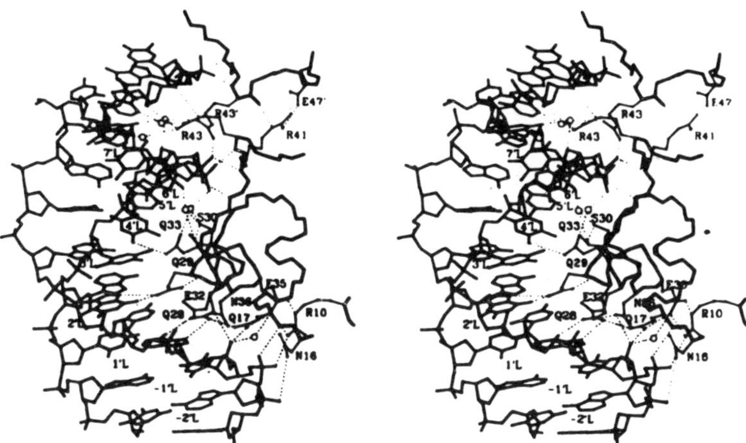

Figure 4.31 DNA operator–phage 434 repressor complex: stereoscopic view of the binding domain between the protein and DNA. Six waters involved in bridging the protein and DNA are represented by small open circles. Maximum resolution 2.5 Å; $R = 0.179$. From Aggarwal *et al.* (1988). Copyright © 1988 AAAS

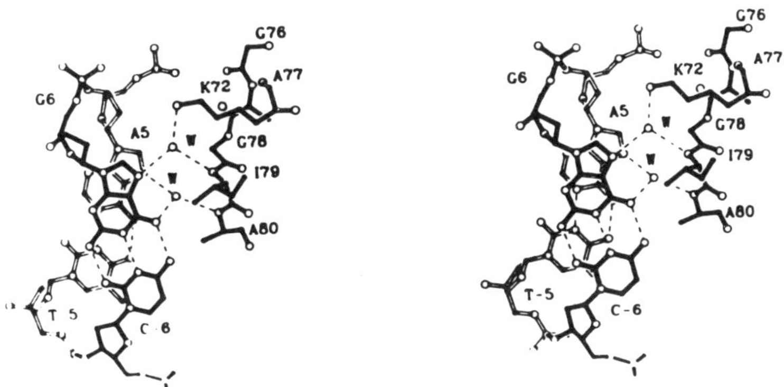

Figure 4.32 *trp* repressor/operator complex: stereoscopic view of details of a portion of the interface, between the operator and the *trp* repressor, which involves waters. The bonds of nucleotide closest to the viewer are solid; those of the protein are semi-filled. Carbon atoms are small filled spheres; oxygen, nitrogen and phosphorus atoms are large open spheres; hydrogen bonds (2.6–3.1 Å) are dashed. 'A' means adenosine or alanine; 'G' means guanosine or glycine; 'W' are waters sites. The solvent interactions with phosphates are omitted for clarity. Maximum resolution 2.4 Å; $R = 0.249$. From Otwinowski *et al.* (1988). Copyright © 1988 Macmillan Magazines Ltd

3 Other Methods

Molecular Dynamics

For the past ten years simulation of the dynamical properties of proteins and of other biological macromolecules has become a major tool to explore

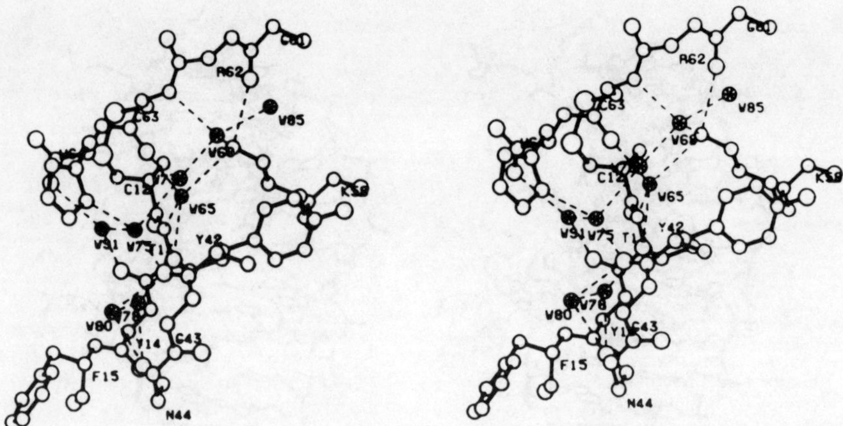

Figure 4.33 *Australis androctonus* scorpion neurotoxin: stereoscopic view of the solvent (starred circles) filled cavity delimited by residues 13, 42, 43, 62 to 64. For clarity, the side-chains of residue 14 and 62 and the possible hydrogen bonds between Lys 58 and Asn 110 are not shown. Maximum resolution 1.8 Å; $R = 0.152$. From Fontecilla-Camps *et al.* (1988). Copyright © 1988 The National Academy of Sciences USA

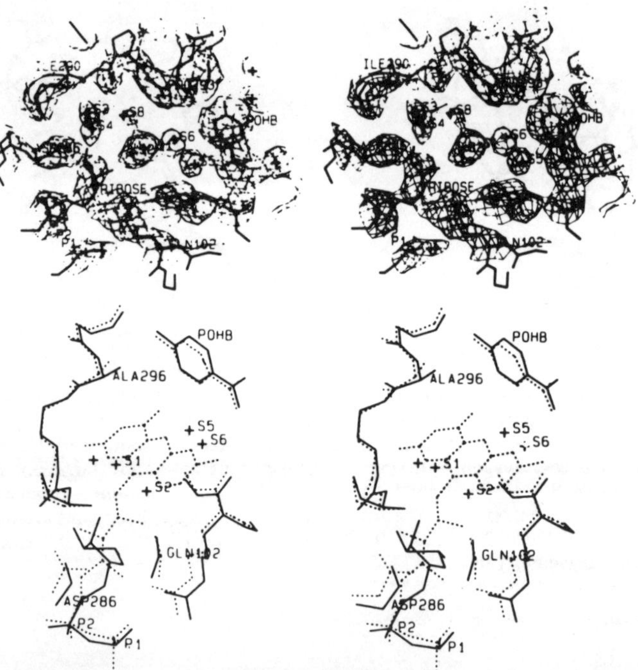

Figure 4.34 *p*-Hydroxybenzoate hydroxylase: stereoscopic view of the electron density map of the solvent molecules (S1 to S8) which occupy the flavin ring site. S3, S4, S8 are present in the structure of the native complex. The remaining S sites are located in the space left by the removal of the flavin ring related to the fixation of the ribose of the analogue. The S1 and S2 bulky density might be occupied by sulphate, phosphate or dithionite. Maximum resolution 2.7 Å; $R = 0.168$. From van der Laan *et al.* (1989). Copyright © 1989 American Chemical Society

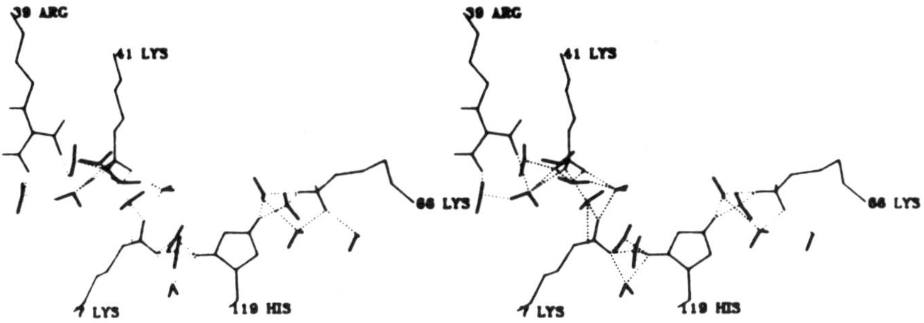

Figure 4.35 Ribonuclease A: dynamics of the active site. Stabilization of positively charged groups by bridging waters in the active site of ribonuclease A without anions. Stereoscopic view of the snapshot of the molecular dynamics trajectory at 15 ps. The picture includes only the residues Lys 7, Arg 39, Lys 41, Lys 66 and His 119 and the bound waters. Hydrogen bonds are represented by dotted lines. From Brünger *et al.* (1985). Copyright © 1985 National Academy of Sciences USA

or to predict the structural and functional properties of these molecules at the atomic level (see, for recent reviews, McCammon and Harvey, 1987; Brooks *et al.*, 1988; Karplus and Petsko, 1990).

Thus, molecular dynamic simulations with stochastic molecular boundary have been performed to study the structure, dynamics and energetics of the solvated active site of ribonuclease A, using a highly refined X-ray model of this enzyme and an equilibrated sample of 'liquid water'. The results indicate that the numerous charged groups, located at the catalytic site, are stabilized in the free enzyme by water networks (Brünger *et al.*, 1985; Figure 4.35), and that the corresponding hydrogen-bonding pattern of waters with protein residues mimics the substrate–enzyme interactions. A similar behaviour has also been found for waters in the active site of lysozyme (Brooks and Karplus, 1989; Figure 4.36). Moreover, it has been shown in this latter case that there is an interdependence between solvent and protein motions, and that waters around apolar groups are less mobile than bulk water or water-solvating polar groups.

MD calculations have also shed light on the contribution of individual amino acid residues and solvent molecules to the free energy changes in proteins. For example, the rather small energy change (a few kcal/mol) due to the mutation of Asp Gl(99)β by an alanine at the α–β interface of the haemoglobin tetramer was seen to result from large energy changes cancelling each other, including those resulting from solvent–protein interactions (Gao *et al.*, 1989; Table 4.2).

It is also worth noting that comparisons between X-ray structures and structures averaged over molecular dynamic simulations have been of major importance in estimation of the validity of the algorithms and force fields. For example, a 40 ps simulation of the structure and dynamics of the

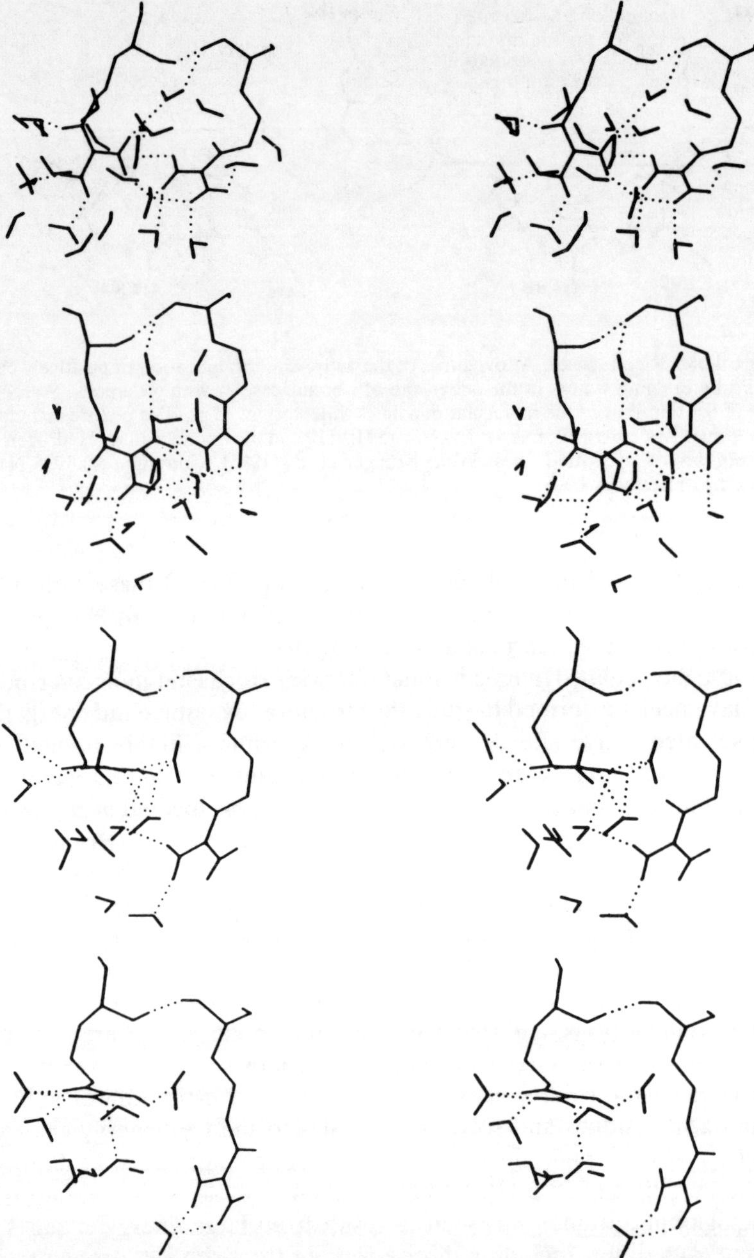

Figure 4.36 Lysozyme: active site region of lysozyme. Formation of a stable water-bridged structure between Arg 61 (*left*) and Arg 73 (*right*) in the solvent stochastic simulation. Stereoscopic views: in (*a*) before equilibrium solvation; in (*b*) initial formation of water-bridged structure 8 ps; in (*c*) 17 ps; in (*d*) 33 ps. From Brooks and Karplus (1989). Copyright © 1989 Academic Press

Table 4.2 Haemoglobin: Molecular dynamic analysis of the free energy (kcal/mol) computed for the mutation Asp G1 (99) β >> Ala for one $\alpha_1\beta_2$ interface; a positive term in ΔG corresponds to the given contribution destabilizing the mutant (Ala) relative to the wild type. $\Delta\Delta G$ is ΔG(oxy)-ΔG (deoxy). From Gao *et al.* (1989). Copyright © 1989 AAAS

Contribution	ΔG*(deoxy)*	ΔG*(oxy)*	ΔΔG
Solvent	46.0	68.5	22.5
Protein[a]	20.0	–8.0	–28.0
Asp G1(99)β_2[b]	–8.8	11.0	–2.2
Inter (α_1)	2.8	–24.4	–27.2
Tyr C7(42)	8.4	4.3	–12.7
Asp G1(94)	–22.0	–44.4	–22.4
Val G3(96)	1.6	7.1	5.5
Asn G4(97)	9.7	13.0	3.3
Intra (β_2)	26.1	27.4	1.3
His FG4(97)	–2.1	–3.3	–1.2
Pro G2(100)	8.2	5.4	–2.8
Glu G3(101)	–11.2	6.9	18.1
Asn G4(102)	14.3	10.1	–4.2
Total	66.0	60.5	–5.5

[a] Residues contributing more than 1.5 kcal/mol to both the deoxy and oxy forms are listed.
[b] Self-energy contribution of the mutant residue.

complete unit cell of BPTI with water molecules has led to root mean square differences of *c.* 1.5 Å (vs. 2–3 Å *in vacuo*) between the average structure generated by molecular dynamics and the observed X-ray structure (van Gunsteren *et al.*, 1983). A recent 60 ps molecular dynamics simulation of *Streptomyces griseus* serine protease A in a crystalline environment consisting of water molecules and of dihydrogen phosphate and sodium ions has led to comparable r.m.s. differences with the X-ray structure: 1.67 Å and 1.25 Å for the two molecules in the asymmetric unit, respectively (Avbelj *et al.*, 1990).

Site-directed Mutagenesis

Parallel to the development of molecular dynamics, site-directed mutagenesis has also become a major tool for exploring the structural and functional properties of proteins, as it is now usual to study the functional and structural role of any residue, or group of residues, through selected changes in the amino acid sequence.

One striking result of the first high-resolution X-ray studies of site-directed mutants has been the demonstration that the protein can remarkably accommodate amino acid substitutions through structural shifts, which may propagate far beyond the zone of mutation. Not unexpectedly, such structural changes are associated with a more or less important

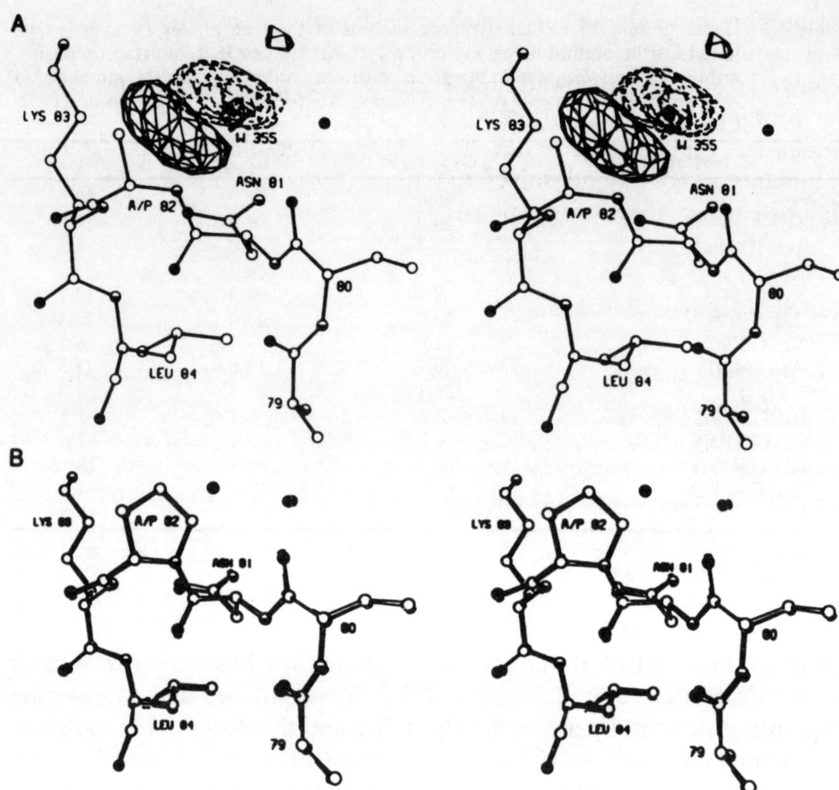

Figure 4.37 Ala82Pro phage T4 lysozyme mutant: stereoscopic views. In (*A*) of the electron density difference map (A82P mutant minus wild type) in the zone of the mutation. The positive peak (full line) indicates the addition of the pyrrolidine ring of the proline. The negative peak corresponds to the site of a water molecule bound to the peptide NH of Ala 82 in the wild type. For clarity part of the side-chains of Leu 79 and Arg 80 have been omitted. In (*B*) of the superposition of the structures of the mutant Ala82Pro lysozyme (open bonds) and of the wild-type lysozyme (filled bonds). Maximum resolution 1.7 Å; R = 0.157. From Matthews *et al.* (1987). Copyright © 1987 National Academy of Sciences USA

reorganization of the adjacent water structure. For example, in thermostable phage T4 lysozyme, the substitutions of Thr157 by various residues can be accommodated not only by localized shifts in the protein structure, but also by inclusions or displacements of water molecules (Matthews *et al.*, 1987; Figure 4.37; Alber *et al.*, 1987). Similarly, for subtilisin BPN', disulphide substitutions result in hydrophobic cavities which are filled with disordered water, and in changes of the water structure around the site of mutation (Katz and Kossiakoff, 1990).

More dramatically, in staphyloccocal nuclease, the replacement of Glu43 by Asp has led to crucial changes of the solvent network structure around the active site. These changes include the probable removal of the nucleophile water molecule, and the destabilization of a loop due to the removal of a constitutive water molecule bridging that loop with the core of the protein (Loll and Lattman, 1990; Figure 4.38). This last example emphasizes the necessity to have at hand the three-dimensional structures of mutants and of 'bound' solvent if one wants to draw reliable conclusions about structure–function relationships.

Nuclear Magnetic Resonance

Progressively, nuclear magnetic resonance (NMR) has become an alternative experimental method to determine the three-dimensional atomic structure of proteins, at least for molecular weights up to *c.* 20 000 (for a review, see, for example, Wühtrich, 1990).

Early NMR experiments confirmed the presence of waters bound to the proteins whose lifetimes were comparable with the overall rotational time of the hydrated protein (*c.* 10^{-9} s). However, NMR has only recently allowed one to identify and characterize the binding sites of individual water molecules: those found at the interior of bovine pancreatic trypsin inhibitor (BPTI: Tüchsen *et al.*, 1987; Otting and Wüthrich, 1989; Otting *et al.*, 1991) and of interleukin 1β (Clore *et al.*, 1990). In each of these cases, the water molecules were found to occupy the same sites on the protein in solution as in the crystal. Moreover, in both BPTI and interleukin 1β, the shortest lifetimes of water molecules, as determined from the same experiments, were found to be 3×10^{-10} s and 2×10^{-10} s, respectively. More recently, for BPTI, an upper lifetime limit of 20 ms was established (Otting *et al.*, 1991), and a protein–water hydrogen bonding network was observed at the protein–water-interface with a lifetime shorter than 3×10^{-10} s.

Thus, NMR studies have unambiguously confirmed (i) that internally 'bound' waters found in crystal structures are also a constitutive element of the protein in solution, and (ii) that the solvent structure is more disordered at the molecular surface of protein structures in solution than in protein crystals.

4 Perspectives

The determination, at the atomic level, of the detailed solvent structure associated with proteins was ten years ago the exception (e.g. Watenpaugh *et al.*, 1978; Blake *et al.*, 1983). This analysis has become common practice for larger and larger molecules. Water is (or should be) considered a constitutive part of X-ray-derived molecular models: the knowledge of the

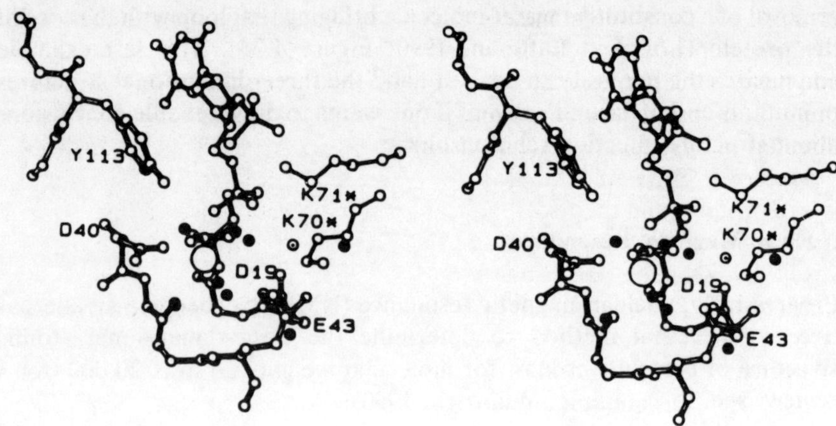

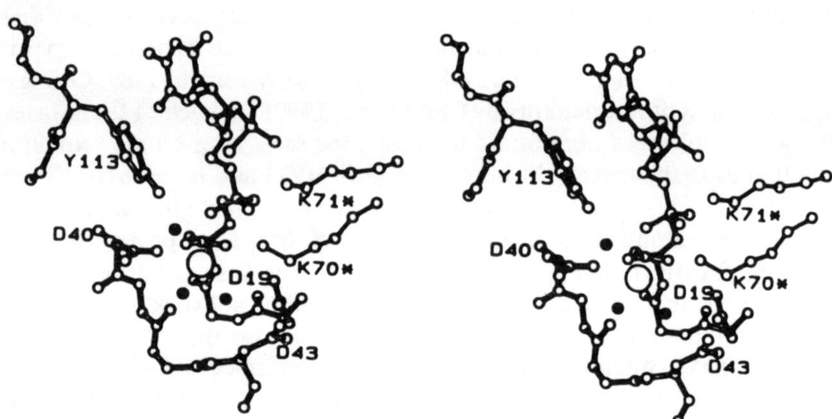

Figure 4.38 Glu43Asp staphylococcal nuclease mutant: stereoscopic views in (WT) of the active site of the wild type; in (E43D) of the active site of the mutant. The calcium atom is represented by a large open circle and waters by small dark circles, except for the putative nucleophile in the wild-type structure, which is represented by a light circle with a dot in the centre. Lys 71 and Lys 72 belong to a symmetry-related molecule. Nine waters can be seen in the wild-type structure: eight in the active site and the ninth bridging the first part of the mobile loop to the main body of the protein. This water is visible below and to the right of E43 in the figure. The bridging water and five of the active site molecules are not seen in the mutant active site. Maximum resolution 1.74 Å; $R = 0.174$. From Loll and Lattman (1990). Copyright © 1990 American Chemical Society

water or solvent structure associated with a protein is not a mere achievement of accurate structural crystallography, but provides an essential element for the understanding of the structural, dynamical and thermodynamical properties of biological macromolecules and, thus, for an adequate perception of the structure–function relationships. This latter point is of particular importance in view of the developments of site-directed mutagenesis.

Moreover, answers to major questions concerning the interactions between proteins and their solvent are still the subject of discussions. Among these, one can mention the proper treatment of electrostatic effects (Harvey, 1989; Schaefer and Froemmel, 1990; Sharp and Honig, 1990; Simonson *et al.*, 1991). New experimental and computational studies on dynamics (e.g. Cusack *et al.*, 1988; Bellissent-Funel *et al.*, 1989; Otting and Wuthrich, 1989; Diamond, 1990; Karplus and Petsko, 1990) should also provide new insights on the solvent structure: in particular, that at the molecular surface which is essential to the understanding of the interactions with other molecular components such as enzyme substrates.

With respect to crystallographic studies, comparison of the solvent structures found in crystals of the same macromolecule grown from different crystallization media are greatly awaited (Walter *et al.*, 1990; Bell *et al.*, 1991). The advances which have been made recently in a more rational search of the crystallization conditions of biological macromolecules (Feigelson, 1986; Giégé *et al.*, 1988; McPherson, 1989; Ward and Gilliland, 1989) suggest the design of crystallization experiments allowing one to obtain a broader view on the solvent structure associated with proteins.

The collection of X-ray diffraction data at high resolutions with monochromatized shorter wavelengths (*c.* 1 Å) will undoubtedly yield more precise electron density maps, and therefore more reliable protein and solvent atomic models. This should provide, for example, an accurate basis for detailed analyses of ordered and of bulk solvent (e.g. Badger and Caspar, 1991; Jensen, 1991).

The recent refinement of the *Desulphovibrio vulgaris* rubredoxin (Adman *et al.*, 1991) has demonstrated that it is now possible to refine stereochemically reasonable macromolecular models to limits comparable to those obtained for small molecules, since in this study the crystallographic *R*-factor was lower than 0.10. However, the modelling and refinement of such an accurate model remains a time-consuming process which still requires ‘a lot of attention and a lot of love’ (James, 1980). Hence, novel graphics and refinement programs, which would treat solvent and multisite conformation in a more automatic way, are now required. For example, software allowing the routine visualization and analysis of the stereochemistry of water molecules and of their possible networks (see, for example, Figure 8 in Smith *et al.*, 1988, or the program WATERPATH, by Steve Sheriff) would be most helpful. Finally, it should be emphasized once more that the final protein and solvent model has to be examined carefully (Bränden and Jones, 1990; Stenkamp *et al.*, 1990), keeping in mind that the inclusion of low-resolution data, obtaining low crystallographic *R*-factors, and a plausible stereochemistry are obligatory to validate a solvent model.

Acknowledgements

I am indebted to Eric Westhof for having suggested this chapter and for many stimulating discussions. I thank my colleagues Juan Carlos Fontecilla-Camps and Frédéric Vellieux for their interest and critical reading of the manuscript. I am also pleased to thank Professor Lyle Jensen and his colleagues for the stimulating times I spent in the Seattle group.

The Centre National de la Recherche Scientifique (CNRS) and the Commissariat à l' Energie Atomique (CEA) are acknowledged for financial support.

Figures 4.1–4.3, 4.5, 4.8–4.14, 4.17–4.19, 4.21, 4.25–4.27, 4.29, 4.30, 4.36 and Table 4.1 have previously appeared in *J. Mol. Biol.* and are reproduced by permission of Academic Press Inc (London) Ltd. Figure 4.4 has been published in *Protein, Structure, Function and Genetics* and is reproduced by permission of Wiley–Liss. Figures 4.6, 4.24, 4.34, 4.38 have been published in *Biochemistry* and are reproduced by permission of the American Chemical Society. Figure 4.7 has been published in *Acta Crystallographica* and is reproduced by permission of the International Union of Crystallography. Figure 4.15 has been published in *Advances in Protein Chemistry* and is reproduced by permission of Academic Press Inc (London) Ltd. Figures 4.16, 4.31 and Table 4.2 have been published in *Science* and are reproduced by permission of the American Association for the Advancement of Science. Figures 4.20, 4.28, 4.32 have been published in *Nature* and are reproduced by permission of Macmillan. Figure 4.22 has been published in *Progress in Biophysics and in Molecular Biology* and is reproduced by permission of Pergamon Press. Figures 4.23, 4.33, 4.35, 4.37 have been published in the *Proceedings of the National Academy of Sciences (USA)* and are reproduced by permission of the National Academy of Sciences (USA).

References

Adman, E. T., Sieker, L. C. and Jensen, L. H. (1991). Structure of rubredoxin from *Desulphovibrio vulgaris* at 1.5 Å resolution. *J. Mol. Biol.*, **217**, 337–352

Aggarwal, A. K., Rodgers, D. W., Drottar, M., Ptashne, M. and Harrison, S. C. (1988). Recognition of a DNA operator by the repressor of phage 434: A view at high resolution. *Science*, **242**, 899–907

Alber, T., Dao-pin, S., Wilson, K., Wozniak, J. A., Cook, S. P. and Matthews, B. W. (1987). Contributions of hydrogen bonds of Thr 157 to the thermodynamic stability of phage T4 lysozyme. *Nature*, **330**, 41–46

Avbelj, F., Moult, J., Kitson, D. H., James, M. N. G. and Hagler, A. T. (1990). Molecular dynamics study of the structure and dynamics of a protein molecule in a crystalline ionic environment, *Streptomyces griseus* Protease A. *Biochemistry*, **29**, 8658–8676

Badger, J. and Caspar, D. L. D. (1991). Water structure in cubic insulin crystals. *Proc. Natl Acad. Sci. USA*, **88**, 622–626

Baker, E. N. (1980). Structure of actinidin, after refinement at 1.7 Å resolution. *J. Mol. Biol.*, **141**, 441–484

Baker, E. N. (1988). Structure of Azurin from *Alcaligenes denitrificans*. Refinement at 1.8 Å resolution and comparison of the two crystallographically independent molecules. *J. Mol. Biol.*, **203**, 1071–1095

Baker, E. N. and Hubbard, R. E. (1984). Hydrogen bonding in globular proteins. *Prog. Biophys. Mol. Biol.*, **44**, 97–179

Barlow, D. J. and Thornton, J. M. (1988). Helix geometry in proteins. *J. Mol. Biol.*, **201**, 601–619

Bartunik, H. D., Summers, L. J. and Bartsch, H. H. (1989). Crystal structure of bovine β-trypsin at 1.5 Å resolution in a crystal form with low molecular packing density. Active site geometry, ion pairs and solvent structure. *J. Mol. Biol.*, **210**, 813–828

Bell, J. A., Wilson, K. P., Zhang, X.-J., Faber, H. R., Nicholson, H. and Matthews, B. W. (1991). Comparison of the crystal structure of bacteriophage T4 lysozyme at low, medium, and high ionic strengths. *Proteins; Structure, Function, and Genetics*, **10**, 10–21

Bellissent-Funel, M.-C., Teixeira, J., Chen, S. H., Dorner, B., Middendorf, H. D. and Crespi, H. L. (1989). Low-frequency collective modes in dry and hydrated proteins. *Biophys. J.*, **56**, 713–716

Bernstein, F. C., Koetzle, T. F., Williams, G. J. B., Meyer, E. F., Brice, M. D., Rodgers, J. R., Kennard, O., Shimanouchi, T. and Tasumi, M. (1977). The protein data bank: A computer-based archival file for macromolecular structures. *J. Mol. Biol.*, **112**, 535–542

Bhat, T. N. (1989). Correlation between occupancy and temperature factors of solvent molecules in crystal structures of proteins. *Acta Cryst.*, **A45**, 145–146

Blake, C. C. F., Pulford, W. C. A. and Artymiuk, P. J. (1983). X-ray studies of water in crystals of lysozyme. *J. Mol. Biol.*, **167**, 693–723

Blevins, R. A. and Tulinsky, A. (1985). Comparison of the independent solvent structures of dimeric α-chymotrypsin with themselves and with γ-chymotrypsin. *J. Biol. Chem.*, **260**, 8865–8872

Blundell, T., Barlow, D., Borkakoti, N. and Thornton, J. (1983). Solvent-induced distortions and the curvature of alpha-helices. *Nature*, **306**, 281–283

Bode, W., Epp, O., Huber, R., Laskowski, M. and Ardelt, W. (1985). The crystal and molecular structure of the third domain of silver pheasant ovomucoid (OMSVP3) *Eur. J. Biochem.*, **147**, 387–395

Bränden, C. I. and Jones, T. A. (1990). Between objectivity and subjectivity. *Nature*, **343**, 687–689

Brooks, C. L. and Karplus, M. (1989). Solvent effects on protein motion and protein effects on solvent motion. Dynamics of the active site region of lysozyme. *J. Mol. Biol.*, **208**, 159–181

Brooks, C. L., Karplus, M. and Pettitt, B. M. (1988). Proteins: A theoretical perspective of dynamics structure and thermodynamics. *Adv. Chem. Phys.*, **71**, 137–248

Brünger, A. T. (1988). Crystallographic refinement by simulated annealing. In Isaacs, N. W. and Taylor, M. R. (Eds). *Crystallographic Computing 4; Techniques and New Technologies*. Clarendon Press, Oxford, pp. 126–140

Brünger, A. T., Brooks, C. L. and Karplus, M. (1985). Active site dynamics of ribonuclease. *Proc. Natl Acad. Sci. USA*, **82**, 8458–8462

Brünger, A. T., Kuriyan, J. and Karplus, M. (1987). Crystallographic R-factor refinement by molecular dynamics. *Science*, **235**, 458–460

Brünger, A. T. and Karplus, M. (1988). Polar hydrogen positions in proteins: Empirical energy placement and neutron diffraction comparison. *Proteins: Structure, Function, and Genetics*, **4**, 148–156

Bushnell, G. W., Louie, G. V. and Brayer, G. D. (1990). High resolution three-dimensional structure of horse-heart cytochrome c. *J. Mol. Biol.*, **214**, 585–595

Chambers, J. L. and Stroud, R. M. (1979). The accuracy of refined protein structures: Comparison of two independently refined models of bovine trypsin. *Acta Cryst.*, **B35**, 1861–1874

Cheng, X. and Schoenborn, B. (1990). Hydration in protein crystals. A neutron diffraction analysis of Carbonmonoxymyoglobin. *Acta Cryst.*, **B46**, 195–208

Chiari, G. and Ferraris, G. (1982). The water molecule in crystalline hydrates studied by neutron diffraction. *Acta Cryst.*, **B38**, 2331–2341

Clore, G. M., Bax, A., Wingfield, P. T. and Gronenborn, A. M. (1990). Identification and localisation of bound internal water in the solution structure of interleukin 1β by heteronuclear three-dimensional ^{1}H rotating-frame Overhauser ^{15}N–^{1}H multiple quantum coherence NMR spectroscopy. *Biochemistry*, **29**, 5671–5676

Cruickshank, D. W. J. (1949). The accuracy of electron-density maps in x-ray analysis with special reference to dibenzyl. *Acta Cryst.*, **2**, 65–82

Cusack, S., Smith, J., Finney, J., Tidor, B. and Karplus, M. (1988). Inelastic neutron scattering analysis of picosecond internal protein dynamics. *J. Mol. Biol.*, **202**, 903–908

Dewar, M. J. S. and Storch, D. M. (1985). Alternative view of enzyme reactions. *Proc. Natl Acad. Sci. USA*, **82**, 2225–2229

Diamond, R. (1990). On the use of normal modes in thermal parameter refinement: Theory and application to the bovine pancreatic trypsin inhibitor. *Acta Cryst.*, **A46**, 425–435

Edsall, J. T. and McKenzie, H. A. (1983). Water and proteins. II. The location and dynamics of water in protein systems and its relation to their stability and properties. *Adv. Biophys.*, **16**, 53–183

Feigelson, R. S. (1986). Proceedings of the First International Conference on Protein Crystal Growth. Stanford University CA USA 14–16 Aug. 1985. *J. Crystal Grwth*, **76**, 535–718

Finney, J. (1979). The organization and function of water in protein crystals. In F. Franks (Ed.), *Water: A Comprehensive Treatise*. Plenum Press, New York, pp. 47–122, and pp. 411–436 for references.

Fontecilla-Camps, J. C., Habersetzer-Rochat, C. and Rochat, H. (1988). Orthorhombic crystals and three-dimensional structure of the potent toxin II from the scorpion *Androctonus australis* Hector. *Proc. Natl Acad. Sci. USA*. **85**, 7443–7447

Frauenfelder, H., Petsko, G. A. and Tsernoglou, D. (1979). Temperature dependent X-ray diffraction as a probe of protein structural dynamics. *Nature*, **280**, 558–563

Frey, M., Genovesio-Taverne, J. C. and Fontecilla-Camps, J. C. (1988). Application of the periodic bond chain (PBC) theory to the analysis of the molecular packing in protein crystals. *J. Cryst. Grwth*, **90**, 245–258

Frey, M., Sieker, L. C., Payan, F., Haser, R., Bruschi, M., Pèpe, G. and LeGall, J. (1987). Rubredoxin from *Desulphovibrio gigas*. A molecular model of the oxidized form at 1.4 Å resolution. *J. Mol. Biol.*, **197**, 525–541

Fujinaga, M., Delbaere, L. T. J., Brayer, G. D. and James, M. N. G. (1985). Refined structure of α-lytic protease at 1.7 Å resolution. Analysis of hydrogen bonding and solvent structure. *J. Mol. Biol.*, **183**, 479–502

Fujinaga, M., Gros, P. and van Gunsteren, W. F. (1989). Testing the method of crystallographic refinement using molecular dynamics. *J. Appl. Cryst.*, **22**, 1–8

Fujinaga, M., Sielecki, A. R., Read, R. J., Ardelt, W., Laskowski, M. and James, M. N. G. (1987). Crystal and molecular structure of the complex of α-chymotrypsin with its inhibitor turkey ovomucoid third domain at 1.8 Å resolution. *J. Mol. Biol.*, **195**, 397–418

Furey, W., Wang, B. C., Yoo, C. S. and Sax, M. (1983). Structure of a Novel Bence-Jones Protein (Rhe) fragment at 1.6 Å resolution. *J. Mol. Biol.*, **167**, 661–692

Gao, J., Kuczera, K., Tidor, B. and Karplus, M. (1989). Hidden thermodynamics of mutant proteins: A molecular dynamics analysis. *Science*, **244**, 1069–1072

Giégé, R., Ducruix, A., Fontecilla-Camps, J. C., Feigelson, R. S., Kern, R. and McPherson, A. (1988). Proceedings of the Second International Conference on Protein Crystal Growth. A FEBS Advanced Lecture Course Bischenberg-Strasbourg, France 19–25 July 1987. *J. Cryst. Grwth*, **90**, 1–374

Hagler, A. T. and Moult J. (1978). Computer simulation of the solvent structure around biological macromolecules. *Nature*, **272**, 222–226

Harrison, S. C. (1991). A structural taxonomy of DNA-binding domains. *Nature*, **353**, 715–719

Harrison, S. C. and Aggarwal, A. K. (1990). DNA recognition by proteins with the Helix-Turn-Helix motif. *Ann. Rev. Biochem.*, **59**, 933–969

Harvey, S. C. (1989). Treatment of electrostatic effects in macromolecular modeling. *Proteins: Structure, Function, and Genetics*, **5**, 78–92

Helliwell, J. R., Moore, P. R., Papiz, M. Z. and Smith, J. M. A. (1984). Measurements of absorption curves for protein single crystals on the oscillation camera with time decaying incident-beam intensity and variable-wavelength synchrotron X-radiation. *J. Appl. Cryst.*, **17**, 417–419

Hendrickson, W. A. (1980). Practical aspects of stereochemically restrained refinement of protein structures. In Machin, P. A.; Campbell, J. W. and Elder, M., *Refinement of Protein Structure*. Daresbury Laboratory, Warrington, pp. 1–8

Hendrickson, W. A. and Konnert, J. H. (1980). Stereochemically restrained least-squares refinement of macromolecule structures. In Srinivasan, R. (Ed.), *Biomolecular Structure, Function, Conformation and Evolution*, Vol. 1. Pergamon Press, Oxford, pp. 43–57

Hope, H. (1990). Crystallography of biological macromolecules at ultra-low temperature. *Ann. Rev. Biophys. Biophys. Chem.*, **19**, 107–126

Jack, A. and Levitt, M. (1978). Refinement of large structures by simultaneous minimization of energy and R factor. *Acta Cryst.*, **A34**, 931–935

James, M. N. G. (1980). Practical aspects of stereochemically restrained refinement of protein structures. In Machin, P. A.; Campbell, J.W. and Elder, M. *Refinement of Protein Structure*. Daresbury Laboratory, Warrington, frontpage.

James, M. N. G. and Sielecki, A. R. (1983). Structure and refinement of penicillopepsin at 1.8 Å resolution. *J. Mol. Biol.*, **163**, 299–361

James, M. N. G., Sielecki, A. R., Brayer, G. D., Delbaere, L. T. J. and Bauer, C.

A. (1980). Structure of product and inhibitor complexes of *Streptomyces griseus* protease A at 1.8 Å resolution. A model for serine proteases catalysis. *J. Mol. Biol.*, **144**, 43–88

Janin, J. and Chothia, C. (1990). The structure of protein-protein recognition sites. *J. Biol. Chem.*, **265**, 16027–16030

Jeffrey, G. A. and Maluszynska, H. (1990). The stereochemistry of the water molecules in the hydrates of small biological molecules. *Acta Cryst.*, **B46**, 546–549

Jensen, L. H. (1990). Solvent models for protein crystals: on occupancy parameters for discrete solvent sites and the solvent continuum. *Acta Cryst.*, **B46**, 650–653

Jensen, L. H. (1991). The water structure in cubic insulin. *Am. Cryst. Assoc. Annual Meeting*, Toledo. Abstract G06

Jensen, L. H. and Watenpaugh, K. D. (1986). Hydrogen bonded water in rubredoxin from *Clostridium pasteurianum*. In Griffin, J. F. (Ed.), *The Hydrogen Bond: New Insights on an Old Story. Trans. Am. Cryst. Assoc.*, **22**, 89–96

Kamphuis, I. G., Drenth, J. and Baker, E. N. (1985). Thiol proteases. Comparative studies based on high resolution structures of papain and actinidin, and of amino acid sequence information for cathepsins B and H, and stem bromelain. *J. Mol. Biol.*, **182**, 317–329

Kamphuis, I. G., Kalk, K. H., Swarte, M. B. A. and Drenth, J. (1984). Structure of papain refined at 1.65 Å resolution. *J. Mol. Biol.*, **179**, 233–256

Karplus, M. and Petsko, G. A. (1990). Molecular dynamics simulations in biology. *Nature*, **347**, 631–639

Karplus, P. A. and Schulz, G. E. (1987). Refined structure of glutathione reductase at 1.54 Å resolution. *J. Mol. Biol.*, **195**, 701–729

Katz, B. and Kossiakoff, A. A. (1990). Crystal structures of subtilisin BPN' variants containing disulfide bonds and cavities: Concerted structural rearrangements induced by mutagenesis. *Proteins: Structure, Function, and Genetics*, **7**, 343–357

Kauzmann, W. (1959). Some factors in the interpretation of protein denaturation. *Adv. Protein Chem.*, **14**, 1–63

Kennard, O., Cruse, W. B. T., Nachman, J., Prangé, T., Shakked, Z. and Rabinovich, D. (1986). Ordered water structure in a A-DNA octamer at 1.7 Å resolution. *J. Biomol. Struct. Dyn.*, **3**, 623–647

Kopfmann, G. and Huber, R. (1968). A method of absorption correction by X-ray intensity measurements. *Acta Cryst.*, **A24**, 348–351

Kossiakoff, A. A. (1985). The application of neutron crystallography to the study of dynamic and hydration properties of proteins. *Ann. Rev. Biochem.*, **54**, 1195–1227

Kundrot, C. E. and Richards, F. M. (1987). Use of the occupancy factor in the refinement of solvent molecules in protein crystal structures. *Acta Cryst.*, **B43**, 544–547

Kuriyan, J., Ösapay, K., Burley, S. K., Brünger, A. T., Hendrickson, W. A. and Karplus, M. (1991). Exploration of disorder in protein structures by x-ray restrained molecular dynamics. *Proteins: Structure, Function, and Genetics*, **10**, 340–358

Kuriyan, J., Wilz, S., Karplus, M. and Petsko, G. A. (1986). X-ray structure and refinement of carbon–monoxy (Fe II)–myoglobin at 1.5 Å resolution. *J. Mol. Biol.*, **192**, 133–154

Lehmann, M. S., Mason, S. A. and McIntyre, G. J. (1985). Study of ethanol-lysozyme interactions using neutron diffraction. *Biochemistry*, **24**, 5862–5869

Lehmann, M. S. and Stansfield, R. F. D. (1989). Binding of dimethyl sulfoxide to lysozyme in crystals, studied with neutron diffraction. *Biochemistry*, **28**, 7028–7033

Loll, P. J. and Lattman, E. E. (1990). Active site mutant Glu-43 → Asp in staphylococcal nuclease displays nonlocal structural changes. *Biochemistry*, **29**, 6866–6873

Louie, G. V. and Brayer, G. D. (1990). High resolution refinement of yeast-iso-1-cytochrome c and comparisons with other eukaryotic cytochromes c. *J. Mol. Biol.*, **214**, 527–555

Luzzatti, V. (1952). Traitement statistique des erreurs dans la détermination des structures cristallines. *Acta Cryst.*, **5**, 802–810

Luzzatti V. (1953). Résolution d'une structure cristalline lorsque les positions d'une partie des atomes sont connues: Traitement statistique. *Acta Cryst.*, **6**, 142–152

McCammon, J. A. and Harvey, S. C. (1987). *Dynamics of Proteins and Nucleic Acids*. Cambridge University Press, Cambridge

McPherson, A. (1989). *Preparation and Analysis of Protein Crystals*, 2nd edn. Krieger, Malabar, Florida

Mason, S. A., Bentley, G. A. and McIntyre, G. J. (1984). Deuterium exchange in lysozyme at 1.4 Å resolution. In Schoenborn, B. P. (Ed.), *Neutron in Biology*, Plenum Press, New York, pp. 335–348

Matthews, B. W. (1968). Solvent content of protein crystals. *J. Mol. Biol.*, **33**, 491–497

Matthews, B. W., Nicholson, H. and Becktel, W. J. (1987). Enhanced protein thermostability from site-directed mutations that decrease the entropy of unfolding. *Proc. Natl Acad. Sci. USA*, **84**, 6663–6667

Messerschmidt, A., Schneider, M. and Huber, R. (1990). ABSCOR: a scaling and absorption correction program for the FAST area detector diffractometer. *J. Appl. Cryst.*, **23**, 436–439.

Moews, P. C. and Kretsinger, R. H. (1975). Refinement of the structure of carp muscle calcium-binding parvalbumin by model building and difference Fourier analysis. *J. Mol. Biol.*, **91**, 201–228

Nar, H., Messerschmidt, A., Huber, R., van de Kamp, M. and Canters, G. W. (1991). X-Ray crystal structure of the two site-specific mutants His35Gln and His35Leu of azurin from *Pseudonomonas aeruginosa*. *J. Mol. Biol.*, **218**, 427–447

North, A. C. T., Phillips, D. C. and Mathews, F. S. (1968). A semi-empirical method of absorption correction. *Acta Cryst.*, **A24**, 351–359

Otting, G., Liepinsh, E. and Wüthrich, K. (1991). Proton exchange with internal water molecules in the protein BPTI in aqueous solutions. *J. Am. Chem. Soc.*, **113**, 4363–4364

Otting, G. and Wühtrich, K. (1989). Studies of protein hydration in aqueous solution by direct NMR observation of individual protein-bound water molecules. *J. Am. Chem. Soc.*, **111**, 1871–1875

Otwinowski, Z., Schevitz, R. W., Zhang, R.-G., Lawson, C. L., Joachimiak, A., Marmorstein, R. Q., Luisi, B. F. and Sigler, P. B. (1988). Crystal structure of *trp* repressor/operator complex at atomic resolution. *Nature*, **335**, 321–329

Pflugrath, J. W. and Quiocho, F. A. (1988). The 2 Å resolution of the sulfate-binding protein involved in active transport in *Salmonella typhimurium*. *J. Mol. Biol.*, **200**, 163–180

Phillips, S. E. V. (1980). Structure and refinement of oxymyoglobin at 1.6 Å resolution. *J. Mol. Biol.*, **142**, 531–554

Pitt, W. R. and Goodfellow, J. M. (1991). Modelling of solvent positions around

polar groups in proteins. *Protein Engng*, **4**, 531–537
Quiocho, F. A., Wilson, D. K. and Vyas, N. K. (1989). Substrate specificity and affinity of a protein modulated by bound water molecules. *Nature*, **340**, 404–407
Rashin, A., Iofin, M. and Honig, B. (1986). Internal cavities and buried waters in globular proteins. *Biochemistry*, **25**, 3619–3625
Read, R. J., Fujinaga, M., Sielecki, A. R. and James, M. N. G. (1983). Structure of the complex of *Streptomyces griseus* protease B and the third domain of the turkey ovomucoid inhibitor at 1.8 Å resolution. *Biochemistry*, **22**, 4420–4433
Read, R. J. and James, M. N. G. (1988). Refined crystal structure of *Streptomyces griseus* trypsin at 1.7 Å resolution. *J. Mol. Biol.*, **200**, 523–561
Richardson, J. (1981). The anatomy and taxonomy of protein structures. In Anfinsen, C. B., Edsall, J. T. and Richards, F. M. (Eds), *Advances in Protein Chemistry*, Vol. 34. Academic Press, New York, pp. 168–340
Rupley, J. A. and Careri, G. (1991). Protein hydration and function. *Adv. Prot. Chem.*, **41**, 37–172
Rupley, J. A., Gratton, E. and Careri, G. (1983). Water and globular proteins. *Trends Biochem. Sci.*, **8**, 18–22
Saenger, W. (1987). Structure and dynamics of water surrounding biomolecules. *Ann. Rev. Biophys. Biophys. Chem.*, **16**, 93–114
Salunke, D. M., Veerapandian, B., Kodandapani, R. and Vijayan, M. (1985). Water-mediated transformation in protein crystals. *Acta Cryst.*, **B41**, 431–436
Savage, H. (1986a). Water structure in vitamin B_{12} coenzyme crystals. I. Analysis of the neutron and X-ray solvent densities. *Biophys. J.*, **50**, 947–965
Savage, H. (1986b). Water structure in vitamin B_{12} coenzyme crystals. II. Structural characteristics of the solvent networks. *Biophys. J.*, **50**, 967–980
Savage, H. and Wlodawer, A. (1986). Determination of water structure around biomolecules using X-ray and neutron diffraction methods. *Methods Enzymol.*, **127**, 162–183
Scanlon, W. J. and Eisenberg, D. (1975). Solvation of crystalline proteins: Theory and its application to available data. *J. Mol. Biol.*, **98**, 485–502
Schaefer, M. and Froemmel, C. (1990). A precise analytical method for calculating the electrostatic energy of macromolecules in aqueous solution. *J. Mol. Biol.*, **216**, 1045–1066
Schirmer, T. and Evans, P. R. (1990). Structural basis of the allosteric behaviour of phosphofructokinase. *Nature*, **343**, 140–145
Schoenborn, B. P. (Ed.) (1984). *Neutrons in Biology*. Plenum Press, New York
Schoenborn, B. P. (1988). Solvent effect in protein crystals. A neutron diffraction analysis of solvent and ion density. *J. Mol. Biol.*, **201**, 741–749
Scott, D. L., White, S. P., Browning, J. L., Rosa, J. J., Gelb, M. H. and Sigler, P. B. (1991). Structures of free and inhibited human secretory phospholipase A from inflammatory exudate. *Science*, **254**, 1007–1010
Scott, D. L., White, S. P., Otwinowski, Z., Yuan, W., Gelb, M. H. and Sigler, P. B. (1990). Interfacial catalysis: The mechanism of phospholipase A_2. *Science*, **250**, 1541–1546
Sharp, K. A. and Honig, B. (1990). Electrostatics interactions in macromolecules: Theory and applications. *Ann. Rev. Biophys. Biophys. Chem.*, **19**, 301–332
Sheriff, S. and Hendrickson, W. A. (1987). Description of overall anisotropy in diffraction from macromolecular crystals. *Acta Cryst.*, **A43**, 118–121
Sielecki, A. R., Hendrickson, W. A., Broughton, C. G., Delbaere, L. T. J., Brayer, G. D. and James, M. N. G. (1979). Protein structure refinement: *Streptomyces griseus* serine protease A at 1.8 Å resolution. *J. Mol. Biol.*, **134**, 781–804

Simonson, T., Perahia, D and Bricogne, G. (1991). Intramolecular dielectric screening in proteins. *J. Mol. Biol.*, **218**, 859–886

Smith, J. L., Corfield, P. W. R., Hendrickson, W. A. and Low, B. W. (1988). Refinement at 1.4 Å resolution of a model of erabutoxin b: Treatment of ordered solvent and discrete disorder. *Acta Cryst.*, **A44**, 357–368

Smith, J. L., Hendrickson, W. A., Honzatko, R. B. and Sheriff, S. (1986). Structural heterogeneity in protein crystals. *Biochemistry*, **25**, 5018–5027

Steitz, T. A. (1990). Structural studies of protein–nucleic acid interaction: the sources of sequence-specific binding. *Q. Rev. Biophys.*, **23**, 205–280

Stenkamp, R. E., Sieker, L. C. and Jensen, L. H. (1990). The structure of rubredoxin from *Desulphovibrio desulfuricans* strain 27774 at 1.5 Å resolution. *Proteins: Structure, Function, and Genetics*, **8**, 352–364

Suguna, K., Bott, R. R., Padlan, E. A., Subramanian, E., Sheriff, S., Cohen, G. H. and Davies, D. R. (1987). Structure and refinement at 1.8 Å resolution of the aspartic proteinase from *Rhizopus chinensis*. *J. Mol. Biol.*, **196**, 877–900

Sundaralingam, M. and Sekharudu, Y. C. (1989). Water-inserted α-helical segments implicate reverse turns as folding intermediates. *Science*, **244**, 1333–1337

Sussmann, J. L., Holbrook, S. R., Church, G. M. and Kim, S.-H. (1977). A structure-factor least-squares refinement procedure for macromolecular structures using constrained and restrained parameters. *Acta Cryst.*, **A33**, 800–804

Takano, T. and Dickerson, R. E. (1981a). Conformation change of cytochrome c. I-Ferrocytochrome c structure refined at 1.5 Å resolution. *J. Mol. Biol.*, **153**, 79–94

Takano, T. and Dickerson, R. E. (1981b). Conformation change of cytochrome c. II. Ferricytochrome c refinement at 1.8 Å and comparison with the ferrocytochrome structure. *J. Mol. Biol.*, **153**, 95–115

Teeter, M. M. (1984). Water structure of a hydrophobic protein at atomic resolution: Pentagon rings of water molecules in crystals of crambin. *Proc. Natl Acad. Sci. USA*, **81**, 6014–6018

Teeter, M. M. (1991). Water–protein interactions: Theory and experiments. *Ann. Rev. Biophys. Biophys. Chem.*, **20**, 577–600

Teeter, M. M. and Kossiakoff, A. A. (1984). The neutron structure of the hydrophobic plant protein crambin. In Schoenborn, B. P. (Ed.), *Neutrons in Biology*. Plenum Press, New York, pp. 335–348

Teeter, M. M. and Whitlow, M. D. (1986). Hydrogen bonding in the high resolution structure of the protein crambin. In Griffin, J. F. (Ed.), *Trans. Am. Cryst. Assoc.*, **22**, 75–88

Teplyakov, A. V., Kuranova, I. P., Harutyunyan, E. H., Vainshtein, B. K., Frömmel, C., Höhne, W. E. and Wilson, K. S. (1990). Crystal structure of thermitase at 1.4 Å resolution. *J. Mol. Biol.*, **214**, 261–279

Thanki, N., Thornton, J. M. and Goodfellow, J. M. (1988). Distribution of water around amino acid residues in proteins. *J. Mol. Biol.*, **202**, 637–657

Thanki, N., Umrania, Y., Thornton, J. M. and Goodfellow, J. M. (1991). Analysis of main-chain solvation as a function of secondary structure. *J. Mol. Biol.*, **221**, 669–691

Tronrud, D. E., Ten Eyck, L. F. and Matthews, B. W. (1987). An efficient general-purpose least-squares refinement program for macromolecular structures. *Acta Cryst.*, **A43**, 489–501

Tüchsen, E., Hayes, J. M., Ramaprasad, S., Copie, V. and Woodward, C. (1987). Solvent exchange of buried water and hydrogen exchange of peptide NH groups hydrogen bonded to buried waters in bovine pancreatic trypsin inhibitor. *Biochemistry*, **26**, 5163–5172

Tulinsky, A. and Blevins, R. A. (1986). Least-squares refinement of two protein molecules per asymmetric unit with and without non-crystallographic symmetry restrained. *Acta Cryst.*, **B42**, 198–200

Tulinsky, A. and Wright, L. H. (1973). An x-ray crystallographic investigation of exchange of localized sulfate ions in crystals of α-chymotrypsin. *J. Mol. Biol.*, **81**, 47–56

van der Laan, J. M., Schreuder, H. A., Swarte, M. B. A., Wierenga, R. K., Kalk, K. H., Hol, W. G. J. and Drenth, J. (1989). The coenzyme analogue adenosine 5-diphosphoribose displaces FAD in the active site of *p*-hydroxybenzoate hydroxylase. An X-ray crystallographic investigation. *Biochemistry*, **28**, 7199–7205

van der Sluis, P. and Spek, A. L. (1990). BYPASS: an effective method for the refinement of crystal structures containing disordered solvent regions. *Acta Cryst.*, **A46**, 194–201

van Gunsteren, W. F., Berendsen, H. J. C., Hermans, J., Hol, W. G. J. and Postma, J. P. M. (1983). Computer simulation of the dynamics of hydrated protein crystals and its comparison with x-ray data. *Proc. Natl Acad. Sci. USA*, **80**, 4315–4319

Veerapandian, B., Cooper, J. B., Šali, A. and Blundell, T. L. (1990). X-ray analysis of aspartic proteinases. III. Three dimensional structure of Endothiapepsin complexed with a transition-state isostere inhibitor of renin at 1.6 Å resolution. *J. Mol. Biol.*, **216**, 1017–1029

Walter, M. H., Westbrook, E. M., Tykodi, S., Uhm, A. M. and Margoliash, E. (1990). Crystallization of Tuna Ferricytochrome c at low ionic strength. *J. Biol. Chem.*, **265**, 4177–4180

Ward, K. B. and Gilliland, G. L. (1989). *Third International Conference on the Crystallization of Biological Macromolecules.* Washington, D.C. 13–18 August 1988. *Program Abstracts*. Laboratory for the Structure of Matter Naval Research Laboratory, Washington, D.C.

Warshel, A., Åqvist, J. and Creighton, S. (1989a). Enzymes work by solvation substitution rather than by desolvation. *Proc. Natl Acad. Sci. USA*, **86**, 5820–5824

Warshel, A., Naray-Szabo, G., Sussmann, F. and Hwang, J.-K. (1989b). How do serine proteases really work? *Biochemistry*, **28**, 3629–3637

Watenpaugh, K. D., Margulis, T. N., Sieker, L. C. and Jensen, L. H. (1978). Water structure in a protein crystal: Rubredoxin at 1.2 Å. *J. Mol. Biol.*, **122**, 175–190

Watt, W., Tulinsky, A., Swenson, R. P. and Watenpaugh, K. D. (1991). Comparison of the crystal structures of a flavodoxin in its three oxidation states at cryogenic temperatures. *J. Mol. Biol.*, **218**, 195–208

Weaver, L. H. and Matthews, B. W. (1987). Structure of bacteriophage T4 lysozyme refined at 1.7 Å resolution. *J. Mol. Biol.*, **193**, 189–199

Westhof, E. (1988). Water: An integral part of nucleic acid structure. *Ann. Rev. Biophys. Biophys. Chem.*, **17**, 125–144

Westhof, E. and Beveridge, D. L. (1990). Hydration of nucleic acids. In Franks, F. (Ed.), *Water Science Reviews*, Vol. 5. Cambridge University Press, Cambridge, pp. 24–136

Wlodawer, A., Borkakoti, N., Moss, D. S. and Howlin, B. (1986). Comparisons of two independently refined models of ribonuclease A. *Acta Cryst.*, **B42**, 379–387

Wlodawer, A., Deisenhofer, J. and Huber, R. (1987). Comparison of two highly refined structures of bovine pancreatic trypsin inhibitor. *J. Mol. Biol.*, **193**, 145–156

Wlodawer, A. and Hendrickson, W. A. (1982). A procedure for joint refinement of macromolecular structures with X-ray and neutron diffraction data from single crystals. *Acta Cryst.*, **A38**, 239–247

Wlodawer, A., Savage, H. and Dodson, G. (1989). Structure of insulin: Result of joint neutron and X-ray refinement. *Acta Cryst.*, **B45**, 99–107

Wlodawer, A., Walter, J., Huber, R. and Sjölin, L. (1984). Structure of pancreatic trypsin inhibitor. Results of joint neutron and x-ray refinement of crystal form II. *J. Mol. Biol.*, **180**, 301–329

Wright, C. S. (1987). Refinement of the crystal structure of wheat germ agglutinin isolectin 2 at 1.8 Å resolution. *J. Mol. Biol.*, **194**, 501–529

Wüthrich, K. (1990). Protein structure determination in solution by NMR spectroscopy. *J. Biol. Chem.*, **265**, 22059–22062

5

Hydration of Protein Secondary Structures—The Role in Protein Folding

C. Y. Sekharudu and M. Sundaralingam

1 Introduction

In general, protein crystals are highly hydrated, with about 27–65% of the volume occupied by solvent.[1] Of the relatively large amount of water in protein crystals, only the ordered water molecules which constitute a small fraction are seen in the electron density maps.[2] These water molecules are generally located in the protein surface, crevices and the intermolecular contact regions, and are involved in hydrogen bonding to the donor and the acceptor atoms of the main-chain or the side-chain atoms, either directly or through other water molecules.[3–10] The water molecules that are directly hydrogen bonded to one or more protein atoms are referred to as first coordination sphere solvent molecules and those that are linked through another water molecule are referred to as second coordination sphere solvents. Water molecules also occupy active sites and catalytic sites, and many of these are displaced upon binding of inhibitors or substrate analogues.[8] Water molecules are also found buried deep inside proteins.[4, 11, 12]

In secondary structures the main-chain N–H donor and C=O acceptor atoms are involved in hydrogen bonds. In addition, they can also be involved in hydrogen bonding to water. The hydrogen bonding to water has generated a great deal of interest since the finding that water molecules can pry open the α-helix secondary structure hydrogen bond and insert between the donor and acceptor atoms and bridge them through hydrogen bonds.[13–17] This has led to an understanding of the role of water in the formation and denaturation of α-helices. Similarly, we find that water molecules can insert into the hydrogen bonds of a β-sheet or a reverse turn.[16] The water-inserted β-sheets provide insights into the assembly of sheets from individual strands. The water-inserted reverse turn provides an important conformational link between the folded α-helix and the extended β-strand.[14, 15] Here we describe the conformational gamut displayed by the water-inserted segments that have provided the pathways for the folding and unfolding of α-helices, β-sheets and reverse turns.

2 Hydration of Alpha-Helices

Three Modes of Backbone Hydrogen Bonding in α-helices

The α-helix is stabilized by sequential 1–5 hydrogen bonds between the main chain CO_i and NH_{i+4} groups. These main-chain groups can also be involved in additional hydrogen bonds particularly to water molecules. The bifunctional nature of the C=O group allows it to participate in an additional hydrogen bond with a water molecule, besides the α-helix hydrogen bond, much more readily than the N−H group. There are three types of hydrogen bonding interactions between water molecules and the helix backbone.[14, 15] In Type I the $C{=}O_i$ group already engaged in the helix hydrogen bonding to NH_{i+4} forms a hydrogen bond with an external water molecule. Therefore, the pentapeptide chain involved in this type of hydrogen bond is referred to as an *externally hydrated* α-helical segment (Figure 5.1a). Such externally hydrated segments are by far the most common,[14] with the water molecule occupying a fairly large area within the constraints of the hydrogen bond geometry.[3, 18] In Type II, besides the carbonyl group, the amide group as well forms a hydrogen bond with the water molecule (Figure 5.1b). Since the amide group is involved in a bifurcated or three-centred hydrogen bond, the pentapeptide segment is referred to as a *three-centred α-helical segment*. The amide group shows a much weaker tendency to form more than one hydrogen bond, unlike the carbonyl oxygen atom; thus the occurrence of the three-centred segments is not as frequent as that of the externally hydrated segments.[14] In the Type III pattern, the water molecule pries open the helix hydrogen bond and inserts itself between the $C{=}O_i$ and NH_{i+4} groups and bridges them through hydrogen bonds. We refer to this pentapeptide segment as a *water-inserted* helical segment (Figure 5.1c). The Type III pattern shows the largest distortion from the α-helix and provides significant information on the conformational transitions between folded and unfolded states of α-helices.

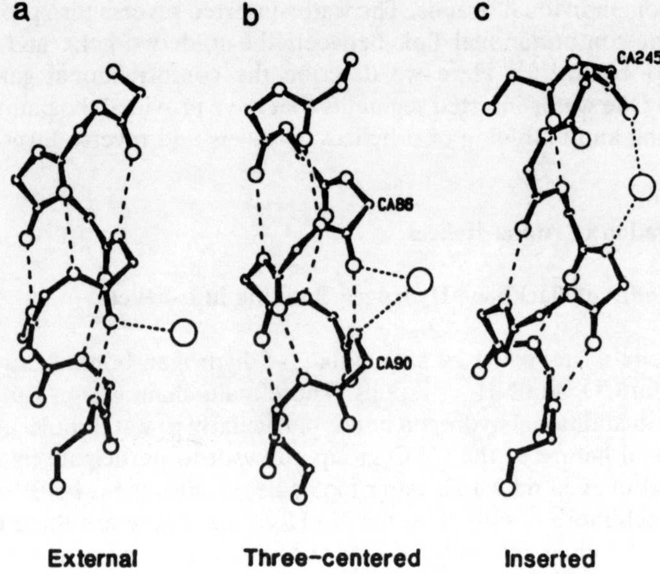

Figure 5.1 Three types of interactions between water molecules and the α-helix backbone: (*a*) externally hydrated α-helical segment; (*b*) three-centred hydrogen-bonded segment; and (*c*) water-inserted α-helical segment. After Reference 14

Hydrogen Bonding Geometries in the Three types of Helix–Water Interactions

The hydrogen bond lengths and angles (Figure 5.2a) involving the water molecule and the $C{=}O_i$ group are the same regardless of whether the water molecule is external, three-centred or inserted. However, in the external and three-centred cases the water molecule is hydrogen bonded to the outer lone-pair orbital of the carbonyl oxygen atom, while it is hydrogen bonded to the inner lone-pair orbital in the inserted cases.[14] In contrast, the hydrogen bonding geometries involving the N−H group are not the same for the three-centred and inserted cases. When both the hydrogen bonds, $N{-}H_{i+4}\cdot\cdot\cdot O_w$ and $N{-}H_{i+4}\cdot\cdot\cdot O{=}C_i$, are present, the hydrogen bonding distances are unequal; the shorter the $N{-}H_{i+4}\cdot\cdot\cdot CO_i$ distance the longer the $N{-}H\cdot\cdot\cdot Ow$ distance and vice versa. Thus, as the three-centred hydrogen bond $N{-}H\cdot\cdot\cdot Ow$ becomes stronger, the helix 5–1 hydrogen bond $N{-}H_{i+4}\cdot\cdot\cdot C{=}O_i$ becomes weaker, until it is ruptured.[14] Therefore, the Type II or three-centred hydrogen bond can be regarded as a transition state between the externally and internally hydrated α-helical segments. A plot showing the relationship between the distances in these two hydrogen bonds is given in Figure 5.2(B).

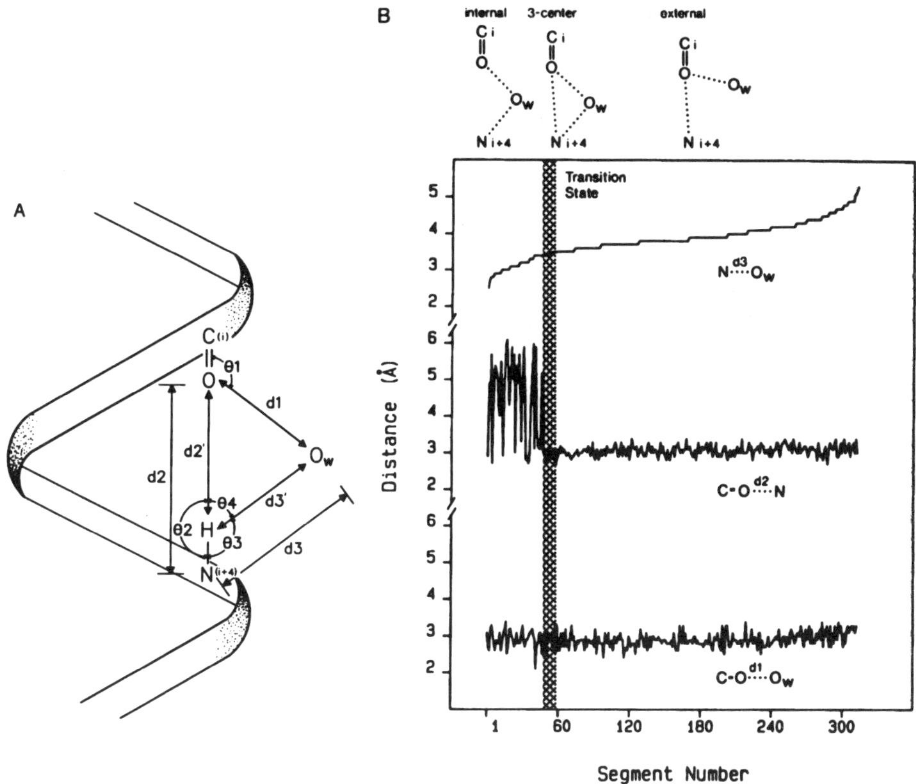

Figure 5.2 (*A*) Geometry of hydrogen bonds between a water molecule and an α-helical segment. (*B*) The relationship between the hydrogen bonds, d2 vs. d3, involving the amide group are shown. Notice that as d3 decreases d2 increases. At the transition state the distances d2 and d3 are the same. After Reference 14

Distortions in the Helix Geometry Due to Water Insertion

All the three types of water interactions induce conformational distortions in the helix, with the least perturbation for the externally hydrated segments[10, 14] and the most perturbation for the inserted segments.[14, 15] In the externally hydrated segments, only small conformational distortions are observed in the Φ, Ψ dihedral angles involving the single bonds adjacent to the $C{=}O_{i+4}$ and $N{-}H_i$ groups involved in the 1−5 hydrogen bond. These small distortions perturb the helix towards the 3_{10}-helix. In the three-centred cases, the distortions are more appreciable and also perturb the helix towards the 3_{10}-helix. In the water-inserted cases the 1−5 helix hydrogen bond is broken and a new 1−4 hydrogen bond is formed involving the penultimate amide group of the pentapeptide segment. The 1−4 hydrogen bonded segments adopt the 3_{10} and reverse turn conformations

(Figure 5.3). Thus, the reverse turn can be accommodated into an α-helix where it is not involved in chain reversal. Therefore it is not unique to the hairpin loop or β-turn. The 1–4 hydrogen bond is not always present, particularly when the inserted water molecule separates the $C{=}O_i$ and $N{-}H_{i+4}$ groups apart by about 6.0 Å, yielding an open-turn conformation.

Water Insertion also Disrupts Adjacent α-Helix Hydrogen Bonds

The insertion of a water molecule into a 1–5 hydrogen bond causes noticeable changes in the local helix structure by disrupting several other adjacent 1–5 helix hydrogen bonds[15] (Figure 5.4). When the water molecule is inserted at the N-terminus of the helix, about 66% of the $CO_{i+1}\cdots NH_{i+5}$ hydrogen bonds, 25% of the $CO_{i+2}\cdots NH_{i+6}$ hydrogen bonds and 5% of the $CO_{i+3}\cdots NH_{i+7}$ hydrogen bonds are disrupted. When the water molecule is inserted at the C-terminus of the helix, the preceding hydrogen bonds, viz. $CO_{i-3}\cdots NH_{i+1}$, $CO_{i-2}\cdots NH_{i+2}$ and $CO_{i-1}\cdots NH_{i+3}$ are disrupted. When the water molecule is inserted into the central segment of the helix, the hydrogen bonds on either side of the pentapeptide are disrupted. The donor and acceptor atoms of the broken 1–5 hydrogen bonds tend to rearrange to form 1–4 hydrogen bonds or combinations of 1–4 and 1–5. In some cases, interaction to additional water molecules and side-chains are observed.

Molecular Dynamics Reveals Insertion of Water into the α-Helices

Recent molecular dynamics simulations on an α-helix in water solution corroborate our observation that the integrity of the helix is disrupted by water molecules penetrating the helix hydrogen bond. DiCapua *et al.*[19] have performed molecular dynamics simulations of Ala_{30} in α-helical conformation both *in vacuo* and in aqueous solution. They found that during the simulations water molecules inserted into the polyalanyl α-helix, inducing a bend in the helix. Analysis of the trajectory from the simulations showed that several water molecules form hydrogen bonds to both the donor NH_{i+4} and acceptor CO_i atoms, thus weakening the helix hydrogen bond. It was found that, in particular, the 1–5 hydrogen bond between residues 22 and 26 at the C-terminal end of the helix is broken with the insertion of a water molecule after 70 ps simulation at 350 K. This study confirms that terminal sites in the helix are more susceptible[14] than others to water and also provides the energetics of water insertion and helix unfolding. They have also performed molecular dynamics simulations on 20- and 40-residue segments of the repressor of primer (ROP).[20] A single long α-helix was constructed with the sequence corresponding to the

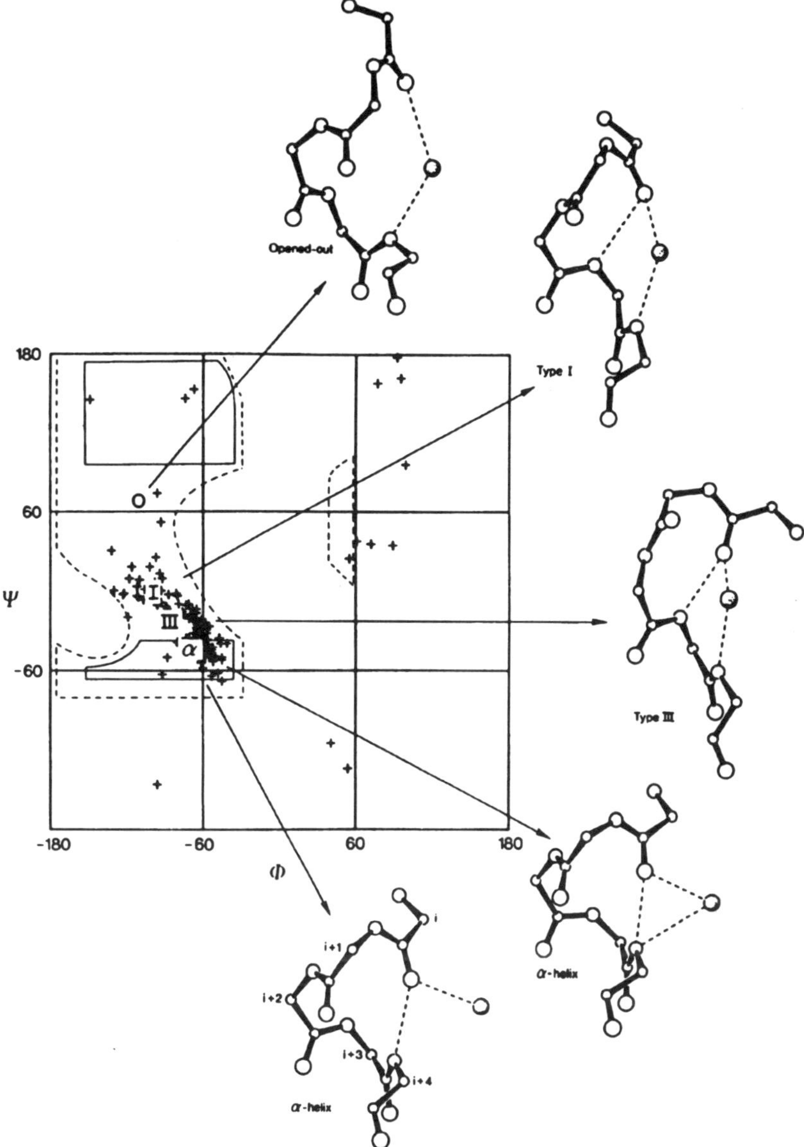

Figure 5.3 Ramachandran Φ, Ψ map showing the conformational angles of the 33 water-inserted α-helical segments shown as crosses (after Reference 15). Notice that the conformations move upwards from the α-helix region towards the β region. Conformations of the representative water-inserted α-helical segments are shown from bottom to top with increasing distortions from the helix. The type-I and type-III reverse turns are indicated in the third and fourth segments from bottom. These reverse turns can be considered as externally hydrated

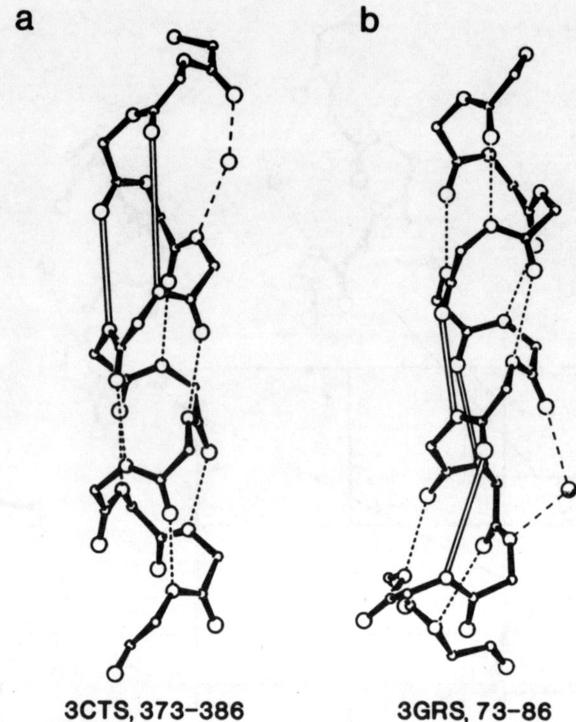

Figure 5.4 Examples of water-inserted helical segments, showing the disruptions in the adjacent 1–5 hydrogen bonds (*a*) amino terminal and (*b*) carboxy terminal. After Reference 15

helix–turn–helix[21] in ROP. Upon simulation of this helix in water, it was found that water molecules inserted into the helix at the region corresponding to the turn in the ROP protein and induced bending, or folding, of the helix. *In vacuo* no such bending was observed, indicating the importance of aqueous medium in protein folding.

3 Hydration of 3_{10} Helices and Reverse Turns

The externally hydrated 1–4 hydrogen bonds are not restricted to water-inserted α-helical segments alone. Protein structures reveal that 3_{10}-helical segments and reverse turns in the hairpin loops of β-sheets are also externally hydrated (Figure 5.5a). In the water-inserted reverse turns or 3_{10}-helical segments, a water molecule pries open the 1–4 hydrogen bond and bridges the CO_i and NH_{i+3} groups (Figure 5.5b) analogous to the 1–5 hydrogen bonds of α-helices. Similarly, water molecules are involved in three-centred and external hydrogen bonds (Figure 5.5c). Thus, both the

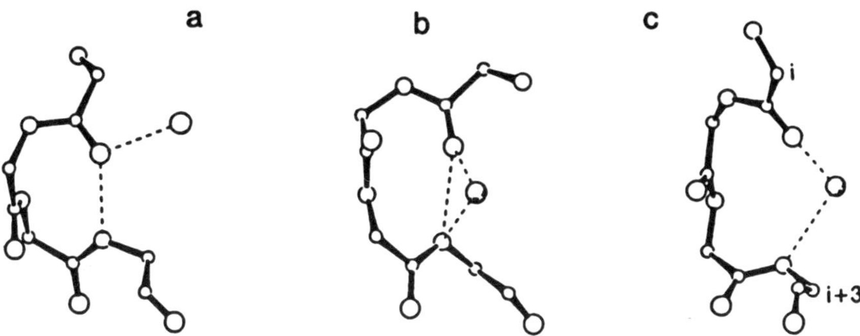

Figure 5.5 Three modes of backbone hydration in reverse turns and 3_{10}-helical tetrapeptide segments: (*a*) externally hydrated; (*b*) three-centred hydrogen-bonded; and (*c*) water-inserted

pentapeptide and tetrapeptide segments show the three types of hydration (Figures 5.1 and 5.5). The 1–4 type of water bridges are the most abundant. There are 373 segments of reverse turns/3_{10}-helices interacting with water molecules in the 40 proteins surveyed.[16] Of these, 37 are internal, 8 are three-centred and the remaining 328 are external. The water-inserted 3_{10} helices or reverse turns are frequently observed at the ends of helices and hairpin loops. The conformations of these water-inserted tetrapeptide segments bridge the gap between the α-helix and the β-sheet region of the Φ, Ψ map. It is found that the water-inserted reverse turns also exhibit conformations similar to the water-inserted pentapeptide segments. Thus, the reverse turn provides an important link between the extended strand and the α-helix. NMR studies have corroborated our analysis that reverse turns are indeed intermediates between the random chain and the α-helix. For instance, the NMR of a synthetic peptide with the sequence corresponding to the C-helix of myohaemerythrin has demonstrated the presence of reverse-turn-like structures interconverting between the folded and unfolded states.[22] Similar NMR studies on shorter synthetic peptides of 4–5 residues have again indicated the presence of reverse turns in aqueous medium.[23]

4 Hydration of β-Sheets

It is known that the C=O and N–H groups at the edges of the β-sheets poking into the solvent with no hydrogen-bonding partners within the protein are commonly engaged in hydrogen bonds with water molecules. Here we present water molecules inserted into the hydrogen bond of the backbone C=O and NH groups of β-sheets. As seen for the α-helices, all the three types of hydration patterns are also found for the sheets, viz.

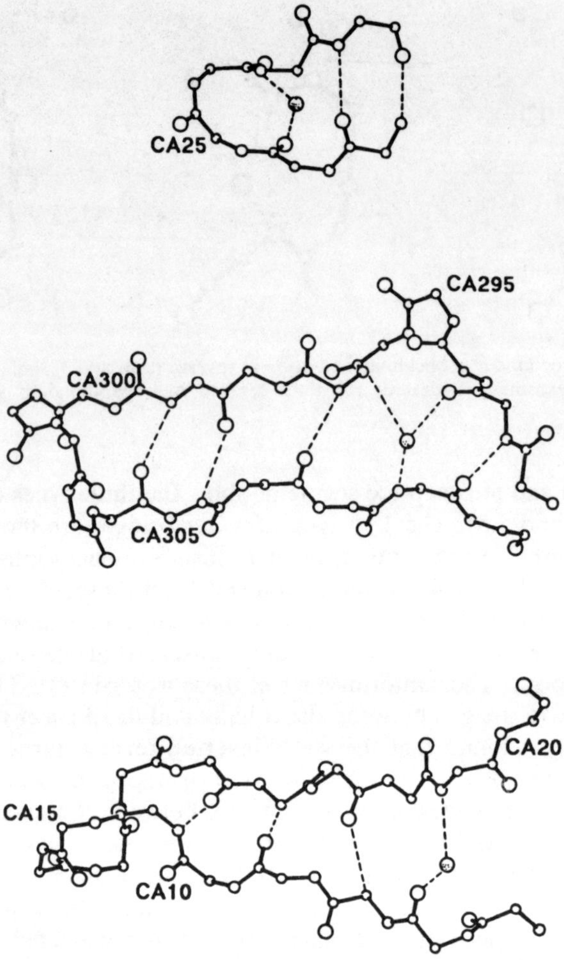

CA25

CA295
CA300
CA305

CA20
CA15
CA10

Figure 5.6 Examples of β-sheets with inserted water molecules. (*Top*) Water molecule inserted in the beginning of the β-sheet, residues 23–28 in aspartic proteinase (2APR). (*Middle*) Water molecule inserted within the β-sheet, residues 292–310 in glyceraldehyde-3-phosphate dehydrogenase (1GD1). (*Bottom*) Water molecule inserted in the terminus of the β-sheet, residues 6–21 in aspartic proteinase (2APR)

external (Type I), three-centred (Type II) and internal (Type III) (Figure 5.6). Here, again, the most interesting class is the Type III, where the water molecule is inserted between the hydrogen-bonding backbone C=O and NH groups of the two β-strands at the ends or the middle.[17] An example of an inserted water bridge at the beginning of a β-sheet, i.e. at the first hydrogen bond, that separates the hairpin loop from an antiparallel β-sheet is shown in Figure 5.6(top). An example of a water molecule trapped at the terminal sites of a β-sheet where the strands are pushed apart is shown in Figure 5.6(bottom). This type of interaction is found in both the parallel and antiparallel β-sheets and is the most common. Water molecules are not found inserted in the middle of the β-sheet, but they have been found between the two strands of a sheet that are not of equal length, as in Figure 5.6(middle), where the water molecule bridges both the constricted region of the bulge or looped-out region of the longer strand and also the shorter strand. Thus, this water molecule is central to the stabilization of this β-sheet segment.

There are some protein structures in which the insertion of the water molecules into the β-sheet is implicated in the functioning of the protein. In the skeletal muscle protein troponin C,[13] there are two calcium binding domains: the low-calcium-affinity N-terminal domain and the high-calcium-affinity C-terminal domain which binds both the calcium ions found in the crystal. In the latter domain the calcium ions are bound by the two helix–loop–helix (EF-hand) motifs, which are themselves connected by a short stretch of antiparallel β-sheet. In the absence of calcium ions in the low-affinity N-terminal domain, the β-sheet is longer, with four direct hydrogen bonds, compared with the high-affinity C-terminal domain, which has only two direct hydrogen bonds (Figure 5.7). Our 1.7 Å resolution structure shows[24] that the two terminal hydrogen bonds of the β-sheet in the high-affinity domain are lost by the insertion of water molecules. Apparently the insertion of water molecules is crucial in orienting the ligands on the two β-strands and the flanking helices to provide the optimal geometry for coordinating to the calcium ions. A striking example of a water-mediated β-sheet structure is seen in the functioning of the allosteric enzyme phosphofructokinase, where the water molecules that separate the two β-strands by inserting into the T-state are stripped in the R-state, bringing the adjacent strands together to form a standard hydrogen bonded β-sheet.[25] Another example of the role of water in β-sheet termination is seen in the dimeric β-protein defensin. Here the β-sheet that bridges the two defensin monomers is terminated at either end by inserted water molecules.[26]

The three sites of hydration, i.e. at the beginning, in the middle and at the end of the sheet discussed above, and the three types of hydration, i.e. external, three-centred and inserted, provide insights into water mediated β-sheet formation. In a fully hydrated β-strand (Figure 5.8), each of the backbone N–H and C=O groups will be at least hydrogen-bonded to one and two water molecules, respectively. When two such chains are brought together to form a β-sheet, three water molecules will be displaced for each NH· · ·OC hydrogen bond that is formed. Thus, a β-sheet with n hydrogen bonds would involve the elimination of $3n$ water molecules. If the β-strands are not in register, the strands will slide with respect to each other, to maximize not only the NH· · ·OC hydrogen bonding interactions but also the side-chain interactions. Maximization of the NH· · ·OC hydrogen

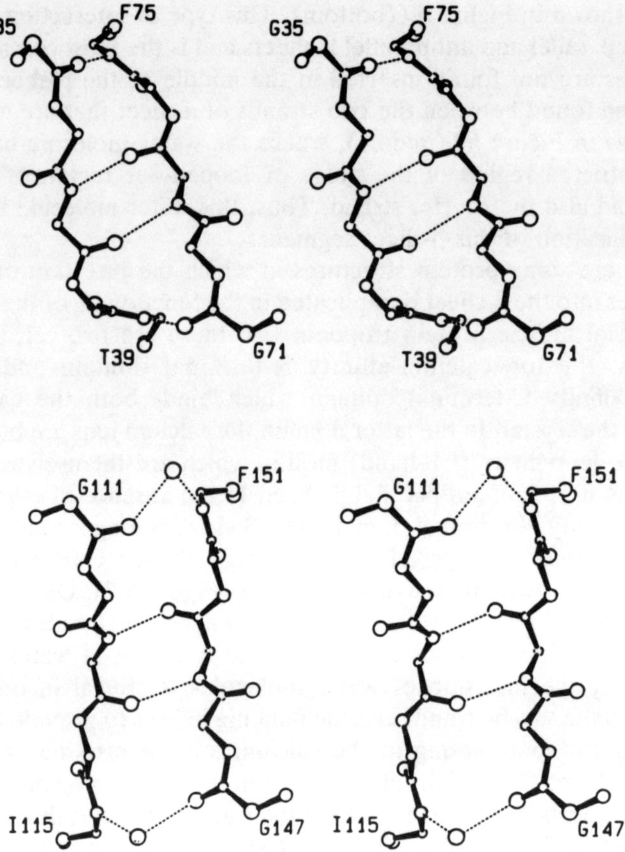

Figure 5.7 The two stretches of short antiparallel β-sheets in chicken skeletal muscle troponin C: (*top*) The N-terminal domain and (*bottom*) the C-terminal domain. Notice that the N-terminal domain has 4 hydrogen bonds, O35· · ·N75, O73· · ·N37, O37· · ·N73 and O71· · ·N39. In contrast, the C-terminal domain has only two hydrogen bonds, O141· · ·N113 and O113· · ·N149. The other two hydrogen bonds at the termini 'N115· · ·0147 and N151· · ·0111' have inserted water molecules[13, 24]

bonding will increase the entropy of the eliminated water molecules. Because of this, it is natural to expect a β-sheet with the highest number of interchain hydrogen bonds. Also, it is possible that the tertiary structure of the protein places additional constraints on the β-sheet, requiring some of the water molecules to remain between the strands to satisfy the geometric and hydrogen bonding of the backbone C=O and N−H groups and to optimize the overall interactions of the β-sheet with the rest of the protein during the folding process.

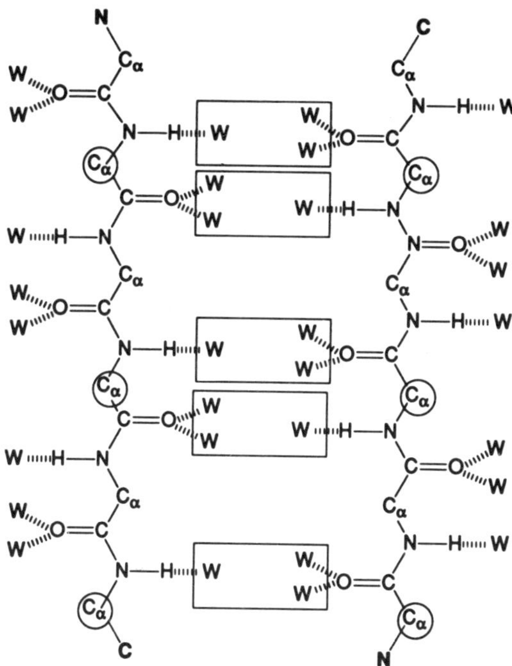

Figure 5.8 Schematic representation of extended polypeptide chains with fully hydrated backbone C=O and N–H groups. In the extended conformation, the alternating hydrophobic residues that are on the same side of the chain aggregate to form a β-sheet. In this process, the formation of each of the β-sheet hydrogen bonds eliminates 3 water molecules

5 Conclusions

Our analysis suggests that the secondary structures are formed from the hydrated unfolded polypeptide chain (Figure 5.9a) by the elimination of water molecules and the formation of C=O· · ·H–N hydrogen bonds (Figure 5.9b) involving either local peptide segments or distal segments. The energetic difference between the folded and unfolded states of the peptide segments slightly favours the folded state from consideration of the sum total of enthalpic and entropic terms of the hydrogen-bonded segments and the eliminated water molecules.[27–31] Protein structures contain numerous examples of the folded and random states of the peptide segment (Figure 5.9a, b) and closely related intermediate states (Figure 5.9c–e). We have found that these states provide useful information on the mechanism and pathway of the folding of the peptide chain in water. It is known that the amino acid sequence determines the folding of a protein[32] and water molecules are needed for this process.[33] Our analysis suggests how water molecules complement the side-chains in bringing the donor and acceptor atoms of the backbone into hydrogen bonding. For instance, in a segment

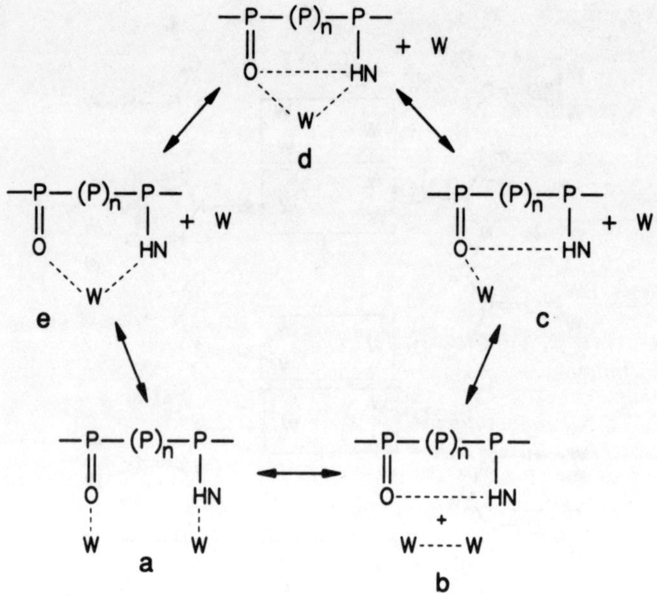

Figure 5.9 Schematic representation of various hydrated states (a–e) of the backbone carbonyl and amide groups of a polypeptide chain. P denotes peptide; n number of units; W water molecule; O carbonyl oxygen atom; and NH the amide group. (*a*) Polypeptide chain in random conformation with all its backbone polar atoms hydrogen bonded to water molecules. (*b*) The backbone NH and CO groups form hydrogen bonds within the chain and the water molecules are eliminated. These two modes of backbone hydration are ubiquitous in protein structures (after Reference 30). (*c–e*) Representations of the three intermediate states of backbone hydration of the secondary structures in proteins, which are referred to as externally hydrated, three-centered and internally hydrated segments

with apolar or polar side-chains at i, $i \pm 3/4$, where the side-chains tend to aggregate to form the well-known hydrophobic triplets[34] or ion-pairs[35, 36] or both, the backbone CO_i and NH_{i+4} groups are brought together through the water bridges, resulting in the formation of the α-helix. Similarly, the β-sheet is formed when extended chains with alternating apolar residues are brought together through water bridges linking the CO and NH groups of opposite strands, while the side-chains of the two strands aggregate. Thus, while the amino acid sequence contains the information for the folding of the protein into secondary and tertiary structures, the above cooperative interactions involving water molecules with the backbone together with the aggregation of side-chains aid in the formation of the final folded state. It has not escaped our mind that a similar mechanism would be involved in the formation of the nucleic acid duplexes from their individual strands. Our analysis of hydration of protein secondary structures has brought out the role of the aqueous medium in the protein folding process.

Acknowledgements

We wish to thank the College of Mathematical and Physical Sciences and the Ohio State University for supporting this work through an endowment to M.S.

References

1. Matthews, B. W. (1968). Solvent content of protein crystals, *J. Mol. Biol.*, **33**, 491
2. Savage, H. F. J. and Wlodawer, A. (1986). Determination of water structure around biomolecules using x-ray and neutron diffraction methods, *Methods Enzymol.*, **127**, 162
3. Baker, E. N. and Hubbard, R. E. (1984). Hydrogen bonding in globular proteins, *Prog. Biophys. Mol. Biol.*, **44**, 97
4. Blake, C. C. F., Pulford, W. C. A. and Artymuik, P. J. (1963). X-ray studies of water in crystals of lysozyme, *J. Mol. Biol.*, **167**, 693
5. Watenpaugh, K. D., Marglus, T. N., Sieker, L. C. and Jensen, L. H. (1978). Water structure in a protein crystal: rubredoxin at 1.2Å resolution, *J. Mol. Biol.*, **122**, 175
6. Karplus, P. A. and Schulz, G. E. (1987). Refined structure of glutathione reductase at 1.54 Angstroms resolution, *J. Mol. Biol.*, **195**, 701–729
7. Teeter, M. M. (1984). Water structure of a hydrophobic protein at atomic resolution: pentagon rings of water molecules in crystals of crambin, *Proc. Natl Acad. Sci. USA*, **81**, 6014–6018
8. James, M. N. G., Sielecki, A. R., Brayer, G. D., Delbare, L. T. J. and Bauer, C. A. (1980). Structures of product and inhibitor complexes of *Streptomyces griseus* protease A at 1.8 Å resolution: a model for serine protease catalysis, *J. Mol. Biol.*, **144**, 43
9. Thanki, N., Thornton J. M. and Goodfellow, J. M. (1988). Distribution of water around amino acids in proteins, *J. Mol. Biol.*, **202**, 637
10. Blundell, T. L., Barlow, D. J., Borkakoti, N. and Thornton, J. M. (1983). Solvent-induced distortions and the curvature of α-helices, *Nature*, **306**, 281
11. Deisenhofer, J. and Steigman, W. (1975). Crystallographic refinement of the structure of the bovine pancreatic inhibitor at 1.5Å resolution, *Acta Cryst.*, **B31**, 238
12. Takano, J. and Dickerson, R. E. (1981). Conformation change of Cytochrome c. I. ferrocytochrome c structure refined at 1.5 Angstroms resolution, *J. Mol. Biol.*, **153**, 79
13. Satyshur, K. A., Rao, S. T., Pyzalska, D., Drendel, W., Greaser, M. and Sundaralingam, M. (1988). Refined structure of chicken skeletal muscle troponin C in the two-calcium state at 2.0Å resolution, *J. Biol. Chem.*, **263**, 1628
14. Sundaralingam, M. and Sekharudu, Y. C. (1989). Water-inserted α-helical segments implicate reverse turns as folding intermediates, *Science*, **244**, 1333–1337
15. Sundaralingam, M. and Sekharudu, Y. C. (1990). Mechanism and pathway of alpha-helix folding and unfolding mediated by water. In Sarma, R. H. and Sarma, M. H. (Eds), *DNA Protein Complexes and Proteins*, Vol. 2. Adenine Press, Schenectady, N.Y., p. 115

16. Sekharudu, Y. C. and Sundaralingam, M. (1989). *ACA Annual Meeting — Abstracts*. Seattle, Washington
17. Karle, I. L., Flippen-Anderson, J., Uma, K. and Balaram, P. (1988). Aqueous channels within apolar peptide aggregates: solvated helix of the α-aminoisobutyric acid (Aib)-containing peptide Boc-(Aib-Ala-Leu)$_3$-Aib-OMe.2H_2O.CH_3OH in crystals, *Proc. Natl Acad. Sci. USA*, **85**, 299
18. Jeffrey, G. A. and Mitra, J. (1984). Three-center (bifurcated) hydrogen bonding in the crystal structures of amino acids, *J. Am. Chem. Soc.*, **106**, 5546–5553
19. DiCapua, F. M., Swaminathan, S. and Beveridge, D. L. (1991). Theoretical evidence for water insertion in α-helix bending: molecular dynamics of Gly_{30} and Ala_{30} *in vacuo* and in solution, *J. Am. Chem. Soc.*, **113**, 6145–6155
20. DiCapua, F. M. and Beveridge, D. L. (private communication)
21. Banner, D. W., Kokkinidis, M. and Tsernoglou, D. (1987). Structure of the Co1E1 Rop protein at 1.7 Å resolution, *J. Mol. Biol.*, **196**, 657
22. Dyson, H. J., Rance, M., Houghten, R. A., Wright, P. E. and Lerner, R. A. (1988). Folding of immunogenic peptide fragments in water solution. I. Sequence requirements for the formation of a reverse turn, *J. Mol. Biol.*, **201**, 217
23. Dyson, H. J., Rance, M., Houghten, R. A., Wright, P. E. and Lerner, R. A. (1988). Folding of immunogenic peptide fragments in water solution. II. The nascent helix, *J. Mol. Biol.*, **201**, 201
24. Satyshur, K. A. and Sundaralingam, M. (unpublished results)
25. Schirmer, T. and Evans, P. R. (1990). Structural basis of the allosteric behavior of phosphofructokinase, *Nature*, **343**, 140
26. Hill, C., Yee, J., Selsted, M. E. and Eisenberg, D. (1991). Crystal structure of the antimicrobial peptide defensin HNP-3: an amphiphilic dimer, *Science*, **251**, 1481–1485
27. Anfinsen, C. B. and Scheraga, H. A. (1975). Experimental and theoretical aspects of protein folding, *Adv. Protein Chem.*, **29**, 205
28. Ptitsyn, O. B. (1972). Thermodynamic parameters of helix-coil transitions in polypeptide chains, *Pure Appl. Chem.*, **31**, 227
29. Snell, C. R. and Fasman, G. D. (1973). Kinetics and thermodynamics of the α↔β transconformation of poly(L-lysine) and L-leucine copolymers: a connection phenomenon, *Biochemistry*, **12**, 1017–1025
30. Brown, J. E. and Klee, W. A. (1971). Helix–coil transition of the isolated terminus of ribonuclease, *Biochemistry*, **10**, 470–476
31. Ptitsyn, O. B. and Finkelstein, A. V. (1980). Why do globular proteins fit the limited set of folding patterns?, *Q. Rev. Biophys.*, **13**, 339.; Finkelstein, A. V. and Ptitsyn, O. B. (1987). Why do globular proteins fit the limited set of folding patterns?, *Prog. Biopys. Molec. Biol.*, **50**, 171.
32. Anfinsen, C. B. (1973). Principles that govern the folding of protein chains, *Science*, **181**, 223 and references therein
33. Rupley, J. A., Gratton, E. and Careri, G. (1983). Water and globular proteins, *Trends Biochem. Sci.*, **8**, 18–22 and references therein
34. Palau, J. and Puigdomenech, P. (1974). The structural code for proteins: zonal distribution of amino acid residues and stabilization of helices by hydrophobic triplets, *J. Mol. Biol.*, **88**, 457
35. Sundaralingam, M., Sekharudu, Y. C., Yahindra, N. and Ravichandran, V. (1987). Ion-pairs in alpha helices, *Proteins*, **2**, 64
36. Maxfield, F. R. and Scheraga, H. A. (1975). The effect of neighboring charges on the helix forming ability of charged amino acids in proteins, *Macromolecules*, **8**, 491

Part 3
Nucleic Acids

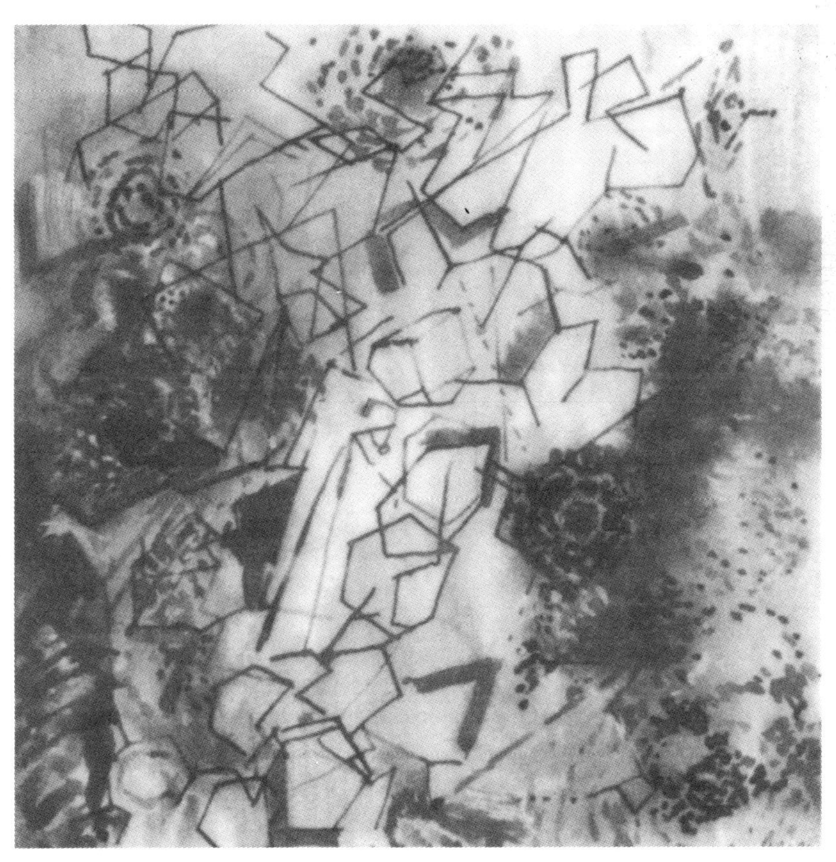

6

Molecular Dynamics Simulations on the Hydration, Structure and Motions of DNA Oligomers

D. L. Beveridge, S. Swaminathan, G. Ravishanker, J. M. Withka, J. Srinivasan, C. Prevost, S. Louise-May, D. R. Langley, F. M. DiCapua and P. H. Bolton

1 Introduction

Nearly 40 years ago, Franklin and Gosling (1953a, b) observed that the diffraction pattern of DNA fibres was sensitive to the relative humidity of the sample. Two forms of DNA were identified, one preferred at lower humidity designated 'A', and the other preferred at high humidity, designated 'B'. Examples of the A- and B-forms of DNA have now been studied extensively, first by fibre diffraction (Arnott *et al.*, 1976) and later by single-crystal X-ray crystallography (Dickerson, 1991). Both are right-handed forms of a DNA double helix, with the B-form corresponding closely to the double helix proposed by Watson and Crick (1953). Another form of DNA was discovered more recently, the left-handed or Z-form (Wang *et al.*, 1979). The A-, B- and Z-forms of DNA, shown in Figure 6.1, now establish the main basis for classification of DNA structures into families (Saenger, 1983). However, crystallographic variations have led to a number of subcategories (Fuller and Mahendrasingam, 1987), and, within each family, the DNA is expected to manifest a certain dynamical range of motions as well as exhibiting sequence-dependent fine structure (Dickerson, 1988; Kennard, 1984; Kennard and Hunter, 1989; Shakked and Rabinovich, 1986) and axis bending (Sundaralingam and Sekharudu, 1988).

While the sensitivity of DNA structure to hydration is now well appreciated, a full explanation of hydration effects on DNA at the molecular level

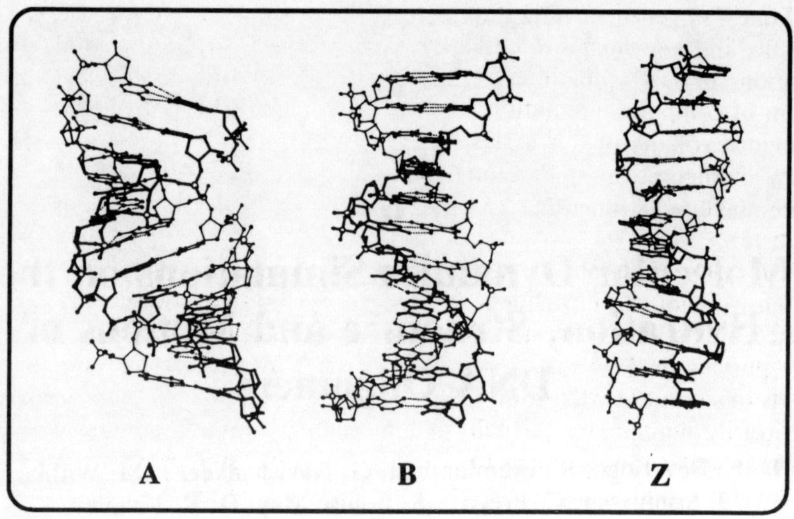

Figure 6.1 Canonical A-, B- and Z-forms of DNA

has not yet been accomplished. The scientific route to this is clear, at least in principle: determine the salient forces in the various forms contributing to the free energy of stabilization. However, this task is difficult to accomplish, since experimental measurements of stabilizing energies and forces due to various components or substructures of the system are not directly feasible. Lacking this information, a number of experimental observations, leading indications and hypotheses have been set forth. Experimental data on the positions assigned to ordered water molecules in the crystals of oligonucleotides figure significantly in this effort (Berman, 1991; Westhof, 1988; Westhof and Beveridge, 1989). However, the ordered waters still comprise only a small fraction of the total hydration, and thus the connection between hydration in a local region of the DNA and the overall thermodynamic stability of a sequence remains necessarily tenuous.

A fundamental approach to 'explaining' stability is to develop a theoretical model for the system which accurately accounts for the relevant observed data, and then examine the stabilizing forces in the model. A general formalism for this type of approach applicable to DNA structure and stability has emerged over the last 15 years from the realm of liquid state molecular physics (Hansen and McDonald, 1976), and involves theoretical calculations based on statistical mechanics and carried out numerically by computer simulation (Abraham, 1986; Allen and Tildesley, 1987). Results on water and nucleic acids based on the probabilistic form of simulation, the Monte Carlo (MC) method, are reviewed elsewhere in this volume (Vovelle and Goodfellow, 1993). In this chapter we review the results and knowledge obtained thus far on the hydration, structure and motions of DNA oligomers and related systems from the deterministic form of computer simulation, molecular dynamics (MD). An article in a previous volume of this series (Olson, 1982c) provided a comprehensive view of theoretical studies on DNA structure about 1982, emphasizing potential energy functions, energy minimization (EM) and configurational statistics.

The MD results on DNA available at this moment in time are not yet extensive enough to provide a critical, comprehensive view of the subject; hence, this review is by necessity fragmentary in places. Also, restricting our purview to MD rather than the whole field of molecular simulation leads to situations where the theoretical and computational perspective is arbitrarily limited; we partially compensate for this with some selected comparisons and references to corresponding MC studies.

Several essential questions emerge as a focus in this chapter. Two of these are:

(1) to what extent can MD simulation accurately describe the water structure around DNA?

(2) what have we learned so far about DNA hydration based on MD simulation?

The framing of additional important questions requires a momentary shift of focus away from hydration to the structure and dynamics of the DNA *per se*. Since DNA structure is experimentally known to be sensitive to hydration, it follows that treating hydration properly in an MD simulation is essential for obtaining an accurate description of the dynamical structure of DNA. However, the first generation of MD simulations on DNA (just as with proteins) were performed on *in vacuo* models, neglecting explicit consideration of water and introducing the effect of hydration via a dielectric screening term in the electrostatic component of the MD force field. This approach arose due to the simple necessity of keeping the dimensionality of MD simulations as low as possible in order to make the computations tractable, conserving computer time and core storage. Obviously, the physics of the *in vacuo* model with dielectric screening is clearly lacking in many ways, including particularly the neglect of specific water–DNA hydrogen bonding involving polar and ionic groups. Recently, with the advent of supercomputers, MD simulations on DNA including water and counterions explicitly have begun to appear. A comparison of recent MD results on fully solvated models of DNA with crystal structure data, NMR results and the results of *in vacuo* simulations now make it possible to approach the question:

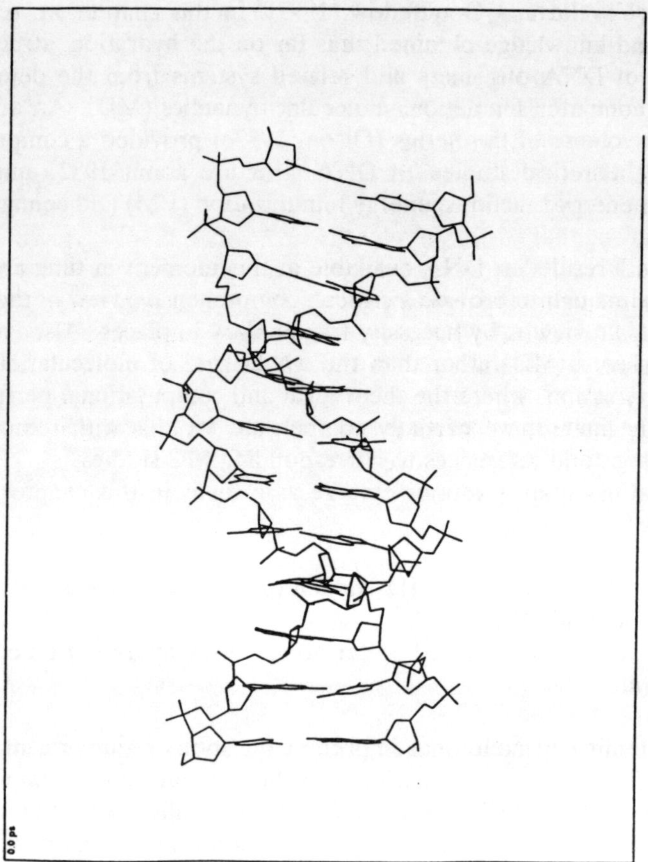

Figure 6.2 Crystal structure of d(CGCGAATTCGCG) (Drew *et al.*, 1981; Wing *et al.*, 1980)

(3) what theoretical treatment of hydration is necessary for the development of an accurate dynamical model of DNA from MD simulation?

The most detailed experimental determination of DNA structure at present comes from X-ray crystallography of oligonucleotides (Dickerson, 1991). Particularly, the dodecamer duplex d(CGCGAATTCGCG), solved by Dickerson and co-workers (Drew and Dickerson, 1981; Drew *et al.*, 1981, 1982; Fratini *et al.*, 1982; Wing *et al.*, 1980) has served thus far as a major testing ground for simulation. This sequence constitutes a full turn of helix, and was the first oligonucleotide example of B DNA to be studied crystallographically at high resolution (Figure 6.2). The crystal structure of d(CGCGAATTCGCG) shows some notable deviations from the structure of canonical B-DNA obtained from fibre diffraction (Arnott *et al.*, 1980; Arnott and Hukins, 1972). An overall bending of the helix axis by some 19°

is observed, correlated with deviations in base-pair roll. Sequence-dependent fine structure of various types is found, including extensive propeller twist, and there is also a pronounced narrowing of the minor groove in the AATT tract. The irregularities detected in the single-crystal structures compared with the regular canonical forms have raised the suggestion that local systematic deformations in the structure could serve as recognition signals for regulatory proteins and drugs (Dickerson, 1983).

A problem that arises in comparing crystallographic and simulation results on DNA sequences is our lack of knowledge as to which effects in the crystal structure are due to packing interactions and which are intrinsic to the DNA. Accurate MD calculations should only reproduce the latter, unless, of course, the calculations are performed on the unit cell. In the $P2_12_12_1$ crystals of the native dodecamer, the DNA helices overlap, with the GC regions from different molecules making crystallographic contacts. A widening of the minor groove is also observed in this region. This could be mechanically responsible for the narrowing of the minor groove observed in the AT region, making this a packing effect rather than a sequence effect. The nature of the problem makes it hard to disentangle intrinsic from packing effects so far (Dickerson, 1990) and thus the crystallographic features we should expect MD simulations on individual molecules to reproduce are not completely clear. MD simulations of nucleic acid crystals reported to date (see below) have been carried out so far only at the dinucleotide level.

The structure of DNA in solution is much less well established than in the crystal, and which details of a crystal structure should carry over to the solution structure is also not predictable. Nuclear magnetic resonance (NMR) provides data related to interproton separations obtained from measurements of nuclear Overhauser effects (NOEs), and on selected torsion angles from scalar coupling analysis. A number of recent reviews deal extensively with these topics (Patel and Shapiro, 1987; Van de Ven and Hilbers, 1988; Wemmer, 1991). Several methodological limitations have been identified. The proton networks in DNA are relatively dense compared with proteins, resulting in more opportunities for spin diffusion and complicating the determination of interproton distances. The isolated spin-pair approximation has thus proved not to be a good approximation in treating DNA NOE data, and matrix relaxation methods have become necessary (Boelens *et al.*, 1989; Borgias and James, 1988; Yip and Case, 1989). Even with accurate distance information, the range of distances obtainable is relatively short, <4.5 Å, and only limited information on sequence-dependent helix fine structure for DNA in solution can be obtained (Metzler *et al.*, 1990). Dynamical effects such as internal motion and the tumbling of the helix in solution are among other complicating features which are just beginning to be treated (Koning *et al.*, 1991; Withka *et al.*, 1991a, b).

NMR studies of d(CGCGAATTCGCG) in solution led to the assignment of all proton resonances (Hare *et al.*, 1983; Patel *et al.*, 1982; Wemmer and Reid, 1985). NMR studies generally indicate that Watson–Crick base pairing is well maintained in solution, with hydrogen exchange occurring only infrequently with respect to the MD time-scale (Moe and Russu, 1990). Ott and Eckstein (1985) assigned the individual ^{31}P NMR resonances of the dodecamer. Anomalous chemical shifts at steps 3 and 9 were observed and interpreted as an indication of a break in conformation at these locations, which correspond to the roll points in the crystal structure.

Recently Nerdal *et al.* (1989) have proposed a solution structure for d(CGCGAATTCGCG) at room temperature based on refinement of NMR-derived distance geometry structure and NOESY spectrum back-calculations on 155 interproton distances. The proposed solution structure appears to be somewhat different from those observed in the crystal. The position of the NMR resonances indicates palindromic symmetry for the dodecamer in solution, whereas the symmetry of the crystal form is not as high. Evidence for kinks in the structure of the C3pG4 (and the symmetry equivalent position C9pG10) and A6pT7 junctions is presented. The former involves an opening into the minor groove caused by base-pair roll, and the latter involves slide and some opening into the major groove. The minor groove is narrowed in the AT region and broadened in the peripheral GC tracts. The proposed structure is also highly underwound with respect to either canonical A- and B-forms of DNA, and markedly so at steps C3pG4 and C9pG10.

Lane *et al.* (1991), in the course of NMR studies on the interaction of berenil with d(CGCGAATTCGCG), reconsidered the structure of the uncomplexed dodecamer in solution. They concluded, in contrast with the results of Nerdal *et al.*, that the structure is much closer to that of a conventional B-form, although the presence of irregularities at the crystallographic roll points is supported. A detailed scalar coupling analysis was carried out and indicates that the sugars are predominantly in the C_2' *endo* form, with a small but significant admixture of other pucker states, particularly at the points of axis deformation observed in the crystal. The helix parameters consistent with the NMR data were determined by extensive parameter space searching. The average twist angle was determined to be $36.5 \pm 5°$, with no particular evidence for serious underwinding. Base-pair rise was within the normal B-form range. Neither propeller twist nor bending characteristics could be accurately determined from the NMR data.

The purview of optical spectra is also limited for the most part to local aspects of the structure. Flow linear dichroism experiments (Dougherty *et al.*, 1983) provide leading evidence that the base pairs are tilted with respect to the helical axis for B-DNA in solution, as opposed to perpendicular, as in the canonical forms. Raman spectra for numerous sequences have been reported (Brahms *et al.*, 1991; Wang *et al.*, 1987). The optical activity of the vibrational exciton transition involving coupled carbonyl groups of DNA bases also provides a possible indication of structural details in solution (Gulotta *et al.*, 1989). These experiments, while differentiating A-, B- and Z-DNA, also do not so far provide enough information to specify DNA fine structure and bending in solution.

In view of the limitations in approaching DNA structure in solution via experimental measurements, there is an obvious need for reliable theoretical methods for *de novo* structure prediction and for use in conjunction with experimental data to develop structural models and carry out structure refinements. MD simulation has recently been used in both of these areas, and the intrinsic accuracy of the MD is a major point of concern. Thus, a further question we address in this chapter is:

(4) How accurate are MD simulations on DNA *per se*, and what can MD simulation contribute to determining and understanding the structure of DNA in aqueous solutions?

Finally, we consider

(5) What applications of MD simulation have been made thus far to DNA oligomers and related systems?

2 MD Simulation

MD simulation is a computer experiment in which the atoms of a postulated system execute Newtonian dynamics on an assumed potential energy surface. The model of the system chosen for study, the assumed energy surface and the simulation protocol are all operational variables in the calculation. The simulation begins with an initial configuration, typically a crystal structure of the macromolecule and an arbitrary arrangement of solvent. Canonical forms and structures obtained from homology mapping are also used as the macromolecular starting point. The starting configuration of the system is first subjected to EM. The initial stage of the MD is a brief period in which the velocities on the particles are increased to the temperature of interest (heating). Then the simulation proceeds to seek out a thermally bounded state (equilibration) and samples it (production). Thus, the simulation locates and then characterizes a thermally bound state in the vicinity of the assumed initial structure. The MD procedures specific for biological macromolecules are described in more detail in several monographs (Karplus and McCammon, 1986; McCammon and Harvey, 1986) and a recent review article emphasizing methodology (van Gunsteren and Berendsen, 1990). Recent developments in the field have extended the purview of simulation into free energy determinations (Beveridge and DiCapua, 1989a, b; McCammon, 1991; van Gunsteren, 1988). Methodological aspects of MD particularly relevant to DNA have also been reviewed recently (Beveridge *et al.*, 1991).

The analysis of MD results involves following the manifold structural changes that occur in the molecule as a function of time. The procedure 'Curves' developed by Lavery and Sklenar (1988) for helicoidal analysis has been used to develop a computer graphics utility called 'Dials and Windows' (Ravishanker *et al.*, 1989), which monitors and displays the time evolution of all of the hundreds of conformational and helicoidal parameters in a DNA oligonucleotide. This allows one to monitor conformational transitions and to report accurately the conformational and helicoidal parameters of the dynamical model over the entire time-course of the simulation. Thus, the stability of the simulation and the nature of the dynamical model can be analysed rigorously.

Examining a succession of MD structural 'snapshots' from the simulation is also informative. The ensemble of MD structures represents the statistical state of the system, and can be presented in panels or else superimposed to convey an idea of the dynamic range of structure covered by the simulation. Plots of root mean square (RMS) deviation against time for the MD structures compared with the initial structure or reference canonical forms convey additional information about stability of the dynamical structure and the location of the thermally bounded state in configuration space. Superimposition of the helical axes from 'Curves' for the various MD snapshots provides a leading idea about axis bending. Most recently, a definition of a persistence-length index based on the statistical theory of chain molecules has been developed for the quantitative characterization of curvature in MD simulations on oligonucleotides (Prevost *et al.*, 1991). The aim is to ascertain quantitatively the degree of linearity, the lateral direction of bending and the flexibility or the lack thereof in interesting local regions of a sequence, such as A-tracts.

The description of the molecular force field is the principal assumption in MD simulations. The MD force fields in AMBER (Weiner *et al.*, 1984), CHARMM (Nilsson and Karplus, 1984) and GROMOS (van Gunsteren and Berendsen, 1986) have each a full set of parameters for nucleic acids, developed on the basis of extensions of polypeptide and protein force fields and supplemented with additional parameters developed from test calculations and some consideration of experimental data on prototype cases. Application of these parameter sets to the simulation of macromolecular duplex DNA has led to a series of MD simulations from various laboratories, particularly on the dodecamer duplex d(CGCGAATTCGCG), for which considerable experimental data are available. The comparison of results from various force fields with each other and with experiment leads to our current state of knowledge of the accuracy of MD simulation on

DNA. Before describing these results, some comments on the truncation of potentials are appropriate.

The truncation of potentials is an issue of particular importance in nucleic acid systems. Truncation is applied in order to avoid extensive computation of forces between particles separated by large distances. A switching function is applied around the cutoff to feather the potentials smoothly off to zero, so that the calculation of the forces remains well conditioned. The truncation error depends on the extent to which the various kinds of interactions fall off with distance. Dispersion energies diminish with distance dependence of r^{-6}, dipole–dipole interactions as r^{-3}, and simple Coulombic electrostatic effects as r^{-1}. The longer-ranged Coulomb forces may still be relatively large at the cutoff, and are thus most affected by truncation effects. As a consequence, it is advisable to implement the cutoff on the basis of electrostatically neutral groups in the molecule (van Gunsteren and Berendsen, 1990), in order that local ionic artefacts with long-range implications are not inadvertently introduced.

The polyanionic character of the DNA duplex and the presence of mobile counterions in the environment raises considerably more concern about truncation effects in MD on nucleic acids than on proteins. One recent approach to this utilizes Ewald sums (Forester and McDonald, 1991), an approach which may obviate the problem. We have been experimenting with a group-by-group approach involving the phosphates and mobile counterions, but attention must also be given to the possibilities for counterion motion, and the extensions of this procedure to divalent counterions becomes problematic. Some methodological aspects of charge grouping in DNA simulations with promising results have recently been described (Beglov and Lipanov, 1991).

3 MD Studies on Prototype systems

The small-crystal hydrates of nucleic acid constituents such as sugars, phosphates, nucleotide bases, nucleosides and nucleotides in many cases have well-defined hydration and can serve as a testing ground for the performance of molecular simulations. Most of the simulation studies on this class of problems to date have been obtained via MC simulation (Vovelle and Goodfellow, 1993). The results of these studies are nevertheless relevant to MD, since the energy functions used in MC are the same as or similar to the potentials used for the calculation of forces in MD. Thus, the level of agreement obtained in the MC studies between calculated and observed quantities serves to validate the functions for use in MD as well.

Nucleotide Bases

The interstrand hydrogen bonding and base-pair stacking interactions are two of the main factors stabilizing the DNA double helix, and have been the subject of a considerable number of experimental investigations and theoretical studies over the years. Experimental evidence indicates that in non-polar solvents and in the gas phase, nucleotide bases associate via hydrogen bonding interactions. In water, stacking interactions predominate. MD studies on this subject have appeared only recently. Cieplak and Kollman (1988) reported a study of the net free energies of association for adenine–thymine and guanine–cytosine complexes based on *in vacuo* MD simulation and thermodynamic cycle perturbation theory. The results were generally in accord with available experimental data, with relatively large statistical errors placed on the calculated values. A subsequent study by Dang and Kollman (1990) used MD to compute potentials of mean force for the case of 9-methyladenine interacting with 1-methylthymine and to determine the preference for stacking relative to hydrogen bonding. The hydrogen bond association, ~12 kcal/mol in the gas phase, is reduced to ~1 kcal/mol in water, whereas the stacking interaction is correspondingly reduced from ~9 kcal/mol to ~2 kcal/mol. These calculations are based on the AMBER force field and the TIP3P model for water. Monte Carlo simulations of absolute free energies of binding for guanine–cytosine and adenine–uracil base pairs in chloroform have also been reported (Pranata and Jorgensen, 1991).

A theoretical study of tautomerism in the gas phase and aqueous solution for 2-oxopyridine, 2-oxopyrimidine and cytosine was carried out using *ab initio* quantum mechanics and free-energy-perturbation methods (Cieplak *et al.*, 1987). The tautomeric equilibria were shown to be very solvent-dependent, switching preferred prototropic states in going from gas phase to solution.

Sugars

The inherent flexibility of the sugar facilitates major structural changes in naturally occurring DNAs and has an important bearing on their chemical reactivity and biological function. Rapid interconversion of puckered states of sugars is observed in NMR studies of nucleosides and nucleotides, indicating that the pseudorotation barrier is low in at least some directions. Opinions on this, as expressed in computational force fields, have ranged from free rotation (Levitt and Warshel, 1978) to fairly rigid models (Nilsson and Karplus, 1984). The flexibility of the furanose ring was the subject of a detailed theoretical study by Olson and Sussman (Olson, 1982a, b), and indicated a low-energy pathway for repuckering from C_2'-*endo* to

C_3'-*endo* via an O_4'-*endo* route, S–E–N on the pseudorotation cycle. The opposite direction on the cycle is indicated to be sterically forbidden at ordinary temperatures. While not an MD paper *per se*, the perspective created in this work makes it still the definitive point of departure for considering the accuracy of the sugar potential in current MD force fields. NMR analysis of sugar conformation has been developed particularly by Altona and co-workers (Rinkel and Altona, 1987).

The cyclodextrins, a prototype for the sugar moieties of the DNA backbone, have been studied with a high degree of precision by neutron diffraction (Saenger, 1987), and the waters of hydration were well determined with respect to both position and orientation. Kohler *et al.* carried out MD calculations on the α-cyclodextrin hexahydrate crystal (Koehler *et al.*, 1987a) and the β-cyclodextrin dodecahydrate crystal (Koehler *et al.*, 1987b) using GROMOS. They obtained excellent agreement with experiment on the structure of the cyclodextrins in the crystalline state. The results of the calculations support the idea of flip-flop hydrogen bond dynamics (Koehler *et al.*, 1988b), and also provide evidence for occurrence of three-centre hydrogen bonds (Koehler *et al.*, 1988a, 1989). Corresponding simulations were carried out on the aqueous hydration of the cyclodextrins, and compared with the results on the crystals. In general, the structure and dynamics of the crystals and aqueous solution were found to be in close correspondence for this system.

Phosphates

The acidic character of the phosphate group confers polyanionic character to DNA, and, hence, a sensitivity of the structure and stability of the DNA to salt concentration. The phosphodiester torsion angles in nucleic acids are another source of conformational flexibility, also sensitive to solvent effects. The structure of hydrated Na^+ ions around a region of A- and B-DNA helix was the subject of an early MD study (Lee *et al.*, 1984). The association of a sodium ion with dimethylphosphate anion was studied by Huston and Rossky (1989), using MD simulation, in which a potential of mean force for the ion pair was determined by use of thermodynamic perturbation theory. The results are summarized in Figure 6.3. For approach along the O−P bond, both contact and solvent-separated minima were found, whereas approach along the bisector led to only a solvent-separated minimum. This latter result is in sharp contrast to the results based on a continuum model for water, where a contact form would be global minimum. Introducing discrete solvent into the model thus creates the possibility of another associated form, with a water intervening between the phosphate group and its associated counterion. The statistical weight of this structure is expected to be large. This result is consistent with

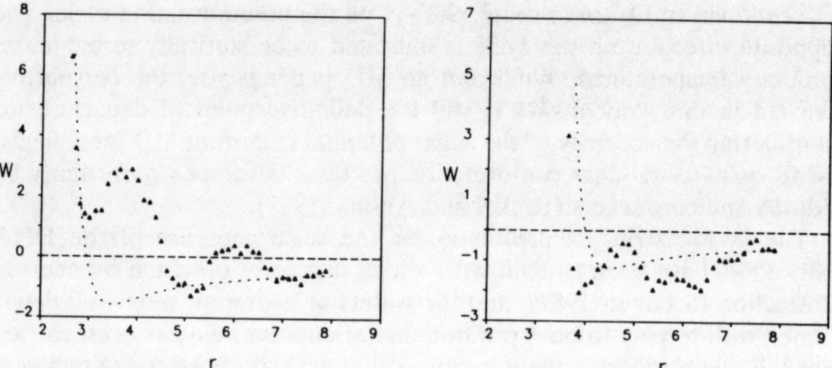

Figure 6.3 Calculated potential of mean force for the interaction of dimethylphosphate anion and Na^+: (left) approach along a P–O bond; (right) approach along the bisector of the O–P–O angle (Huston and Rossky, 1989)

the model of Manning (Manning, 1978), in which ions are 'condensed' on the DNA as a shroud, but within this region have considerable lateral mobility rather than forming ion pairs. The sensitivity of results to truncation of the potential was investigated, with spherical truncation found to introduce non-physical behaviour. The solvent-averaged potential computed with spherical truncation was found to be purely repulsive at large separations. Ion–ion interaction in this case should be attractive, with the extent of interaction attenuated by a factor of the dielectric constant. Results based on the minimum image convention behaved more normally.

Nucleosides and Nucleotides

Pearlman and Kollman (1991) recently reported MD studies in conjunction with thermodynamic perturbation techniques to study the conformational stability of nucleosides. Free energy conformational maps were generated and compared with corresponding adiabatic potential energy maps computed by standard minimization techniques. While qualitatively similar, the two types of maps display significant quantitative differences. The sensitivity of results to all-atom as opposed to united atom options and to explicit as opposed to implicit treatment of solvent indicates that significant changes can result from small, seemingly innocuous differences in simulation protocol. The parametrization of force fields on the basis of experimental data to date has been based for the most part on direct comparisons with potential energies and not free energies. This study indicates that considerable complications lie in the way of subsequent developments in simulation energy functions and force fields. On an optimistic note, the advent of free energy techniques and the explicit treatment of solvent in

MD simulation carries the computational approach to a higher level of sophistication.

The dinucleoside–proflavine (dCpG/Pf) crystal hydrate (Shieh *et al.*, 1980) has received considerable theoretical attention, since it features the most extensive experimentally defined hydration network of any nucleic acid system, and thus serves as a convenient critical test of potential functions and simulation methodology. The ordered water network in this crystal (Figure 6.4(top)), consists of ~100 molecules per unit cell organized into an elaborate system of edge-connected pentagons and a higher-order polygon disc (Neidle *et al.*, 1980). A recent review of water in crystals (Savage, 1986) questioned whether or not all these contacts really correspond to hydrogen-bonded molecules. Water in this crystal is essentially all in a fairly close (first solvation shell) relationship to one or other of the four dCpG/Pf adducts present as asymmetric units in each cell.

Following up earlier Monte Carlo studies (Mezei *et al.*, 1983), the dCpG proflavine crystal hydrate was studied via EM and MD simulation in considerably more detail by Swaminathan *et al.* (1990), using the GROMOS force field on a system of three crystallographic unit cells. The EM calculations resulted in a hydration structure recognizable in comparison with the crystallographic results, but all the 'contacts' did not turn out to be hydrogen-bonded. However, this result only applies to 0 K. The MD calculations at 300 K were subsequently analysed by means of a superposition of snapshots taken at various intervals along the time-line of the simulation to provide a dynamical, time-averaged view of the structure. The results (Figure 6.4(bottom)) indicate that the complete network as proposed from the crystallographic analysis is essentially intact at 300 K. The calculated internal geometry for dCpG/proflavine in the simulation also agreed well with experiment, maintaining the mixed sugar pucker observed in the crystal. The results obtained from GROMOS MD simulation on both cyclodextrins and dCpG/Pf generally indicate the force field to be performing reliably on these prototype cases.

MD simulations have recently carried out on rGpC and dCpG proflavine crystal hydrates, using the AMBER force field, together with time-correlational analysis of the trajectories (Herzyk *et al.*, 1991). The results show significant differences in mobilities of the sugar, phosphate and base groups for the uncomplexed dinucleoside, decreasing in that order. This contrasts with experimental indications that the phosphate group is most mobile, but the calculations are consistent with the behaviour expected for C3′-*endo* sugars of the A-family and observed in tRNA. Motions observed in the crystal were seen to be uniformly larger than those observed in the simulation, attributed to a combination of thermal effects in the crystals and force field problems in the simulation. The discrepancies with experiment are carefully detailed, and contribute to understanding of the study of mobility by MD.

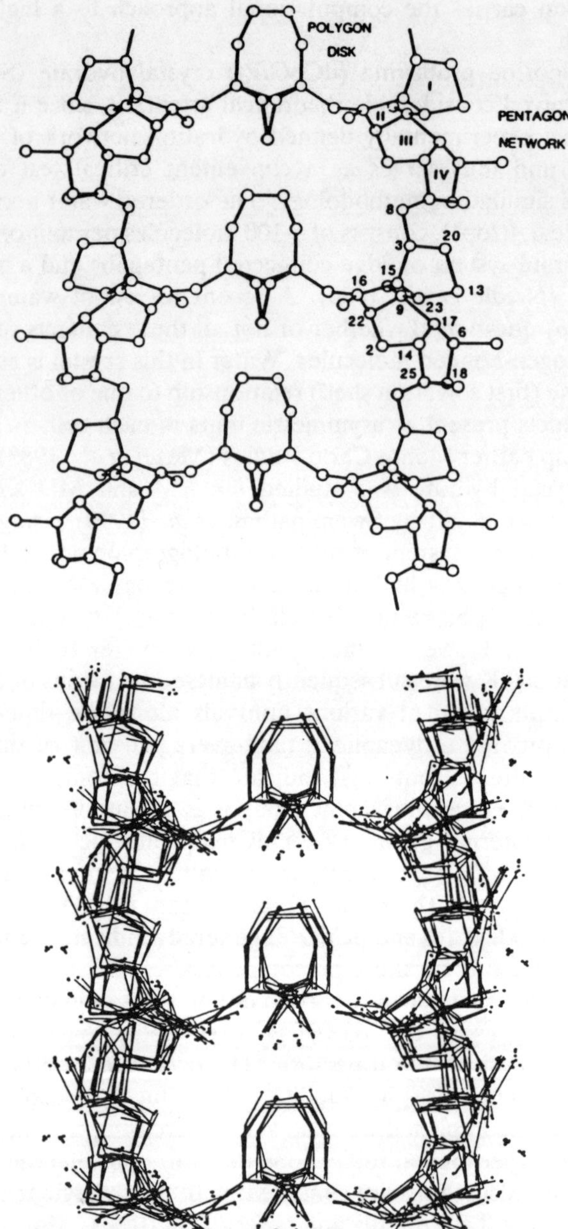

Figure 6.4 Water networks in the dCpG/proflavin crystal hydrate: results of the crystallographic determination (top) (Neidle *et al.*, 1980; Shieh *et al.*, 1980); results from the ensemble of structures obtained from MD simulation (bottom) (Swaminathan *et al.*, 1990)

MD Studies on DNA Hydration

There have recently been a number of MC simulation studies of the hydration of DNA. MC was used to study the dodecamer duplex d(CGCGAATTCGCG) together with 1777 water molecules in a hexagonal cubic cell under periodic boundary conditions (Subramanian and Beveridge, 1989; Subramanian *et al.*, 1988, 1990). The results are of interest in discussing MD simulations on a related system (see below). The calculated total first coordination number of 10.4 waters per nucleotide compared closely to the critical level of hydration necessary for DNA stability (Falk *et al.*, 1962, 1963a,b; Falk, 1970) and supported the idea that the experimental number of 10 waters per nucleotide corresponds to an intact first hydration shell. The addition of the second shell coordination of the minor and major groove gives a total of 17.4 waters/nucleotide, close to the value of 20 required in Falk's experiments for the stability of the B-form. Considering that slightly in excess of first and second hydration shells would be necessary to completely fill the grooves of a B-DNA structure, the calculated results correspond closely to the estimates from solvent accessibility calculations by Alden and Kim (1979).

Results on the calculated hydration density in the minor groove are shown in Figure 6.5(top). The localization of hydration density occurs at A–N3 and G–N2 sites and describes essentially monodentate hydrogen bonding. The crystallographic sites are essentially limited to the AT tract, and indicate the existence of A–N3· · ·W· · ·T–O2 bridges which comprise the first shell of the 'spine of hydration'. The inherently greater propeller twist in AT compared with GC base pairs orients the bases such that A–N3 and T–O2 are 3.7 Å apart, a propitious distance for bridging by a single water molecule. Comparison of the calculated results with the observed crystallographic ordered water positions revealed a close correspondence between the calculated and observed values (Subramanian *et al.*, 1990).

The MC simulation results predicted that the CG as well as the AT region can support an ordered water structure. The CG region, with the floor of the minor groove lined with N2 donor groups and cytosine O2 acceptor sites, appears in the simulation to nucleate localized hydration density as readily as the AT region. The calculated ordered water network in the minor groove in the theoretical model thus extends completely from one end of the structure to the other, with the interface between the CG and AT regions dealt with readily by the geometric versatility of water–water hydrogen bonding. The penetration of water into the DNA is, of course, clearly greater for AT than for GC tracts, and this feature distinguishes the 'spine of hydration', as defined by Dickerson and co-workers, from the generic minor groove ordered water network. The calculation also predicts that the spine of hydration branches into a double ribbon motif in the

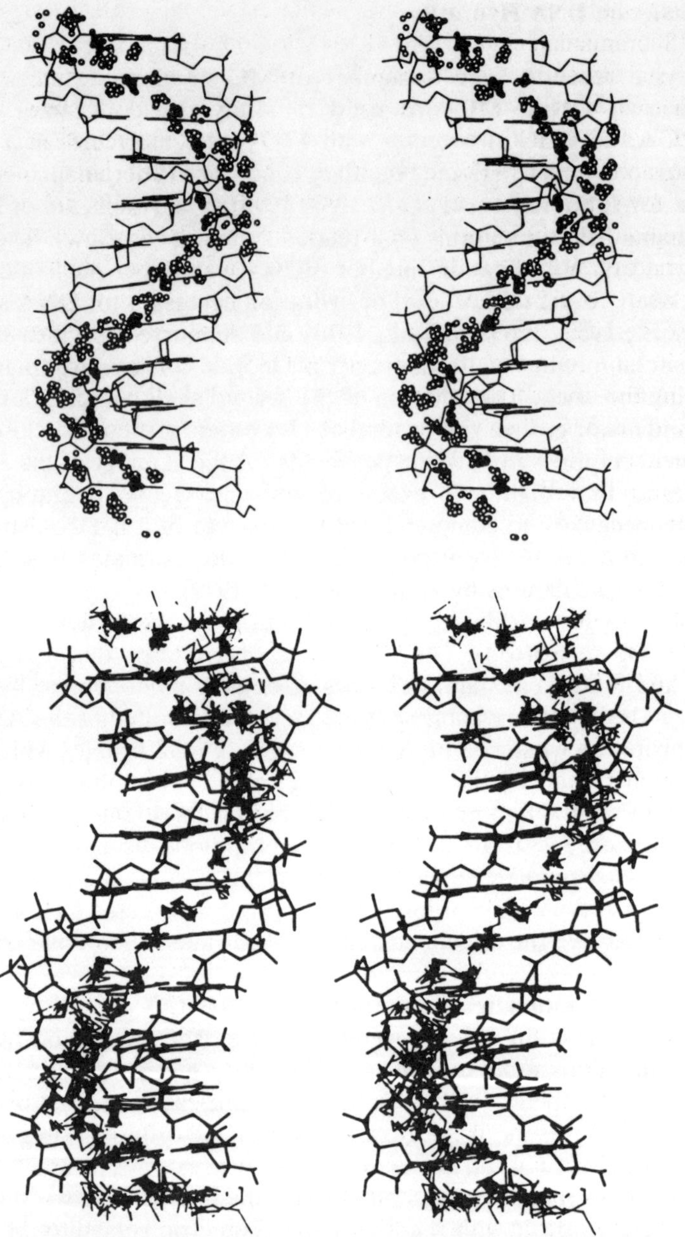

Figure 6.5 Calculated water networks in the minor groove of DNA oligomers (stereo): results from an MC simulation on d(CGCGAATTCGCG) (top) (Subramanian *et al.*, 1988); results from an MD simulation on d(CCAACGTTGG) (bottom) (Chuprina *et al.*, 1991)

GC tracts, where the minor groove in the crystal structure is wider than normal (Subramanian *et al.*, 1990). The hydration of this region in the dodecamer crystal structure is blocked by helix–helix packing.

For the major groove, the calculated saturation of hydrophilic binding sites with waters of hydration in the calculation corresponds closely to the crystallographic results. The 16 K structure and the MPD7 structure show evidence for some various types of water bridges as well as in the major groove region. At the phosphates, the calculated results show the characteristic triad of waters in the 'cone of hydration' motif (Pullman *et al.*, 1978) at nearly every anionic oxygen. The phosphate hydration is observed to be disordered in the room temperature structure, but extensive ordered solvent sites are found in the 16 K and MPD7 structures of the dodecamer. Examining the crystallographic results with the idea of cones of hydration in mind, the observed solvation sites can nearly all be interpreted in terms of fragments of this motif in which only one or two of the three waters has turned out to be ordered. Only rarely does one see a fully ordered triad of waters of a single cone in the crystal structure. The hydrophobic hydration around the thymine methyl group and the non-polar atoms of the furanose sugar ring is not highly ordered. The water 'wedged' between T−5CH_3 and a phosphate occurs in the simulation with a frequency of 51%.

The crystal structure of the B-DNA decamer of sequence d(CCAACGTTGG) (Prive *et al.*, 1991) indicates that the helices pack end-to-end rather than overlapping, as in the dodecamer, and the minor groove hydration was thus more fully revealed. The results indicate a spine of hydration in the central region, and a double ribbon motif for the hydration of the flanking sequence. A 40 ps MD trajectory for the hydration of d(CCAACGTTGG) was recently reported by Chuprina *et al.* (1991) based on the AMBER force field, in an article which focuses mainly on the calculated pattern of hydration in the minor groove. The results on the calculated hydration density are shown in Figure 6.5(bottom), and display the two motifs of hydration expected from the crystal structure. The water molecule between the central base pairs C*G and G*C fluctuated rapidly during the simulation, with hydrogen bonds occurring at the acceptors and donor sites of G6 and G16. Contrary to the results from the crystal structure, this water never bridges the N3 atoms of G6 and G16 in the length of trajectory performed. The major conclusion is the idea that the pattern of hydration in the minor groove may depend more on groove width than base sequence as originally surmised on the basis of only the ordered water positions in the dodecamer. Chuprina *et al.* suggest that the spine of hydration may be a particular property of overwound variants of B-DNA such as the C- and D-forms.

An MD simulation focused on the structure and dynamics of both water and counterions around the canonical B-form duplex of d(CGCGCGCG) was also recently reported (Forester and McDonald, 1991). The interactions of water molecules was described using the SPC potential, with the interactions of water, ions and DNA treated consistently. Five different simulations were carried out, one on fully charged, polyanionic DNA in pure water and four others on electrically neutral systems containing different combinations of Na^+, Ca^{2+} and Cl^- ions and water. Periodic boundary conditions were applied in all directions, making the results applicable to an infinite chain. The pair potentials were truncated at 9.0 Å, except for the Coulomb forms, which were treated by Ewald summation. The ions were introduced near the periphery of the simulation cell, and their diffusion during the course of the simulation was monitored.

The MD hydration was analysed using radial distribution functions, and the results indicate a first-shell coordination of 17.5 waters, 8–9 of which are strongly coordinated for the DNA via hydrogen bonds. Each of the anionic oxygens of the phosphate groups is separately solvated, with few occurrences of bridges. This is consistent with the 'cone of hydration' model for phosphate hydration proposed earlier based on quantum mechanical calculations (Pullman *et al.*, 1978) and seen in corresponding MC simulations (Subramanian and Beveridge, 1989). The dynamics of the various shells of hydration around the DNA was monitored via autocorrelation functions. The first-shell waters were found to have considerably hindered reorientation times and translational motion. The ions were found to diffuse somewhat into the region of the second hydration shell, leaving the number of first-shell waters essentially unaltered, but significantly affecting their orientational ordering. No direct site binding of the ions on the DNA was observed, but some solvent-separated associations were identified for divalent Ca ions.

An MD simulation of 70 ps on a dodecamer of poly(dG*dC) in the Z-conformation including water and counterions was reported by Laaksonen *et al.* (1989) and the surrounding solvent structure was subjected to a detailed analysis based on distribution functions and related dynamical indices of structure. Different techniques for treating the long-range electrostatic interactions were tested with respect to indices of stability in the simulation, with the Ewald summation technique found to give the best results. Cones of hydration were found for the phosphate oxygens, but little order beyond the first hydration shell was observed. Structured water at the sugar O4′ and a bridging water in the ribophosphate backbone were noted. Counterions were found to preferentially coordinate with the nucleotide bases, which are pushed to the surface of the Z-form helix. The rotational and translational dynamics of the solvent were studied, and translational diffusion of water was found to be retarded close to the helix. Counterion diffusion was found to be lowered to a third of that in corresponding aqueous solutions of ions.

Experimental measurements of the force between DNA molecules have been reported (Parsegian *et al.*, 1985; Rau *et al.*, 1984), and indicate the existence of significant repulsions, contrary to the predictions of double-layer theory. An explanation of this phenomenon has been advanced in terms of specific effects in the waters of hydration, a so-called 'hydration force' (Israelchavelli, 1985). A series of MD simulations were initiated to determine the range of influence of the DNA on the surrounding water (Reddy and Berkowitz, 1989). Two DNA helices, fixed in the canonical B-form, and placed at a particular interaxial separation, were solvated and MD was performed on the water only. The results indicated that when the DNAs are so far apart that the influence of bulk water between the surfaces is large, the polarization of water next to the DNA surface is low. As the helices approach each other, the polarization of water layers next to the surface is increased, effectively resulting in a net increase in repulsion between the DNA surfaces.

4 MD Studies on DNA *in vacuo*

The first MD simulations reported on a DNA helix were reported in 1983 (Levitt, 1983; Tidor *et al.*, 1983). Trajectories of 90 ps were performed by Levitt on *in vacuo* models of the duplexes of d(CGCGAATTCGCG) and a dA·dT 24-mer (Levitt, 1983). In these simulations, the electrostatic terms in the non-bonded interactions were omitted. The simulations demonstrated large-amplitude bending and twisting in the model DNA, affecting particularly the major groove. The propeller-like twisting in base pairs, later to be recognized as a common feature in oligomer structures, was quite evident in these simulations. Some estimates of electrostatic effects were made, but when uncompensated charges were added to the model at an internal dielectric of unity, electrostatic repulsions caused the duplex to unwind.

MD and normal mode analysis of the duplexes d(CGCGCG) and normal mode analysis for d(CGCGCG) and d(TATATA) were reported by Karplus and co-workers at about the same time (Tidor *et al.*, 1983), with the electrostatics treated by partially reducing the negative charge on each phosphate to −0.32 on the basis of a counterion screening model, and a distance-dependent dielectric constant used to mimic solvent shielding. The simulation produced a stable B-form with amplitudes of atomic fluctuation somewhat smaller than those reported in crystallography on the dodecamer, possibly indicative of some static disorder in the crystal. Some preliminary comparisons with results on NMR order parameters and fluctuations inferred from fluorescence depolarization were given. A series of MD simulations on t-RNA was also reported at about this same time (McCammon and Harvey, 1986; Prabhakaran *et al.*, 1983), also using a screened-charge model for the phosphates. The overall shape and exposed surface area remained near the X-ray value for some 32 ps of dynamics, giving average atomic fluctuations reasonably close to corresponding X-ray values.

MD on d(CGCGA) with alternative treatments of electrostatics was then carried out by Kollman and co-workers (Singh *et al.*, 1985), considering first a fully anionic form and then explicitly including 'hydrated' counterions, a species with a charge of +1 and a mass and size of the cation of hexahydrated sodium. The simulation was based on the AMBER force field with a non-bonded cutoff of 12 Å and a distance-dependent dielectric screening function of $\varepsilon = R$. The duration reported was 83 ps. The average structural parameters obtained for the anionic and hydrated counterion dynamics were essentially within a standard deviation of each other. A number of internal correlations between conformational variables were pointed out, and atomic motions were found to be of the order of 1 Å. The average MD values of twist and tilt angles were close to the average obtained from the dodecamer crystallography, although local differences were noted. The anionic model gave 9 bp/turn as opposed to 10 for the hydrated counterion model, the latter being closer to the canonical *B*-value of 10.6. Most hydrated counterions remained in the vicinity of the backbone, but one, in the last 10 ps, slipped into the minor groove region equidistant between phosphates.

Nilsson and Karplus (1984), in the course of studies on force field development for nucleic acids, report a series of MD simulations, 5–20 ps in length, of the hexamer duplex d(CGTACG) with different electrostatic options in a CHARMM and AMBER force field. The CHARMM calculation proved to be stable at 9.1 bp/turn, a little low, with $\varepsilon = R$. The helix became somewhat distorted when $\varepsilon = 2$ was used or when the van der Waals radii of extended carbons were increased by 0.3–0.5 Å. Also, the particular treatment of electrostatic truncation was demonstrated to have a significant influence on the overall behaviour of the system.

Westhof *et al.* (1986) have reported a temperature-dependent MD and restrained X-ray refinement simulations for a Z-DNA hexamer, d(CGCGCG), with each cytosine brominated in the 5-position. The MD calculations were based on the AMBER force field, using a distance-dependent dielectric and neglecting counterions. The MD analysis revealed evidence for sequence-dependent flexibility, with the sugar and phosphate groups internal to the GpC base-pair step more mobile than those for CpG. The molecule at 300 K exhibited a dynamical range of motion spanning the crystallographic Z-I and Z-II structures. The phosphate groups clearly displayed a higher degree of mobility than the bases and sugars, and a larger RMS difference from the molecular geometry in the corresponding crystal structure.

Stretches of adenine nucleotides in DNA have been implicated in unusual conformations and anomalous behaviour of DNA oligomers, including bending phenomena (Dickerson, 1988; Sundaralingam and Sekharudu,

1988). One line of thought holds that the particular conformational preferences of A-tracts is due to the high propeller twist at each A–T base pairing, resulting in the formation of cross-chain inter-residue hydrogen bonds in the major groove. Fritsch and Westhof (1991c) have carried out *in vacuo* MD simulations on poly (dA)* poly (dT) and examined the structure and dynamics of these interactions. They concluded on the basis of calculated correlation functions that the three-centre hydrogen bonds in the dynamical structure appear more as a consequence of the anomalous A-tract structural characteristics than as determining factors. A comparison of results based on simple distance-dependent and sigmoidal dielectric screening constants is reported in this and related studies (Fritsch and Westhof, 1991a, b), and the results indicate that the sigmoidal function, with a behaviour reflecting electrostriction effects, behaves more reasonably.

The structural features of the backbone of poly (dA)* poly (dT) and of the alternating poly d(Br5 U–A) and poly d(A–T) have been recently studied by Raman spectroscopy and MD simulation (Brahms *et al.*, 1991). Two characteristic Raman bands have been assigned to two different furanose conformations, C2′ *endo* and O4′ *endo*. The authors conclude from their data that alternating and non-alternating A- and T-containing DNA oligomers present a dynamical behaviour based on differently fluctuating sugar populations, and this suggests dynamical differences at the level of the A- and T-nucleotide backbone. This could imply particular functional properties for protein binding and recognition of the backbone in A- and T-nucleotides in DNA.

A GROMOS MD simulation was configured with a screened charge model for the phosphates (−0.25) and sodium-sized solvatons of equal and opposite charge for electrical neutrality, and used as a demonstration case for 'Dials and Windows' analysis of conformational and helicoidal parameters (Ravishanker *et al.*, 1989). This calculation was extended in parallel with simulations on solvated models using GROMOS, as discussed further in the following section.

The characteristics of 100 ps of molecular dynamics on d(CGCGAATTCGCG) at 300 K based on *in vacuo* AMBER were reported recently (Srinivasan *et al.*, 1990). Some snapshots from the simulation are shown in Figure 6.6. The simulation is based on an *in vacuo* model and the 3.0 force field of Weiner *et al.* (1984) configured for DNA in the manner of Singh *et al.* (1985), with counterions of hydrated radii and a distance-dependent screening function for water. The starting structure was a canonical B-form of DNA. The results indicate this dynamical model to be a provisionally stable double helix which lies at ~3.2 Å RMS deviation from the canonical B-form. There is, however, a persistent non-planarity in the base-pair orientations which resembles that observed in canonical A-DNA. The major groove width is seen to narrow during the

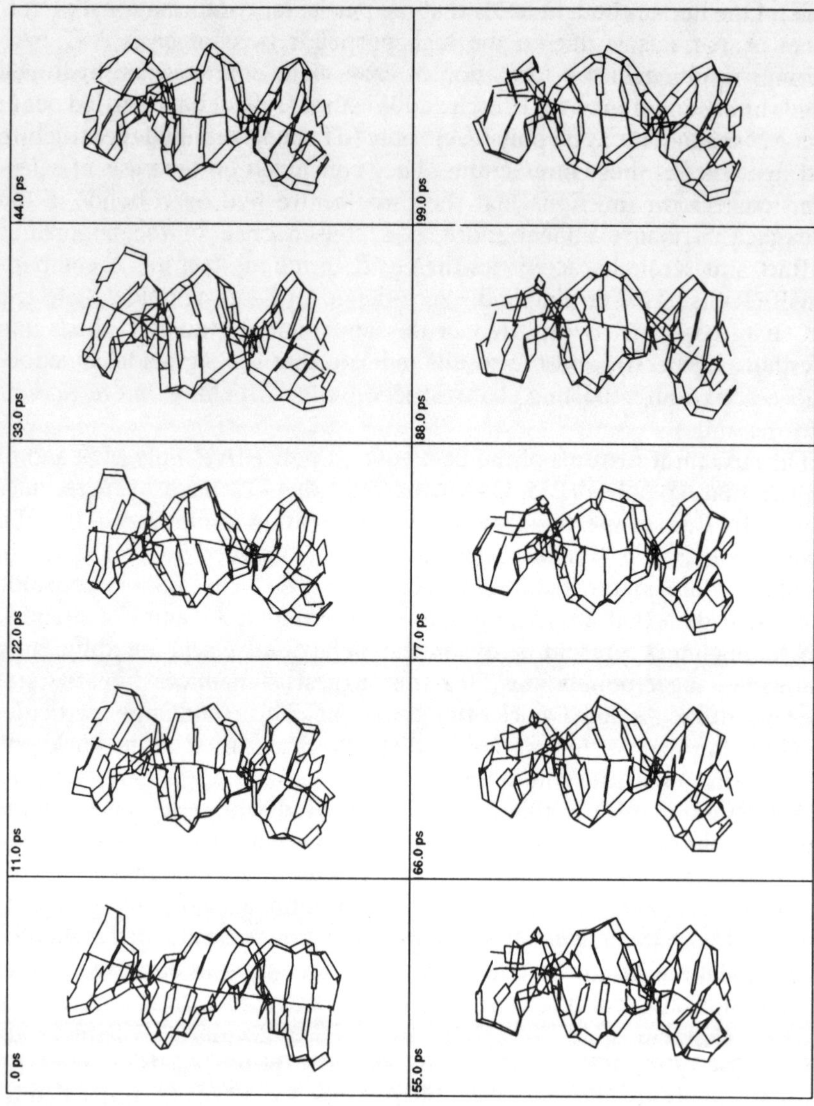

Figure 6.6 A series of snapshots from an MD simulation of 100 ps on d(CGCGAATTCGCG) based on an *in vacuo* model and the AMBER 3.0 force field (Srinivasan *et al.*, 1990). The initial structure (lower left-hand panel) was a canonical B-form of DNA

course of the simulation and the minor groove expands, contravariant to the alterations in groove width seen in the crystal structure of the native dodecamer. The propeller twist in the bases, the sequence dependence of the base-pair roll and aspects of bending in the helix axis are in some degree of agreement with the crystal structure. A second AMBER simulation on the dodecamer was initiated from a canonical A-structure, with all other characteristics of the simulation carried over directly from the previous B-form simulation. The dynamical structure in the A-DNA simulation was observed after ~100 ps to reside in the same state obtained in the B-form simulation, indicating that *in vacuo* AMBER does not support independent A- and B-forms of DNA. Since experimentally it is known that the relative stability of A- and B-DNA is a consequence of water activity, this does not necessarily mean that the force field is deficient.

Rao and Kollman (1990) carried out at this same time their own AMBER-based MD simulations on duplex d(CGCGAATTCGCG) and d(CGCGCGCGCG), both simulations initiated from either canonical or crystallographic B-form structures. The dynamical structures of both sequences take on a markedly kinked form in the course of 100 ps of MD, and the authors pointed out that the absence of solvent in the simulation could exaggerate the bend. A comparison of results from all-atom and united-atom versions of the AMBER force field indicate that the helix repeat is more consistent with experiment in the all-atom case, but the overall evidence was not sufficient to say that the united-atom model is deficient. Comparison with the corresponding results described above (Srinivasan *et al.*, 1990) shows essentially complete agreement between these two independent AMBER runs, and, in particular, the tendency in the MD to produce a considerably wider minor groove in the AATT region than that observed in the crystal structure.

Kumar and co-workers (Kumar, 1990; Kumar *et al.*, 1991), using AMBER, carried out MD on the dodecamer *in vacuo*, commencing with the crystal structure of the native form. The result ends up midway between A- and B-forms of DNA, with the base pairs displaced ~3 Å from the helix axis. Some sensitivity of results to the treatment of the counterions was noted. A simulation on the dodecamer commencing from the kinked form of DNA found in the crystal structure of the complex with EcoRI endonuclease quickly relaxed back to the same state obtained in the previous simulation, indicating that the kinked form is not an altered equilibrium state of DNA but is induced by the interaction with protein. The deformation occurs at the roll points in the structure, consistent with the idea that there is a propensity to bend at these positions. A 100 ps MD simulation on d(TATCACC) using AMBER has also been reported (Miriganik and Kothekar, 1991).

In vacuo simulations have also been carried out on d(CGCGAATTCGCG) using the CHARMM force field (Louise-May and Beveridge, 1991). A discussion of at least a precursor of this parametrization has been given by Nilsson and Karplus (1984) in their previous study involving the hexanucleotide duplex d(CGTACG). This calculation uses a distance-dependent dielectric screening constant in the electrostatic interactions, and the phosphate charges are set to −0.34 to represent the

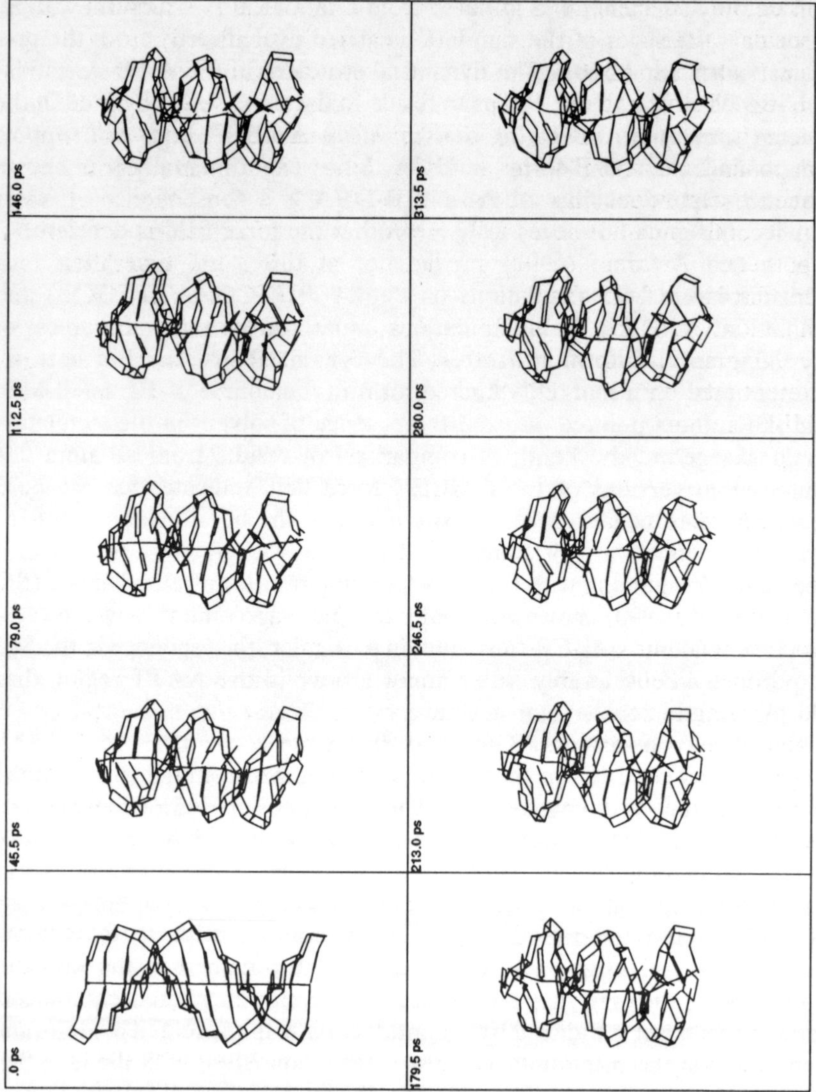

Figure 6.7 Snapshots from an MD simulation of 347 ps on d(CGCGAATTCGCG) based on an *in vacuo* model and the CHARMM version 21.2.7b force field (Louise-May and Beveridge, 1990). The initial structure (lower left-hand panel) was a canonical B-form of DNA

effect of condensed counterions. The trajectory was extended to ~350 ps to check convergence behaviour. Some snapshots from the trajectory are shown in Figure 6.7. The dynamical structure is somewhat constricted as compared with the corresponding canonical form, a feature commonly observed in *in vacuo* simulational models. Further analysis of the simulation indicates that the CHARMM force field results in a model with a considerably smaller range of dynamical motion than that of AMBER. In particular, the sugar puckers in CHARMM are tightly pinned at B-form values, whereas AMBER permits the possibility of repuckering. The CHARMM structure overall maintains the base pairs centred on the helix axis in the manner of B-DNA, but there is considerable base-pair inclination to an extent similar to that seen in AMBER. Also, considerable sequence-dependent fine structure is observed.

A second simulation using CHARMM was initiated, commencing with a hypothetical A-form structure for d(CGCGAATTCGCG), with the base pairs displaced from the helix axis by 5.4 Å and tilted by 19.1°. The simulation, even after 72 ps (Figure 6.8), was observed to be quite stable in the A-form, indicating that *in vacuo* CHARMM supports well-differentiated forms of DNA, in contrast to the results obtained from AMBER.

5 MD Studies on DNA with Solvent

The first MD simulation on a DNA oligomer including solvent water explicitly was performed on the sequence d(CGCGA), treating the duplex structure together with 8 Na^+ counterions and 830 TIP3P water molecules in a droplet (Siebel *et al.*, 1985). The MD was carried out for a reported 106 ps using a non-bonded cutoff of 10 Å. The simulation produced a duplex of 10 bp/turn, and the average dihedral angles remained in the same range as that found in a previous *in vacuo* simulation (Singh *et al.*, 1985). More tilt and twist of central base pair was found when water was included, and the values of glycosidic torsion angles were altered for the central C and G. However, phosphate motions were damped by a factor of 2 by the explicit water as compared with the implicit treatment. Sugar puckers were mixed in a statistical ratio (70–80% C2'-*endo*), similar to NMR results. All but two Na^+, initially at 3.1 Å, remained close to phosphates. One diffused to the edge of the droplet, and another ended up in the major groove. The waters were observed to hydrogen bond to the expected hydrophilic sites along the base pairs in both grooves, but detailed analysis of the water network was not attempted.

An MD simulation on the octamer duplex d(CGCAACGC) including 14 Na^+ and 1231 SPC water molecules under periodic boundary conditions was reported in 1986 (van Gunsteren *et al.*, 1986) based on the GROMOS force field. The Na^+ initial positions were chosen as favourable with respect to electrostatic potential. The length of the simulation was 80 ps, the dielectric shielding $\varepsilon = 1$, and a twin range cutoff in which the >8 Å interactions were updated less frequently was used. The simulated structure turned out to be 2.2 Å RMS different from the canonical B-form and 3.5 Å RMS different from the A-form, i.e. intermediate between the RMS

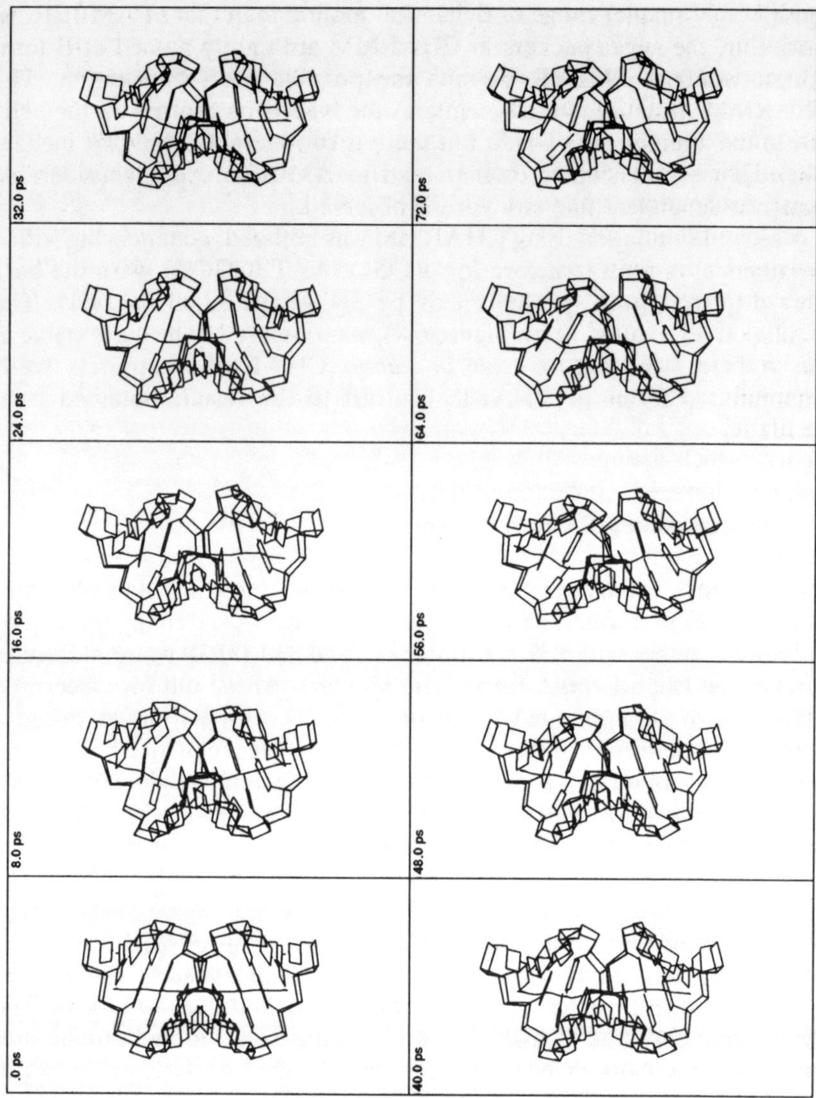

Figure 6.8 Snapshots from an MD simulation of 72 ps on d(CGCGAATTCGCG) based on an *in vacuo* model and the CHARMM version 21.2.7b force field (Louise-May and Beveridge, 1990). The initial structure (lower left-hand panel) was a canonical A-form of DNA

of the B-form Dickerson dodecamer to canonical B. Some 80% of a set of 2D-NMR NOESY distances were satisfied by the computed average structure. No Na^+ ions were found to end up within 3.0 Å of any phosphate anion, leaving the sodiums and phosphates completely hydrated and DNA solvated by water and hydrated Na^+ ions.

Swamy and Clementi (1987) also reported an early MD on B- and

Z-DNA in the presence of water and counterions. Calculations were carried out on G–C and A–T dodecamer sequences in B-form with 1500 H_2O and 22 Na^+, and on a G–C dodecamer in the Z-conformation with 1851 waters and 12 counterions. The simulations in the two cases were performed for 4.0 ps and 3.5 ps, respectively, after equilibration. The water–DNA interactions were discussed on the basis of the calculated radial distribution functions, with the gross features of a composite of ionic, hydrophilic and hydrophobic hydration effects presented in the analysis. Counterion structure defined in terms of localized solvation sites relative to the phosphate groups was presented and were said to confirm previous results from Monte Carlo simulations. However, in 3–4 ps none of the ions has any significant chance to move, and so this point remains essentially unsubstantiated. The conclusions on site binding are contrary to the findings of van Gunsteren *et al.* (1986) and to counterion condensation theory, which assumes that a large percentage of counterions are fully hydrated and freely diffusing in the vicinity of the DNA.

Several instances of nuclease activity specific for G-tracts led Zielinski and Shibata (1990) to carry out an MD simulation on the $dG_6{*}dC_6$ duplex and develop a model for $dG_n{*}dC_n$. The simulation was carried out on the hexamer and 10 sodium counterions in a droplet of 292 water molecules, and based on the GROMOS force field. Base-pair hydrogen bonds were observed to fluctuate with time, and analysis revealed some unusual patterns in them. The hexamer maintained a B-form conformation over the course of 60 ps of MD, but featured high propeller twist and a surprisingly narrow minor groove. If this feature is accurately predicted in the simulation, it provides a possible explanation for the resistance of G-tract sequences to DNAase I.

Experimental measurements indicate that the 5-methyl cytosine analogue of poly (dG–dC) is considerably more prone to adopting the Z-conformation than the underivitized polymer, and undergoes the B-to-Z transition at near-physiological pH (Behe *et al.*, 1985). Kollman *et al.* (1982) had previously proposed on the basis of energy minimization that this was due to intramolecular effects, whereas others (Ho *et al.*, 1988), on the basis of solvent-accessibility calculations, had ascribed it to differential solvation. Pearlman and Kollman (1990) used MD simulation and free-energy-perturbation methods to develop a more extensive theoretical account of this phenomenon, considering both *in vacuo* and solvated cases. The calculations indicated that, in solution, methylation makes the B-to-Z transition more favourable by about 0.4 kcal/mol of base pairs as opposed to 0.3 kcal/mol observed experimentally. Analysis of the results shows that the preferential stabilization of the Z-form has comparable contributions from intramolecular and solvent effects. However, it was noted that the calculated structures in this study exhibited certain non-standard stacking and base pairing, although the overall morphology of the Z-form was maintained in the final state.

An MD simulation on d(CGCGAATTCGCG) including water and counterions was recently carried out (Swaminathan *et al.*, 1991). The simulation involved the dodecamer and 1927 water molecules and 22 Na^+ counterions treated under periodic boundary conditions in a hexagonal prism elementary cell. The force field for the simulation was GROMOS supplemented with a restraint potential for maintaining Watson–Crick (WC) base pairing to be consistent with NMR results (Moe and Russu, 1990) on the lack of base-pair opening on the picosecond time-scale. Extensive Monte Carlo equilibration of the solvent was used to prepare the system in a suitable state to perform a stable dynamical trajectory. The initial structure in the simulation was a canonical B-form, leading us to designate this calculation WC/B. The MD trajectory was extended for 140 ps; and the results are shown in Figure 6.9. The structure at the termination of the trajectory resides clearly in the B-DNA family, 2.3 Å RMS deviation from the corresponding canonical form and 2.5 Å RMS deviation from the corresponding crystal structure. This is significantly closer agreement with experiment than that obtained in previous *in vacuo* simulations. The MD simulations including solvation also show a smaller range of dynamical fluctuations than the *in vacuo* case, indicating that the explicit presence of water has a damping effect on the DNA motions.

The analysis of the simulation reveals good accord with a number of detailed features seen in the X-ray crystal structure of the dodecamer, including local axis deformation near the GC–AT interfaces in the sequence, and large propeller twist in the base pairs. The DNA base pairs show a consistent inclination in the simulation, in accord with the interpretation of results obtained from flow dichroism studies. The negative propeller twist found in the crystal structure is clearly observed in the MD, but is not so far different for AT and GC, in spite of the expectation that it should be larger for the AT base pairs. Zhurkin theory (Zhurkin, 1985) in base-pair roll is followed in some instances but not in others. Narrowing of the minor groove in the AT region of the crystal structure is not observed over the time-course of the simulation, nor is there any evidence for the spontaneous narrowing of the minor groove in AT regions of d(CGCGAATTCGCG) in any MD simulations so far. On the other hand, DNA footprinting studies provide quite strong evidence for sequence-dependent fine structure, and Tullius and co-workers suggest it to be linked to minor groove narrowing (Burkhoff and Tullius, 1987). This is a point of considerable interest and will no doubt be the subject of further studies on various DNA oligonucleotide sequences.

In an additional simulation carried out here without the WC restraint function, referred to as the electrostatic (ES) model, more pronounced axis deformations and a base-pair opening is observed at or near the primary roll point. This provides a dynamical model for how base-pair opening might occur via water insertion. A corresponding *in vacuo* simula-

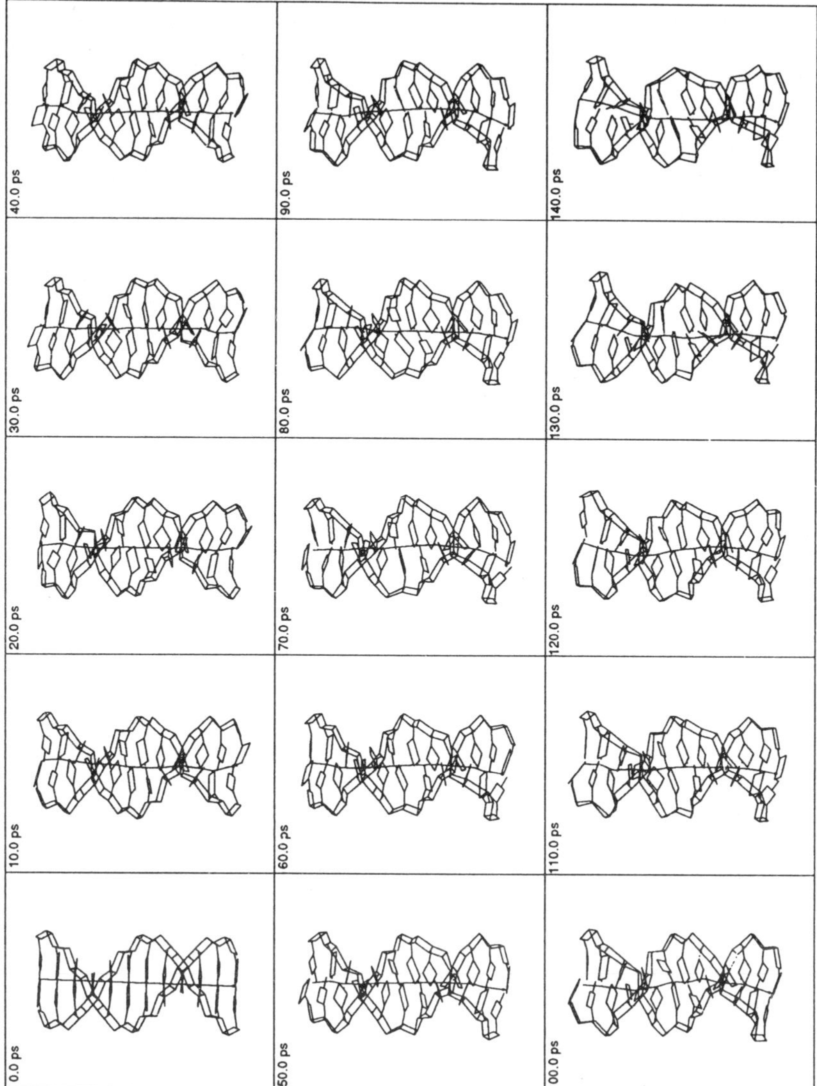

Figure 6.9 Snapshots from an MD simulation of 140 ps on d(CGCGAATTCGCG) including counterions and water based on the GROMOS force field with WC restraints (Swaminathan and Beveridge, 1990). The initial structure (lower left-hand panel) was a canonical B-form of DNA

tion shows that explicit inclusion of the water molecules is necessary to properly support the major and minor groove structure of the DNA helix. Overall, the RMS results particularly indicate that including explicit solvation significantly improves the agreement of the calculated dynamical structure with crystal structures, and maintains it closer to the canonical form.

A GROMOS MD simulation including counterions and water, commencing with the crystal geometry of the native dodecamer and using substantially longer cutoffs, has been carried out for 500 ps by Srinivasan *et al.* (1992). The results of this simulation, called WC/X, indicate that a stable B-form structure is well established, and the narrow minor groove in the AATT region becomes substantially wider in the simulation. The results correspond closely but not exactly to those obtained within the WC/B MD simulation described above. Srinivasan *et al.* (1991) have subsequently calculated the theoretical scalar coupling constants for the sugar protons of the dodecamer from the WC/B, ES/B and WC/X GROMOS simulations described above and compared them with the corresponding observed values. The results, based particularly on the WC/B model, are in best agreement with experiment, predominately $C_{2'}$-*endo* sugar puckers, with a distribution of other forms, particularly in the vicinity of the roll points.

Three groups (Kumar, 1990; Miaskiewicz *et al.*, 1992; Singh, 1991) have recently carried out MD simulations on d(CGCGAATTCGCG) including counterions and water using the AMBER force field, all as yet unpublished. Kumar (1990) reports a simulation of 81 ps in which a disruption in Watson–Crick base pairing occurred after 70 ps, effectively a hydrogen bond exchange with solvent water. The structure maintains a good B-like form, with base-pair displacements of ~0.1 Å, but increasing up to 1 Å at the point of base-pair opening. Singh (1991) carried out 100 ps of AMBER MD on the dodecamer, but found the RMS displacement still to be drifting when the simulation was terminated. The structure was bent into a C-shaped form. Kumar discusses a number of the methodological problems in equilibrating DNA simulations in his thesis (Kumar, 1990). Miaskiewicz *et al.* (1992) carried out 150 ps of AMBER MD on the dodecamer in a droplet of 1431 water molecules, beginning from a canonical B-form (Figure 6.10). The dynamical structure was observed to make transitions at two points, 60 ps and 100 ps into the run, and ended up 4.6 Å RMS from canonical B. The final structure was sharply kinked at two places, and showed evidence of considerable minor deformations. The results were discussed in the context of distortions thermally accessible to B-DNA in solution, and with respect to the Reid group's NMR structure.

An MD study of left-handed d(CGCGCGCGCGCG)2 including water and sufficient counterions to create electrical neutrality in the solution has recently been carried out based on the AMBER force field (Erikksson and Laaksonen, 1992). This is an extension of work described in a previous article focusing on hydration (Laaksonen *et al.*, 1989), and discussed in a preceding section. The MD on the DNA *per se* (Figure 6.11), based primarily on the AMBER force field, produced a model with considerable kinking and base-pair opening. Most of the residues ended up in the Z_{II} conformation, stabilized to some extent by ion coordination between one

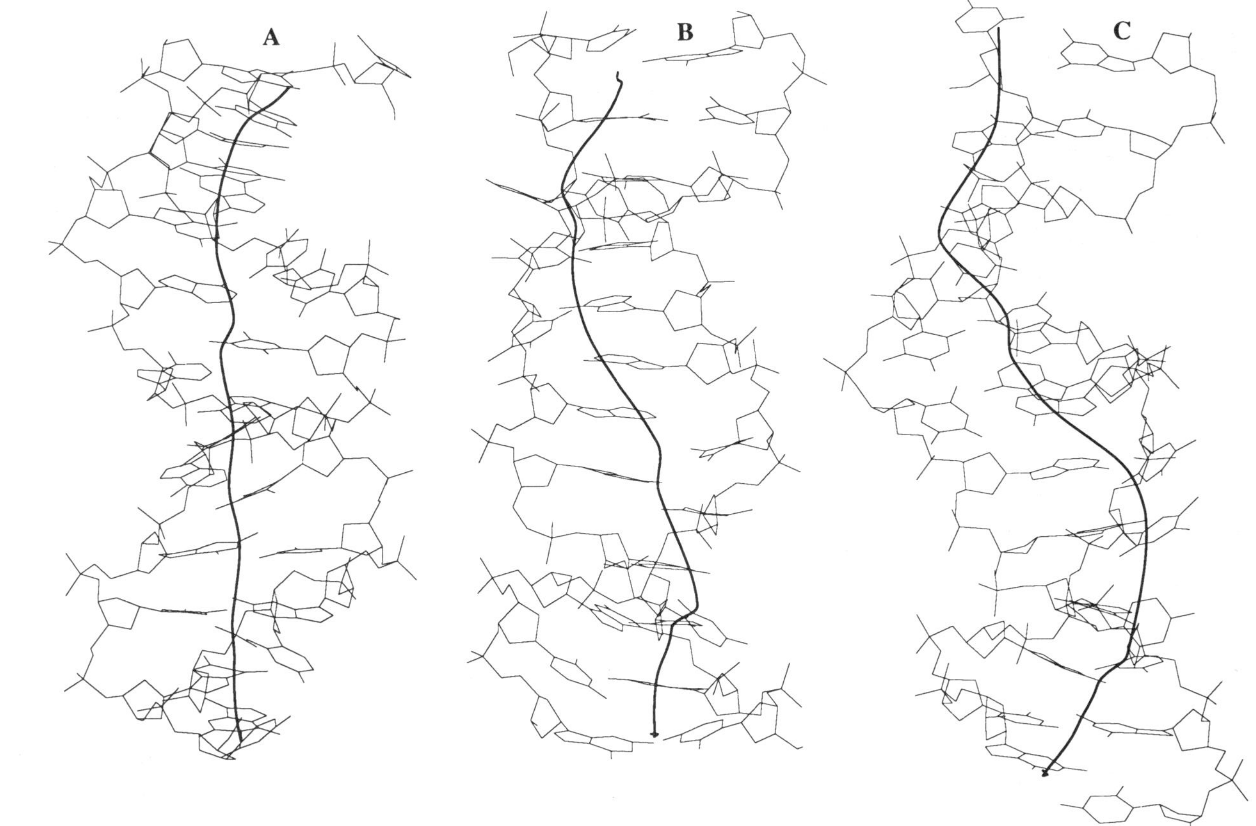

Figure 6.10 Snapshots from an MD simulation of 140 ps on d(CGCGAATTCGCG) including counterions and water based on the AMBER force field (Miaskiewicz *et al.*, 1992), starting from canonical B-DNA. (A) at 32 ps; (B) at 72 ps; (C) at 142 ps

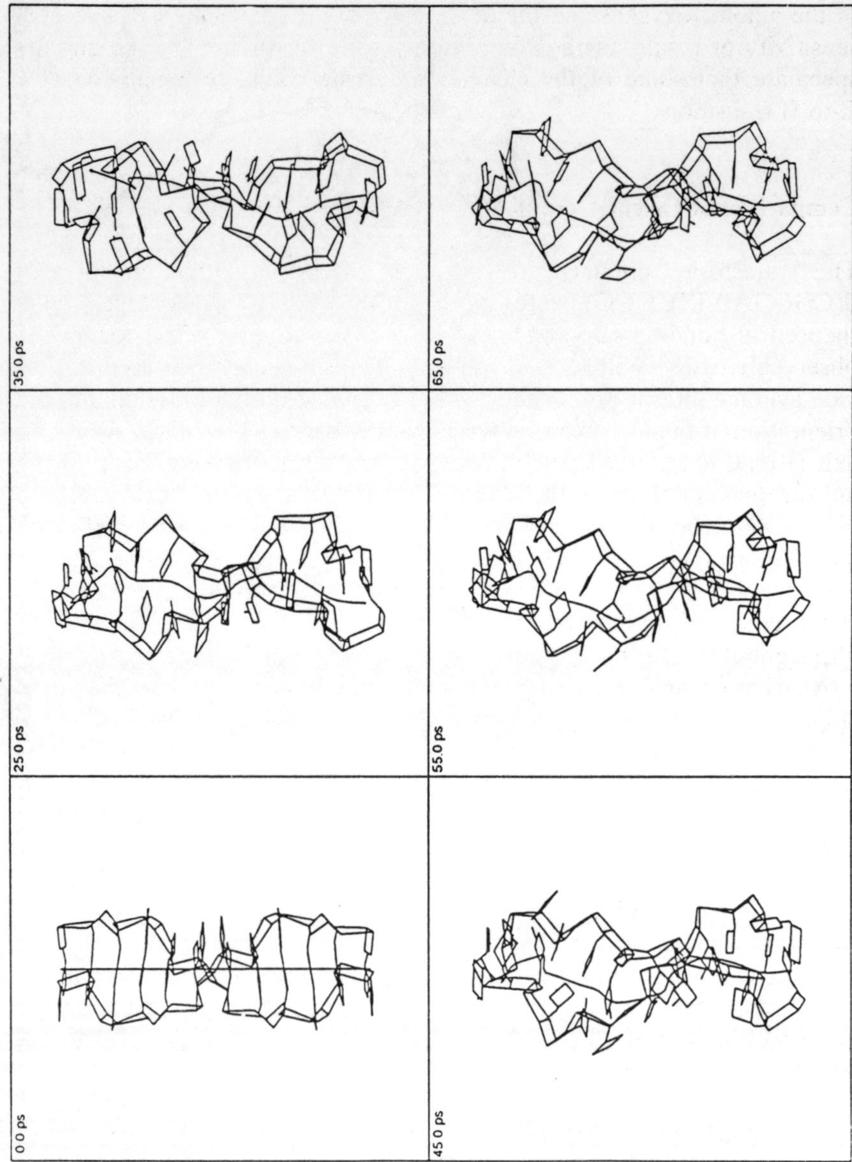

Figure 6.11 Snapshots from an MD simulation of 70 ps on d(CGCGCGCGCGCG) based on AMBER (Erikksson and Laaksonen, 1992). The initial configuration in this simulation was a Z-form of DNA

of the anionic oxygens and the next G–N7 in the sequence. Considerable sensitivity of results to the counterion positions was noted. The authors speculate that some of the changes occurring could be precursors of a Z-to-B transition.

Comparison of Calculated and Observed NOEs

The trajectories obtained for the GROMOS simulation on solvated d(CGCGAATTCGCG) were subsequently used to calculate *de novo* theoretical build-up curves for the nuclear Overhauser effect for all the observable cases (Withka *et al.*, 1991a). The simulation was used to provide average interproton distances, the extent of internal motion and the orientation of the interproton vectors with respect to the helical symmetry axis (Eimer *et al.*, 1982, 1990). Independent estimates were made of the anisotropic tumbling of the duplex from hydrodynamic theory. These results were combined to calculate correlation times for each interproton pair. Transition probabilities were computed by standard methods, and the time dependence of the NOEs was determined by solving the generalized Bloch equations by numerical integration. This includes consideration of all magnetization transfer pathways.

Analysis of the calculated NOE build-up curves confirms that the initial slopes are not necessarily proportional to the inverse of the sixth power of the internuclear distance, as in the isolated two-spin approximation, and have magnitudes and lag phase behaviour which arise from extensive spin diffusion. This is indicated to occur even at the shortest experimentally accessible mixing times. Inclusion of the orientation effects and the internal motions of the interproton vectors were found to have significant effects on the calculated NOE intensities and the extent of spin diffusion. Comparing the results based on the WC and ES simulations described above, the build-up curves for distant sites were found to be well differentiated in the two cases (Withka *et al.*, 1991b), indicating that the method has a high degree of sensitivity to the underlying dynamical model. Some selected results from this study are shown in Figure 6.12.

NMR measurements of the build-up curves of NOE cross-peak volumes for the dodecamer were then carried out, and the overall agreement between theoretically calculated and experimentally measured intensities was determined for each residue of each of the different models. Calculations were also made for the duplex dodecamer modelled as the canonical A-form, the canonical B-form and the crystal structure (single structure models), as well as for the ensemble of structures comprising the dynamical models obtained from the various MD simulations. The results are shown in Figure 6.13. The crystal structure yielded a better fit to the experimental

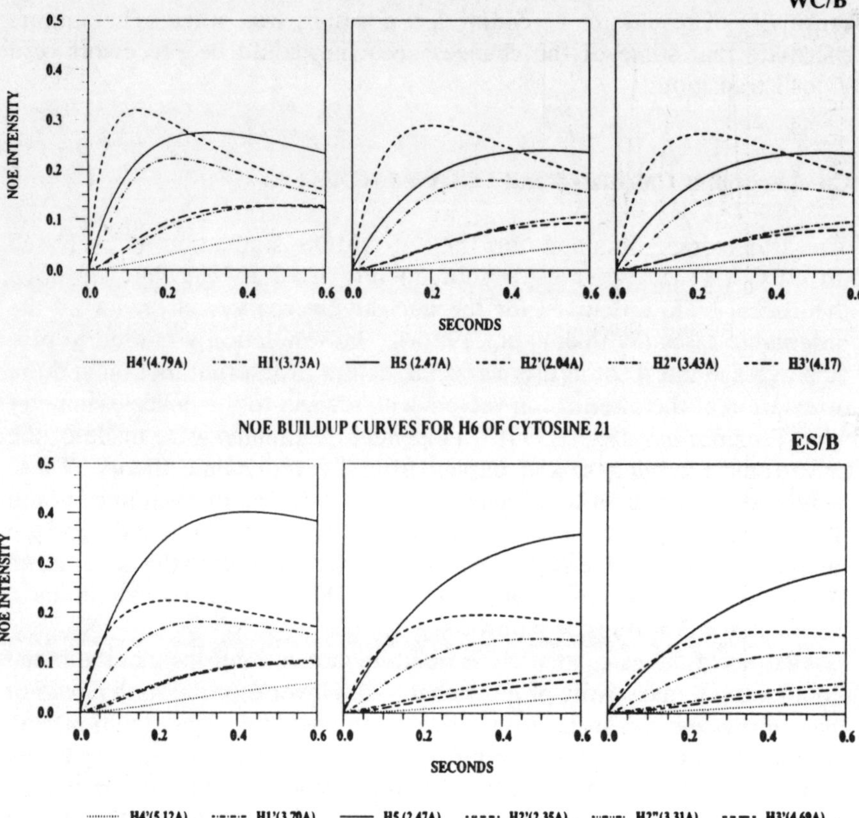

Figure 6.12 Theoretical NOE build-up curves for the aromatic proton H6 of a cytosine residue C-21 in d(CGCGAATTCGCG), based on the WC simulation (top) and the ES simulations (bottom). The left panel of each set contains build-up curves which depend only on interproton distance. The middle panels include effects of the orientation and anisotropy of the duplex. The rightmost panels includes effects of orientation, internal motion and anisotropy. Average interproton distances are included at the bottom

data than the canonical B-form structure, and both, as expected, are clearly better than the results based on the canonical A-form. However, the results show that none of the single-structure models gives a particularly good fit to the experimental data.

The corresponding results for the WC/B, ES/B and WC/X dynamical models, including the effects of anisotropic and internal motion, demonstrate that each of the simulations leads to a significantly better agreement than any of the single-structure models, canonical or crystallographic. The dynamical model intraresidue results are within experimental error, and the inter-residue results tend to correspond to internuclear distances of no

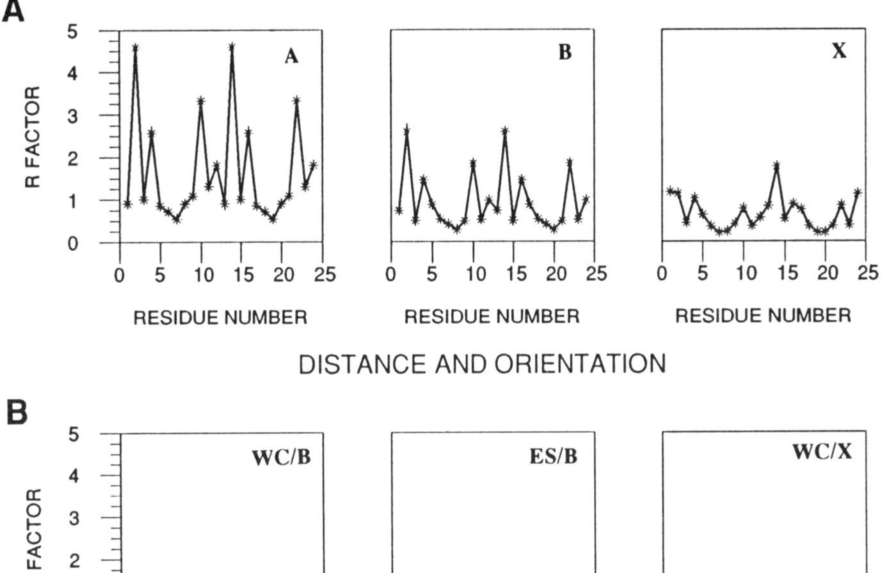

Figure 6.13 RMS deviation between calculated and observed NOE data on d(CGCGAATTCGCG) (Withka *et al.*, 1991b): (A) results based on single structure, reading left to right, for canonical A, canonical B and the native dodecamer; (B) results based on MD ensembles of structures, reading left to right, for WC/B, ES/B and WC/X models

more than 0.1–0.2 Å less than the observed values. Therefore, the dynamical structure for DNA in solution obtained from MD simulation agrees significantly better with experiment than any of the single-structural models. With respect to *R*-values, the three distinct MD dynamical models all turn out to have about the same overall quality of fit to the corresponding observed results. Detailed inspection reveals results that differ in the fit from site to site, and an improved definition of agreement factors which are sensitive to these effects has been devised (Withka *et al.*, 1992).

The comparison between the theoretical and experimental results including motional considerations demonstrates that both the anisotropic tumbling and local motions of the DNA need to be explicitly included for accurate modelling of DNA in solution. The improved agreement obtained from the MD models over that of single structures indicates that the dynamical models are a better description of DNA structure in solution than the crystallographic or canonical models. The combined scalar coupling and NOE results in comparison with experiment indicate that the ensemble of structures comprising the WC/B MD model is our best current prediction of the structure of d(CGCGAATTCGCG) in solution.

6 Restrained MD and Refinement

Obtaining the structure of a DNA oligomer from experimental data is now recognized to be an underdetermined problem. Crystal structures of oligonucleotides are typically obtained at *ca.* 2.0 Å resolution, leaving considerable detail to be filled in during the crystallographic refinement process. The NMR data from 2D-NOESY intensity measurements provide information related to interproton separations of <4.5 Å, sufficient only to define certain local aspects of structure, such as the pseudorotation angle of the sugar pucker and the *syn* or *anti* orientation of the nucleotide bases with respect to the sugar ring. Spin-coupling analysis is primarily used only for the determination of sugar puckers.

In recent years theoretical calculations on the corresponding structures have been incorporated into a refinement process to compensate for the limitations in the experimental data base. However, the calculations introduce additional approximations of their own that must be considered. Early work in this area used empirical energy functions and minimization procedures, with experimental data introduced as restraints via a non-physical energy penalty function. Now MD simulation based on force calculations augmented with restraints has become the method of choice, since it provides a means of sampling the restrained energy hypersurface more broadly.

The use of restrained MD (RMD) based on the CHARMM force field along with 2D-NMR NOESY distance information was first introduced in 1986 by Nilsson *et al.* (1986) in a structure refinement of the oligonucleotide d(CGTACG). First, a model set of canonical A and B conformational constraint distances (< 4.5 Å to mimic NOE distances) was formed, and applied to MD simulations of the hexamer, beginning in the B- and A-conformations, respectively. The authors demonstrated that the MD calculation beginning in A with B restraints applied transits expeditiously to B, and vice versa. Application of the observed NOE distances as restraints to separate MD calculations beginning at A- and in B-forms, respectively, showed reasonable convergence to a proposed solution structure generally in the B-family, but with considerable deviations from a canonical B-type structure.

There is now a considerable literature developing on structural prediction of DNA oligonucleotides based on NMR data and restrained MD protocols. A diverse number of cases have been studied, including d(GCATGC) (Nilges *et al.*, 1987a), d(CTGGATCCAG) (Nilges *et al.*, 1987b), d(CGCGPATTTCGCG) (Clore *et al.*, 1988), the RNA–DNA hybrid duplex r(GCA) d(TGC) (Scalfi Happ *et al.*, 1988), d(CGCIAAT-)*d(ATTAGCG) (Fujii *et al.*, 1988), d(CTGAAATTCAG) (Nordlund *et al.*, 1989), an extrahelical adenosine tridecamer oligodeoxyribonucleotide duplex (Nikonowicz *et al.*, 1989), d(CGCGPATTCGCG) (Gronenborn and Clore, 1989), d(GCGTTGCG) (Boelens *et al.*, 1989), hairpin structures in DNA-containing arabinofuranosylcytosine (Pieters *et al.*, 1990), the phage lambda half-operator DNA sequence (Baleja *et al.*, 1990), d(GGAAATTTCC) (Katahira *et al.*, 1990), a mismatched GA decamer (Nikonowicz and Gorenstein, 1990), d(CGCTTAAGCG) (Powers and Gorenstein, 1990; Powers *et al.*, 1990), $d(AC)_4$*$d(GT)_4$ (Gochin and James, 1990), d(CCTTAAGG) (Ito *et al.*, 1991), d(GTACGTAC) and d(CATGCATG) (Baleja *et al.*, 1990), d(ATATATAUAT) (Kerwood *et al.*, 1991), d(GCGTTCGC) (Koning *et al.*, 1991), junction sequences between two A/T tracts (Luo *et al.*, 1991) and d(GTATATAC) (Schmitz *et al.*, 1991).

The main points of conceptual and theoretical development in this area in the course of the studies referenced above as well as other theoretical work (Yip and Case, 1989) have been the recognition that (a) a matrix relaxation approach is necessary to account for spin-diffusion effects and (b) the use of NOE intensities (cross-peak volumes) as restraints is preferred over using interproton distances in order to maintain a closer contact with experimental data. It is also now standard practice to iterate the results to self-consistency, whereas in the initial Nilsson *et al.* study (1986) it was not determined whether the resulting structural forms produced predicted NOEs which matched experiment. Kaptein and co-workers (Koning *et al.*, 1991) recently introduced the consideration of local motions of the interproton vectors into an iterative RMD protocol, and found improved agreement with experiment, as did Withka *et al.* (1991a, b) in free MD simulations on DNA. Some recent refinement studies have begun to incorporate scalar coupling data from COSY experiments, as well as NOESY data (Gochin *et al.*, 1990; Lane *et al.*, 1991; Srinivasan *et al.*, 1991).

There are additional issues of concern in RMD calculations beyond the assumptions in the force field. From a purely theoretical point of view, in an RMD calculation the experimental restraints may distort to some extent the intentions of the force field *per se*, pulling the atoms away from the natural equilibrium regions of the potential surface. The resulting dynamics will not necessarily be a good theoretical description of the motions of the system. This is especially true in cases where the electrostatic interactions are removed from the MD force field in refinements. Also, the procedure is typically not unique; recent studies have shown that a number of individual structures may be equally consistent with the experimental data (Wemmer, 1991). The idea of refining experimental data on structure in solution to a single static form belies the fact that the dynamical structure in solution is, in fact, an ensemble. This procedure is currently being addressed by RMD methods which involve interproton distance or NOE intensity restraints based on time-average information obtained for an ensemble of structures over a long MD trajectory (Pearlman, 1991; Torda *et al.*, 1990).

The fundamental question being addressed at the present moment is how well the DNA structure can *ever* be determined from NMR and RMD procedures. Pardi *et al.* (1988) performed simulations geared to determine how accurately and precisely DNA structures can be determined from a set of interproton distances that mimic an optimal NMR data set. The backbone torsion angles turned out to be rather loosely determined, although χ could be defined. Metzler *et al.* (1990), extending this work, note that in using available techniques that most parameters representative of the fine structure such as helix twist, rise, propeller, roll, the α, β, γ, ε and ζ torsions and probably the tilt, δ torsion and displacement cannot be determined precisely enough by NMR and RMD refinement to justify comparisons of local structure variations with sequence. They concluded that in many cases a parameter in DNA can only be defined by NMR with a precision similar to the whole range of values that has ever been observed for this parameter in X-ray structures of right-handed DNAs. Gorenstein and co-workers (Kaluarachi *et al.*, 1991) also studied this problem and provide evidence that it is possible to determine sequence-specific variations in some of the local helical parameters and also in the sugar–phosphate backbone using a ‘hybrid matrix relaxation’ methodology that involves the substitution of NOE cross-peak volumes determined from a model structure for overlapped or suspect experimental volumes. The extent to which dynamical fine structure can be determined for DNA in solution thus remains an active area of research investigation.

The RMD method may also be applied to refinement of structure with respect to crystallographically determined electron densities, and the procedure developed by Brünger known as X-PLOR has been used in several structure determinations of DNA oligonucleotides — for example, d(CGCATATATGCG) (Yoon *et al.*, 1988). Once again, the risk that assumptions in the underlying force field could bias the resulting structure is unavoidable; however, the advantage introduced by the procedure over fitting by hand and eye in front of a graphics terminal is considerable, and has made the X-PLOR approach currently very popular. Brünger (1990) has recently published a good critical introduction to the field of both NMR and crystallographic RMD refinement.

7 MD Studies of DNA Variants and Complexes

Perturbed and Damaged DNA

Rao *et al.* (1986) reported MD simulation studies of sequence dependence and the role of mismatch pairs in the DNA helix. Ideas about the solution structure of two DNA tridecamer sequences containing inserted adenosines were developed based on molecular modelling and MD simulation by Hirshberg *et al.* (1988), and keyed to NMR and calorimetric results. The adenosines were found to act as wedges and kink the structure by some 30° at the insertion site.

A molecular dynamics study of the effect of GT mispairs on the conformation of DNA in solution has been reported based on MD simulation using the GROMOS force field (Shibata *et al.*, 1991). Trajectories of 60 ps were performed on the d(G_3C_3) normal duplex as described previously and on the d(G_3TC_2) mispair duplex, treating the duplexes along with counterions in a droplet of water. Incorporation of the wobble mispairs in the middle of the mini-helix did not alter the helix structure to a major extent in the simulation, which remained in the B-family with some admixture of characteristics of the A-form. This contrasts with the general behaviour of GC-rich sequences in the crystalline state, but is consistent with NMR and CD results in solution. Good intrastrand stacking was maintained, with nevertheless high propeller twist. The MD predicts the mismatched duplex to be dynamically more rigid than the corresponding normal duplex.

Two-dimensional proton and phosphorus-31 NMR spectra and restrained MD were used to work out a proposed structure of a mismatched GA decamer oligodeoxyribonucleotide duplex (Nikonowicz and Gorenstein, 1990). Preliminary results of an MD study on a DNA dodecamer in which one of the thymines was replaced with a 5-hydroxy-6-thymidine radical have been described (Osman *et al.*, 1991). This is a main product of the attack of a OH radical as produced by energy deposition due to irradiation of DNA in aqueous media. Quantum chemical methods were used to study the reaction mechanism and product energetics, and MD was used to study possible conformational changes produced by the local damage. The major groove was narrowed and the minor groove widened in the simulated form, and an impressive global kink in the helix was induced by the presence of the altered thymidine. This article also contains a useful review of the field of radiation-damaged DNA.

DNA Triplexes

The ability of DNA to form a triple helix by adding a single strand of nucleic acid to the major groove of a duplex has been known for some time

(Felsenfeld *et al.*, 1957), and recently the inhibition of gene promoter sites by triplex formation has been demonstrated (Cooney *et al.*, 1988). This raises the possibility of a whole new class of pharmaceutical agents, and is the subject of considerable current research interest. The experimental information currently available on triplex structure is mainly the fibre diffraction study of poly(dT)–poly(dA)–poly(dT) (Arnott and Selsing, 1974), and recent NMR studies (Rajagopal and Feigon, 1989; Sklenar and Feigon, 1990). Hausheer *et al.* (1990) reported MD simulations on DNA triplexes with oligonucleotide dimethyl phosphonates (MP) and naturally occurring oligodeoxynucleotides (ODN) as the third strand, and the results are consistent with the expectation that the uncharged MP is more favourable to triplex formation as well as having an advantage in pharmaceutical bioavailability. The initial *in vacuo* MD studies were followed up with MD and electrostatic potential calculations considering solvated forms (Hausheer, 1990), and the results indicate that charge neutralization of the backbone has relatively small effects on the conformational properties of DNA.

Pettitt and co-workers (van Vlijnen *et al.*, 1990) carried out a number of EM and MD studies of triplexes, and found that the experimentally observed destabilizing effect of base mismatches between duplex and ligand strand was not reproduced satisfactorily using *in vacuo* CHARMM simulations. Conformational deformations and local energetic destabilizations were found in all triplexes with mismatches. Cheng and Pettitt (1991) subsequently used modelling, solubility estimates and some MD simulation to explore further the nucleotide base-pairing schemes of triplex formation, focusing particularly on the structure and energetics of Hoogsteen as opposed to reverse Hoogsteen pairing schemes. For TAT triplexes, the calculations indicate that the additional strand prefers Hoogsteen base pairing and parallel orientation. Hoogsteen base pairing was also predicted for the CGG case, but a parallel orientation of the additional strand, consistent with the results on the c-*myc* gene promoter, gives the most favourable energetics. The orientation effect was linked to forces of solvation in the complex. This subject, along with antisense RNA technology, is expected to be an area critically important to new directions in drug design, and is likely to be the subject of a considerable number of MD and modelling studies in the near future.

Drug–DNA Complexes

MD simulation is emerging as a valuable methodology in the development of models for the structure and action of drugs which interact with DNA. The accuracy of these calculations must, of course, be considered in the context of what we know to be the accuracy of MD simulations on DNA alone, as described in the preceding sections. However, useful ideas can sometimes be gleaned from even very exploratory calculations considered in the proper context. Since developing pharmaceutical agents is enormously expensive, any good lead on a drug may result in considerable efficiencies and economies. As a result, most major drug companies have now established groups which specialize in theoretical and computational studies, and MD simulations involving DNA figure prominently in their efforts.

Prabhakaran and Harvey (1985, 1988) were the first to use MD in the study of drug–DNA interactions. A preliminary work describing the formation of an intercalation site in DNA *in vacuo* was followed up by a more extensive MD study of an intercalation site in both B- and Z-DNA, and on the binding of actinomycin D. The formation of an adjacent site was found to be energetically more costly, accounting for the empirically observed neighbour exclusion rule in intercalation complexes. Rudolf and Case (1989) reported harmonic dynamics simulation bound to d(CGCGCG) on the ethidium. The intercalation was found to significantly lower the mobility of base pairing adjacent to the drug, affecting the guanosines more than the cytosines. The wobble of bound ethidium was smaller than that inferred from fluorescence anisotropy measurements, but qualitative features were correct. Out-of-plane mobility in the ethidium–DNA complex has also been considered (Haerd, 1987).

The binding of actinomycin D to the oligonucleotide d(ATGCAT) duplex was the subject of a subsequent NMR and *in vacuo* RMD study by Creighton *et al.* (1989), subject to the two-spin approximation and R^{-6} weighted interproton distances. Some substantial discrepancies between the inferred and calculated interproton distances turned up, understandable in light of the concerns about spin diffusion. Some general features of the crystallographic model obtained for the complex were, however, confirmed. Feuerstein *et al.* (1989) used MD to model the interaction of spermine, a physiologically important polyamine, with $dG_{10}{\cdot}dC_{10}$ and $(dGdC)_5{\cdot}(dGdC)_5$, including water and counterions. In the initial energy minimization step, spermine binding induced a bend in the heteropolymer but not in the homopolymer. In the subsequent 90 ps of MD, the spermine moved out of the major groove in $dG_{10}{\cdot}dC_{10}$, ending up interacting 'non-specifically'. The bend in the heteropolymer was maintained in the heteropolymer complex, and the spermine remained in position. Although the MDs are rather short, this was interpreted as evidence for sequence-specific binding properties.

Cieplak *et al.* (1990) used *in vacuo* MD based on the AMBER force field in conjunction with free-energy-perturbation theory to study the binding of daunomycin and 9-aminoacridine to B-form DNA. The results support the observed preference of the two complexing agents to bind to a specific base sequence. The pure intercalator 9-aminoacridine prefers CG base pairs, while daunomycin, part intercalator and part groove binder, prefers AT base pairs. The results suggested that one could obtain hybrid molecules that optimize intercalation and groove binding preferences independently, and also provided some leading ideas about the design of sequence-specific binding agents.

Minimization and MD studies of guanosine and Z-DNA modified by *N*-2-acetylaminofluorene (AAF), a potent rat liver carcinogen which covalently links to nucleic acids, have been reported by Fritsch and Westhof (1990, 1991a). The calculations were carried out on *in vacuo* models, and sensitivity of results to force field parameters and various treatments of the electrostatics was considered. The results were found to show a pronounced dependence on the choice of these terms, and provide an informative perspective on the fragility of the computational models in the area of drug–DNA interactions that apply critically to all of the studies described in this section. A dielectric screening function featuring a sigmoidal dependence on distance, which provides a better model for dielectric saturation effects, shows some improved features over a simple distance-dependent screening term. The AAF appears from this study to prefer a conformation in which the fluorene ring stacks on a sugar–phosphate backbone of the following 5′ C–G bases.

Boehncke *et al.* (1991) investigated the molecular dynamics of the complex of distamycin with DNA. Langley *et al.* (1991) used MD based in the CHARMM force field together with DNA-affinity cleavage data to develop a model for the dynemicin–DNA intercalation complex. Particular attention was given to the stability of the complex in the simulation. Only the minor groove intercalated enantiomers produced a dynamically stable model consistent with affinity cleavage results (Figure 6.14). The results suggested ideas on how dynemicin is activated into a DNA cleavage, the mode of binding, the absolute stereochemistry of the drug and the cleavage patterns and specificity. An MD modelling study for the dynemycin complex was also reported by Wender *et al.* (1991), supporting the three-base-pair offset cleavage.

O'Handley *et al.* (1991) have recently reported on the use of MD in conjunction with energy minimization and NMR data to propose a structure for a nine-base-pair duplex containing an N-(deoxyguanosine-8-yl)-2-(acetylamino) fluorene adduct, introducing a lesion opposite to a normal cytidine, as it would be prior to replication. The AAF–G5 residue adopted a *syn* conformation, with the fluorene protruding into the minor groove, while also stacking with a nearby guanine. There is a flexible bend at the site of adduct formation, consistent with NMR data. It is speculated that the distortion may be responsible for initiating adduct-induced repair *in vivo*, as well as providing a model for template-induced mutagenesis during replication.

Singh *et al.* (1991) studied the structures of mirror-image *trans-anti* adduct of benzopyrene adducts, using wide-scale conformational searches

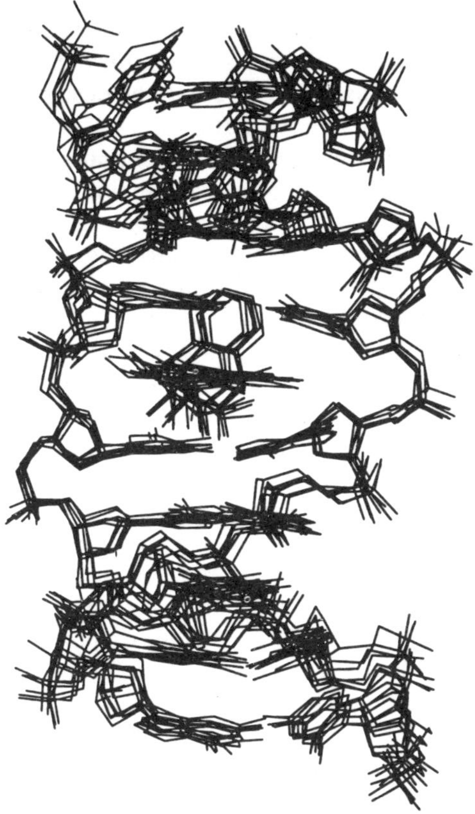

Figure 6.14 A dynamical model for the dynemycin–DNA complex developed from MD simulation based on the CHARMM force field (Langley *et al.*, 1991)

followed up by ~100 ps MD using AMBER and including water and counterions. Groove binding was found to be preferred over intercalation in these complexes. Comparison with experiment was favourable for *trans-anti* adducts, but *cis* forms are indicated to intercalate.

Lane *et al.* (1990), in the paper discussed previously in which the characteristics of the EcoRI dodecamer were worked out, used MD refinement in conjunction with NMR to propose a solution structure for the complex between the dodecamer and berenil, a drug used in the treatment of bovine trypanosomiasis.

Protein–DNA Complexes

The nature of protein–DNA interactions and the molecular basis of regulatory processes in the genome is currently a subject of considerable research interest in molecular biophysics. Crystal structures of the protein–nucleic acid complexes reported in the recent literature (Steitz, 1990) and the concurrent 2D-NMR studies (Russu, 1991) have contributed essential details on the molecular basis of association. In addition to these structural studies and others now in progress, a parallel understanding of the dynamical stability and functional energetics will be required to develop a complete picture of the specificity of protein interactions with operator sequences on DNA, as well as non-specific associations.

Following earlier simulation studies of the structure and thermodynamics of the counterion atmosphere of DNA based on MC computer simulations (Jayaram and Beveridge, 1990; Jayaram *et al.*, 1990), a theoretical calculation of the ion atmosphere contribution to the free energy of association for a protein–DNA complex was carried out using MC simulation and thermodynamic perturbation theory (Jayaram *et al.*, 1991). The system considered in this project was the dimer of the amino-terminal fragment of the λ repressor in a complex with a 17 base pair oligonucleotide of DNA, based on the crystal structure (Pabo and Sauer, 1984). The λ repressor fragment binds to DNA via the 'helix-turn-helix' motif.

Recently this system became the focus of a full MD simulation of the protein–DNA complex, including water and counterions (DiCapua, 1991; DiCapua and Beveridge, 1991). Separate, individual simulations of the DNA sequence (20 base pairs) and the λ repressor protein were carried out as well. The protein dynamics was analysed by means of a variance–covariance analysis of atomic fluctuations. The results, reported as a dynamical cross-correlation map (Swaminathan *et al.*, 1991), showed evidence for considerable internal correlations in the protein dimer. However, one place where correlated motions are *not* seen is in the recognition helices. Independent motions may be an advantage in this region in order for the dynamical recognition process of the operator sequence on the DNA to occur.

The corresponding simulation on the complex is the first MD simulation of an intact protein–DNA complex including solvation; 7300 water molecules and several counterions are involved in the calculation. The complex is found to be dynamically quite rigid as compared with the uncomplexed constituents (Figure 6.15), and the results support the idea that the tight binding to the DNA is primarily due to hydrogen bonding. These interactions remain very stable over the course of the simulation, with very little dynamical fluctuation in strength. Internal correlations in the protein are observed to be much weaker in the complex than in the uncomplexed form, suggesting that large-scale correlated motions are overcome by the necessity of maintaining hydrogen bond contacts.

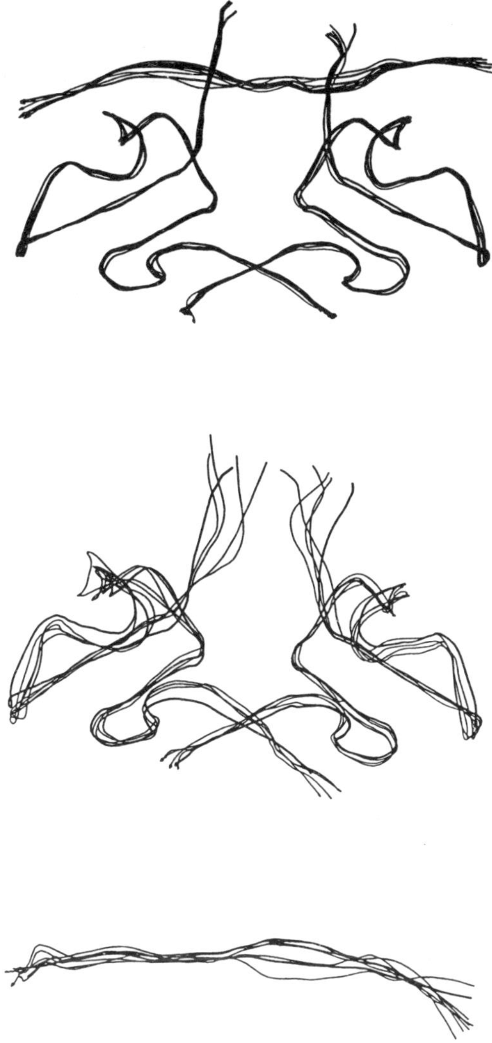

Figure 6.15 MD studies of the solvated λ repressor–operator complex based on the GROMOS force field (DiCapua, 1991; DiCapua and Beveridge, 1991): helicoidal axis dynamics of the DNA oligomer (bottom) and the protein components treated separately (middle); helicoidal axis dynamics of the protein DNA complex (top)

8 Brownian Dynamics Simulations

Computer simulations described up to this point have been based on Newtonian dynamics, treating all or essentially all atoms explicitly and carried out on the picosecond time-scale. The longest DNA simulation carried out to date with solvent included is ~500 ps (Srinivasan *et al.*,

1991). While a number of internal motions and some local bending of the DNA do occur in this frame of time, other motions of interest, such as large-scale bending motions as well as rotational and translational diffusion processes, are intrinsically slower, and thus outside the range of dynamical simulations discussed up to this point. Similar problems arise in the solvent part of the problem, when counterion diffusion is too slow to be accessible to comprehensive study via Newtonian dynamics.

A strategy for computer simulation on a longer time-scale is to simplify the system and use a correspondingly adjusted simulation methodology. Both counterion motions and DNA dynamics have recently been studied in highly simplified representations of the system by the method of Brownian dynamics, with the motion of the particles of the system described by the Langevin equation, with an added term representing the external solvent force. While a considerable methodological history underlies this process, today the Ermak–McCammon algorithm (Ermak and McCammon, 1978) is often the method of choice for integrating the equations of motion and calculating the time evolution of the system.

Several recent studies have used Brownian dynamics to study counterion motions in DNA systems, treating the DNA as lines or helices of charges and the ions as hydrated charged spheres. Reddy *et al.* (1987) performed Brownian dynamics simulations on DNA solutions at several concentrations, and considered the nature of the quadrupolar relaxation mechanism of the ^{23}Na nucleus as observed in NMR studies. The simulation results indicate that quadrupolar relaxation effects observed occur via ion motions in the vicinity of the polyion electrostatic potential, with negligible contributions originating from radial diffusion.

Guildbrand (1989) used Brownian dynamics to perform simulations of 10–50 ns on the motions of hydrated counterions in the field of charges in both a linear and a helical array of DNA charges. The static distributions, average residence times, macroscopic diffusion coefficients and quadrupolar splitting for ^{23}Na were investigated. The results for the distribution of mobile ions around the macroionic charges were generally consistent with the concepts of counterion condensation theory (Manning, 1978). For Li-DNA, for which data were available for detailed comparisons, the calculations predicted diffusion constants an order of magnitude higher than experiment for Li^+-DNA. Raising the concentration of DNA lowers the diffusion constants, presumably owing to obstructive effects. Results on quadrupole coupling constants were found to be higher than experiment, but roughly of the same order of magnitude.

A Brownian dynamics study of base-pair opening kinetics in DNA has been described recently (Briki *et al.*, 1991), considering the $(dA_5)^*(dT_5)$oligomer and assuming that the base is allowed to rotate into the major groove. The results are in good agreement with experiment, supporting the model and the stochastic nature of the process.

The application of Brownian dynamics to the study of DNA flexibility has been developed particularly by Allison (Allison, 1991; Allison *et al.*, 1989, 1990; Lewis *et al.*, 1988). The systems treated are simplified representations of DNAs consisting of 150–2350 base pairs, with a force field that allows for static bends and anisotropic bending, and involves elastic constants for bending and twisting along the chain. Trajectories produced in the Brownian dynamics simulations ranged from nanoseconds to microseconds. An ensemble of trajectories is generated in order to provide sufficient statistics for the determination of accurate correlation functions, appropriate for the time-scale of a particular experiment. The calculated results were analysed by obtaining the distribution of underlying relaxation times and choosing models which best account for observed data. This approach has been used to study flexibility of DNA as probed by fluorescence depolarization (Allison, 1986; Allison and McCammon, 1984; Song *et al.*, 1990), triplet anisotropy decay (Allison *et al.*, 1989), transient electric birefringence (Lewis *et al.*, 1988) and dynamic light scattering (Allison, 1991; Allison *et al.*, 1990). One interesting problem which emerges from these studies is the discrepancy of a factor of 2 between the static persistent length of DNA, ~500 Å, and the dynamic persistence length. The DNA seems to behave as an elastic body at short times, but somewhat more sluggishly over a long time-period. The physical basis for this is suggested to be the existence of multiple stable substates. Even so, the flexibility of DNA is likely to be of considerable importance in understanding regulatory processes in the genome at the molecular level, since crystal structures of protein–DNA complexes reveal substantial changes in twist or bending of the DNA at or near regulatory binding sites (Steitz, 1990).

In conclusion, the area of DNA supercoiling (Hao and Olson, 1989) has recently been broached in dynamical simulations. Schlick and Olson (1991) performed Langevin dynamics on a circular piece of duplex DNA, modelled by a B-spline ribbon with a number of control vertices. The energy function includes bending and twisting integrals, a Lennard-Jones description of intrachain contacts, all of which are parameterized to effectively include hydration. New techniques were designed to integrate the equations of motion (Schlick *et al.*, 1990). The results, based on trajectories of the order of nanoseconds, reveal rapid folding of the unstable circular state into supercoiled interwound forms, and are generally consistent with electron microscopy data. Significant bending and twisting motions of the interwound structures in the model are observed about the helix axis (Figure 6.16). The picture emerging from these studies is that of a highly dynamic supercoiled DNA, with considerable bending and twisting continuously occurring. These dynamical features of supercoiled DNA are expected ultimately to be of primary importance in understanding the biophysical nature and the fundamental molecular mechanisms involved in DNA recombination, transcription and replication.

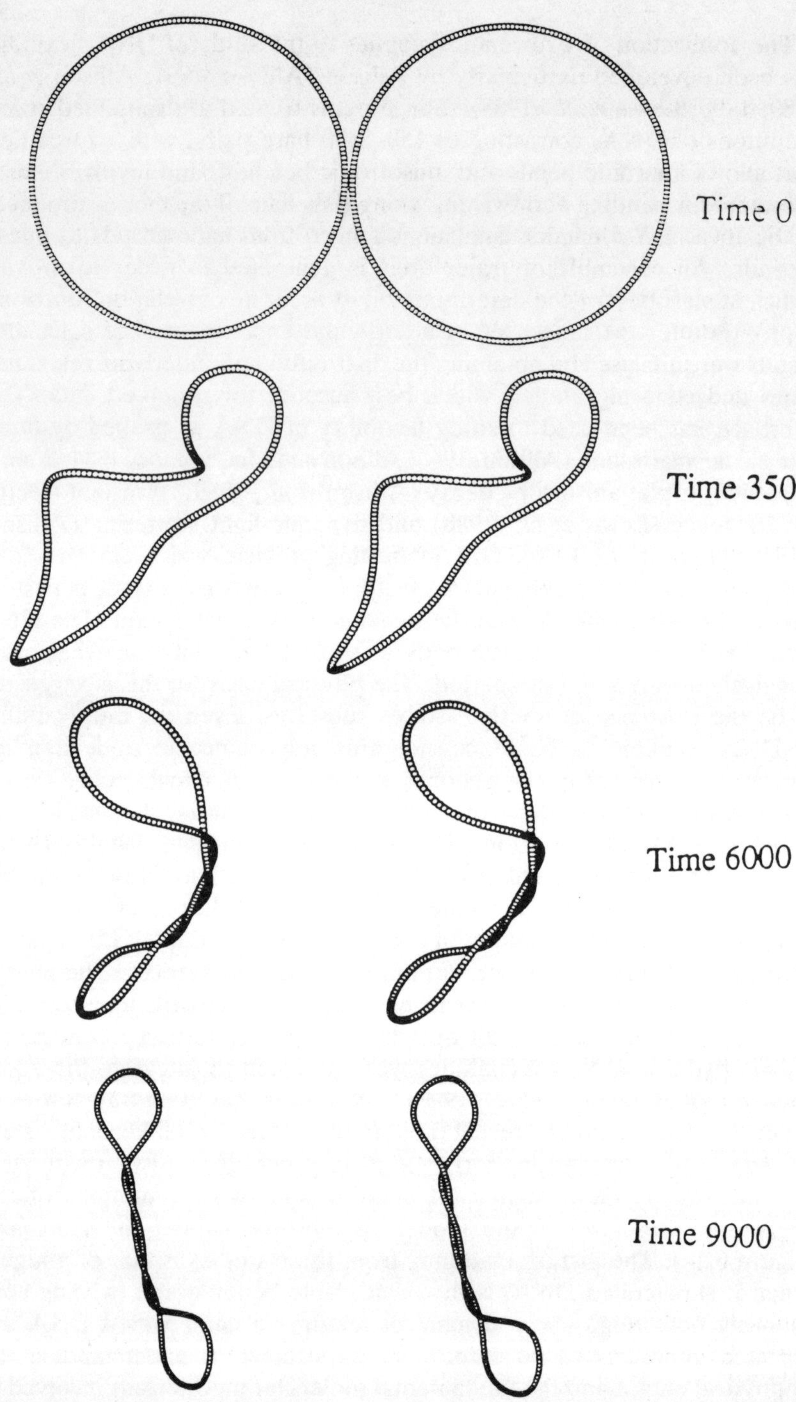

Figure 6.16 Snapshots from a stochastic dynamics simulation of DNA supercoiling (Schlick and Olson, 1991)

Acknowledgements

This study was supported by Grant No. GM-37909 from the National Institutes of Health, a Cooperative High Technology Research and Development Grant from the State of Connecticut with Bristol-Myers Squibb, and an NSF Grant for the Nucleic Acids Data Bank (H. M. Berman, PI). F. DiCapua, J. Withka and S. Louise-May are recipients of NIH Traineeships in Molecular Biophysics, Grant No. GM-08271. Interaction and advice from Dr Richard Lavery, Professor Helen Berman, Professor Eric Westhof, and colleagues participating in the series of CECAM Workshops in Paris and Orsay on nucleic acid topics are also gratefully acknowledged.

References

Abraham, F. F. (1986). Computational statistical mechanics: methodology, applications and supercomputing. *Adv. Phys.*, **35**, 1–111

Alden, C. J. and Kim, S. H. (1979). Solvent accessible surfaces of nucleic acids. *J. Mol. Biol.*, **132**, 411–434

Allen, M. P. and Tildesley, D. J. (1987). *Computer Simulation of Liquids*. Clarendon Press, Oxford

Allison, S. A. (1986). Brownian dynamics simulation of wormlike chains. Fluorescence depolarization and depolarized light scattering. *Macromolecules*, **19**, 118–124

Allison, S. A. (1991). A Brownian dynamics algorithm for arbitrary rigid bodies. *Macromolecules* (in press)

Allison, S., Austin, R. and Hogan, M. (1989). Bending and twisting dynamics of short linear DNAs. Analysis of the triplet anisotropy decay of a 209 base pair fragment by Brownian simulation. *J. Chem. Phys.*, **90**, 3843–3854

Allison, S. A. and McCammon, J. A. (1984). Multistep Brownian dynamics: Application to short wormlike chains. *Biopolymers*, **23**, 363–375

Allison, S. A., Sorlie, S. S. and Pecora, R. (1990). Brownian dynamics simulations of wormlike chains: Dynamic light scattering from a 2311 base pair fragment. *Macromolecules*, **23**, 1110–1118

Arnott, S., Campbell-Smith, P. J. and Chandrasakaran, R. (1976). In Fasman, G. (Ed.), *CRC Handbook of Biochemistry and Molecular Biology*. CRC Press, Cleveland

Arnott, S., Chandrasekeharan, R., Birdsall, D. L., Leslie, A. G. W. and Ratliffe, R. L. (1980). Left-handed DNA helices. *Nature*, **283**, 743.

Arnott, S. and Hukins, D. W. L. (1972). Optimized parameters for A and B DNA. *Biochem. Biophys. Res. Commun.*, **47**, 1504–1510

Arnott, S. and Selsing, E. (1974). Structures for the polynucleotide complexes poly(dA)–poly(dT) and poly(dA)–poly(dT)–poly(dT). *J. Mol. Biol.*, **88**, 509–521

Baleja, J. D., Germann, M. W., Van de Sande, J. H. and Sykes, B. D. (1990). Solution conformation of purine-pyrimidine DNA octamers using NMR, restrained molecular dynamics and NOE-based refinement. *J. Mol. Biol.*, **215**, 411–428

Baleja, J. D., Pon, R. T. and Sykes, B. D. (1990). Solution structure of phage lambda half-operator DNA by use of NMR, restrained molecular dynamics, and NOE-based refinement. *Biochemistry*, **29**, 4828–4839

Beglov, D. B. and Lipanov, A. A. (1991). Charge grouping approaches to calculation of electrostatic forces in molecular dynamics of macromolecules. *J. Biomol. Struct. Dyn.*, **9**, 205–214

Behe, M. J., Felsenfeld, G., Szu, S. C. and Charney, E. (1985). Temperature-dependent conformational transitions in poly(dG-dC) and poly (dG-m5dC). *Biopolymers*, **24**, 289–300

Berman, H. M. (1991). Hydration of DNA. *Curr. Opinion Struct. Biol.*, **1**, 423–427

Beveridge, D. L. and DiCapua, F. M. (1989a). Free energy via molecular simulation: A primer. In van Gunsteren, W. F. and Weiner, P. (Eds), *Computation of Free Energy for Molecular Systems*, (Eds), ESCOM, Leiden

Beveridge, D. L. and DiCapua, F. M. (1989b). Free energy via molecular simulation: Applications to chemical and biomolecular systems. *Ann. Rev. Biophys. Biophys. Chem.*, **18**, 431–492

Beveridge, D. L., Swaminathan, S., Ravishanker, G., Withka, J., Srinivasan, J., Prevost, C., Louise-May, S., DiCapua, F. M. and Bolton, P. H. (1991). Methodological considerations on molecular dynamics simulations on DNA oligonucleotides. In Lavery, R., Rivail, J.-L. and Smith, J. (Eds), *Advances in Biomolecular Simulations*. American Institute of Physics, New York

Boehncke, K., Nonella, M., Schulten, K. and Wang, A. H.-J. (1991). Molecular dynamics investigation of the interaction between DNA and distamycin. *Biochemistry*, **30**, 5465–5475

Boelens, R., Koning, T. M. G., van der Marel, G. A., Van Bloom, J. H. and Kaptein, R. (1989). Iterative procedure for structure determination from proton-proton NOe's using a full matrix relaxation approach. Application to a DNA octamer. *J. Mag. Res.*, **82**, 290

Borgias, G. A. and James, T. L. (1988). COMATOSE: a method for constrained refinement of macromolecular structure based on two-dimensional Nuclear Overhauser Spectra. *J. Mag. Res.*, **79**, 493–512

Brahms, S., Fritsch, V., Brahms, J. G. and Westhof, E. (1992). Investigations on the dynamic structures of adenine and thymine containing DNA. *J. Mol. Biol.*, **223**, 455–476

Briki, F., Ramstein, J. and Lavery, R. (1991). Evidence for the stochastic nature of base pair opening in DNA: A Brownian dynamics simulation. *J. Am. Chem. Soc.*, **113**, 2490–2493

Brünger, A. T. (1990). Refinement of three-dimensional structures of proteins and nucleic acids. In Goodfellow, J. M. (Ed.) *Molecular Dynamics: Applications in Molecular Biology*. Macmillan Press, London

Burkhoff, A. M. and Tullius, T. D. (1987). The unusual conformation adopted by the adenine tracts in kinetoplast DNA. *Cell*, **48**, 935–943

Cheng, Y. K. and Pettitt, B. M. (1992). Hoogsteen vs reversed-Hoogsteen base pairing: DNA triple helices. *J. Amer. Chem. Soc.*, **114**, 4465–4474

Chuprina, V. P., Heinemann, U., Nurislamov, A. A., Zielenkiewicz, P., Dickerson, R. E. and Saenger, W. (1991). Molecular dynamics of the hydration shell of a B-DNA decamer reveals two main types of minor groove hydration depending on groove width. *Proc. Natl Acad. Sci. USA*, **88**, 593–597

Cieplak, P., Bash, P., Singh, U. C. and Kollman, P. A. (1987). A theoretical study of tautomerism in the gas phase and aqueous solution: a combined use of state-of-the-art *ab initio* quantum mechanics and free energy-perturbation methods. *J. Am. Chem. Soc.*, **109**, 6283–6289

Cieplak, P. and Kollman, P. A. (1988). Calculation of the free energy of association of nucleic acid bases *in vacuo* and in water solution. *J. Am. Chem. Soc.*, **110**, 3734–3939

Cieplak, P., Rao, S. N., Grootenhuis P. D. J. and Kollman, P. A. (1990). Free energy calculation on base specificity of drug–DNA interactions: application to daunomycin and acridine intercalation into DNA. *Biopolymers*, **29**, 717–727.

Clore, G. M., Oschkinat, H., McLaughlin, L. W., Benseler, F., Happ, C. S., Happ, E. and Gronenborn, A. M. (1988). Refinement of the solution structure of the DNA dodecamer 5′d(CGCGPATTCGCG)2 containing a stable purine-thymine base pair: combined use of nuclear magnetic resonance and restrained molecular dynamics. *Biochemistry*, **27**, 4185–4197

Cooney, M., Czernuszewicz, G., Postel, E. H., Flint, J. and Hogan, M. E. (1988). Site-specific oligonucleotide binding represses transcription of the human c-myc gene *in vitro*. *Science*, **241**, 456–459

Creighton, S., Rudolph, B., Lybrand, T., Singh, U. C., Shafer, R., Brown, S., Kollman, P., Case, D. A. and Andrea, T. (1989). A combined 2D-NMR and molecular dynamics analysis of the structure of the actinomycin D: d(ATGCAT)2 complex. *J. Biomol. Struct. Dyn.*, **6**, 929–969

Dang, L. X. and Kollman, P. A. (1990). Molecular dynamics simulations study of the free energy of association of 9-methyladenine and 1-methylthymine bases in water. *J. Am. Chem. Soc.*, **112**, 503–507

DiCapua, F. (1991). *Molecular Dynamics and Monte Carlo Studies of Protein Stability and Protein–DNA Interactions*. PhD Thesis, Wesleyan University

DiCapua, F. M. and Beveridge, D. L. (1991). Molecular dynamics studies of the lambda repressor–operator protein–DNA complex. *Biochemistry* (submitted)

Dickerson, R. E. (1983). The DNA helix and how it is read. *Sci. Am.*, **249**, 94–111

Dickerson, R. E. (1988). Usual and unusual DNA structures: A summing up. In Wells, R. D. and Harvey, S. C. (Eds), *Unusual DNA Structures*. Springer Verlag, New York

Dickerson, R. E. (1991). DNA structure from A to Z (preprint)

Dickerson, R. E. (1990). What do we really know about B-DNA? In Sarma, R. H. and Sarma, M. H. (Eds), *Structure and Methods*, Vol. 3: *DNA and RNA*. Adenine Press, Schenectady, N.Y.

Dougherty, A. M., Causley, G. C. and Johnson Jr, W. C. (1983). Flow dichroism evidence for tilting of the bases when DNA is in solution. *Proc. Natl Acad. Sci. USA*, **80**, 2193–2195

Drew, H. R. and Dickerson, R. E. (1981). Structure of a B-DNA dodecamer. III. Geometry of hydration. *J. Mol. Biol.*, **151**, 535–556

Drew, H. R., Samson, S. and Dickerson, R. E. (1982). Structure of a B-DNA dodecamer at 16 K. *Proc. Natl Acad. Sci. USA*, **79**, 4040

Drew, H. R., Wing, R. M., Takano, T., Broka, C., Tanaka, S., Itikura, K. and Dickerson, R. E. (1981). Structure of a B DNA dodecamer. I. Conformation and dynamics. *Proc. Natl Acad. Sci. USA*, **78**, 2179–2983

Eimer, W., Williamson, J. R., Boxer, S. G. and Pecora, R. (1982). Kinetics for exchange of imino protons in the d(CGCGAATTCGCG) double helix, and in two similar helices that contain a GT base pair. *Biochemistry*, **21**, 6567

Eimer, W., Williamson, J. R., Boxer, S. G. and Pecora, R. (1990). Characterization of the overall and internal dynamics of short oligonucleotides by depolarized dynamic light scattering and NMR relaxation measurements. *Biochemistry*, **29**, 799–811

Erikksson, M. A. L. and Laaksonen, A. (1992). A molecular dynamics study of conformational changes and hydration of left-handed d(CGCGCGCGCGCG)2 in a non-salt solution. *Biopolymers*, **32**, 1035–1059

Ermak, D. L. and McCammon, J. A. (1978). Brownian dynamics with hydrodynamic interactions. *J. Chem. Phys.*, **69**, 1352

Falk, M. K. A., Hartman, J. and Lord, R. C. (1962). Hydration of deoxyribonucleic acid. I. A gravimetric study. *J. Am. Chem. Soc.*, **84**, 3843–3846

Falk, M. K. A., Hartman, J. and Lord, R. C. (1963a). Hydration of deoxyribonucleic acid. II. An infrared study. *J. Am. Chem. Soc.*, **85**, 387–391

Falk, M. K. A., Hartman, J. and Lord, R. C. (1963b). Hydration of deoxyribonucleic acid. III. A spectroscopic study of the effect of hydration on the structure of deoxyribonucleic acid. *J. Am. Chem. Soc.*, **85**, 391–394

Falk, M., Poole, A. G. and Goyman, C. G. (1970). IR study of the state of water in the hydration shell of DNA. *Can. J. Chem.*, **48**, 1536–1542

Felsenfeld, G., Davies, D. R. and Rich, A. (1957). Formation of a three-stranded polynucleotide molecule. *J. Am. Chem. Soc.*, **57**, 2023–2024

Feuerstein, B. G., Pattabiraman, N. and Marton, L. J. (1989). Molecular dynamics of spermine–DNA interactions: sequence specificity and DNA bending for a simple ligand. *Nucleic Acids Res.*, **17**, 6883–6892

Forester, T. R. and McDonald, I. R. (1991). Molecular dynamics studies of the behavior of water molecules and small ions in concentrated solutions of polymeric B-DNA. *Mol. Phys.*, **72**, 643–660

Franklin, R. E. and Gosling, R. G. (1953a). Molecular configuration in sodium thymonucleate. *Nature*, **171**, 740–741

Franklin, R. E. and Gosling, R. G. (1953b). The structure of sodium thymonucleate fibers. I. The influence of water content. *Acta Cryst.*, **6**, 673–677

Fratini, A. V., Kopka, M. L., Drew, H. R. and Dickerson, R. E. (1982). Reversible bending and helix geometry in a B-DNA dodecamer: CGCGAATT(Br)CGCG. *J. Biol. Chem.*, **257**, 14686–14707

Fritsch, V. and Westhof, E. (1990). Minimization and molecular dynamics of Z-DNA modified by acetylaminofluorene. *Stud. Phys. Theoret. Chem.*, **71**, 627–634

Fritsch, V. and Westhof, E. (1991a). Minimization and molecular dynamics studies of guanosine and Z-DNA modified by N-2-acetylaminofluorene. *J. Comput. Chem.*, **12**, 147–166

Fritsch, V. and Westhof, E. (1991b). Molecular dynamics simulations of DNA oligomers under various electrostatic parameters. In Lavery, R., Rivail, J. L. and Smith, J. (Eds), *Advances in Biomolecular Simulations*. American Institute of Physics, New York

Fritsch, V. and Westhof, E. (1991c). Three center hydrogen bonds in DNA: Molecular dynamics of poly (dA)*poly (dT). *J. Am. Chem. Soc.*, **113**, 8271–8277

Fujii, S., Oda, Y., Uesugi, S., Ohtsuka, E. and Tomita, K. (1988). Molecular dynamics simulations of d(CGCIAAT)*d(ATTAGCG) with anti-anti and anti-syn orientations in I-A base pair. *Nucleic Acids Symp. Ser.*, **20**, 127–128

Fuller, W. and Mahendrasingam, A. (1987). X-Ray fibre diffraction studies of DNA: Recent results and future possibilities. In Neidle, S. and Fuller, W. (Eds), *Nucleic Acid Structure*. Macmillan Press, London

Gochin, M. and James, T. L. (1990). Solution structure studies of d(AC)4*d(GT)4 via restrained molecular dynamics simulations with NMR constraints derived from two-dimensional NOE and double-quantum-filtered COSY experiments. *Biochemistry*, **29**, 11172–11180

Gochin, M., Zon, G. and James, T. L. (1990). Two-dimensional COSY and two-dimensional NOE spectroscopy of d(AC)4*d(GT)4: Extraction of structural constraints. *Biochemistry*, **29**, 11161–11171

Gronenborn, A. M. and Clore, G. M. (1989). Analysis of the relative contributions of the nuclear Overhauser interproton distance restraints and the empirical energy function in the calculation of oligonucleotide structures using restrained molecular dynamics. *Biochemistry*, **28**, 5978–5984

Guldbrand, L. (1989). The distribution and dynamics of small ion in simulations of ordered polyelectrolyte solutions. *Mol. Phys.*, **67**, 217–237

Gulotta, M., Goss, D. J. and Diem, M. (1989). IR vibrational CD of model deoxyribonucleotides: observation of the B to Z transition and extended coupled oscillator calculations. *Biopolymers*, **28**, 2047–2058

Haerd, T. (1987). Out-of-plane mobility in the ethidium/DNA complex. *Biopolymers*, **26**, 613–618

Hansen, J. P. and McDonald, I. R. (1976). *Theory of Simple Liquids*. Academic Press, New York

Hao, M. H. and Olson, W. K. (1989). The global equilibrium configurations of supercoiled DNA. *Macromolecules*, **22**, 3292–3303

Hare, D. R., Wemmer, D. E., Chou, S. H., Drobny, G. and Ried, B. R. (1983). Assignment of the non-exchangeable proton resonances of d(CGCGAATTCGCG) using 2D-NMR methods. *J. Mol. Biol.*, **171**, 319–336

Hausheer, F. H. (1990). Dynamic properties and electrostatic potential surfaces of neutral DNA heteropolymers. *J. Am. Chem. Soc.*, **112**, 9468–9474

Hausheer, F. H., Singh, U. C., Saxe, J. D., Colvin, O. M. and T'so, P. O. P. (1990). Can oligonucleoside methylphosphonates form a stable triplet with a double DNA helix? *Anti-Cancer Drug Design*, **5**, 159–167

Herzyk, P., Goodfellow, J. M. and Neidle, S. (1991). Molecular dynamics simulations of dinucleoside and dinucleoside drug crystal hydrates. *J. Biomol. Struct. Dyn.*, **9**, 363–386

Hirshberg, M., Sharon, R. and Sussman, J. L. (1988). A kinked model for the solution structure of DNA tridecamers with inserted adenosines: Energy minimization and molecular dynamics. *J. Biomol. Struct. Dyn.*, **5**, 965–979

Ho, P. S., Quigley, G. J., Tilton, R. F. J. and Rich, A. (1988). Hydration of methylated and non-methylated B-DNA and Z-DNA. *J. Phys. Chem.*, **92**, 939–945

Huston, S. E. and Rossky, P. J. (1989). Free energies of association for the sodium-dimethyl phosphate ion pair in aqueous solution. *J. Phys. Chem.*, **93**, 7888–7895

Israelchavelli, J. (1985). Solvation forces and liquid structure, as probed by direct force measurements. *Acc. Chem. Res.*, **20**, 415–421

Ito, N., Nakamura, H., Sumikawa, H., Nagashima, N., Arata, Y. and Nishimura, Y. (1991). Structure of a DNA octamer, d(CCTTAAGG) obtained by restrained molecular dynamics based on Raman and NMR data. *J. Mol. Struct.*, **242**, 119–123

Jayaram, B. and Beveridge, D. L. (1990). Grand canonical Monte Carlo simulations on aqueous solutions of NaCl and NaDNA: Excess chemical potentials and sources of non-ideality in electrolyte and polyelectrolyte solutions. *J. Phys. Chem.*, **95**, 2506–2516

Jayaram, B., DiCapua, F. M. and Beveridge, D. L. (1991). A theoretical study of polyelectrolyte effects in protein–DNA interactions: Monte Carlo free energy simulations on the ion atmosphere contribution to the thermodynamics of lambda repressor–operator complex formation. *J. Am. Chem. Soc.*, **113**, 5211–5215

Jayaram, B., Swaminathan, S., Beveridge, D. L., Sharp, K. and Honig, B. (1990). Monte Carlo simulation studies on the structure of the counterion atmosphere of B-DNA. Variations on the primitive dielectric model. *J. Phys. Chem.*, **23**, 3156–3165

Kaluarachi, K., Meadows, R. P. and Gorenstein, D. G. (1991). How accurately can oligonucleotide structures be determined from the hybrid relaxation rate matrix/NOESY distance restrained molecular dynamics approach? *Biochemistry*, **30**, 8785–8797

Karplus, M. and McCammon, A. J. (1986). The dynamics of proteins. *Sci. Am.*, **254**, 42–51

Katahira, M., Sugeta, H., Kyogoku, Y. and Fujii, S. (1990). Determination of the conformation of d(GGAAATTTCC)2 in solution by use of proton NMR and restrained molecular dynamics. *Biochemistry*, **29**, 7214–7222

Kennard, O. (1984). DNA from A-Z: A survey of oligonucleotide structures. *Pure and Appl. Chem.*, **56**, 989–1004

Kennard, O. and Hunter, W. N. (1989). Oligonucleotide structure: A decade of results from single crystal X-ray diffraction studies. *Q. Rev. Biophys.*, **22**, 327–329

Kerwood, D. J., Zon, G. and James, T. L. (1991). Structure determination of d(ATATATAUAT) via 2D-NOE spectroscopy and molecular dynamics calculations. *Eur. J. Biophys.*, **197**, 583–595

Koehler, J. E. H., Saenger, W. and van Gunsteren, W. F. (1987a). A molecular dynamics simulation of crystalline alpha-cyclodextrin hexahydrate. *Eur. J. Biophys.*, **15**, 197–210

Koehler, J. E. H., Saenger, W. and van Gunsteren, W. F. (1987b). A molecular dynamics simulation of crystalline beta-cyclodextrin dodecahydrate at 293 K and 120 K. *Eur. J. Biophys.*, **15**, 211–224

Koehler, J. E. H., Saenger, W. and van Gunsteren, W. F. (1988a). Conformational differences between alpha-cyclodextrin in aqueous solution and in crystalline form: A molecular dynamics study. *J. Mol. Biol.*, **203**, 241–250

Koehler, J. E. H., Saenger, W. and van Gunsteren, W. F. (1988b). The flip–flop hydrogen bond phenomena: A molecular dynamics simulation of beta-cyclodextrin. *Eur. J. Biophys.*, **16**, 153–168

Koehler, J. E. H., Saenger, W. and van Gunsteren, W. F. (1989). On the occurrence of three center hydrogen bonds in cyclodextrins in crystalline form and in aqueous solution: Comparison of neutron diffraction and molecular dynamics results. *J. Biomol. Struct. Dyn.*, **6**, 181–198

Kollman, P. A., Weiner, P., Quigley, G. and Wang, A. (1982). Molecular mechanical studies of Z-DNA: A comparison of the structural and energetic properties of Z and B DNA. *Biopolymers*, **21**, 1945–1969

Koning, T. M. G., Boelens, R., van der Marel, J. H. and Kaptein, R. (1991). Structure determination of a DNA octamer in solution by NMR spectroscopy: Effect of fast local motions. *Biochemistry*, **30**, 3787–3797

Kumar, S. (1990). *Dynamical Behavior of DNA*. PhD Thesis, University of Pittsburgh

Kumar, S., Kollman, P. A. and Rosenberg, J. M. (1991). Dynamical behavior of kinky and straight DNA. *J. Biomol. Struct. Dyn.*, **8**, a114

Laaksonen, A., Nilsson, L. G., Joensson, B. and Teleman, O. (1989). Molecular dynamics simulation of double helix Z-DNA in solution. *Chem. Phys.*, **129**, 175–183

Lane, A. N. (1990). The determination of conformational properties of nucleic acids in solution from NMR data. *Biochim. Biophys. Acta*, **1049**, 189–204

Lane, A., Jenkins, T. C., Brown, T. and Neidle, S. (1991). Interaction of Berenil with the EcoRI dodecamer d(CGCGAATTCGCG) in solution studied by NMR. *Biochemistry*, **30**, 1372–1385

Langley, D. R., Doyle, T. W. and Beveridge, D. L. (1991). The Dynemicin-DNA intercalation complex. A model based on DNA affinity cleavage and molecular dynamics simulation. *J. Am. Chem. Soc.*, **113**, 4395–4403

Lavery, R. and Sklenar, H. (1988). The definition of generalized helicoidal parameters and of axis curvature for irregular nucleic acids. *J. Biomol. Struct. Dyn.*, **6**, 63–91

Lee, W. K., Gao, Y. and Prohofsky, E. W. (1984). Structure of hydrated Na^+ ions around a region of A- or B-DNA helix. *Biopolymers*, **23**, 257–270

Levitt, M. (1983). Computer simulation of DNA double-helix dynamics. *Cold Spring Harbor Symp. Quant. Biol.*, **47**, 251–262

Levitt, M. and Warshel, A. (1978). Extreme conformational flexibility of the furanose ring in DNA and RNA. *J. Am. Chem. Soc.*, **100**, 2607–2613

Lewis, R. J., Allison, S. A., Eden, D. and Pecora, R. (1988). Brownian dynamics simulations of a three subunit and a ten subunit worm-like chain: Comparison of results with Trumbell theory and with experimental results from DNA. *J. Chem. Phys.*, **89**, 2490–2503

Louise-May, S. and Beveridge, D. L. (1990). Molecular dynamics of d(CGCGAATTCGCG) based on the CHARMM force field. *Biopolymers* (in preparation)

Louise-May, S. and Beveridge, D. L. (1991). Molecular dynamics of d(CGCGAATTCGCG) based on the CHARMM force field. *Biopolymers* (ms in preparation)

Luo, J., Sarma, M., Gupta, G. and Sarma, R. H. (1991). DNA bending studied by MD and NOESY simulations: Role of the junction sequence between two A/T tracts. Preprint

McCammon, J. A. (1991). Free energy from simulations. *Current Opinion in Structural Biology*, **1**, 196–200

McCammon, A. J. and Harvey, S. C. (1986). *Dynamics of Proteins and Nucleic Acids*. Cambridge University Press, Cambridge

Manning, G. S. (1978). The molecular theory of polyelectrolyte solutions with applications to the electrostatic properties of polynucleotides. *Q. Rev. Biophys.*, **11**, 179–246

Metzler, W. J., Wang, C., Kitchen, D. B., Levy, R. M. and Pardi, A. (1990). Determining local conformational variations in DNA. Nuclear magnetic resonance structures of the DNA duplexes d(CGCCTAATCG) and d(CGTCACGCGCG) generated using back calculation of the nuclear Overhauser effect, a distance geometry algorithm and restrained molecular dynamics. *J. Mol. Biol.*, **214**, 711–736

Mezei, M., Beveridge, D. L., Berman, H. M., Goodfellow, J. M., Finney, J. L. and Neidle, S. (1983). Monte Carlo studies on water in the dCpG/proflavin crystal hydrate. *J. Biomol. Struct. Dyn.*, **1**, 287–297

Miaskiewicz, K., Osman, R. and Weinstein, H. (1992). Molecular dynamics simulation of the hydrated d(CGCGAATTCGCG) dodecamer. *J. Am. Chem. Soc.* (submitted)

Miriganik and Kothekar, V. (1991). 100 ps molecular dynamics of d(TATCACC). *J. Biomol. Struct. Dyn.*, **8**, 1147–1167

Moe, J. G. and Russu, I. M. (1990). Proton exchange and base pair opening kinetics in 5′-d(CGCGAATTCGCG)-3′ and related dodecamers. *Nucleic Acids Res.*, **18**, 821–827

Neidle, S., Berman, H. M. and Shieh, H. S. (1980). Highly structured water network in crystals of a deoxydinucleside–drug complex. *Nature*, **288**, 129–133

Nerdal, W., Hare, D. R. and Ried, B. R. (1989). Solution structure of the Eco RI DNA sequence. Refinement of NMR derived distance geometry structures by

NOESY spectrum back calculations. *Biochemistry*, **28**, 10008–10021

Nikonowicz, E. P. and Gorenstein, D. G. (1990). Two-dimensional proton and phosphorus-31 NMR spectra and restrained molecular dynamics structure of a mismatched GA decamer oligodeoxyribonucleotide duplex. *Biochemistry*, **29**, 8845–8858

Nikonowicz, E., Roongta, V., Jones, C. R. and Gorenstein, D. G. (1989). Two-dimensional proton and phosphorus-31 NMR spectra and restrained molecular dynamics structure of an extrahelical adenosine tridecamer oligodeoxyribonucleotide duplex. *Biochemistry*, **28**, 8714–8725

Nilsson, L., Clore, G. M., Gronenborn, A. M., Brunger, A. J. and Karplus, M. (1986). Structure refinement of oligonucleotides by molecular dynamics with nuclear Overhauser effect interproton distance restraints: Application to 5′ d(CGTACG). *J. Mol. Biol.*, **188**, 455–475

Nilges, M., Clore, G. M., Gronenborn, A. M., Brunger, A. T., Karplus, M. and Nilsson, L. (1987a). Refinement of the solution structure of the DNA hexamer d(GCATGC): Combined use of nuclear magnetic resonance and restrained molecular dynamics. *Biochemistry*, **26**, 3718–3733

Nilges, M., Clore, G. M., Gronenborn, A., Piel, N. and McLaughlin, L. W. (1987b). Refinement of the solution structure of the DNA decamer d(CTGGATCCAG): Combined use of nuclear magnetic resonance and restrained molecular dynamics. *Biochemistry*, **26**, 3734–3744

Nilsson, L. and Karplus, M. (1984). Energy functions for energy minimization and dynamics of nucleic acids. *J. Comp. Chem.*, **1**, 591–616

Nordlund, T. M., Andersson, S., Nilsson, L., Rigler, R., Graeslund, A. and McLaughlin, L. W. (1989). Structure and dynamics of a fluorescent DNA oligomer containing the EcoRI recognition sequence: fluorescence, molecular dynamics, and NMR studies. *Biochemistry*, **28**, 9095–9103

O'Handley, S. F., Sanford, D. G., Xu, R., Lester, C. C., Hingerty, B. R., Broyde, S. and Krugh, T. R. (1991). Structure of an acetylaminofluorene modified DNA oligomer (ms submitted)

Olson, W. K. (1982a). How flexible is the furanose ring? 1. A comparison of experimental and theoretical studies. *J. Am. Chem. Soc.*, **104**, 270–278

Olson, W. K. (1982b). How flexible is the furanose ring? An updated potential energy estimate. *J. Am. Chem. Soc.*, **104**, 278–286

Olson, W. (1982c). Theoretical studies of nucleic acid conformation: Potential energies, chain statistics, and model building. In Neidle, S. (Ed.), *Topics in Nucleic Acid Structure*. Macmillan, London

Osman, R., Miaskiewicz, K. and Weinstein, H. (1991). Structure–function relations in radiation damaged DNA. In Varma, M. and Glass, W. (Eds), *Structure–Function Relations in Radiation Damaged DNA*. Plenum Press, New York

Ott, J. and Eckstein, F. (1985). Phosphorous NMR spectral analysis of the dodecamer d(CGCGAATTCGCG). *Biochemistry*, **24**, 2530–2535

Pabo, C. O. and Sauer, R. T. (1984). Protein-DNA recognition. *Ann. Rev. Biochem.*, **53**, 293–321

Pardi, A. R. and Wang, C. (1988). Determination of DNA structures by NMR and distance geometry techniques: A computer simulation. *Proc. Natl Acad. Sci. USA*, **85**, 8785–8789

Parsegian, V. A., Rand, R. P. and Rau, D. C. (1985). Hydration forces: what next? *Chem. Scripta*, **25**, 28–31

Patel, D., Pardi, A. and Itakura, K. (1982). DNA conformation, dynamics, and interactions in solution. *Science*, **216**, 581–590

Patel, D. J. and Shapiro, L. (1987). Nuclear magnetic resonance and distance geometry studies of DNA structures in solution. *Ann. Rev. Biophys. Biophys. Chem.*, **16**, 423–54

Pearlman, D. A. (1991). Are time-averaged restraints necessary for nuclear magnetic resonance refinement? *J. Mol. Biol.*, **18**, 457–479

Pearlman, D. A. and Kollman, P. A. (1990). The calculated free energy effects of 5-methyl cytosine on the B to Z transition in DNA. *Biopolymers*, **29**, 1193–1209

Pearlman, D. A. and Kollman, P. A. (1991). Evaluating the assumptions underlying force field development and application using free energy conformation maps for nucleosides. *J. Am. Chem. Soc.*, **113**, 7167–7177

Pieters, J. M. L., De Vroom, E., Van der Marel, G. A., Van Boom, J. H., Koning, T. M. G., Kaptein, R. and Altona, C. (1990). Hairpin structures in DNA containing arabinofuranosylcytosine. A combination of nuclear magnetic resonance and molecular dynamics. *Biochemistry*, **29**, 788–799

Powers, R. and Gorenstein, D. G. (1990). Two-dimensional proton and phosphorus-31 NMR spectra and restrained molecular dynamics structure of a covalent CPI–CDPI2–oligodeoxyribonucleotide decamer complex. *Biochemistry*, **29**, 9994–10008

Powers, R., Jones, C. R. and Gorenstein, D. G. (1990). Two-dimensional proton and phosphorus NMR spectra and restrained molecular dynamics structure of an oligodeoxyribonucleotide duplex refined via a hybrid relaxation matrix. *J. Biomol. Struct. Dyn.*, **8**, 253–294

Prabhakaran, M. and Harvey, S. C. (1985). Molecular dynamics anneals large-scale deformations of model macromolecules: stretching the DNA double helix to form an intercalation site. *J. Phys. Chem.*, **89**, 5767–5769

Prabhakaran, M. and Harvey, S. C. (1988). Molecular dynamics of structural transitions and intercalation in DNA. *Biopolymers*, **27**, 1239–1248

Prabhakaran, M., Harvey, S. C., Mao, B. and McCammon, J. A. (1983). Molecular dynamics of phenylalanine transfer RNA. *J. Biomol. Struct. Dyn.*, **1**, 357–369

Pranata, J. and Jorgensen, W. L. (1991). Monte Carlo simulations yield absolute free energies of binding for guanine–cytosine and adenine–uracil base pairs in chloroform. *Tetrahedron*, **47**, 2491–2501

Prevost, C., Louise-May, S., Ravishanker, G., Lavery, R. and Beveridge, D. L. (1991). Persistence analysis of the static and dynamical helix deformations of DNA oligonucleotides: Application to the crystal structure and molecular dynamics simulations of d(CGCGAATTCGCG)2. *Biopolymers* (in press)

Prive, G. G., Yanagi, K. and Dickerson, R. E. (1991). The structure of the B-DNA decamer d(CCAACGTTGG), and comparison with the isomorphic decamers d(CCAAGATTGG) and d(CCAGGCCTGG). *J. Mol. Biol.*, **217**, 177–199

Pullman, A., Pullman, B. and Berthod, H. (1978). An SCF *ab initio* investigation of the 'through-water' interaction of the phosphate-anion with the Na^+ cation. *Theoret. Chim. Acta (Berlin)*, **47**, 175–192

Rajagopal, P. and Feigon, J. (1989). NMR studies of triple-strand formation from the homopurine-homopyrimidine deoxyribonucleosides d(GA)4 and d(TC)4. *Biochemistry*, **28**, 7859–7870

Rao, S. N. and Kollman, P. A. (1990). Simulations of the B-DNA molecular dynamics of d(CGCGAATTCGCG) and d(CGCGCGCGCGCG): An analysis of the role of initial geometry and a comparison of united and all-atom models. *Biopolymers*, **29**, 517–532

Rao, S. N., Singh, U. C. and Kollman, P. A. (1986). Molecular dynamics simulations of DNA double helices: studies of sequence dependence and the role of mismatch pairs in the DNA helix. *Israel J. Chem.*, **27**, 189–197

Rau, D. C., Lee, B. and Parsegian, V. A. (1984). Measurement of the repulsive force between polyelectrolyte molecules in ionic solution: Hydration forces between parallel double helices. *Proc. Natl Acad. Sci. USA*, **81**, 2621–2625

Ravishanker, G., Swaminathan, S., Beveridge, D. L., Lavery, R. and Sklenar, H. (1989). Conformational and helicoidal analysis of 30 psec of molecular dynamics on the d(CGCGAATTCGCG) double helix. *J. Biomol. Struct. Dyn.*, **6**, 669–699

Reddy, M. R. and Berkowitz, M. (1989). Hydration forces between parallel DNA double helices: Computer simulations. *Proc. Natl Acad. Sci. USA*, **86**, 3165–3168

Reddy, M. R., Rossky, P. J. and Murthy, C. S. (1987). Counterion spin relaxation in DNA solutions: A stochastic dynamics simulation study. *J. Phys. Chem.*, **91**, 4923–4933

Rinkel, L. J. and Altona, C. A. (1987). Conformational analysis of the deoxyribofuranose ring in DNA by means of sums of proton–proton coupling constants. *J. Biomol. Struct. Dyn.*, **4**, 621–649

Rudolph, B. R. and Case, D. A. (1989). Harmonic dynamics of a DNA hexamer in the absence and presence of the intercalator ethidium. *Biopolymers*, **28**, 851–871

Russu, I. M. (1991). Studying protein–DNA interactions using NMR. *TIBTech.*, **9**, 96–104

Saenger, W. (1983). *Principles of Nucleic Acid Structure*. Springer, New York

Saenger, W. (1987). Structure and dynamics of water surrounding biomolecules. *Ann. Rev. Biophys. Biophys. Chem.*, **16**, 93

Savage, H. F. (1986). Water structure in crystalline solids: Ices to proteins. In Franks, F. (Ed.), *Water Science Reviews*, Vol. 2. Cambridge University Press, Cambridge, pp. 67–148

Scalfi Happ, C., Happ, E., Clore, G. M. and Gronenborn, A. M. (1988). Refinement of the solution structure of the RNA-DNA hybrid 5′-[r(GCA)d(TGC)]$_2$. Combined use of nuclear magnetic resonance and restrained molecular dynamics. *FEBS Lett.*, **236**, 62–70

Schlick, T., Hingerty, B., Peskin, C. S., Overton, M. L. and Broyde, S. (1990). Search strategies, minimization algorithms, and molecular dynamics simulations for exploring conformational spaces of nucleic acids. In Beveridge, D. L. and Lavery, R. (Eds), *Theoretical Chemistry and Molecular Biophysics*. Adenine Press, Schenectady, N.Y.

Schlick, T. and Olson, W. (1991). Supercoiled DNA energetics and dynamics by computer simulation (submitted for publication)

Schmitz, U., Pearlman, D. A. and James, T. L. (1991). Solution structure of d(GTATATAC) via restrained molecular dynamics simulations with NMR constraints derived from relaxation matrix analysis of 2D NOE experiments. *J. Mol. Biol.* (in press)

Shakked, Z. and Rabinovich, D. (1986). The effect of base sequence on the fine structure of the DNA double helix. *Prog. Biophys. Mol. Biol.*, **47**, 159–195

Shibata, M., Zielinski, T. J. and Rein, R. (1991). A molecular dynamics study of the effect of GT mispairs on the conformation of DNA in solution. *Biopolymers*, **31**, 211–232

Shieh, H. S., Berman, H. M. and Neidle, S. (1980). The structure of a drug–deoxynucleoside phosphate complex: Generalized conformational behaviour of intercalation complexes with DNA and RNA fragments. *Nucleic Acids Res.*, **8**, 85–97

Siebel, G. L., Singh, U. C. and Kollman, P. A. (1985). A molecular dynamics simulation of double-helical B-DNA including counterions and water. *Proc. Natl Acad. Sci. USA*, **82**, 6537–6540

Singh, S. (1991). *Dynamics of DNA and Drug–DNA Complexes*. PhD Thesis, New York University

Singh, S. B., Hingerty, B. E., Singh, U. C., Greenberg, J. P., Geacintov, N. E. and Broyde, S. (1991). Structures of the (+) and (−) - trans BPDE adducts to guanine-N2 in a duplex dodecamer. *Cancer Res.* (in press)

Singh, U. C., Weiner, S. J. and Kollman, P. A. (1985). Molecular dynamics simulations of d(CGCGA)* d(TCGCG) with and without 'hydrated' counterions. *Proc. Natl Acad. Sci. USA*, **82**, 755–759

Sklenar, V. and Feigon, J. (1990). Formation of a stable triplex from a single DNA strand. *Nature*, **345**, 836–838

Song, L., Allison, S. A. and Schurr, J. M. (1990). Normal mode theory for the Brownian dynamics of a weakly bending rod. Comparison with Brownian dynamics simulations. *Biopolymers*, **29**, 1773

Srinivasan, J., Withka, J. M. and Beveridge, D. L. (1990). Molecular dynamics of an *in vacuo* model of duplex d(CGCGAATTCGCG) in the B form based on the amber force field. *Biophys. J.*, **58**, 533–547

Srinivasan, J., Withka, J. M. and Beveridge, D. L. (1992). Ms in preparation

Srinivasan, J., Withka, J. M., Swaminathan, S., Beveridge, D. L. and Bolton, P. H. (1991). Dynamical structure of DNA in solution: Comparison of theoretical molecular dynamics and experimental NMR coupling constant results. *J. Am. Chem. Soc.* (ms in preparation)

Steitz, T. A. (1990). Structural studies of protein–nucleic acid interactions: The sources of sequence-specific binding. *Q. Rev. Biophys.*, **23**, 205–280

Subramanian, P. S. and Beveridge, D. L. (1989). A theoretical study of the aqueous hydration of canonical B d(CGCGAATTCGCG): Monte Carlo simulation and comparison with crystallographic ordered water sites. *J. Biomol. Struct. Dyn.*, **6**, 1093–1122

Subramanian, P. S., Ravishanker, G. and Beveridge, D. L. (1988). Theoretical considerations on the 'spine of hydration' in the minor groove of d(CGCGAATTCGCG)*d(GCGCTTAAGCGC): Monte Carlo computer simulation. *Proc. Natl Acad. Sci. USA*, **85**, 1836–1840

Subramanian, P. S., Swaminathan, S. and Beveridge, D. L. (1990). Theoretical account of the 'spine of hydration' in the minor groove of duplex d(CGCGAATTCGCG). *J. Biomol. Struct. Dyn.*, **7**, 1161–1165

Sundaralingam, M. and Sekharudu, Y. C. (1988). Sequence dependent bending and curvature. An overview. In Olson, W. K., Sarma, M. H., Sarma, R. H. and Sundaralingam, M. (Eds), *Structure and Expression*, Vol. 3: *DNA Bending and Curvature*. Adenine Press, Schenectady, N.Y.

Swaminathan, S. and Beveridge, D. L. (1990). Molecular dynamics of B-DNA including water and counterions: A 140 psec trajectory for d(CGCGAATTCGCG) based on the GROMOS force field. *J. Am. Chem. Soc.* (submitted)

Swaminathan, S., Beveridge, D. L. and Berman, H. M. (1990). Molecular dynamics simulation of a deoxydinucleoside drug intercalation complex: d(CpG/ Proflavine. *J. Phys. Chem.*, **92**, 4660–4665

Swaminathan, S., Harte Jr., W. E. and Beveridge, D. L. (1991). Identification of domain structure in proteins via molecular dynamics simulation: Application to HIV-1 protease. *J. Am. Chem. Soc.*, **113**, 2717–2721

Swaminathan, S., Ravishanker, G. and Beveridge, D. L. (1991). Molecular dynamics of B-DNA including counterions and water: A 140 psec trajectory for d(CGCGAATTCGCG) based on the GROMOS force field. *J. Am. Chem. Soc.*, **113**, 5027–5040

Swamy, K. and Clementi, E. (1987). Hydration structure and dynamics of B- and Z-DNA in the presence of counterions via molecular dynamics simulations. *Biopolymers*, **26**, 1901–1927

Tidor, B., Irikura, K. K., Brooks, B. R. and Karplus, M. (1983). Dynamics of DNA oligomers. *J. Biomol. Struct. Dyn.*, **1**, 231

Torda, A. E., Sheek, R. M. and Van Gunsteren, W. F. (1990). Time-averaged distance restraints in molecular dynamics simulations. *Chem. Phys. Lett.*, **157**, 289–294

Van de Ven, J. M. and Hilbers, C. W. (1988). Nucleic acids and nuclear magnetic resonance. *Eur. J. Biochem.*, **178**, 1–38

van Gunsteren, W. F. (1988). Methods for calculation of free energies and binding constants: successes and problems. In van Gunsteren, W. F. and Weiner, P. K. (Eds), *Computer Simulation of Biomolecular Systems: Theoretical and Experimental Applications*. ESCOM, Leiden

van Gunsteren, W. F. and Berendsen, H. J. C. (1986). *GROMOS86: Gröningen Molecular Simulation System*. PhD Thesis, University of Gröningen

van Gunsteren, W. F. and Berendsen, H. J. C. (1990). Computer simulation of molecular dynamics: Methodology, applications, and perspectives in chemistry. *Angew. Chem. Int. Edn Engl.*, **29**, 992–1023

van Gunsteren, W. F., Berendsen, H. J., Guersten, R. G. and Zwinderman, H. R. (1986). A molecular dynamics computer simulation of an eight base pair DNA fragment in aqueous solution: Comparison with experimental 2d NMR data. *Ann. N. Y. Acad. Sci.*, **482**, 287–303

van Vlijnen, H. W. T., Rame, G. L. and Pettitt, B. M. (1990). A study of model energetics and conformational properties of polynucleotide triplexes. *Biopolymers*, **30**, 517–532

Vovelle, F. and Goodfellow, J. (1993). Hydration sites and hydration bridges around DNA helices. This volume

Wang, A. H., Quigley, G. J., Kolpak, F. J., Crawford, J. L., van Boom, J. H., van der Marel, G. and Rich, A. (1979). Molecular structure of a left-handed double helical DNA fragment at atomic resolution. *Nature*, **283**, 743–745

Wang, Y., Thomas, G. A. and Peticolas, W. (1987). Sequence dependent conformation of oligomeric DNAs in aqueous solution and in crystals. *J. Biomol. Struct. Dyn.*, **5**, 249–274

Watson, J. D. and Crick, F. H. C. (1953). A structure for deoxyribonucleic acid. *Nature*, **171**, 737–738

Weiner, S. J., Kollman, P. A., Case, D. A., Singh, U. C., Ghio, C., Alagona, G. S., Profeta, J. and Weiner, P. (1984). A new force field for molecular mechanical simulation of nucleic acids and proteins. *J. Am. Chem. Soc.*, **106**, 765–784

Wemmer, D. E. (1991). The applicability of NMR methods to the solution structure of nucleic acids. *Current Opinion in Biology*, **1**, 452–458

Wemmer, D. and Reid, B. (1985). High resolution NMR studies of nucleic acids and proteins. *Ann. Rev. Phys. Chem.*, **36**, 105–137

Wender, P. A., Kelly, R. C., Beckham, S. and Miller, B. L. (1991). Studies on DNA cleaving agents: computer modeling analysis of the mechanism of activation and cleavage of dynemycin oligonucleotide complexes. *Proc. Natl Acad. Sci. USA*, **88**, 8835–8839

Westhof, E. (1988). Water: An integral part of nucleic acid structure. *Ann. Rev. Biophys. Biophys. Chem.*, **17**, 125

Westhof, E. and Beveridge, D. L. (1989). Hydration of nucleic acids. In Franks, F. (Ed.), *The Molecules of Life*. Cambridge University Press, Cambridge

Westhof, E., Chevrier, B., Gallion, S. L., Weiner, P. K. and Levy, R. M. (1986). Temperature-dependent molecular dynamics and restrained X-ray refinement simulations of a DNA hexamer. *J. Mol. Biol.*, **190**, 699–712

Wing, R. M., Drew, H. R., Takano, T., Broka, C., Tanaka, S., Itakura, I. and

Dickerson, R. E. (1980). Crystal structure analysis of a complete turn of B-DNA. *Nature*, **287**, 755–758

Withka, J. M., Srinivasan, J. and Bolton, P. H. (1992). Problems with, and alternatives to the NMR R factor. *J. Mag. Res.* (in press)

Withka, J. M., Swaminathan, S., Beveridge, D. L. and Bolton, P. H. (1991a). Time dependence of nuclear Overhauser effects in duplex DNA from molecular dynamics trajectories. *J. Am. Chem. Soc.*, **113**, 5041–5049

Withka, J. M., Swaminathan, S., Beveridge, D. L. and Bolton, P. H. (1991b). Towards a dynamical structure of duplex DNA in solution: Comparison of theoretical and experimental NOE intensities of d(CGCGAATTCGCG). *Science* (in press)

Yip, P. and Case, D. A. (1989). A new method for refinement of macromolecular structures based on nuclear Overhauser spectra. *J. Mag. Res.*, **83**, 643–648

Yoon, C., Prive, G. G., Goodsell, D. S. and Dickerson, R. E. (1988). Structure of an alternating-B DNA helix and its relationship to A tract DNA. *Proc. Natl Acad. Sci. USA*, **85**, 6332–6336

Zhurkin, V. B. (1985). Sequence dependent bending of DNA and phasing of nucleosomes. *J. Biomol. Struct. Dyn.*, **2**, 785–804

Zielinski, T. J. and Shibata, M. (1990). A molecular dynamics simulation of the $(dG)_6$:$(dC)_6$ minihelix including counterions and water. *Biopolymers*, **29**, 1027–1044

Note Added in Proof

Our group has just completed several new MD simulations on the d(CGCGAATTCGCG) duplex including couterions and water on the nanosecond timescale.The calculations reveal some artifacts due to cutoffs applied to the potential functions, and which manifest themselves only very slowly, i.e. $\approx$500 ps into the simulation. They are serious enough, however, to materially affect the results. A full documentation of our results is currently being prepared for publication. We urge those undertaking MD on DNA of any kind to proceed with extreme caution in using truncated potentials.

7
Structural Water Bridges in Nucleic Acids

E. Westhof

1 Introduction

The tertiary structures of nucleic acids result from equilibria between (1) electrostatic forces due to the negatively charged phosphates; (2) stacking interactions between the bases due to hydrophobic and dispersion forces as well as to hydrogen bonding interactions between the polar atoms of the bases and water molecules; and (3) the conformational energy of the sugar–phosphate backbone. In its preferred conformations, the polynucleotide backbone exposes the negatively charged phosphates to the dielectric screening by the solvent and promotes the stacked helical arrangement of adjacent bases. In this way, a hydrophobic core is created where hydrogen bond formation between the nucleic acid bases as well as additional sugar–base and sugar–sugar interactions are favoured. Further, via variations in torsion angles of the sugar–phosphate backbone and through reorientations of the bases, nucleic acids adapt their structures so that their polar hydrophilic atoms form favourable interactions with the molecules of the solvent. This interdependence between solvent and nucleic acid structure constitutes the physicochemical basis for DNA polymorphism. In such helical structures, only the internal atoms involved in hydrogen bonding between the bases are protected from solvent, while most of the other atoms are accessible to water. Thus, water molecules contribute to the overall stability of helical conformations of nucleic acids by (1) screening the charges of the phosphates; (2) bonding to and bridging between the polar exocyclic atoms of the bases; and (3) influencing the conformations of residues with methyl groups via hydrophobic interactions. Besides, owing to the periodicity of the helical structures of nucleic acids, water sites and water bridges involving polar base atoms or phosphate oxygens lead to structured arrangements of water molecules, called columns, chains, filaments (Clementi and Corongiu, 1981), or spines (Drew and Dickerson, 1981).

The discovery by Franklin and Gosling (1953) of the effects of humidity and salts on DNA conformations was decisive for the establishment of the double-helical structure of DNA (Watson and Crick, 1953). Soon afterwards, the possibility that hydration plays a role in the stability of nucleic acid helices was suggested by Geiduschek and Gray (1956). Later, base-stacking forces together with hydrogen bonding between complementary bases were held responsible for double-helical structures in solution. In 1967 Lewin propounded the concept that water bridges contribute greatly to the stability of the DNA double helix in solution on the basis of model building and theoretical considerations. However, the evidence was indirect and the lack of direct crystallographic experimental evidence held the paper in respectable obscurity. A new impetus was given about 15 years later by the pioneering work of Drew and Dickerson (1981). Following that work, a wealth of information on nucleic acid hydration has been gathered from crystal structures of nucleic acid oligomers. Extended reviews have appeared on nucleic acid hydration describing the water molecules found in nucleic acid structures on the basis of crystallographic data (Saenger, 1987; Westhof, 1987a, 1988) as well as of simulation results (Westhof and Beveridge, 1990; see also the other chapters in this book).

Here the various hydrogen-bonded bridges mediated by water molecules observed in crystal structures of nucleic acids will be reviewed in more detail than in a previous work (Westhof, 1990). It will be emphasized that similar water binding sites and water bridges are found repeatedly in small as well as in large nucleic acid crystals and that they should, consequently, play an important part in the overall stability not only of helical conformations but also of conformations which are either non-helical or present non-Watson–Crick base pairings. In short, the analysis tries to convey that water is an integral part of nucleic acid structure and that water does not play a mere space-filling role. Nucleic acid structures, in other words, result from an interplay between solvent and nucleic acid molecules with both together constituting the functional structural system. Thus, descriptions like 'a water bridge stabilizes the nucleic acid conformation' or 'the nucleic acid conformation nucleates a water bridge' are meaningful, at best, if stated simultaneously.

The frequently observed water bridges between polar atoms belonging to the same nucleotide or to neighbouring ones are summarized in Tables 7.1, 7.2. With X-ray crystallography, one can identify localized hydration sites, that is sites close to the macromolecule which are locally stable so that these positions are, on average, frequently visited and occupied by solvent molecules (for discussion and references, see the chapters by H. Savage and M. Frey in this book). For helical conformations, energy minimization studies have reproduced the one-water bridges seen by X-ray crystallography (Vovelle *et al.*, 1989; see also the chapter by Vovelle and Goodfellow). Recently the roles of the solvent-induced interactions

Table 7.1 Structural water-mediated bridges in nucleic acids

HYDRATION AROUND PHOSPHATE GROUPS

Cones of hydration (B– and Z-DNA)
O1P (*i*)· · ·W· · ·O2P (*i* + 1) bridges (A- and Z-DNA, RNA)

5′ -PHOSPHATE–WATER–BASE BRIDGES

B-DNA: methyl· · ·Ow· · ·phosphate
A-DNA and RNA: N7 (purines)· · ·Ow· · ·phosphate
Purine in *syn*:
External G*syn*–C*anti* base pairs in Z-DNA
G*anti*–A*syn* and G*syn*–A*anti* in B-DNA

3′-PHOSPHATE–WATER–SUGAR BRIDGES

Helical conformations : never seen
Present in non-helical conformations (turns,· · ·)

3′-PHOSPHATE–WATER–BASE BRIDGES

Right-handed helical structures: never seen
Purine in *syn*
Internal G*syn*–C*anti* base pairs in Z-DNA

OTHER PHOSPHATE–WATER–BASE BRIDGES

Unusual pairs : A–A, A–G (tRNAs, RNA–drug complexes)
Z-DNA : N7(G)· · ·(hydrated ion)· · ·5′-phosphate of C

SUGAR–WATER–BASE BRIDGES

In minor groove of all helical forms (A, B, and Z)
Periodicity → Spines of hydration (B and Z)

BASE–WATER–BASE BRIDGES

Purines : N6/O6· · ·Ow· · ·N7 and N2· · ·Ow· · ·N3
Inter-residue (inter- and intrastrand) versatile

Table 7.2 Water bridges observed in the minor groove of B- and Z-DNA forms as well as in the shallow groove of A-DNA, in helical RNA fragments and tRNAs

Type of bridge	*Helical form*
Intrastrand	
N2[*i*]· · ·Ow· · ·N3[*i*][a]	RNA, A-DNA
N3[*i*]· · ·Ow· · ·O4′[*i* + 1]	A-, B-DNA
O2[*i*]· · ·Ow· · ·O4′[*i* + 1]	A-, B-DNA
O2′[*i*]· · ·Ow· · ·N3[*i*]	RNA
O2′[*i*]· · ·Ow· · ·O2[*i*]	RNA
O2′[*i*]· · ·Ow· · ·O4′[*i* + 1]	RNA
N2[*i*]· · ·Ow· · ·O1P[*i*, *i* + 1][b]	Z-DNA
Interstrand	
O2[*i*]· · ·Ow· · ·O2[*j* + 1]	B-DNA
N3[*i*]· · ·Ow· · ·O2[*j* + 1]	B-DNA
O2[*i*]· · ·Ow· · ·O2[*j* + 1]	Z-DNA

[a] Brackets contain residue number, with positive sign for the 5′- to 3′-OH direction; *j* refers to a residue on the other strand, with positive sign for the 5′-to 3′-OH direction.
[b] The water involves the phosphate of the same residue when it is a terminal residue. In such cases the sugar pucker is C(2′)-*endo* instead of C(3′)-*endo*.

between two hydrophilic groups via water-bridged H-bonds in molecular recognition processes has been stressed and examined within the framework of classical statistical mechanics (Ben-Naim *et al.*, 1990).

2 Hydration around Phosphate Groups

Around each anionic phosphate oxygen, the motif most frequently seen, in nucleotides as well as in oligonucleotides, is the 'cone of hydration' built of three water molecules. This motif was predicted by Pullman and co-workers using quantum mechanical calculations (Langlet *et al.*, 1979). Recently Monte Carlo simulations of the B-dodecamer d(CGCGAATTCGCG) revealed the extensive presence of 'cones of hydration' around anionic phosphate oxygen atoms (Subramanian and Beveridge, 1989). Crystallographically, 'cones of hydration' are also seen in the B-dodecamer referred to above (Westhof and Beveridge, 1990). Another example coming from the Z-DNA hexamer d(5BrCG5BrCG5BrCG) (Chevrier *et al.* 1986) shown in Figure 7.1. Up to now, such 'cones of hydration' have not been seen in large structures of RNA molecules. It is not yet clear whether this observation is genuine or due to lack of resolution or crystal packing. For example, in several dinucleotide structures the binding of sodium ions perturbs such arrangements, since direct ion binding and through-water binding, which are of the same order of magnitude (Pullman *et al.*, 1978), have been observed (Rosenberg *et al.*, 1976; Seeman *et al.*, 1976). Also, helical RNA structures (or A-DNA-type structures of deoxyoligomers) present instead frequently a water bridge between anionic phosphate oxygen atoms of successive phosphate groups on the same strand (Saenger *et al.*, 1986; Westhof *et al.*, 1988). Figure 7.2 displays examples of 'cones of hydration' in the extended dinucleotide CpA intercalated by proflavine (Westhof *et al.*, 1980). Monte Carlo simulations on a A-DNA dodecamer do show the presence of such 'cones of hydration' (Subramanian and Beveridge, in preparation) together with one-water bridges between successive phosphate groups. It is interesting to note on Figure 7.2 the presence of 'cones of hydration' around O(2′) and terminal O(3′) ribose oxygen atoms as well. Similarly, binding of phosphate groups in protein complexes involves frequently 'cones of H-bonding' in which the water molecules have been replaced by protein backbone amide N–H or side-chain donor atoms, as, for example, in the complex between carbamoyl phosphate and aspartate carbamoyltransferase (Gouaux and Lipscomb, 1988) or in that between phosphate and phosphate-binding protein (Luecke and Quiocho, 1990). In nucleic acids amino groups have a strong tendency for binding directly to phosphate anionic oxygen atoms as well as to ribose hydroxyl groups. The former situation is dramatically displayed in the structure of a non-palindromic dodecamer (Timsit *et al.*, 1989).

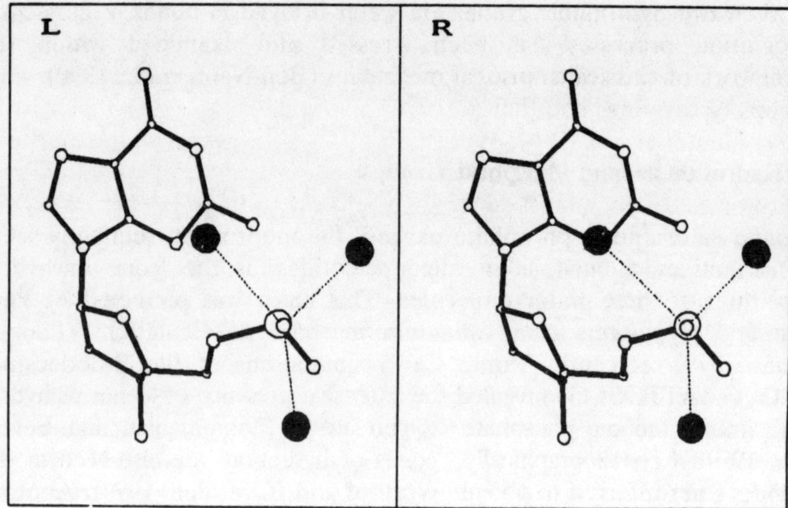

Figure 7.1 Stereo view of three water molecules around an anionic phosphate oxygen forming a 'cone of hydration' in the crystal structure of Z-DNA hexamer d(5BrC–G–5BrC–G–5BrC–G) (Chevrier *et al.*, 1986). The resolution of the data is 1.4 Å and refinement gave an *R*-factor of 12.5%

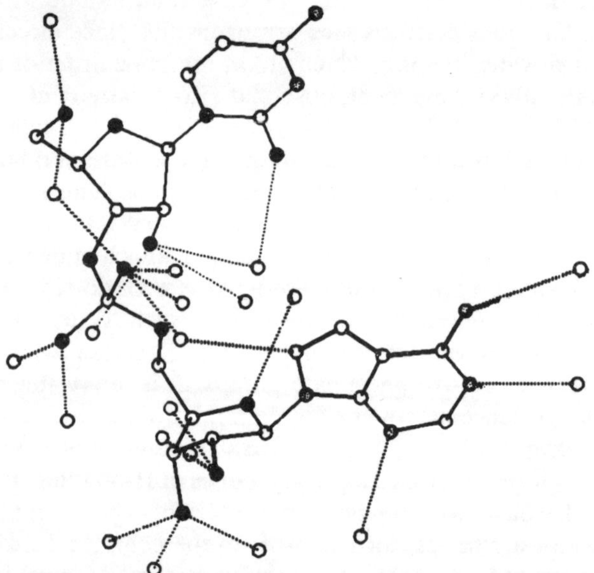

Figure 7.2 Some water molecules binding to cytidilyl-3′, 5′-adenosine in its complex with proflavine (Westhof *et al.*, 1980). Notice the 'cones of hydration' around the phosphate anionic oxygen atoms as well as around the hydroxyl groups of the riboses. There are also three water bridges: one linking C8(A) to an anionic oxygen; the same anionic oxygen is linked to O5′(C) via a water molecule; and the frequent bridge between the ribose O2′ and O2(C). The resolution of the data is 1.1 Å and refinement gave an *R*-factor of 11.0%

3 5′-Phosphate–Water–Base Bridges

In DNA structures such bridges are not frequent, except between methyl groups of thymines and their attached phosphates (Drew and Dickerson, 1981; Hunter *et al.*, 1986). With purine bases in the *syn* orientation with respect to the sugar, water bridges between N3 (if adenine: Hunter *et al.*, 1986) or N2 (if guanine: Leonard *et al.*, 1990) and a 5′-phosphate anionic oxygen occur in B-DNA structures. In Z-DNA the water bridges between N2 of the *syn* guanine residues and an anionic phosphate oxygen are made towards the 5′-phosphate in terminal residues (in such instances, the sugar pucker is C(2′)-*endo* instead of C(3′)-*endo*: Wang *et al.*, 1984). Such a situation was seen also in the crystal structure of the nucleotide 2-amino-8-methyl-adenosine 5′-monophosphate (Silverton *et al.*, 1982), in which the base adopts the *syn* conformation about the glycosyl bond (Figure 7.3). Water bridges (with one or two water molecules) between the N7 atoms of purines and their attached 5′-phosphate are very frequent in A-form DNA structures or RNA structures (Westhof, 1987a). Figure 7.4 shows one example of such a water bridge in the crystal structure of the complex between glutaminyl-tRNA and its cognate synthetase (Rould *et al.*, 1991) and a similar one between the ring N−H of a pseudouridine and its 5′-phosphate, a common occurrence (Westhof *et al.*, 1988).

4 3′-Phosphate–Water–Sugar Bridges

In RNA structures a water bridge between the O2′ hydroxyl group and an anionic oxygen atom of the 3′-phosphate group is seen only when the sugar–phosphate backbone makes sharp turns and adopts non-helical conformations. Two examples are shown in Figure 7.5. In one drawing the phosphate between A58 and U59 of yeast tRNA–Asp (Westhof *et al.*, 1988) adopts the *trans-gauche plus* conformation; and in the other the phosphate between C20 and A21 adopts the *gauche minus-gauche plus* conformation.

5 3′-Phosphate–Water–Base Bridges

In right-handed helical structures (either RNA or DNA), such bridges are not observed. However, the *syn* conformation of the guanine base in left-handed Z-DNA is stabilized by an intramolecular water bridge between its amino group N2 and one of its anionic phosphate oxygens (Wang *et al.*, 1979). Thus, in right-handed helical structures water molecules do not link the nucleoside and its 3′-phosphate, most probably because of the local compaction resulting from the right-handedness. However, the

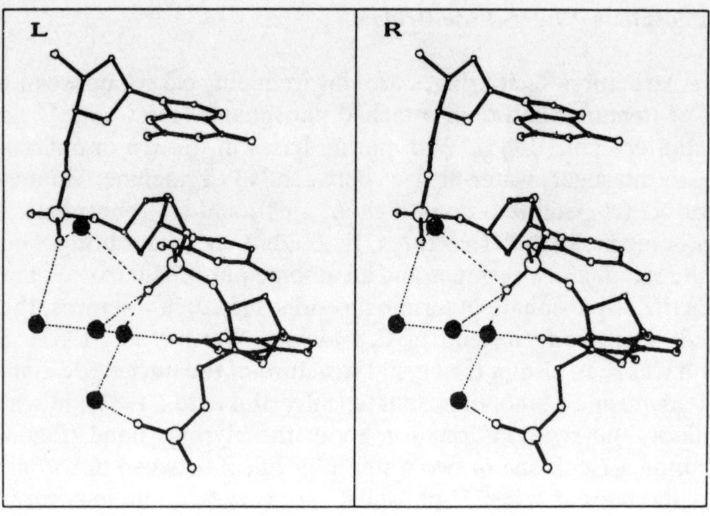

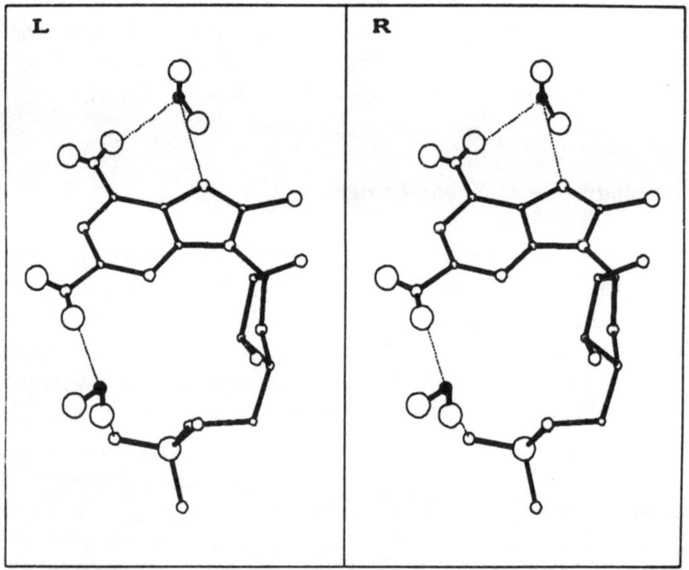

Figure 7.3 (*Top*) Stereo view of some water molecules around part of the Z-DNA hexamer d(5BrC–G–5BrC–G–5BrC–G) (Chevrier *et al.*, 1986). Notice the water bridge linking the amino group of the guanine residue and an anionic oxygen of its 3′-phosphate and the similar bridge made via two water molecules with an anionic oxygen of its 5′-phosphate. Note also the one-water bridge linking successive anionic oxygen atoms belonging to the 3′- and 5′-phosphate of the cytosine residue. The resolution of the data is 1.4 Å and refinement gave an *R*-factor of 12.5% (*Bottom*) Stereo view of two one-water bridges in the crystal structure of 2-amino-8-methyl-adenosine 5′-monophosphate (Silverton *et al.*, 1982): one is between N6(A) and N7(A) and the second is between the amino group and the 5′-phosphate. The resolution of the data is 0.54 Å and refinement gave an *R*-factor of 2.3%

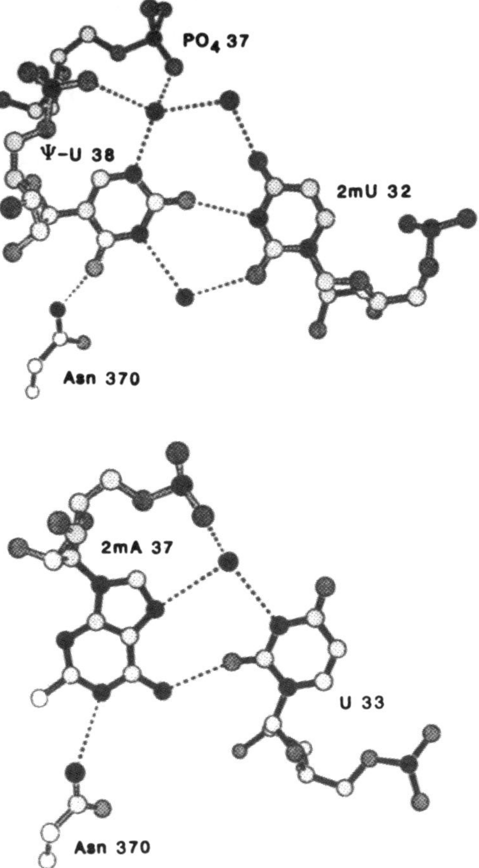

Figure 7.4 Some water molecules around two unusual base pairs in the anticodon loop of *Escherichia coli* tRNA–Gln complexed with its cognate synthetase (Rould *et al.*, 1991). Two frequent water bridges, those between N7(A37) and its 5′-phosphate and between N1(pseudouridine 38) and its 5′-phosphate, are used for binding with the other base of the pair. The resolution of the data is 2.5 Å and refinement gave an *R*-factor of 21.0%

structural and stabilizing importance of such water bridges should be kept in mind when computing and model-building unusual and non-helical conformations of nucleic acids.

6 Other Phosphate–Water–Base Bridges

Crystal structures of DNA oligomers and of tRNAs show that, around non-standard Watson–Crick base pairs or unusual base pairs, water molecules tend to cluster and compensate for the loss in hydrogen bonds (Ho *et al.*, 1985; Kennard, 1986; Westhof *et al.*, 1988). This can already be noticed

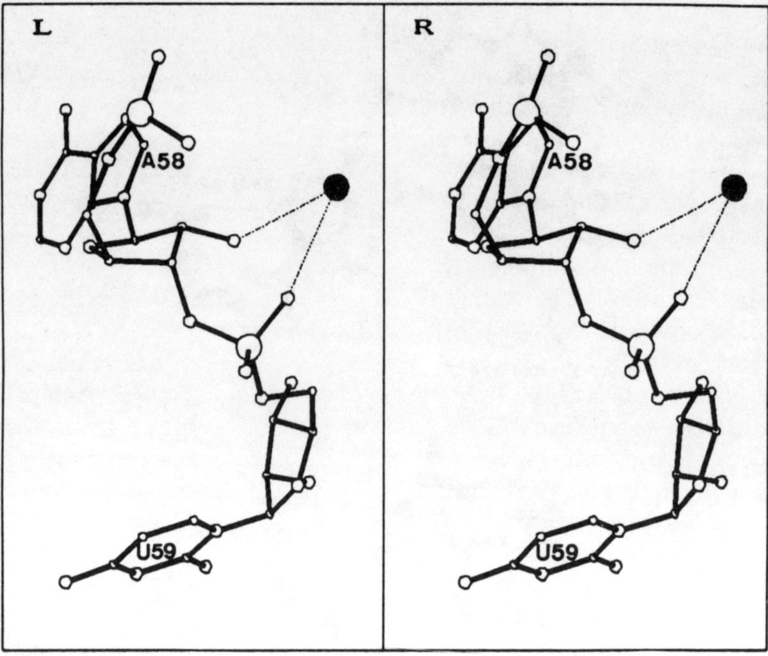

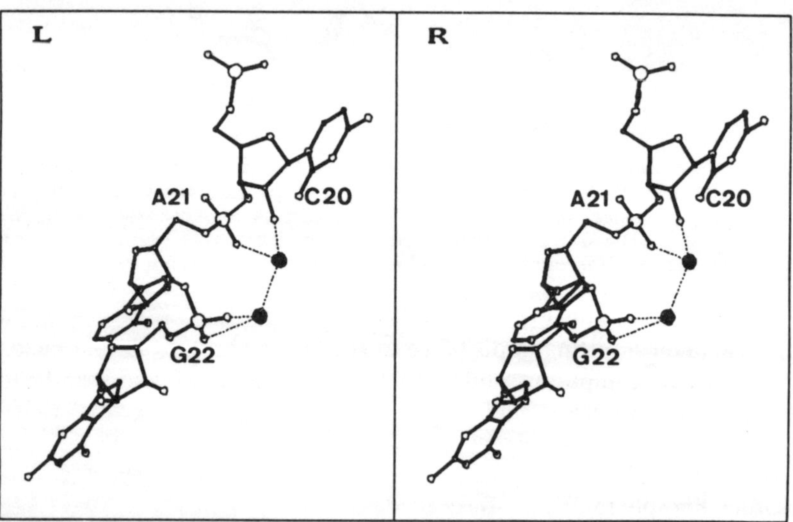

Figure 7.5 Stereo view of 3′-phosphate–water–sugar bridges in two extended conformations of the sugar–phosphate backbone of yeast tRNA–asp (Westhof *et al.*, 1988). The resolution of the data is 3.0 Å and refinement gave an *R*-factor of 17.2%

in the crystal structure of 8-bromoguanosine (Tavale and Sobell, 1970), where the symmetrical G–G base pair (via N1 and O6 of the guanine bases) is strengthened by a water bridge linking the amino group N2 of one base and the O6 of the other. Besides those additional base–water–base bridges across the pair, unusual base pairs also present new phosphate–water–base bridges. Thus, in A–A self-pairs the amino N6 of one adenine is linked via a water molecule to an anionic phosphate oxygen of the opposite base (Figure 7.6). In the ApApA structure the N6 atom is linked directly to the opposite phosphate, with the N1 atom linked to the same phosphate group via a water molecule (Suck *et al.*, 1976). A similar situation occurs in the adenosine–proflavine sulphate complex, in which sulphate groups play the role of phosphates (Swaminathan *et al.*, 1982). One example of a similar hydration pattern around the G22–A46 pair in yeast tRNA–Asp (Westhof *et al.*, 1988) is shown also in Figure 7.6. Figure 7.4 contains two further examples of water bridges between non-standard pairs involving a phosphate group.

7 Other Phosphate–Water–Sugar Bridges

Again, these bridges are observed mainly around unusual base pairs or non-helical conformations and the structures of tRNA molecules offer several examples. For instance, around the *trans* Watson–Crick base pair G15–C48 in yeast tRNA–Phe (Westhof and Sundaralingam, 1986), three water molecules are observed: (1) one links N2(G15) to N4(C48); (2) one bridging O2′(G15) and N3(G15) hydrogen bond also to O6 of neighbouring G20; (3) a third one links N7(G15) to an anionic phosphate oxygen of neighbouring U7. At the same time, such water molecules fill cavities and notches formed by the tertiary folding of the tRNA. Further, they have been found in different tRNA crystals despite the rather low resolution (Westhof *et al.*, 1988). Thus, they can be regarded as constituent of the tRNA structure itself. Two examples are shown in Figure 7.7. Note how in one instance the water bridge between an hydroxyl O(2′) and an anionic phosphate oxygen is on the ridge of the macromolecule, while in the other instance it is deep inside the tRNA fold.

8 Sugar–Water–Base Bridges

Such bridges occur in the minor groove of helical nucleic acids (see Table 7.2). They are the least dependent on the nucleic acid form (A, B or Z), on the sequence or on the nature of the sugar. Water molecules participating

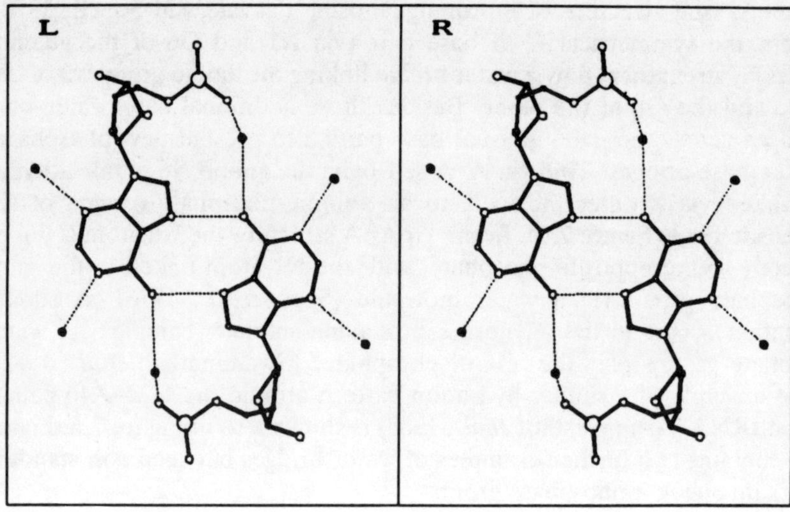

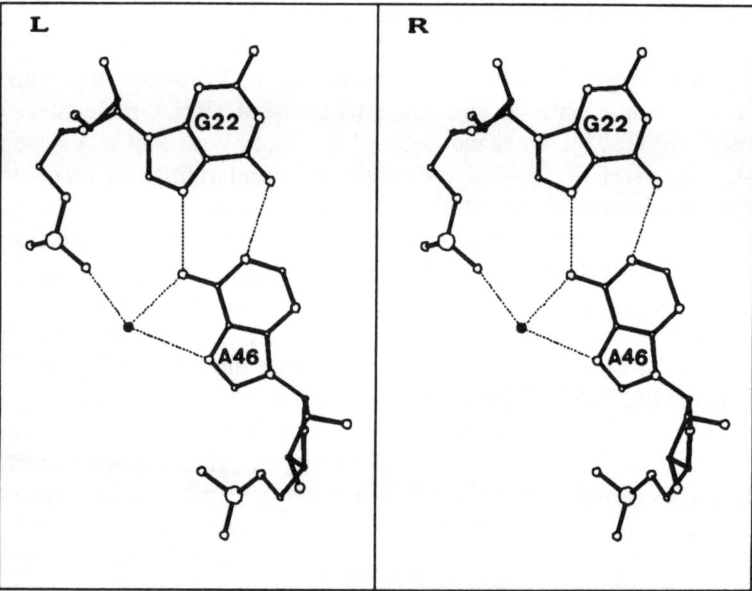

Figure 7.6 (*Top*) Stereo view of some water molecules around the unusual A–A base pair in the complex between cytidilyl-3′, 5′-adenosine and proflavine (Westhof *et al.*, 1980). A similar water bridge between the amino group of the adenosine residues and an opposite phosphate anionic oxygen atom exists in the A9–A23 unusual interaction of tRNAs (Westhof *et al.*, 1988). The resolution of the data is 1.1 Å and refinement gave an *R*-factor of 11.0%. (*Bottom*) Stereo view of a one-water bridge linking the adenosine residue of the unusual interaction A46–G22 to the 5′-phosphate of the guanine residue in yeast tRNA–Asp (Westhof *et al.*, 1988). The resolution of the data is 3.0 Å and refinement gave an *R*-factor of 17.2%

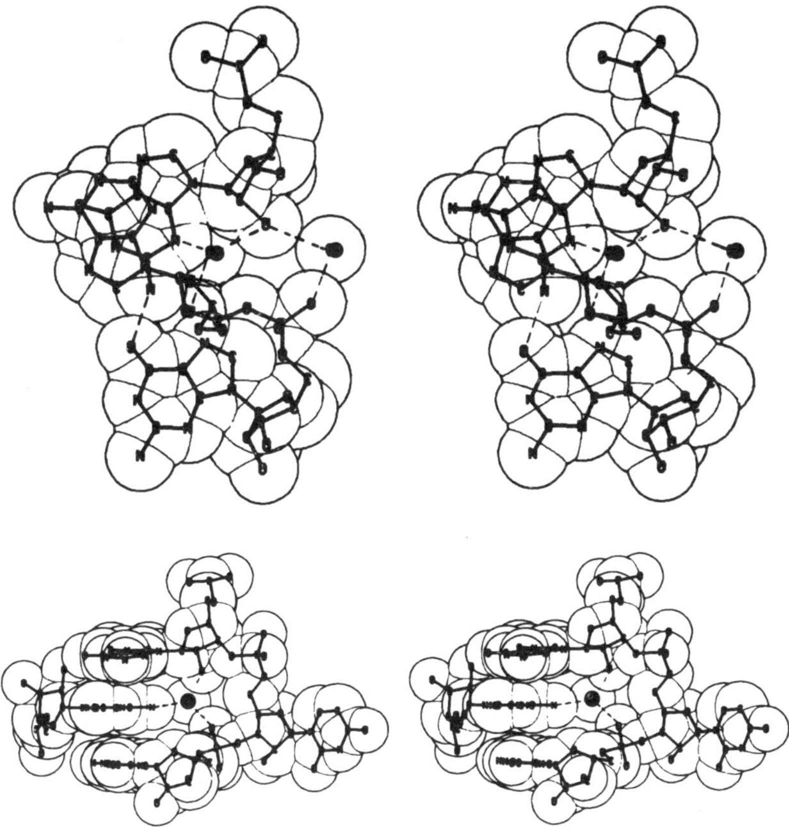

Figure 7.7 (*Top*) Stereo view of the turn between A9 and G10 in yeast tRNA–Asp (Westhof *et al.*, 1988), with G45 stacked on top of A9, showing an internal water molecule linking O2′(A9) to 02′(G45) and N3(G45). The O2′(G45) is itself linked to an anionic oxygen atom of G10 by a water molecule at the ridge of the macromolecule boundary. The resolution of the data is 3.0 Å and refinement gave an *R*-factor of 17.2%. (*Bottom*) Stereo view of the stack G46–A21–C48 in the monoclinic form of yeast tRNA–Phe (Westhof and Sundaralingam, 1986). Notice the water molecule which fills a cavity and forms hydrogen bonds with N6(A21), O2′(G46), and O1P (C48). The resolution of the data is 3.0 Å and refinement gave an *R*-factor of 19.4%

in such bridges are tightly bound and appear clearly in electron density maps. They should, therefore, contribute to the structural integrity of all helical nucleic acids. In B- and Z-helices the minor grooves form canyons particularly devoid of hydrogen bond donors (Quigley, 1986), so that water molecules, by binding to it, fulfil the hydrogen-bonding potential and complete the H-bond network. The most frequent sugar–water–base bridge in RNA structures links the O2′ hydroxyl group to the exocyclic O2 atom of pyrimidines or to the ring nitrogen N3 of purines (Westhof, 1988). In DNA structures the O2/N3· · ·W· · ·O4′ bridge within the same residue

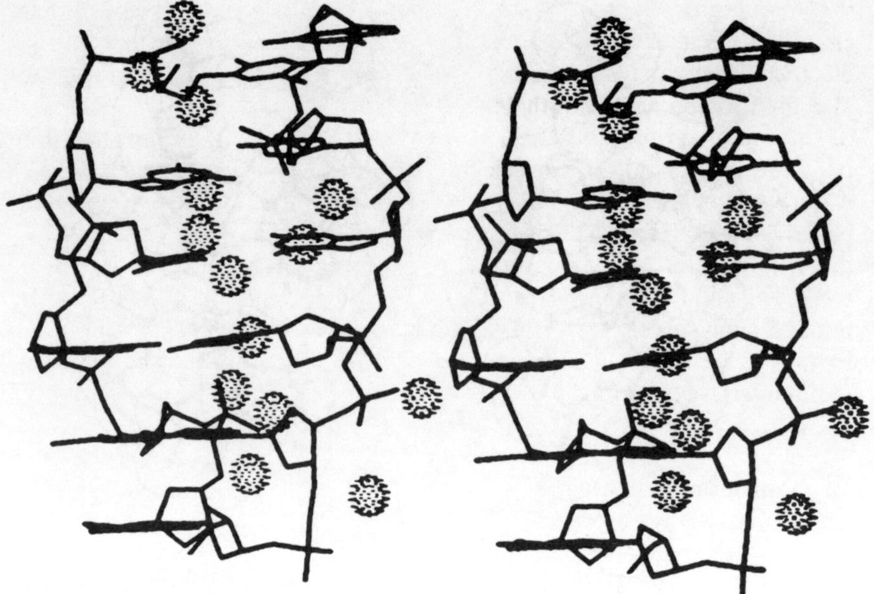

Figure 7.8 Stereo view of the Z hydration spine in the crystal structure of the hexamer d(5BrC–G–5BrC–G–5BrC–G) (Chevrier *et al.*, 1986). The view is down the minor groove and shows the string of water molecules bound to the O2 exocyclic oxygen atoms of the cytosine residues. Examples of two other types of water bridges already mentioned can also be seen. The resolution of the data is 1.4 Å and refinement gave an *R*-factor of 12.5%

occurs but is rare (example in the decamer structure d(CCAAGATTGG): Privé *et al.*, 1987). Otherwise, in DNA and in RNA, the O2/N3· · ·W· · ·O4′ water bridge occurs between successive residues on the same strand. The sequences rich in A–Ts, together with the helical periodicity, lead to spines of hydration running down the central canyon of the minor groove of B-DNA (Drew and Dickerson, 1981; Privé *et al.*, 1991). Theoretical simulations give good account of the experimental data (Subramanian *et al.*, 1988, 1990; Chuprina *et al.*, 1991; see also the chapter by Beveridge in this book). In Z-DNA also, the periodicity between the CpG dimers leads to a spine of hydration (Figure 7.8) formed by water molecules binding to the exocyclic oxygen atoms of the cytosine residues and another group of bridging water molecules (Wang *et al.*, 1984).

9 Base–Water–Base Bridges

Such bridges, when intraresidue, occur only around purine bases; in the major groove, between N6/O6 and N7, and in the minor groove, between N2 and N3 in guanine residues. Inter-residue base–water–base bridges, interstrand as well as intrastrand, are versatile and variable, especially in the major groove (see Westhof, 1987a; Vovelle *et al.*, 1989; and the chapter by Vovelle and Goodfellow in this book). Their localization in electron density maps is often patchy, revealing not only their mobility but also the protean nature of their arrangements around the hydrophilic sites of the major groove of helical nucleic acids. Those around unusual base pairs are, instead, often very clearly seen in electron density maps. Base–water–base bridges should play an important role not only in defining the preferred stacking geometries (Vovelle and Goodfellow, 1990) but also in determining the local fine structure around base pairs, as described by twist, roll, and propeller-twist angles (Raghunathan *et al.*, 1990). A systematic analysis of water molecules around DNA bases based on crystallographic data has revealed how remarkably well hydrated the bases are (Schneider *et al.*, 1992).

10 Conclusions

On a souvent besoin d'un plus petit que soi.
La Fontaine

The most frequent water bridges appear mainly in the minor groove of helical nucleic acids. The systematic use of the sugar–water–base and base–water–base bridges in the minor groove contrasts with the versatility and mobility of the base–water–base bridges and settlements in the major groove of helical nucleic acids. The hydration around phosphate groups of helical structures is characterized by 'cones of hydration' centred on each anionic phosphate oxygen, by water bridges between anionic phosphate oxygen atoms of successive residues on the same strand and by 5′-phosphate–water–base bridges. Those water molecules mediating structural bridges between atoms of the nucleic acid should be regarded as an integral constituent of helical nucleic acids in aqueous solution and, consequently, be considered also as responsible for fine-structural parameters such as stacking geometries, twist and roll angles between base pairs as well as some propeller-twist angles of base pairs.

The 3′-phosphate–water–base and 3′-phosphate–water–sugar water bridges around unusual nucleotide conformations (e.g. *syn* conformation of the base or the *gauche minus-gauche plus* conformation of the sugar–phosphate backbone) stress the role of water bridges in non-standard conformations. Around non-canonical base pairs (e.g. A–G, A–A, U–G pairs) as well, new phosphate–water–base bridges appear systematically.

As advocated by Lewin (1967), water bridges at different sites of the nucleic acids may not be of the same strength. If the frequency of occurrence of a water bridge is related to its strength, this proposal may be correct. Preliminary investigations based on the crystallographically determined occupancies and temperature factors indicated differences in those two parameters for the water molecules building up the B-spine and the Z-spine (Westhof, 1987b). However, the establishment of such a suggestion would still require extensive computer simulation studies. Theoretical works in that direction are beginning to appear (Laaksonen *et al.* 1989; Forester and McDonald, 1991). It is further suggested that the dynamics of the hydration structure around helical nucleic acids in aqueous solution is only slightly perturbed, except for the water bound to phosphate groups and for the structurally important water fraction around repetitive sequences (spines of A–T stretches), unusual conformations (phosphate–water–base bridges of *syn* bases), and non-Watson–Crick base pairs (G–T, A–A, G–A, . . .).

In short, water molecules constitute an integral part of helical as well as of non-helical structures of nucleic acids. Clearly, the development of an adequate description of the solvation and electrostatic properties of the highly charged nucleic acids is essential for correct modelling and simulation of nucleic acid structures.

References

Ben-Naim, A., Ting, K. L. and Jernigan, R. L. (1990). Solvent effects on binding thermodynamics of biopolymers. *Biopolymers*, **29**, 901–919

Chevrier, B., Dock, A. C., Hartmann, B., Leng, M., Moras, D., Thuong, M. T. and Westhof, E. (1986). Solvation of the left-handed hexamer d(5BrC–G–5BrC–G–5BrC–G) in crystals grown at two temperatures. *J. Mol. Biol.*, **188**, 707–719

Chuprina, V. P., Heinemann, U., Nurislamov, A. A., Zielenkiewicz, P., Dickerson, R. E. and Saenger, W. (1991). Molecular dynamics simulation of the hydration shell of a B-DNA decamer reveals two main types of minor-groove hydration depending on groove width. *Proc. Natl Acad. Sci. USA*, **88**, 593–597

Clementi, E. and Corongiu, G. (1981) Solvation of DNA at 300 K: Counter-ion structure, base-pair sequence recognition and conformational transitions. A computer experiment. In Sarma, R. H. (Ed.), *Biomolecular Stereodynamics*, Vol. 1. Adenine Press, Schenectady, N.Y., pp. 209–259

Drew, H. R. and Dickerson, R. E. (1981). Structure of a B-DNA dodecamer. III. Geometry of hydration. *J. Mol. Biol.*, **151**, 535–556

Forester, T. R. and McDonald, I. R. (1991). Molecular dynamics studies of the behaviour of water molecules and small ions in concentrated solutions of polymeric B-DNA. *Mol. Phys.*, **72**, 643–660

Franklin, R. E. and Gosling, R. G. (1953). The structure of sodium thymonucleate fibres. I. The influence of water content. *Acta Cryst.*, **6**, 673–677

Geiduschek, E. P. and Gray, I. (1956). Non-aqueous solutions of sodium desoxyribose nucleate. *J. Am. Chem. Soc.*, **78**, 879–880

Gouaux, J. E. and Lipscomb, W. N. (1988). Three-dimensional structure of carbamoyl phosphate and succinate bound to aspartate carbamoyltransferase. *Proc. Natl Acad. Sci. USA*, **85**, 4205–4208

Ho, P. S., Frederick, C. A., Quigley, G. J., van der Marel, G. A., van Boom, J. H., Wang, A. H.-J. and Rich, A. (1985). G.T. wobble pairing in Z-DNA at 1.0 Å atomic resolution; the crystal structure of d(CGCGTG). *EMBO Jl*, **4**, 3617–3623

Hunter, W. N., Brown, T. and Kennard, O. (1986). Structural features and hydration of d(CGCGAATTAGCG): a double helix containing two G·A mispairs. *J. Biomol. Struct. Dyn.*, **4**, 173–191

Kennard, O. (1986). Structural studies of DNA fragments: The G·T wobble base pair in A, B, and Z-DNA. The G·A base pair in B-DNA. *J. Biomol. Struct. Dyn.*, **3**, 205–226

Laaksonen, A., Nilsson, L. G., Jönsson, B. and Teleman, O. (1989). Molecular dynamics simulation of double helix Z-DNA in solution. *Chem. Phys.*, **129**, 175–183

Langlet, J., Claverie, P., Pullman, B. and Piazzola, D. (1979). Studies of solvent effects. IV. Study of hydration of the dimethyl phosphate anion (DMP-) and the solvent effect upon its conformation. *Int. J. Quant. Chem.*, **6**, 409–437

Leonard, G. A., Booth, E. D. and Brown, T. (1990). Structural and thermodynamic studies on the adenine–guanine mismatch in B-DNA. *Nucleic Acids Res.*, **18**, 5617–5623

Lewin, S. (1967). Some aspects of hydration and stability of the native state of DNA. *J. Theoret. Biol.*, **17**, 181–212

Luecke, H. and Quiocho, F. A. (1990). High specificity of a phosphate transport protein determined by hydrogen bonds. *Nature*, **347**, 402–406

Privé, G. G., Heinemann, U., Chandrasegaran, S., Kan, L. S., Kopka, M. L. and Dickerson, R. E. (1987). Helix geometry, hydration and G·A mismatch in a B-DNA decamer. *Science*, **238**, 498–504

Privé, G. G., Yanagi, K. and Dickerson, R. E. (1991). Structure of the B-DNA decamer C–C–A–A–C–G–T–T–G–G and comparison with the isomorphous decamers C–C–A–A–G–A–T–T–G–G and C–C–A–G–G–C–C–T–G–G. *J. Mol. Biol.*, **217**, 177–199

Pullman, A., Pullman, B. and Berthod, H. (1978). An SCF *ab initio* investigation of the 'through-water' interaction of the phosphate anion with the Na+ cation. *Theoret. Chim. Acta (Berlin)*, **47**, 175–192

Quigley, G. (1986). The roles of hydrogen bonding in nucleic acid conformation. *Trans. Am. Crystallogr. Assoc.*, **22**, 121–130

Raghunathan, G., Jernigan, R.L., Ting, K.L. and Sarai, A. (1990). Solvation effects on the sequence variability of DNA double helical conformation. *J. Biomol. Struct. Dyn.*, **8**, 187–198

Rosenberg, J. M., Seeman, N. C., Day, R. O. and Rich, A. (1976). RNA double helical fragment at atomic resolution. II. The crystal structure of sodium guanylyl-3′, 5′-cytidine nonahydrate. *J. Mol. Biol.*, **104**, 145–167

Rould, M. A., Perona, J. J. and Steitz, T. A. (1991). Structural basis of anticodon loop recognition by glutaminyl-tRNA synthetase. *Nature*, **352**, 213–218

Saenger, W. (1987). Structure and dynamics of water surrounding biomolecules. *Ann. Rev. Biophys. Biophys. Chem.*, **16**, 93–114

Saenger, W., Hunter, W. N. and Kennard, O. (1986). DNA conformation is determined by economics in the hydration of phosphate groups. *Nature*, **324**, 385–388

Schneider, B., Cohen, D. and Berman, H. M. (1991). Hydration of DNA bases: Analysis of crystallographic data. *Biopolymers* (in press)

Seeman, N. C., Rosenberg, J. M., Suddath, F. L., Kim, J. J. P. and Rich, A. (1976). RNA double-helical fragment at atomic resolution. I. The crystal and molecular structure of sodium adenylyl-3′, 5′-uridine hexahydrate. *J. Mol. Biol.*, **104**, 109–144

Silverton, J. V., Limm, W. and Miles, H. T. (1982). 2-Amino-8-methyladenosine 5′-monophosphate dihydrate. A nucleotide with syn C4′-exo conformation and 'triple-stranded' packing. *J. Am. Chem. Soc.*, **104**, 1082–1087

Subramanian, P. S. and Beveridge, D. L. (1989). A theoretical study of the aqueous hydration of canonical B d(CGCGAATTCGCG): Monte Carlo simulation and comparisons with crystallographically ordered water sites. *J. Biomol. Struct. Dyn.*, **6**, 1093–1122

Subramanian, P. S., Ravishanker, G. and Beveridge, D. L. (1988). Theoretical considerations on the 'spine of hydration' in the minor groove of d(CGCGAATTCGCG): d(CGCGAATTCGCG): Monte Carlo computer simulation. *Proc. Natl Acad. Sci. USA*, **85**, 1836–1840

Subramanian, P. S., Swaminathan, S. and Beveridge, D. L. (1990). Theoretical account of the 'spine of hydration' in the minor groove of duplex d(CGCGAATTCGCG). *J. Biomol. Struct. Dyn.*, **7**, 1161–1165

Suck, D., Manor, P. C. and Saenger, W. (1976). The structure of a trinucleotide diphosphate: adenylyl-3′, 5′-adenylyl-3′, 5′-adenosine hexahydrate. *Acta Cryst.*, **B32**, 1727–1737

Swaminathan, P., Westhof, E. and Sundaralingam, M. (1982). Structure of a 1:2 sandwich complex of proflavine and adenosine with an unusual puckering disorder and a site shared by sulfate and water molecules. *Acta Cryst.*, **B38**, 515–522

Tavale, S. S. and Sobell, H. M. (1970). Crystal and molecular structure of 8-bromoguanosine and 8-bromoadenosine, two purine nucleosides in the *syn* conformation. *J. Mol. Biol.*, **48**, 109–123

Timsit, Y., Westhof, E., Fuchs, R. P. P. and Moras, D. (1989). Unusual helical packing in crystals of DNA bearing a mutation hot spot. *Nature*, **341**, 459–462

Vovelle, F., Elliott, R. J. and Goodfellow, J. M. (1989). Solvent bridging sites in A- and B-DNA helices. *Int. J. Biol. Macromol.*, **11**, 39–42

Vovelle, F. and Goodfellow, J. M. (1990). Sequence dependent hydration of DNA. *Int. J. Biol. Macromol.*, **12**, 369–373

Wang, A.H.-J., Hakoshima, T., van der Marel, G. A., van Boom, J. H. and Rich, A. (1984). A·T base pairs are less stable than G·C base pairs in Z-DNA: The crystal structure of d(m5CGTAm5CG). *Cell*, **37**, 321–331

Wang, A.H.-J., Quigley, G. J., Kolpak, F. J., Crawford, J. L., van Boom, J. H., van der Marel, G. A. and Rich, A. (1979). Molecular structure of a left-handed double helical DNA fragment at atomic resolution. *Nature*, **282**, 680–686

Watson, J. D. and Crick, F. H. C. (1953) A structure for deoxyribose nucleic acid. *Nature*, **171**, 737–738.

Westhof, E. (1987a). Hydration of oligonucleotides in crystals. *Int. J. Biol. Macro.*, **9**, 186–192.

Westhof, E. (1987b). Re-refinement of the B-dodecamer d(CGCGAATTCGCG) with a comparative analysis of the solvent structure in it and in the Z-hexamer d(5BrCG5BrCG5BrCG). *J. Biomol. Struct. Dyn.*, **5**, 581–600

Westhof, E. (1988). Water: An integral part of nucleic acid structure. *Ann. Rev. Biophys. Biophys. Chem.*, **17**, 125–144

Westhof, E. (1990). Structural water of nucleic acids. In *Water and Ions in Biomolecular Systems*. Birkhäuser Verlag, Basel, pp. 11–18

Westhof, E. and Beveridge, D. L. (1990). Hydration of nucleic acids. *Water Science Reviews*, **5**, 24–135

Westhof, E., Dumas, P. and Moras, D. (1988). Hydration of transfer RNA molecules: a crystallographic study. *Biochimie*, **70**, 145–165

Westhof, E., Rao, S. T. and Sundaralingam, M. (1980). Crystallographic studies of drug-nucleic acid interactions: Proflavine intercalation between the noncomplementary base-pairs of cytidilyl-3′, 5′-adenosine. *J. Mol. Biol.*, **142**, 331–361

Westhof, E. and Sundaralingam, M. (1986) Restrained refinement of the monoclinic form of yeast phenylalanine transfer RNA. Temperature factors and dynamics, coordinated waters, and base-pair propeller twist angles. *Biochemistry*, **25**, 4868–4878.

8

Hydration Sites and Hydration Bridges around DNA Helices

F. Vovelle and J. M. Goodfellow

1 Introduction

It has been well known, since the early studies of Franklin and Gosling (1953), that solvent influences the helical conformations of nucleic acids. Since then a variety of experimental techniques have been used to study the interactions of solvent with nucleic acids, including infrared spectroscopy (Falk *et al.*, 1963), gravimetric studies (Falk *et al.*, 1962) and fibre diffraction (Leslie *et al.*, 1980). These studies have been reviewed by Texter (1978) and Saenger (1984).

From these studies, it appears that 20 water molecules per nucleotide are found in the first hydration shell of a B-DNA, of which 11 or 12 are directly bound to DNA. These water molecules bind (in increasing order of strength) to anionic oxygens of the phosphate group, to the ester oxygens of the phosphodiester linkage, to the O_4' oxygen of the furanose ring and to the electronegative atoms of the base pairs, with A–T base pairs having one or two more water molecules than G–C base pairs.

Structural information on the organization of the solvent around the different helical forms of DNA has come from X-ray diffraction studies on oligomeric fragments of A-, B- and Z-DNA (for reviews see Kennard, 1984; Berman, 1986; Saenger, 1987; Westhof, 1987a, b, 1988; Westhof and Beveridge, 1990). These studies provide evidence for the stabilizing role of water and the importance of the bridging sites which lead through the helical periodicity of DNA to regular motifs of water molecules, such as 'spine of hydration' in the AATT region of the minor groove of B-DNA (Drew and Dickerson, 1981; Kopka *et al.*, 1983) and in the minor groove of Z-DNA (Westhof, 1987a, b; Wang *et al.*, 1984); 'filament' of water molecules in the minor groove of A-DNA (Cruse *et al.*, 1986) and of B-DNA

(Prive *et al.*, 1987); and pentagons of water molecules in the major groove of the A- form of DNA (Kennard *et al.*, 1986).

However, the number of crystal structures in which the solvent region has been resolved is relatively low. Moreover, the solvent, especially around the phosphate groups, is found to be disordered in several nucleic acid crystal hydrate structures. Therefore, several theoretical approaches have been used in an attempt to provide a more detailed understanding of the DNA–solvent interactions.

Three main theoretical techniques have been used, depending on the type of results required, and have been reviewed recently by Westhof and Beveridge (1990). They are as follows:

(1) *The hydration site approach*, in which one looks at water molecules closely bound to one or several polar or charge groups of the molecule,

(2) *The Monte Carlo computer simulation technique*, which is a numerical approximation based on statistical mechanics and can be used to predict structural and energetic properties of the solvent network around biomolecules.

(3) *The molecular dynamics technique*, which provides insight into solvent fluctuations and will be reviewed elsewhere in this book.

Most of these approaches require a set of potential energy functions to model each type of atom–atom interaction. The analytical formulae describing the potential functions are based on several approximations (Price and Goodfellow, 1992). It is, therefore, important to test the ability of any potential used to reproduce experimental structures.

2 Hydration Site Approach

What Is a Hydration Site?

Clementi (1983) gives the definition of a hydration site as 'a region of space with a volume and a shape such as to closely contain a water molecule'. The site position is relative to a particular atom of the solute molecule and corresponds to an energy minimum of the interaction energy between a water molecule and the solute. Thus, a molecule has several hydration sites which are related to the number of polar and charged groups of the molecule. Two alternative approaches can be used to determine these solvation sites. The first is based on quantum mechanical molecular orbital calculations and the second on empirical potential energy functions.

Molecular Orbital Calculations

The molecular orbital approach has been extensively used by B. and A. Pullman and collaborators to study the hydration scheme of the constituents of DNA. These calculations are achieved within the supermolecule model, in which the solute and solvent molecules are considered as an unique entity. The *ab initio* calculations are performed by using a contracted set of Gaussian orbitals.

Initially the hydration sites of purine and pyrimidine bases and of the complementary base pairs of DNA are calculated (Pullman and Pullman, 1975; Pullman and Perahia, 1978; Goldblum *et al.*, 1978; Pullman *et al.*, 1979). For AT and CG base pairs it is found that the more stable sites are those that bridge between the N_7 and N_6H_6 atoms of adenine or between the O_6 and N_7 atoms of guanine. Other strong binding sites are found on the O_2 atom of thymine (two sites), the N_3 atoms of adenine and guanine. In contrast, cytosine presents only weak binding sites. From the comparison of the stability of the different hydration sites, it is concluded that the hydration preferences of the bases in a base pair are in the order G > A > T > C and that AT base pairs are more hydrated than GC pairs in nucleic acid duplexes. This is consistent with experimental data on base-pair hydration.

For the ribose unit three main hydration sites are calculated as O_2', O_5' and O_3' atoms in decreasing order of stability (Berthod and Pullman, 1978). The hydration scheme of the phosphodiester backbone has been studied using the dimethylphosphate anion as a model (Pullman *et al.*, 1975 a, b). These studies show that 6 water molecules can bind to the anion; 3 water molecules are bound to each anionic oxygen in symmetrical positions forming a 'cone of hydration'.

For larger entities such as base pairs, *ab initio* SCF calculations become prohibitive. As it has been shown that the electrostatic contribution of the interaction energy plays the dominant role, the hydration scheme for larger fragments may be obtained using an electrostatic approach (Pullman and Perahia, 1978). Therefore, the hydration sites of purine and pyrimidine bases and base pairs were also calculated by using this electrostatic approach (known as OMTP).

More recently (de Neto Oliveira, 1986a, b) used this electrostatic procedure based on molecular orbital calculation to determine the hydration sites and the lability of the water molecule around these sites for the principal components of nucleic acids and for a model segment of B DNA poly (dA)·(dT) and poly (dG)·(dC). The results of this newer study reproduce satisfactorily the results obtained previously in the Pullmans' laboratory. It was also shown that many bound waters can be displaced slightly from their optimal binding positions, giving rise to only small variations in their interaction energy, i.e. they are very labile. For example, strong binding sites are found around the anionic oxygens of the phosphate groups but they are very labile.

Pure quantum mechanical studies on nucleic acid hydration have also been performed by Del Bene (1981, 1982), on thymine and uridine.

Calculations Based on Empirical Potentials

Chronologically, this second approach has been introduced by B. Pullman and co-workers (Perahia *et al.*, 1977). They used a very sophisticated empirical potential energy function (Caillet and Claverie, 1974, 1975) to determine the hydration scheme of a mini-helix constituted of three G–C pairs in B-conformation. Five water molecules are found on each of the phosphate groups. One of them bridges the two anionic oxygens of the phosphate group and two more water molecules are strongly bound to the N_7 atom of guanine, one of them forming another hydrogen bond with the O_6 atom. Bridging waters are also found between the O_4' oxygen of the deoxyriboses and the O_2 oxygen of cytosine or the N_3 of guanine (of the same residue). Individual hydration sites occurred on N_4 of cytosine. All these results corroborated the results of the molecular orbital calculations.

During the same period Clementi and collaborators have developed an analytical pair potential for a water molecule interacting with DNA (Scordamaglia *et al.*, 1977; Corongiu and Clementi, 1978; Clementi and Corongiu, 1979a). Using these potential energy functions, they studied the hydration of the bases (Scordamaglia *et al.*, 1977), of base pairs (Clementi and Corongiu, 1980) and of nucleic acid included in A-DNA (Clementi and Corongiu, 1979a, c), B-DNA (Clementi and Corongiu, 1979b, c), Z-DNA (Clementi and Corongiu, 1982b) single and double helices.

In these studies, isoenergy contour maps are drawn for selected planar sections of the biomolecule or for cylindrical surfaces enclosing the DNA helix. This approach gives a picture of DNA hydration which differs from the previous one. The isoenergy contour maps represent the interaction energy of a single water molecule at any position on the surface. From these maps, one can distinguish four regions, as follows: (1) the molecular hard core region (i.e. repulsive contours); (2) the hydrophobic region, which constitutes the part of the first hydration shell around hydrophobic groups; (3) the hydrophilic zone, characterized by negative contours; and (4) the bulk region. The cylindrical contour maps reflect the helical periodicity of DNA. The attraction of water molecules to the phosphate group is very strong and may affect water not only in the first layer but also in two additional layers. These results were found to hold for DNA in A-, B- and Z-conformation, but a detailed examination of these maps shows large differences for the three canonical forms of DNA corresponding to different hydration patterns.

Another simple atom–atom potential function has been developed by Poltev and co-workers (1984) to study the behaviour of nucleic acids in aqueous solution. They tested the reliability of their potential by calculating the preferential hydration sites of the nucleic acid bases before using more sophisticated Monte Carlo techniques. The most favourable sites were found to be sites bridging between hydrophilic groups of the base. These water positions are consistent with the available crystallographic data. However, sites are found on the N_1 atoms of thymine, cytosine and guanine and N_9 of the adenine. These latter sites would *not* be accessible to a water molecule when the bases are paired and linked to the sugar phosphodiester backbone. In general, these studies on hydration of the constituents of nucleic acids lead to a large number of hydration sites, but these sites are reduced when the components are formed into a DNA fragment. Conversely, other sites bridging different parts of the molecule may appear.

Solvent Bridging Sites

More recently we have used energy minimization studies to search for the possible hydration sites bridging two polar atoms for oligonucleotides (Vovelle *et al.*, 1989; Vovelle and Goodfellow, 1990, 1991). We first considered the canonical A- and B-forms of DNA based on Arnott data and alternating purine–pyrimidine sequences AT and CG. The interaction energy between water and the nucleic acid has been modelled using two different potential energy functions, which are: (1) our EMPWI model (Vovelle and Ptak, 1979) combined with Huckel Del Re charges (Lavery *et al.*, 1984) and (2) TIPS2 water model (Jorgensen, 1982) combined with non-bonded coefficients and charges from Weiner *et al.* (1984).

Both sets of potentials led to almost the same results. A comparison of the hydration site for A- and B-form shows that major differences appear in the hydration of the phosphate groups: the A- form allows for a bridging site between two adjacent phosphate groups (Figure 8.1), while in the B-form each phosphate group is individually hydrated. On the other hand, the hydration sites of the bases are bridging sites for both A- and B-forms. These sites are reported in Tables 8.1 (intrastrand solvent bridging sites) and 8.2 (interstrand solvent bridging sites) together with the water bridges calculated for the ZI- and ZII-forms of DNA (Wang *et al.*, 1979, 1981). In the Z-conformation, few water bridges are found on the bases which can be individually hydrated.

The analysis has been extended to look at further sequences in A- and B-conformation $d(G)_n{\cdot}d(C)_n$, $d(AAT)_n{\cdot}d(TTA)_n$, $d(AC)_n{\cdot}d(GT)_n$ and $d(AG)_n{\cdot}d(CT)_n$ (Vovelle and Goodfellow, 1990). Among these sequences several are known to adopt the A-conformation. From this study and our

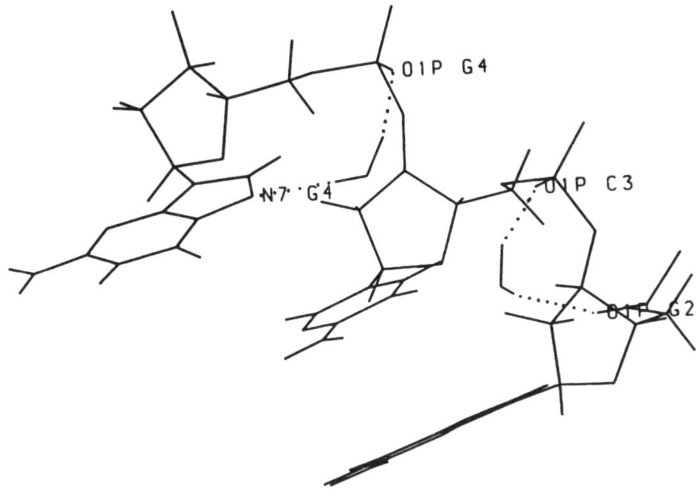

Figure 8.1 Diagram showing the positions of solvent bridging sites between adjacent intrastrand phosphate groups and between phosphate and base atoms in the sequence d(CG)$_n$ in the A-conformation

Table 8.1 Intrastrand solvent bridging sites

A	*B*	*Z*
Phosphate–W–Phosphate		
$O1P_i\cdots W\cdots O1P_{i+1}$	None	None
Phosphate–W–Base		
$O1P_i\cdots W\cdots N7(PU)_i$	None	$O1P(C)_{i+1}\cdots W\cdots N2(G)_i$ [a]
		$O1P(C)_i\cdots W\cdots O6(G)_{i+1}$ [b]
		$O1P(C)_i\cdots W\cdots N7(G)_{i+1}$ [b]
Base–W–Sugar		
$O2_i\cdots W\cdots O4'_{i+1}$	$O2_i\cdots W\cdots O4'_{i+1}$	None
$N3_i\cdots W\cdots O4'_{i+1}$	$N3_i\cdots W\cdots O4'_{i+1}$	None
$Base_i$–W–$Base_i$		
N6H6(A)· · ·W· · ·N7(A)	N6H6(A)· · ·W· · ·N7(A)	None
O6(G) · · ·W· · ·N7(G)	N7(G)· · ·W· · ·O6(G)	None
$Base_i$–W–$Base_{i+1}$		
N7(A)· · ·W*· · ·N7(A)	N7(A)· · ·W· · ·O4(T)	N7(G)· · ·W· · ·N4(C)
· · ·N6(A)	N7(A)· · ·W· · ·N4(C)	
N7(G)· · ·W*· · ·N7(G)	N7(A)· · ·W· · ·O6(G)	
· · ·N6(G)	O4(T)· · ·W· · ·N7(G)	
N7(A)· · ·W · · ·N4(C)		
O6(G)· · ·W · · ·N4(C)		
O4(T)· · ·W · · ·N6(A)		
O4(T)· · ·W · · ·O6(G)		
N4(C)· · ·W · · ·O4(T)		
O6(G)· · ·W*· · ·N6(A)		
N7(G)· · ·		

[a] in ZI form.
[b] in ZII form.
* This indicates sites with three hydrogen bonds.

Table 8.2 Interstrand solvent bridging sites

A	*B*	*Z*
Minor groove		
O2(T)· · ·W· · ·O2(T)	O2(T)· · ·W· · ·O2(T)	None
O2(C)· · ·W· · ·O2(C)		
O2(C)· · ·W· · ·O2(T)		
N3(A)· · ·W· · ·N3(A)		
N3(A)· · ·W· · ·O2(C)		
N3(A)· · ·W· · ·O2(T)		
Major groove		
O6(G)· · ·W· · ·N4(C)	O6(G)· · ·W· · ·O6(G)	None
	N4(C)· · ·W· · ·O4(T)	
	O6(G)· · ·W· · ·O4(T)	
	O4(T)· · ·W· · ·O4(T)	

previous analysis some general features emerge on the form and sequence dependence of DNA hydration. The results have been summarized in Tables 8.1 and 8.2:

(1) In the A-conformation bridging sites occur between two adjacent phosphate oxygens if the second residue is cytosine (Figure 8.1), while no bridging site is found on the phosphate groups of the B-form.

(2) Bridging sites between phosphate groups and the base atoms ($P_i \cdots B_i$) occur only between O_1P and N_7 of purine bases only in the A-conformation (Figure 8.1).

(3) Whatever the sequence and the form a water molecule bridges the purine N_3 or the pyrimidine O_2 atom to O_4' oxygen of the following residue in the minor groove (Figure 8.2).

(4) A large number of base· · ·base bridging sites are found in A- and B-form, but the A-form favours water bridges in which a water molecule is involved in three hydrogen bonds (Figure 8.3).

(5) The interstrand bridging sites occur mainly in the minor groove of the A-form (Figure 8.4), while they are mainly in the major groove of B-DNA (Figure 8.5).

Another significant result that emerges from these studies, concerns the number of bridging sites for a given sequence in the A- compared with the B-form. We have calculated this number for a sequence of 12 base pairs for each conformational form. It appears that the sequences which are found experimentally in the A-conformation have more bridging sites in this form than in the B-conformation. This confirms and extends the 'economy of hydration hypothesis' of Saenger *et al.* (1986). This hypothesis maintains that the main difference in the hydration of the A- and B-forms of DNA was the presence of water bridges between adjacent phosphate groups in the low-hydration A-form but not in the B-form; thus, fewer waters are

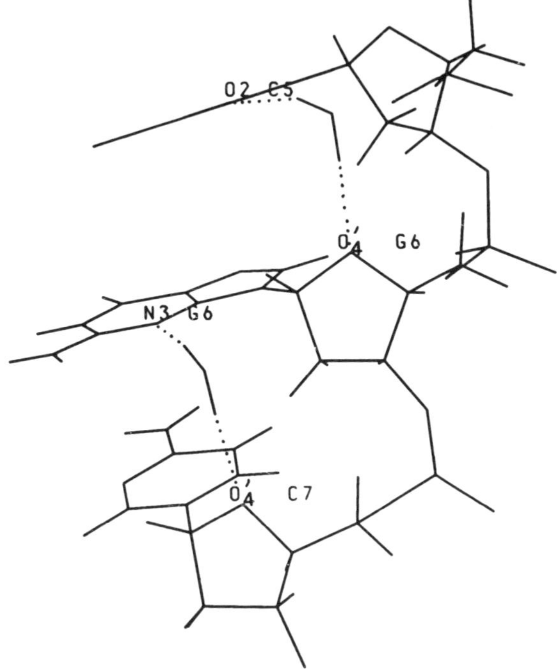

Figure 8.2 Diagram showing the position of solvent bridging sites between base and sugar atoms on the same strand in the sequence d(CG)$_n$ in the A-conformation

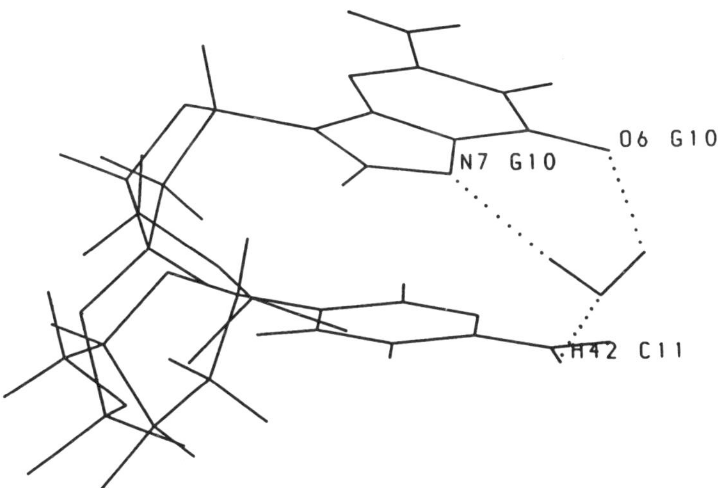

Figure 8.3 Diagram showing the position of solvent bridging sites between bases on the same strand in the sequence d(CG)$_n$ in the A-conformation

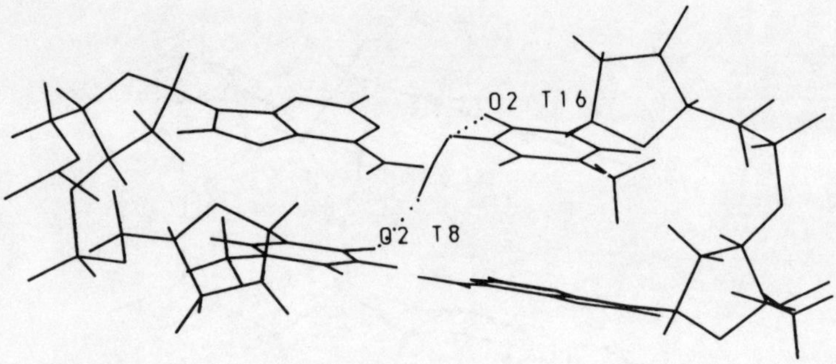

Figure 8.4 Diagram showing the position of solvent bridging sites between bases in different strands in the sequence $d(AT)_n$ in the B-conformation

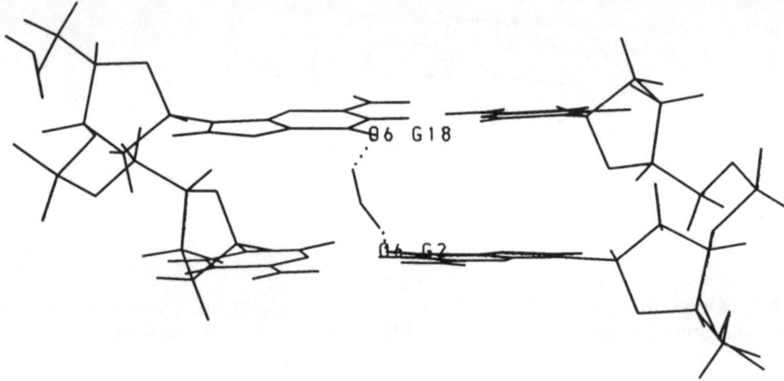

Figure 8.5 Diagram showing the position of solvent bridging sites between bases in different strands in the sequence $d(CG)_n$ in the B-conformation

necessary to satisfy all the hydrogen bond possibilities of the phosphate groups in the A-form than in the B-form. Our work shows that this concept is sequence-dependent and needs to include all the polar groups of DNA.

Our results described above were found using the canonical forms of DNA. However, we can compare calculations using the B-DNA canonical form with those using the crystallographic structure determined by Dickerson and colleagues (Drew and Dickerson, 1981; Kopka *et al.*, 1983) for the dodecamer d(CGCGAATTCGCG). The hydration sites calculated for the crystallographic structure differ notably from those found using standard coordinates (Vovelle and Goodfellow, 1990), especially in the minor groove. In the canonical form the only bridging sites found are between the O_2 of thymines. In the narrow minor groove of the crystal structure, interstrand bridges are found from O_2 of thymines to N_3 or O_2 of the base on the opposite strand and also to the O_4' atom of the next sugar. These calculated positions correspond to experimentally observed water mol-

Table 8.3 Comparison of bridging sites from Monte Carlo and energy minimization calculations for B-dodecamer

	EM	*MC*
Intrastrand		
Minor Groove		
O2(T)· · ·W· · ·O2(C)	No	Yes
N2(G)· · ·W· · ·O4′	No	Yes
O2(T)· · ·W· · ·O4′	Yes	Yes
N3(G)· · ·W· · ·O4′	Yes	No
N3(A)· · ·W· · ·O4′	Yes	No
O2(C)· · ·W· · ·O4′	Yes	No
Major Groove		
N7(G)$_i$· · ·W· · ·O6(G)$_i$	Yes	Yes
N7(A)$_i$· · ·W· · ·N6(A)$_i$	Yes	Yes
N6(A)· · ·W· · ·N7(G)	Yes	Yes
N4(C)· · ·W· · ·C7(T)	No	Yes
N6(A)· · ·W· · ·N7(A)	Yes	Yes
N7(G)· · ·W· · ·O1P	Yes	Yes
C7(T)· · ·W· · ·O1P	Yes	Yes
N7(A)· · ·W· · ·O1P	Yes	Yes
Interstrand		
O2(T)· · ·W· · ·N3(T)	No	Yes
O2(T)· · ·W· · ·O2(T)	Yes	No

ecules which constitute the ‘spine of hydration’. These results confirm the importance of water molecules in *inducing* DNA structure (Westhof, 1988): in the minor groove of the dodecamer, the bases adopt a conformation in which they optimize their interactions with water.

3 Monte Carlo Calculations

Methodology

The previous sections have focused on the use of quantum mechanics and energy minimization methods to describe the most favourable interactions between solvent molecules and oligonucleotides. Some of the limitations of these methods can be overcome with the use of more sophisticated simulation methods, such as the Monte Carlo technique. First, the simulations are carried out at a given temperature, usually room temperature for simulation of macromolecular ensembles. Second, many water molecules can be included, such that the solute is surrounded by solvent and the water· · ·water contribution to the energy can be calculated as well as the water· · ·solute contribution. Third, it is now possible to estimate the more realistic free-energy difference rather than potential energy differences between related conformations. The results from these calculations lead to a more realistic but complex view of nucleotide hydration.

The Monte Carlo simulation technique depends on generating a Boltzmann weighted distribution of states for a given temperature, T. An advantage of this method compared with molecular dynamics techniques is that no derivatives are required. One of the most frequently used methods is due to Metropolis (Metropolis *et al.*, 1953), in which sampling is considered as part of a Markov chain of configurations. In this method an initial molecular ensemble is generated for a given number of molecules within a specified volume and at a given temperature. The potential energy, V_0, is calculated using the potential-energy functions described previously.

One particle in the system is chosen at random and moved a random amount translationally and rotationally, and the potential energy is recalculated, V_1. The difference in energy between the two states can be calculated as $V_1 - V_0$. If this difference is negative (i.e. if the new state is energetically more favourable), the move is accepted and the cycle of random moves is repeated. If the difference in energy is unfavourable, then the probability of this occurring is calculated as $\exp -(V_0 - V_1)/k_B T$ (where k_B is the Boltzmann constant and T the temperature) and compared with a random number between 0 and 1. If the probability is less than the random number, the move is again accepted. On the other hand, if the probability is greater than the random number, then the move is rejected and the system returns to the initial state.

This procedure is repeated many times, so that a large number of energetically accessible configurations of the system are generated. In the initial stages, the total energy is usually seen to decrease fairly rapidly as the system equilibrates. After this equilibration period, properties such as the average potential energy tend to oscillate around an average value. Data from this latter stage are used in the structural and energetic analysis of the solvent–solute interactions.

In practice, it has only been possible to calculate potential energies from computer simulations. However, recently methods have been evolved which allow one to estimate free-energy differences between two related systems.

The Gibbs free energy, ΔG, is often described as

$$\Delta G = \Delta H - T\Delta S$$

where ΔH is the change in enthalpy and $T\Delta S$ the entropic component.

The free-energy difference can be estimated from computer simulations using the following equation which is derived from statistical mechanics

$$\Delta G = -k_B T \ln \langle \exp-(V_1 - V_0)/k_B T \rangle_0$$

where V_1 and V_0 are the potential energies of two closely related states and the angled brackets imply averaging over the initial state. This is the perturbation approach. There are also a number of other methods involving changes to the sampling procedures used in the Monte Carlo algorithm. These methods have been reviewed by Beveridge and Di Capua (1989) and van Gunsteren (1989).

Hydration of DNA Constituents

Although it is possible to model large systems, many initial simulations involved small fragments from nucleic acids. Some of the earliest Monte Carlo simulations on macromolecules in solution were undertaken by Clementi and co-workers (Clementi and Corongiu, 1980) after they had established solvent–solvent and solvent–solute potential energy functions from *ab initio* quantum mechanical calculations. Their initial simulations involved individual bases enclosed in a cluster of 40 water molecules and base pairs surrounded by 50 water molecules, and led to the identification of the positions and orientations of water molecules in the first hydration shell.

The studies of Beveridge and co-workers (1984) extended these simulations to include the hydration not only of the individual bases but also of ribose and deoxyribose sugars and the dimethylphosphate anion. A detailed analysis procedure was developed using proximity criteria in order to provide a detailed analysis of the hydration shells around each of the components and the framework for discussing the stability of ordered networks of solvent molecules such as the 'spine of hydration' in the minor groove of a B-like dodecamer found experimentally in the crystal structure (Kopka *et al.*, 1983) and the solvent filaments found in the pioneering simulations of Clementi and co-workers (Corongiu and Clementi, 1981a).

A more detailed study (Jayaram *et al.*, 1987) of the hydration of a dimethylphosphate anion and its complex with a Na^+ cation was undertaken to predict the most stable hydrated phosphate group conformation. The first hydration shells of the three conformations (distinguished by the alpha and zeta backbone torsion angles — i.e. *gauche-gauche* (*gg*), *gauche-trans* (*gt*) and *trans-trans* (*tt*)) were similar, each containing between 23 and 26 water molecules around the whole molecule and with approximately 6 water molecules associated with the 2 free phosphate oxygens. This description of the first hydration shell is essentially similar to that found in a previous study by Alagona *et al.* (1985) using different potential energy functions. The effect of changes in conformation is primarily to change the interactions of these 6 water molecules with the anionic oxygens and with the ester oxygens as well. In the absence of the Na^+, the *gg* and *gt* conformations are favoured. In the presence of the counterion, the *tt* conformation is most stable. Calculations using umbrella and adaptive sampling techniques, in order to estimate free-energy changes (Jayaram *et al.* 1988), indicated that the *gg* conformation of the dimethylphosphate anion is the most stable, but the *gt* and *tt* conformations are also likely to be accessible at room temperature.

Similar methods have been used to study the relative stability of a single uracil base, stacked uracil bases and coplanar uracil bases in solution (Danilov *et al.*, 1984). These studies indicated that the base-stacking interaction is the most favourable in solution due to an increase in water–water interactions around the dimer. Larger stacks of bases and base pairs have been simulated by Teplukhin *et al.* (1989), who showed that the number of hydrogen bonds between water molecules and bases changed as the stack of bases was changed from that of an 'A'-like to a 'B'-like helix.

The hydration of methylated derivatives of uracil and thymine has also been investigated using Monte Carlo simulations (Danilov and Tolokh, 1990). This study shows that the interaction energy between the water molecules and the bases increases (i.e. becomes less favourable) on the introduction of methyl groups, while the water–water interaction energy remained very similar. Lower water–water interaction energies and higher number of hydrogen bonds in the water structure are seen in the simulations of thymine and its methylated derivatives compared with those of uracil and its derivatives. The authors believe that this indicates greater regulation of water molecules around hydrophobic groups.

The influence of the type of solvent on planar and stacked bases has been investigated using Monte Carlo simulations of each type of base in a non-polar solvent, CCl_4 (Porhille *et al.*, 1984). It is found that solute–solvent interactions, which are primarily determined by dispersion forces, are most favoured when there is the maximum contact area between the base and the non-polar solvent. Given the planar shape of the bases, solute–non-polar solvent interactions are not favoured by stacking. Analysis of the solvent–solvent interactions shows that the introduction of the bases causes significant changes in the solvent structure. In contrast, in aqueous solution, the electrostatic forces rather than the dispersion forces dominate, and the most stable structures are stacked associations, which allows for maximum exposure of polar groups to solvent.

Dinucleotides

Simulations on nucleotides in solution are important, as they often add to the information available from X-ray crystallography on nucleotide crystal hydrates. In the experimental structural studies the solvent around oligonucleotides, especially close to the phosphate groups, may be disordered, and thus individual sites could not be refined into specific electron density. Simulations, whether of the Monte Carlo or molecular dynamics type, allow one to see ordered and disordered solvent sites.

Simulations on a dinucleotide duplex dCpG in the canonical B-DNA form were carried out in solution using the TIPS2 water model (Goodfellow *et al.*, 1986, 1987). The results from these simulations showed that all polar or charged groups along the phosphate backbone and accessible polar base atoms are hydrated. Solvent molecules often formed complex networks consisting of loops or chains linking polar groups at different parts of the nucleotide.

Monte Carlo simulations on a dinucleotide duplex rGpC surrounded by 562 water molecules using the TIP4P water model together with the AMBER force field has shown that the calculated hydration sites are generally inconsistent with the experimental data (Subramanian *et al.*, 1990a). The hydration of this molecule has also been simulated in the crystal environment using the AMBER molecular dynamics package (Herzyk *et al.*, 1991). Simulations with and without restrained solvent molecules showed that although restraining solvent oxygen atoms to their experimentally determined positions had little effect on the nucleotide atomic fluctuations, it did have a significant effect on the backbone conformation of the nucleotide. Thus, underlying the importance of establishing realistic potential energy functions which accurately model the solvent–nucleotide and solvent–solvent interactions.

Hydration of Oligonucleotides

The first simulations to include a complete turn of the B-DNA helix together with 447 water molecules and also Na^+ counterions to balance the charge on the phosphate groups were carried out by Clementi and coworkers (Corongiu and Clementi, 1981a, b; Clementi and Corongiu, 1982a). These pioneering simulations predicted the existence of *trans*-groove filaments of hydrogen-bonded water molecules connecting phosphate groups on opposite sides of the major groove as well as interphosphate filaments connecting successive phosphate groups on the same strand.

Subsequent simulations have often used the dodecamer sequence d(CGCGAATTCGCG)·d(GCGCTTAAGCGC), as this was the first oligonucleotide sequence in the B-like conformation to be solved experimentally using crystallographic techniques. This structure is of considerable interest to those studying DNA hydration, as the spine of hydration in the minor groove is first seen in this conformation (Drew and Dickerson, 1981). A simulation of this sequence together with 1777 water molecules has indicated considerable localization of solvent site in the minor groove, which would correspond to the spine of hydration seen in the central AATT sequence (Subramanian *et al.*, 1988). However, further minor groove hydration is seen in the flanking CG sequence in the simulation, which has specific implications for stability of DNA sequences in water. A more detailed analysis of the same simulation (Subramanian and Beveridge, 1989) has shown hydration sites not only in the minor groove but also in the major groove and along the sugar–phosphate backbone. A number of one- and two-water strand bridges are also seen in this simulation. Comparison of simulations of the same sequence using canonical and crystal structures as well as two different potential energy functions has shown that both have an effect on the location of the hydration sites (Subramanian *et al.*, 1990). The anomalous structure and properties of poly dA·dT have also been explained in terms of this minor groove hydration (Chuprina, 1987). More recently, a molecular dynamics study on a decamer (C–C–A–A–C–G–T–T–G–G) has shown that there are two hydration patterns that can exist in the minor groove, depending on its width (Chuprina *et al.*, 1991). In the narrow regions the hydration pattern corresponds to the spine, but in the wider regions each base pair is hydrated individually.

Simulation of Crystal Hydrates

Although attempts have been made to compare the results of simulations of nucleotide hydration in solution with the experimentally determined water molecule sites from crystallography, there is obviously the problem that the environments are not identical. However, simulations can be undertaken in a crystalline environment in which the asymmetric units are replicated to produce the crystallographic unit cell and, if necessary, the unit cells can be replicated to produce a cell dimension larger than twice the standard cutoff distance for the energy calculations. The necessary prerequisite for such calculations is the existence of a high-resolution well-refined nucleotide structure in which many of the water molecule sites have been determined.

Early attempts using Monte Carlo computer simulations and a number of different potential energy functions have studied the well-defined solvent structure in the crystal of dCpG–proflavine (a dinucleotide–intercalating drug complex (Neidle *et al.*, 1980)). This crystal has a distinctive solvent network consisting of a polygonal disc of and a pentagonal network of water molecules. Initial analysis showed that although the structure was of high accuracy, there appeared to be one or two water molecules missing per asymmetric unit. This is within the error of experimental crystal density measurements needed to determine the total number of water molecules with unit occupancy.

Monte Carlo calculations are carried out with fixed solute positions and with periodic boundary conditions to represent the condensed phase environment. The intermolecular potential energy functions of Clementi and co-workers were used in one simulation as well as those obtained from the literature and used in other crystal systems by Goodfellow and collaborators (Mezei *et al.*, 1983). The results are analysed in terms of hydrogen bond topology and distances as well as mean water positions. Considerable agreement is found with the experimental solvent molecule positions, although there was some evidence that the solvent–solute energies in the simulation using the MCY model of water may be overestimating these interactions at the expense of the water–water interactions.

Further simulations on the same system by Kim *et al.* (1983) and Kim and Clementi (1985a, b) found good agreement with the experimental solvent positions, especially when they assumed the existence of an extra 20–30 water molecules per unit cell over that found in the original X-ray refinement. This extra number of water molecules is unlikely, given the experimentally determined unit cell volume and crystal density. The authors conclude that this discrepancy may be due to the MCY water model and the level of detail used in the quantum mechanical calculations used to establish the solute–solvent interactions in the hydrated crystal.

The dCpG–proflavine crystal system has also been simulated using molecular dynamics techniques (Swaminathan *et al.*, 1990; Herzyk *et al.*, 1991, 1992). In the latter study the importance of solvent interactions was underscored by monitoring the effect of restraining water molecules. This had the effect of improving the conformation (i.e. leading to better agreement with experiment) and having some effect on the atomic fluctuations, especially of the phosphate groups.

Other small-crystal systems have been simulated using Monte Carlo techniques in order to compare the predictive power of the simulation with the experimentally determined solvent structures (Goodfellow, 1984). The use of explicit hydrogen atom sites compared with united atom models was studied by Howell and Goodfellow (1984). They found that differences in the precise location of water molecules were found which depended on the exact location of the hydrogen atoms (which are not normally known from X-ray or neutron crystallography for large solutes). However, the use of explicit hydrogen atoms improved the agreement in the water–solute distances when compared with those found experimentally.

Further simulations using different starting positions for the water oxygen atoms (Howell and Goodfellow, 1988) and extending the type of interactions to include counterions (Elliott and Goodfellow, 1987a, b) showed that the agreement between the water molecule sites from the simulation compared with the experimentally determined positions was within the error of the simulations, and thus it seemed feasible to extend the simulations from the condensed phase to solution conditions and to study larger fragments of more direct biological interest.

4 Discussion

In this review we have tried to present much of the data on the hydration of DNA helices obtained from *ab initio* quantum mechanics, energy minimization and Monte Carlo computer simulations. We should like to be able to compare results from these different techniques as well as from molecular dynamics simulations and the experimental results from X-ray crystallography. This is a complicated task, given the variety of environments (crystal, solution or *in vacuo*), variety of sequences (both length and type of bases) and the number of helical families (A, B and Z).

Comparison between computationally predicted sites and experimentally determined solvent structures in nucleotide crystals is limited, partly because it is not possible to determine experimentally the disordered sites in the crystal structures. We have also seen that energy calculations indicate that solvent bridging sites are dependent on sequence and the relevant sequence may not have been crystallized. The early calculations on components of nucleotides may not be valid for larger fragments, because some of the sites may become inaccessible to solvent molecules.

However, there is one system which has been studied by many workers — namely the 'B'-dodecamer d(CGCGAATTCGCG)·d(GCGCTTAAGCGC). Both the canonical and the crystal atomic coordinates have been used in both energy minimization and Monte Carlo simulations. The crystallographic data showed narrowing of the minor groove in the central AATT region (compared with the canonical B-form) and the existence of the now well-characterized spine of hydration. Energy minimization and Monte Carlo studies of the crystal conformation have confirmed the existence of solvent molecule sites which can bridge between bases on different strands and thus lead to the spine of hydration. Differences are seen if the canonical 'B'-conformation is used in the calculations, as this has a uniform minor groove width. In the energy minimization calculations, solvent bridging sites are found between the thymine bases when the canonical 'B'-conformation is used. Moreover, Monte Carlo calculations have shown that the hydration network extends into the CG regions in the minor groove and along the phosphate backbone.

In summary, a number of techniques have been and are still being used to elucidate the role of water molecules in stabilizing the helical conformations of DNA. No one technique alone has led to the complete picture. No doubt, it will be the partnership between high-resolution X-ray structures and a variety of simulation techniques which will lead to a full understanding of DNA hydration.

Acknowledgements

JMG would like to thank the SERC for support from the Molecular Recognition Initiative and the Computational Science Initiative (GR/F/81071). JMG gratefully acknowledges a Research Leave Fellowship from The Wellcome Trust.

References

Alagona, G., Ghio, C. and Kollman, P. A. (1985). Monte Carlo simulations of the solvation of the dimethyl phosphate anion. *J. Am. Chem. Soc.*, **107**, 2229–2239

Berman, H. M. (1986). Hydration of nucleic acid crystals. *Ann. Rev. N. Y. Acad. Sci.*, **482**, 166–178

Berthod, H. and Pullman, A. (1978). Quantum mechanical exploration of the properties of sugar. *Theoret. Chim. Acta*, **47**, 59–66

Beveridge, D. L. and Di Capua, F. M. (1989). Free energy via molecular simulation: A primer. In van Gunsteren, W. F. and Weiner, P. (Eds), *Computer Simulation of Biomolecular Systems*. ESCOM, Leiden, pp. 1–26

Beveridge, D. L., Maye, P. V., Jayaram, B., Ravishanker, G. and Mezei, M. (1984). Aqueous hydration of nucleic acid constituents: Monte Carlo computer simulation studies. *J. Biomol. Struct. Dyn.*, **2**, 261–270

Caillet, J. and Claverie, P. (1974). Differences of nucleotide stacking patterns in a crystal and in binary complexes. *Biopolymers*, **13**, 601–614

Caillet, J. and Claverie, P. (1975). Theoretical evaluation of the intermolecular interaction energy of a crystal: Application to the analysis of crystal geometry. *Acta Cryst.*, **A31**, 448–461

Chuprina, V. P. (1985). Regularities in formation of the spine of hydration in the DNA minor groove and its influence on the DNA structure. *FEBS*, **186**, 98–102

Chuprina, V. P. (1987). Anomalous structure and properties of poly dA·dT computer simulation of the polynucleotide structure with the spine of hydration in the minor groove. *Nucleic Acids Res.*, **15**, 293–310

Chuprina, V. P., Heinemann, U., Nurislamov, A. A., Zielenkiewicz, P., Dickerson, R. E. and Saenger, W. (1991). Molecular dynamics simulation of the hydration shell of a B-DNA decamer reveals two main types of minor-groove hydration depending on groove width. *Proc. Natl Acad. Sci. USA*, **88**, 593–597

Clementi, E. (1983). Structure of water and counterions for nucleic acids in solutions. In Clementi, E. and Sarma, R. H. (Eds), *Structure and Dynamics: Nucleic Acids and Proteins*. Adenine Press, Schenectady, N. Y., pp. 321–365

Clementi, E. and Corongiu, G. (1979a). Interaction of water with DNA single helix in the A conformation. *Biopolymers*, **18**, 2431–2450

Clementi, E. and Corongiu, G. (1979b). Interaction of water with DNA single and double helices in the B conformation. *Int. J. Quant. Chem.*, **116**, 897–915

Clementi, E. and Corongiu, G. (1979c). Iso-energy contour maps for the interaction of water with a DNA double helix in the B conformation. *Chem. Phys. Lett.*, **60**, 175–178

Clementi, E. and Corongiu, G. (1980). A theoretical study on the water structure for nucleic acids bases and base pairs in solution at t=300 K. *J. Chem. Phys.*, **72**, 3979–3992

Clementi, E. and Corongiu, G. (1982a). B-DNA structural determination of Na^+ counterions at different humidities, ionic concentrations and temperature. In

Lowdin, P. O. (Ed.), *Proceedings Int. Symp. in Quantum Biology and Quantum Pharmacology*, Vol. 9. Wiley, New York, pp. 213–221

Clementi, E. and Corongiu, G. (1982b). Simulations of the solvent structure for macromolecules. III. Determination of the Na^+ counterion structure. *Biopolymers*, **21**, 763–777

Corongiu, G. and Clementi, E. (1978a). Intramolecular and intermolecular interactions for deriving chemical formulae and for simulating complex chemical systems. *Gazz. Chim. Ital.*, **108**, 273–306

Corongiu, G. and Clementi, E. (1978b). Interaction of water with dimethyl phosphate ion. *Gazz. Chim. Ital.*, **108**, 687–691

Corongiu, G. and Clementi, E. (1981a). Simulations of the solvent structure for macromolecules. I. Solvation of B-DNA double helix at $T = 300$ K. *Biopolymers*, **20**, 551–571

Corongiu, G. and Clementi, E. (1981b). Simulations of the solvent structure for macromolecules. II. Structure of water solvating Na^+ B-DNA at 300 K and a model for conformational transitions induced by solvent variations. *Biopolymers*, **20**, 2427–2483

Cruse, W. B. T., Salisbury, S. A., Brown, T., Cosstick, R., Eckstein, F. and Kennard, O. (1986). Chiral phosphorothioate analogues of B-DNA. The crystal structure of Rp-d Gp(S)CpGp(S)CpGp(S)C. *J. Mol. Biol.*, **192**, 891–905

Danilov, V. I., Tolokh, I. S., Poltev, V. I. and Malenkov, G. G. (1984). Nature of the stacking interaction of nucleotide bases in water: a Monte Carlo study of the hydration of uracil molecule associates. *FEBS*, **167**, 245–248

Danilov, V. I. and Tolokh, I. S. (1990). Hydration of uracil and thymine methyl derivatives: a Monte Carlo simulation. *J. Biomol. Struct. Dyn.*, **7**, 1167–1183

Del Bene, J. (1981). Molecular orbital theory of hydrogen bonding. 24 Ground state water–uracil complexes. *Comp. Chem.*, **2**, 188–199

Del Bene, J. (1982). Molecular orbital theory of the hydrogen bond XXIX water–thymine complexes. *J. Chem. Phys.*, **76**, 1058–1063

Drew, H. and Dickerson, R. E. (1981). Structure of a B-DNA dodecamer. III. Geometry of hydration. *J. Mol. Biol.*, **151**, 535–556

Elliott, R. J. and Goodfellow, J. M. (1987a). Monte Carlo computer simulation of nucleotide crystal hydrates and their counter-ions. *J. Theoret. Biol.*, **127**, 403–412

Elliott, R. J. and Goodfellow, J. M. (1987b). Ammonium Ion representation in Monte Carlo simulation of crystal hydrates. *J. Theoret. Biol.*, **128**, 121–125

Falk, M., Hartman, J. K. A. and Lord, P. G. (1962). Hydration of deoxyribonucleic acid. I. A Gravimetric study. *J. Am. Chem. Soc.*, **84**, 3843–3846

Falk, M., Hartman, J. K. A. and Lord, P. G. (1963). Hydration of deoxyribonucleic acid. II. A spectroscopic study of the effect of hydration on the structure of deoxyribonucleic acid. *J. Am. Chem. Soc.*, **85**, 391–394

Franklin, R. E. and Gosling, R. G. (1953). The structure of sodium thymonucleate fibres. I. The influence of water content. *Acta Cryst.*, **6**, 673–677

Goldblum, A., Perahia, D. and Pullman, A. (1978). Hydration scheme of the complementary base-pairs of DNA. *FEBS Lett.*, **91**, 213–215

Goodfellow, J. M. (1984). Solvent interactions in nucleic acid hydrates. *J. Theoret. Biol.*, **107**, 261–274

Goodfellow, J. M., Howell, P. L. and Elliott, R. J. (1987). Water at biomolecule interfaces. In Moras, D., *et al.* (Eds), *Crystallography of Molecular Biology*. Plenum Press, New York, pp. 167–177

Goodfellow, J. M., Howell, P. L. and Vovelle, F. (1986). Monte Carlo studies of water in crystal hydrates. *Ann. N.Y. Acad. Sci.*, **482**, 179–194

Herzyk, P., Goodfellow, J. M. and Neidle, S. (1991). Molecular dynamics simulations of dinucleoside and dinucleoside–drug crystal hydrates. *J. Biomol. Struct. Dyn.*, **9**, 363–386

Herzyk, P., Neidle, S. and Goodfellow, J. M. (1992). Conformation and dynamics of drug/DNA interactions. *J. Biomol. Struct. Dyn.* (in press)

Howell, P. L. and Goodfellow, J. M. (1984). Solvent networks in nucleotide crystal hydrates. *J. Phys.* (*Paris*), **45**, C7, 211–218

Howell, P. L. and Goodfellow, J. M. (1988). Effect of initial positions on the simulation of water networks in crystal hydrates. *Mol. Simulation*, **1**, 333–345

Jayaram, B., Mezei, M. and Beveridge, D. L. (1987). Monte Carlo hydration of the aqueous hydration of dimethylphosphate conformations. *J. Comp. Chem.*, **8**, 917–942

Jayaram, B., Mezei, M. and Beveridge, D. L. (1988). Conformational stability of dimethyl phosphate anion in water: liquid state free energy simulations. *J. Am. Chem. Soc.*, **110**, 1691–1694

Jorgensen, W. (1982). Revised TIPS for simulation of liquid water and aqueous solutions. *J. Chem. Phys.*, **77**, 4156

Kennard, O. (1984). DNA from A-Z. A survey of oligonucleotide structure. *Pure and Appl. Chem.*, **56**, 989–1004

Kennard, O., Cruse, W. B. T., Nachman, J., Prange, T., Shakked, Z. and Rabinovich, D. J. (1986). Ordered water structure in a A-DNA octamer at 1.7Å resolution. *J. Biomol. Struct. Dyn.*, **3**, 623

Kim, K. S. and Clementi, E. (1985a). Energetics and pattern analysis of crystals of proflavine deoxydinucleotide phosphate complex. *J. Am. Chem. Soc.*, **107**, 227–234

Kim, K. S. and Clementi, E. (1985b). Hydration analysis of the intercalated complex of deoxydinucleotide phosphate and proflavin: Computer simulations. *J. Chem. Phys.*, **89**, 3655–3663

Kim, K. S., Corongiu, G. and Clementi, E. (1983). Networks of water molecules in a proflavine deoxydinucleotide phosphate complex. *J. Biomol. Struct. Dyn.*, **1**, 263–285

Kopka, M. L., Fratini, A. V., Drew, H. R., Drew, R. E. and Dickerson, R. E. (1983). Ordered water structure around a B-DNA dodecamer: A quantitative study. *J. Mol. Biol.*, **163**, 129–146

Lavery, R., Zakrzwska, K. and Pullman, A. (1984). Optimized monopole expansions for the representation of the electrostatic properties of nucleic acids. *J. Comp. Chem.*, **9**, 363

Leslie, A. G. W., Arnott, S., Chandrasekaran, R. and Ratcliff, R. L. (1980). Polymorphism of DNA double helices. *J. Mol. Biol.*, **143**, 49–72

Metropolis, M., Metropolis, A. W., Rosenbluth, M. N., Teller, A. H. and Teller , E. (1953). Equation of state calculations by fast computing machines. *J. Chem. Phys.*, **21**, 1087–1092

Mezei, M., Beveridge, D. L., Berman, H. M., Goodfellow, J. M., Finney, J. L. and Neidle, S. (1983). Monte Carlo studies on water in the dCpG/proflavin crystal hydrate, *J. Biomol. Struct. Dyn.*, **1**, 287–297

Neidle, S., Berman, H. M. and Shieh, H. S. (1980). Highly structured water network in crystals of a deoxyribonucleoside–drug complex. *Nature*, **288**, 129–133

de Oliveira Neto, M. (1986a). Rapid location of the preferred interaction sites between small polar molecules and macromolecules. I. Binding of water to the component units of nucleic acids. *J. Comp. Chem.*, **7**, 617–628

de Oliveira Neto, M. (1986b). Rapid location of the preferred interaction sites between small polar molecules and macromolecules. II. Binding of water to a model segment of B-DNA. *J. Comp. Chem.*, **7**, 629–639

Perahia, D., Jhon, M. S. and Pullman, B. (1977). Theoretical study of the hydration of B-DNA. *Biochim. Biophys. Acta*, **474**, 349–362

Poltev, V. I., Grokhlina, T. I. and Malenkov, G. G. (1984). Hydration of nucleic acid bases studied using novel atom–atom potential functions. *J. Biomol. Struct. Dyn.*, **2**, 413–429

Porhille, A., Burt, S. K. and MacElroy, R. D. (1984). Monte Carlo simulation of the influence of solvent on nucleic acid base associations. *J. Am. Chem. Soc.*, **106**, 402–409

Price, S. L. and Goodfellow, J. M. (1992). Potential energy functions. In Moss, D. S. and Goodfellow, J. M. (Eds), *Computer Modelling of Biomolecular Processes*. Ellis Horwood, Chichester (in press)

Prive, G. G., Heinemann, U., Chendrasegaran, U., Kan, L. S., Kopka, M. M. and Dickerson, R. E. (1987). Helix geometry, hydration and G. A. mismatch in a B-DNA decamer. *Science*, **203**, 498–504

Pullman, A., Berthod, H. and Gresh, N. (1975a). Quantum mechanical studies of environmental effects on biomolecules. An *ab initio* study of the hydration of dimethylphosphate. *Chem. Phys. Lett.*, **33**, 11–15

Pullman, A. and Perahia, D. (1978). Hydration scheme of uracil and cytosine. *Theoret. Chim. Acta*, **48**, 29–36

Pullman, A. and Pullman, B. (1975). New paths in the molecular orbital approach to solvation of biological molecules. *Q. Rev. Biophys.*, **7**, 505–566

Pullman, B., Miertus, S. and Perahia, D. (1979). Hydration scheme of the purine and pyrimidine bases and base pairs of the nucleic acids. *Theoret. Chim. Acta*, **50**, 317–325

Pullman, B., Pullman, A., Berthod, H. and Gresh, N. (1975b). Quantum mechanical studies of environmental effects on biomolecules VI. *Theoret. Chim. Acta*, **40**, 93–111

Saenger, W. (1984). *Principles of Nucleic Acid Structure*. Springer Verlag, New York, pp. 368–384

Saenger, W. (1987). Structure and dynamics of water surrounding biomolecules. *Ann. Rev. Biophys. Biophys. Chem.*, **16**, 94–114

Saenger, W., Hunter, W. N. and Kennard, O. (1986). DNA conformation is determined by economics in the hydration of phosphate groups. *Nature*, **324**, 385–388

Scordamaglia, R., Cavallone, F. and Clementi, E. (1977). Analytical potentials from '*ab initio*' computation for the interaction between biomolecules. 2. Water with nucleic acids. *J. Am. Chem. Soc.*, **99**, 5545–5549

Subramanian, P. S. and Beveridge, D. L. (1989). A theoretical study of aqueous hydration of canonical B d(CGCGAATTCGCG): Monte Carlo simulation and comparison with crystallographic ordered water sites. *J. Biomol. Struct. Dyn.*, **6**, 1093–1122

Subramanian, P. S., Ravishanker, G. and Beveridge, D. L. (1988). Theoretical considerations on the 'spine of hydration' in the minor groove of d(CGCGAATTCGCG)·d(GCGCTTAAGCGC): Monte Carlo simulation. *Proc. Natl Acad. Sci. USA*, **85**, 1836–1840

Subramanian, P. S., Swaminathan, S. and Beveridge, D. L. (1990). Theoretical account of the 'spine of hydration' in the minor groove of duplex d(CGCGAATTCGCG). *J. Biomol. Struct. Dyn.*, **7**, 1161–1165

Swaminathan, S., Beveridge, D. L. and Berman, H. M. (1990). Molecular dynamics simulation of a deoxydinucleoside–drug intercalation complex: dCpG/proflavine. *J. Phys. Chem.*, **94**, 4660–4665

Teplukhin, A. V., Poltev, V. I., Shulyuprina, N. V. and Malenkov, G. G. (1989). Monte Carlo simulation of hydration of nucleic acid fragments. *J. Biomol. Struct. Dyn.*, **7**, 75–89

Texter, J. (1978). Nucleic acid–water interactions. *Prog. Biophys. Mol. Biol.*, **33**, 83

van Gunsteren, W. F. (1989). Methods for calculations of free energies and binding constants: successes and problems. In van Gunsteren, W. F. and Weiner, P. (Eds), *Computer Simulation of Biomolecular Systems*. ESCOM, Leiden, pp. 27–59

Vovelle, F., Elliott, R. J. and Goodfellow, J. M. (1989). Solvent bridging sites in A and B DNA helices. *Int. J. Biol. Macromol.*, **11**, 39–42

Vovelle, F. and Goodfellow, J. M. (1990). Sequence dependent hydration of DNA. *Int. J. Biol. Macromol.*, **12**, 369–373

Vovelle, F. and Goodfellow, J. M. (1991). Modelling of DNA interactions. In Beveridge, D. L. and Lavery, R. (Eds), *Theoretical Chemistry and Molecular Biophysics*, Adenine Press, Schenectady, N.Y., pp. 389–398

Vovelle, F. and Ptak, M. (1979). A fractional charge model for empirical calculations of peptide water interactions. *Int. J. Peptide Protein Res.*, **13**, 435–446

Wang, A. N. J., Hakoshima, T. H., van der Marel, G., van Boom, J. H. and Rich, A. (1984). AT base pairs are less stable than GC base pairs in Z-DNA: the crystal structure of d(m^5CG7Am^5CG). *Cell*, **37**, 321–331

Wang, A. N. J., Quigley, G. J., Kolpak, F. J., Crawford, J. L., Van Boom, J. H., Van der Marel, G. and Rich, A. (1979). Molecular structure of a left-handed double helical DNA fragment at atomic resolution. *Nature*, **282**, 680–686

Wang, A. H. J., Quigley, F. J., Kolpak, F. J., van der Marel, G., Van Boom, H. and Rich, A. (1981). Left handed double helical DNA: variations in the backbone conformation. *Science*, **211**, 171–176

Weiner, J., Kollman, P., Case, D. A., Chandra Singh, U., Ghio, C., Alagona, G., Profeta, S. and Weiner, P. (1984). A new force field for molecular mechanical simulation of nucleic acids and proteins. *J. Am. Chem. Soc.*, **106**, 765–784

Westhof, E. (1987a). Hydration of oligonucleotides in crystals. *Int. J. Biol. Macromol.*, **9**, 186–192

Westhof, E. (1987b). Re-refinement of the B-dodecamer d(CGCGAATTCGCG) with a comparative analysis of the solvent in it and in the Z-hexamer d(5BrC–G–5BrC–G–5BrC). *J. Biomol. Struct. Dyn.*, **5**, 581–599

Westhof, E. (1988). Water: an integral part of nucleic acid structure. *Ann. Rev. Biophys. Biophys. Chem.*, **17**, 125–144

Westhof, E. and Beveridge, D. L. (1990). Hydration of nucleic acids. In Franks, F. (Ed.), *Water Science Reviews*, Vol. 5. Cambridge University Press, Cambridge, pp. 24–136

9
Light Scattering Spectroscopy Studies of the Water Molecules in DNA

N. J. Tao

1 Introduction

The water molecules surrounding DNA are usually described in terms of a primary hydration shell and a secondary hydration shell (Saenger, 1984; Texter, 1978). Various experiments indicate that the primary hydration contains 20–25 water molecules per nucleotide pair. The size of the secondary hydration shell is less clear (Saenger, 1984; Texter, 1978). To understand the roles water plays in DNA properties at a microscopic level, it is essential to know the structure of the DNA hydration shells. There have been many studies of the structure of the DNA hydration shells, although a complete understanding of the structure is still lacking. The most direct method is probably X-ray crystallography, and it has been used to locate the water molecules in DNA single crystals. For example, Kopka *et al.* (1983) and Kennard *et al.* (1986) studied water structure in various oligonucleotide single crystals using high-resolution X-ray diffraction. Various water bridges that depend on DNA conformation have been observed in DNA. However, this method can only locate a small fraction of the water molecules in the hydration shells, since most water molecules diffuse too rapidly, and it is also limited to DNA single crystals. Infrared spectroscopy has been used by many workers (e.g. Blinska and Wilczk, 1976; Falk *et al.*, 1970; Pilet *et al.*, 1975). Falk *et al.* (1970) studied the infrared spectrum of DNA and water in the region of 400–4000 cm^{-1} (1 cm^{-1} = 30 GHz) as a function of relative humidity (r.h.). They were able to determine the possible water binding sites at various r.h. Other methods, such as water self-diffusion measurements (Wang, 1955), buoyant density measurements (Hearst and Vinograd, 1961; Tunis and Hearst 1968), NMR (Kuntz *et al.*, 1969), isopiestic measurements (Hearst, 1965), also contributed to our understanding of the structure of DNA hydration shells. We do not describe them here, but some of them will be referred to later. In addition to these experimental investigations, theoretical considerations, such as by Lewin (1967), and computer simulations (Corongiu and Clementi, 1981) have also been carried out.

We shall show in this chapter that, using Raman spectroscopy, we can determine the number of water molecules in the DNA primary hydration shell, obtain information about the structure of the DNA hydration shells and get the water content of a hydrated DNA sample (Tao *et al.*, 1989).

While most studies have been focused on the structures of the hydration shells (mainly the primary hydration shell), the dynamics of the hydration shells are much less well studied. We all know that DNA is rich in movement. It vibrates, rotates and translates. It is quite obvious that, in order to understand the biological functions of DNA in terms of atoms and molecules, it is necessary to understand these motions.

Eyster and Prohofsky (1974) calculated all vibration modes in DNA using lattice dynamics. It has been argued that the interesting frequencies of motions lie between a few GHz and a few thousand GHz (Lindsay *et al.*, 1985; Lindsay and Tao, 1988). The lower limit of the frequency range is set by the viscous damping from the medium. Part of this chapter will show that using Brillouin scattering we can determine the lower limit. The upper limit is fixed by considering the fact that the smallest biologically meaningful blocks of DNA are nucleotides, and the primary interactions involved in the motions of nucleotides are van der Waals and hydrogen bonds. Using typical values for the interactions and the mass of a nucleotide, the frequency is estimated to be ~ 2000 GHz. High-frequency motions involve vibrations of local chemical bonds and they are the same from one biomolecule (or one conformation of the same molecule) to another.

Both bulk water and water molecules in the hydration shells have many motions, with frequencies lying in this frequency window, so one may expect that these motions of water may couple with the vibrations of DNA at similar frequencies. Indeed, as we shall discuss below, the acoustic vibrations (sound waves) of DNA couple strongly to the water molecules in the hydration shells (Tao *et al.*, 1987, 1988). The lowest-frequency optical phonons are also believed to couple with DNA hydration shells (Tominaga *et al.*, 1985). Striking correlations between the motions of other biomolecules, such as myoglobins (Singh *et al.*, 1981) and membranes (Nimtz, 1986) and the water molecules in the hydration shells have also been observed.

Despite the importance of water, a proper theory of DNA dynamics which takes into account the water molecules in the hydration shells has not been developed. Crude models such as a harmonic oscillator (Tominaga *et al.*, 1985) or a collective acoustic mode (Tao *et al.*, 1987, 1988) coupled to a relaxation mode have been used to analyse experimental data.

A hydrodynamic theory has been developed by Van Zandt (1986), Van Zandt and Davis (1986) and Sokoloff (1988). A molecular dynamics simulation of the dynamical properties of water molecules in the DNA hydration shells has also appeared (Swamy and Clementi, 1987).

Some dielectric (Cross and Pethig, 1983; Wittlin *et al.*, 1986) and inelastic neutron scattering (Schreiner *et al.*, 1987) measurements have been used to study the hydration shells of DNA in this frequency range. Brillouin spectroscopy enables us to investigate not only the dynamics of the hydration shell, but also the dynamic interactions between DNA and the hydration shells at GHz frequencies.

This chapter is arranged as follows: in Section 2, we shall give a brief introduction to the theory and experimental methods of light scattering (Raman and Brillouin spectroscopies); in Section 3, we shall discuss the structure of the water molecules in DNA as probed by Raman spectroscopy; in Section 4, we shall describe the dynamic properties of the water molecules in DNA and the dynamic interaction between the water molecules and DNA. Section 5 is a summary.

2 Raman and Brillouin Spectroscopies

Raman and Brillouin spectroscopies are both inelastic light scattering techniques. Many good review articles and books (e.g. Berne and Percora, 1976) have been published on the basic theory and experimental techniques of light scattering, as well as the applications. Here we outline some basic results which we shall refer to in later sections. Light scattering is a result of dielectric constant fluctuation. The dielectric constant, ε, which describes the interaction between light and the material can be separated into two parts:

$$\varepsilon(\mathbf{r}, t) = \varepsilon 0 + \delta\varepsilon(\mathbf{r}, t) \tag{9.1}$$

where the first term is the average static dielectric constant, which is related to the refractive index as $n = \varepsilon_0^{1/2}$, and the second term is the fluctuating part, which can be caused by many physical processes, such as molecular vibrations and density vibrations. For a typical experimental arrangement, the scattered light intensity (spectrum) is

$$I(\mathbf{q}, \omega) \sim \langle \delta\varepsilon(\mathbf{q}, 0)\delta\varepsilon^*(\mathbf{q}, \omega)\rangle \tag{9.2}$$

where $\langle\ \rangle$ indicates a thermal average, $\delta\varepsilon(\mathbf{q}, \omega)$ is the spatial and temporal Fourier transform of $\delta\varepsilon(\mathbf{r},t)$, $\omega = \omega_f - \omega_i$($\omega_i$ and ω_f are the frequencies of incident and scattered light), and $\mathbf{q} = \mathbf{k}_f - \mathbf{k}_i$ ($\mathbf{k}_i$ and $\mathbf{k}_f$ are the momenta of incident and scattered light). The frequency change, ω, in the scattering process can be either positive (called anti-Stokes scattering) or negative (Stokes scattering). Note for a scattering angle, θ, $|\mathbf{q}| = (4\pi n/\lambda)\sin(\theta/2)$, where λ is the wavelength of light.

Suppose the fluctuation of the dielectric constant is the result of fluctuations in a quantity y (for example, density) and using the quantum mechanical fluctuation-dissipation theorem (Kubo, 1966), the light scattering spectrum can be expressed as

$$I(\mathbf{q}, \omega) \sim n(\omega)\,\mathrm{Im}\{\chi(\mathbf{q}, -\omega)\} + \{n(\omega) + 1\}\,\mathrm{Im}\{\chi(\mathbf{q}, \omega)\} \quad (9.3)$$

where $n(\omega) = 1/[\exp(\hbar\omega/k_BT) - 1]$ and $\chi(\mathbf{q},\omega)$ is the response function of the quantity y, defined as $y(\mathbf{q}, \omega)/F(\omega)$ ($F(\omega)$ is a generalized driving force). So the spectrum can be easily calculated if $\chi(\mathbf{q}, \omega)$ is known.

Though the basic principles of Raman and Brillouin spectroscopies are the same, they are distinguished because Raman spectroscopy normally studies optical phonons, molecular vibrations, etc., which are in the frequency range of a few cm^{-1} to thousands of cm^{-1}, while Brillouin spectroscopy usually measures acoustic phonons (sound waves) or other low-frequency excitations, which are in the range of 0.01 cm^{-1} to a few cm^{-1}. Since different frequency ranges are studied, different experimental techniques are used to analyse the spectroscopies. Grating spectrometers are normally used in Raman spectroscopy. Instead of using optical gratings as in Raman spectroscopy, Fabry–Perot interferometers are used in Brillouin spectroscopy. To overcome the limited contrast of single Fabry–Perot interferometers, multipass tandem Fabry–Perot systems are employed (Sandercock, 1982; Lindsay *et al.*, 1981). High contrast is extremely important for studying hydrated biopolymers, such as DNA, because the strong elastic scattering could block the weak signal. We used a 9-pass tandem Fabry–Perot system (Lindsay *et al.*, 1981), which has a transmission higher than 10%, contrast observed to be better than 10^{17} and an effective finesse which is found to be about 60. A scanning range can be as low as 1 GHz and as high as 1500 GHz.

3 Structure of DNA Hydration Shell Studied by Raman Spectroscopy

O–H Stretching Vibrations in Liquid Water

Raman scattering of O–H stretching vibrations of liquid water has been studied by many workers (e.g. Walrafen, 1971). A depolarized spectrum of the mode shows a peak at ~ 3400 cm^{-1} (the top one in Figure 9.3), while the polarized spectrum resolves not only the 3400 cm^{-1} peak but also a peak at 3200 cm^{-1} (the top one in Figure 9.2). The depolarized 3400 cm^{-1} mode has been assigned to the O–H stretching vibration. The polarized band (3200 cm^{-1}) was interpreted earlier in terms of a Fermi resonance with bending overtones (Walrafen, 1964), but this interpretation is not consistent with data for ice I and amorphous ice (Whally, 1977). Wong and Whally (1977) assigned the polarized band in ice (crystalline and amorphous) to an in-phase collective stretching band. Green *et al.* (1986) extended the new interpretation to liquid water, and pointed out that the polarized band arises from the strongly hydrogen-bonded patches. This new interpretation is consistent not only with experimental results but also with theoretical studies (Sceats and Rice, 1983). On the basis of this interpretation, the intensity of the mode can be regarded as a measure of how well the hydrogen bond network is formed. Green *et al.* (1987a,b) studied this mode by introducing hydrogen-bond-breaking defects such as LiCl and H_2O_2, and indeed found that the intensity decreases as the impurity (LiCl and H_2O_2) concentrations increase. They also observed an increase in the intensity when introducing 'network-enhancing' impurities (Green, 1987c). Two interesting questions are raised immediately about DNA: (1) Is DNA a 'network-enhancing' or 'network-breaking' impurity? (2) How many water molecules surrounding DNA have altered hydrogen bonding? The answer to the first question provides information about the hydration structure, while the answer to the second one gives the number of water molecules in the primary hydration shell.

Experiments and Results

Wet-spun films (Rupprecht and Forslind, 1970) of Na-DNA (1% excess NaCl in weight) and Li-DNA (4.5% excess LiCl in weight) were used. The Na-DNA is in a B-like conformation when the relative humidity (r.h.) is lower than 45% r.h., A-form between 45% and 92% r.h. and B-form above 92% r.h. The Li-DNA is in the B-form. The advantage for using the highly ordered films is that the conformations of DNA can be determined easily by X-ray diffraction. A list of the water content of these films as a function of r.h. can be found in Lee *et al.* (1988).

Figure 9.1 shows a typical Raman spectrum of DNA samples. We can see a broad strong fluorescence band at ~ 2000 cm^{-1}. Vibrational modes are superimposed on the fluorescence. Detailed assignments of these are given by Prescott *et al.* (1984). The fluorescence is quite similar in shape and position from sample to sample, and it can be fitted approximately by a gaussian at ~ 2100 cm^{-1} with a width of ~ 2000 cm^{-1}, so we can subtract it from the spectra.

The spectrum of D_2O hydrated Na-DNA at 75% r.h. was also studied (Tao *et al.*, 1989), which showed a mode at ~ 2960 cm^{-1}, and several other weak modes at ~ 2893, 3025 and 3100 cm^{-1}. They have been assigned as the C–H stretching bands of various groups (Prescott *et al.*, 1984). There

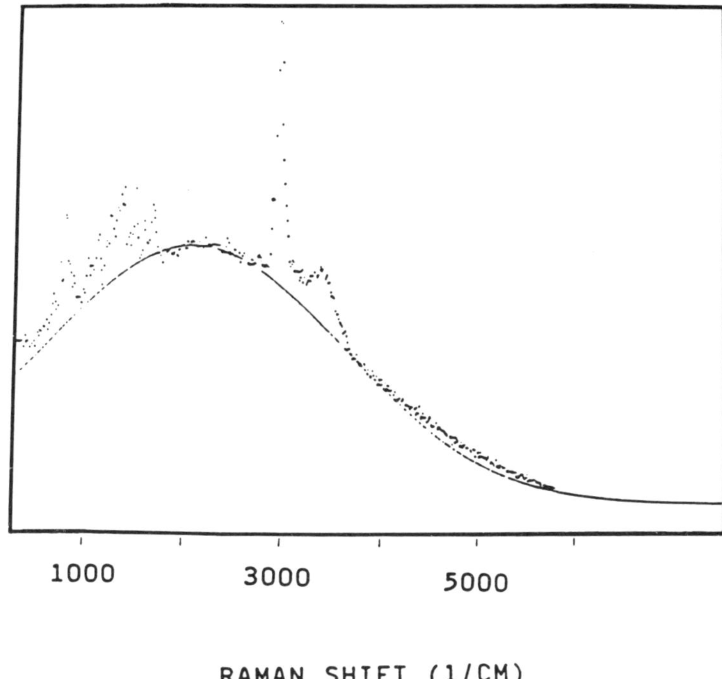

Figure 9.1 Raman spectrum of Li-DNA at 0% r.h. The dots are the experimental points, and the solid line is a Gaussian fitting of the fluorescence background. After Tao *et al.* (1989)

are no DNA bands around 3200 cm^{-1} and 3400 cm^{-1}. Thus, we conclude that all the intensity of 3223 cm^{-1} and 3430 cm^{-1} bands arises from water in the sample.

Figure 9.2 shows the VV (vertically polarized scattered light with vertically polarized incident light) polarized spectra of Li-DNA at various water contents. The top one is pure water, and the following spectra are for DNA from 98% to 0% r.h. The corresponding concentration of DNA is from 30% to nearly 100% by weight. Note that the intensity of the local O–H stretching band is not zero at 0% r.h., which indicates that there are some water molecules in the 0% r.h. sample. If the intensity (at x% r.h.) of the local O–H stretching band ($I^{oh}(x\%\text{r.h.})$) and the 2960 cm^{-1} C–H stretching band ($I^{ch}(x\%\text{r.h.})$) are proportional to the number of water molecules and nucleotide pairs per unit volume, respectively, then

$$\frac{I^{oh}\,(x\%\text{r.h.})}{I^{ch}\,(x\%\text{r.h.})} = Kn\,(x\%\text{r.h.}) \tag{9.4}$$

where $n(x\%\text{r.h.})$ is the number of water molecules per nucleotide pair at x%r.h. K is a constant. At 0% and 80% r.h., the intensity ratios were

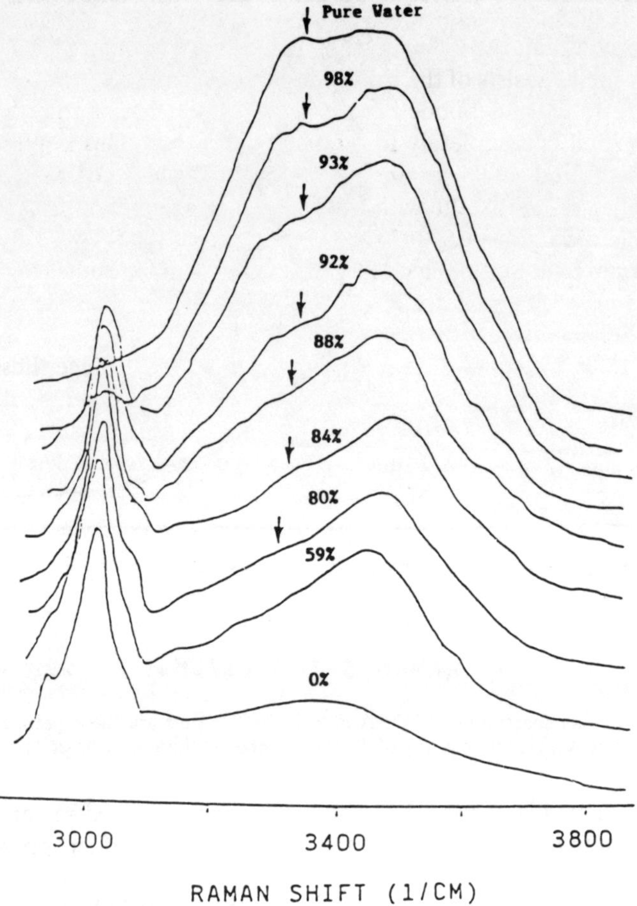

Figure 9.2 VV Raman spectra of (from top to bottom) pure water, Na-DNA at 98%, 93%, 92%, 88%, 84%, 80%, 59% and 0% r.h. The arrows mark the location of the 3200 cm^{-1} band, which is due to tetrahedrally bonded water patches. After Tao *et al.* (1989)

determined from the spectra with an uncertainty of ~ 5%. Substituting the ratios into Equation (9.4) and using $n(80\%\text{r.h.}) - n(0\%\text{r.h.}) = 21$ water molecules per nucleotide pair (from gravimetric measurements), the residual O–H stretching band intensity at 0% r.h. is then accounted for by 5.5 ± 0.5 water molecules per nucleotide pair. Falk *et al.* (1963) reached a similar conclusion from fits to their gravimetric data using BET theory (Brunauer *et al.*, 1938).

The intensity of the 3200 cm^{-1} O–H mode drops quickly from the value for pure water to almost zero at 80% r.h., as the DNA concentration increases. This means that DNA disrupts the hydrogen bonded structure of free water. The vanishing of this band's intensity when the humidity is reduced to 80% indicates that the water tetrahedral hydrogen bonded network is completely destroyed at 80% r.h. If we assume that the primary hydration shell consists of the water molecules forming a structure different from the tetrahedrally bonded network, we can conclude that the primary hydration shell is completely formed at ~ 80% r.h. This implies that the primary hydration shell contains ~ 27 water molecules per nucleotide pair. This number is consistent with most literature values if we include the $6H_2O$/nucleotide pair at 0%r.h.

Tunis and Hearst (1968) have suggested that Falk's results from the films and fibres may not apply to DNA in solutions. Indeed, it has been found that interhelical interactions are very strong in the films and fibres (DeMarco *et al.*, 1986; Lee *et al.*, 1988; Lindsay *et al.*, 1988). Since these interactions soften strongly as the water content increases (Lee *et al.*, 1988), it is of interest to see whether the number of water molecules in the primary hydration shell changes as a function of water content. We can extract the primary hydration number at different water contents from the intensity of the ~ 3200 cm^{-1} (Green *et al.*, 1987a). Let C_w and C_x be the collective mode intensities of pure water and the DNA sample at x%r.h. (the definition and extraction procedure for the intensities, C, are described by Tao *et al.*, 1988); then we have

$$1 - C_x/C_w = P_d \tag{9.5}$$

where P_d is the defect density in the pure water hydrogen bond network created by DNA and the ions. Suppose that N_0 O–H oscillators per nucleotide pair are removed from the hydrogen-bonded tetrahedral network in the primary hydration shell; then the defect probability is $N_0 f_x$(*where* f_x is number of nucleotide pairs at x% r.h.), so

$$1 - C_x/C_w = N_0 f_x, \text{ or } N_0 = (1 - C_x/C_w)/f_x \tag{9.6}$$

Values of f_x were obtained from gravimetric data (see Table 9.1), amended to account for the 6 water molecules which remain at 0% r.h. Derived values for N_0 are listed in Table 9.1. We can see that N_0 is basically constant. This indicates that the interhelical interactions and the formation of the secondary hydration have little influence on the size of the primary hydration shell.

VH (horizontally polarized scattered light with vertically polarized incident light) polarized spectra are shown in Figure 9.3 (decreasing water content from top to bottom). VH spectra only show the ~ 3400 cm^{-1} O–H stretching band. The peak frequency of the ~ 3400 cm^{-1} O–H stretching band decreases quite significantly as water content decreases. This behaviour was also observed by Falk (1966) and Falk *et al.* (1970) using infrared spectroscopy. The decrease in the frequency from that in pure

Table 9.1 The number of water molecules in the primary hydration shell of Na- and Li-DNA as determined by Raman spectroscopy. Also listed are the water contents from the gravimetric data (Lee *et al.*, 1988) and the ratio of the intensities, C_x/C_w, of the ~3200 cm^{-1} mode in DNA at a given r.h. to the same band in pure water. These intensity ratios were obtained as outlined by Tao *et al.* (1988)

	Na-DNA			*Li-DNA*		
r.h.%	f_x	C_x/C_w	N_0	f_x	C_x/C_w	N_0
80	0.037	0	28 ± 3	0.038	0	27 ± 3
88	0.022	–	–	0.025	0.18 ± 0.09	33 ± 4
92	0.022	0.44 ± 0.06	26 ± 3	0.024	0.31 ± 0.07	29 ± 3
93	0.019	0.54 ± 0.06	24 ± 3	0.020	0.43 ± 0.06	29 ± 3
95	0.015	0.62 ± 0.05	25 ± 3	0.014	0.58 ± 0.04	29 ± 3
98	0.010	0.66 ± 0.05	35 ± 5	0.008	0.72 ± 0.05	35 ± 5

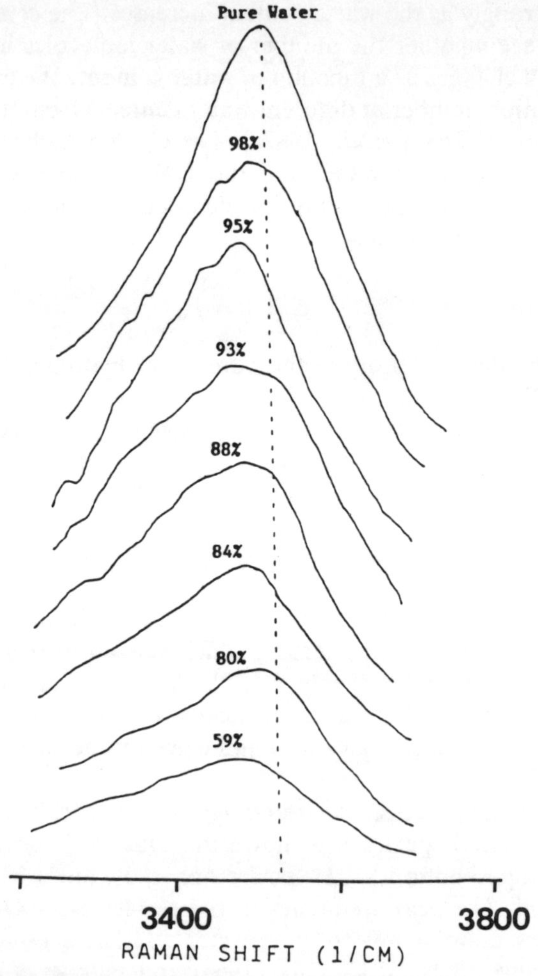

Figure 9.3 VH Raman spectra of (from top to bottom) pure water, Na-DNA at 98%, 95%, 93%, 88%, 84%, 80% and 59%. The dashed line marks the peak position of the 3400 cm^{-1} band of pure water. After Tao *et al.* (1989)

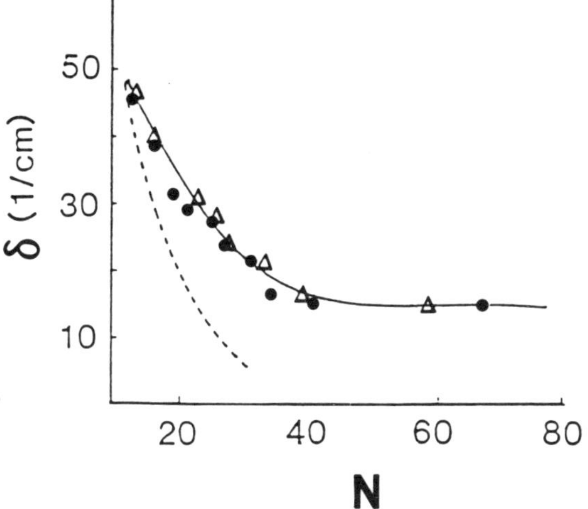

Figure 9.4 Frequency shift of the local O–H stretching mode from its value in pure water as a function of number of water molecules per nucleotide pair (• for Li-DNA, Δ for Na-DNA). The solid line is a guide for the eye. After Tao *et al.* (1989)

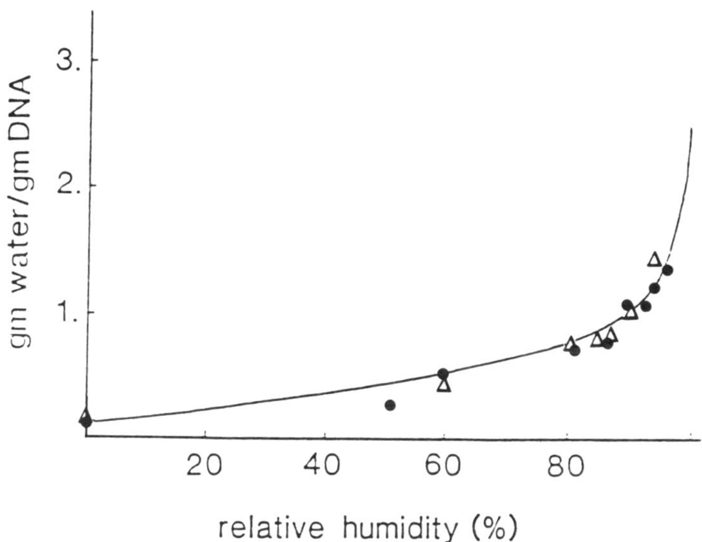

Figure 9.5 Water content of Li- (•) and Na-DNA (Δ) from the intensity of the water local O–H stretching band as a function of r.h. The solid line summarizes gravimetric data. After Tao *et al.* (1989)

water is plotted against the number of water molecules per nucleotide pair in Figure 9.4.

As r.h. decreases, the intensity of the band decreases. Using Equation (9.4), water content at each r.h. was determined from the ratio, I^{oh}/I^{ch}. The results are plotted in Figure 9.5 (open triangles for Na-DNA and filled circles for Li-DNA). The solid line in Figure 9.5 is obtained from Na-DNA gravimetric data (modified to include the 6 water molecules per nucleotide pair at 0% r.h.). Values for Li-DNA are similar below 98% r.h. The data measured by these two methods are in agreement (with the exception of the 98% r.h. datum), indicating that laser dehydration is minimal. The agreement of these data shows that Raman spectroscopy can be used to determine the water content of DNA at a given r.h.

Discussion

Our hydration numbers also include the hydration of the counterions and the excess ions, but the correction from the excess ions is not large at these salt concentrations in our sample. Our Li-DNA contains 4.5% excess LiCl (0.7 LiCl per nucleotide pair), so the correction is only ~ 3.5 water molecules per nucleotide pair using hydration number of 5 for LiCl (Green *et al.*, 1987b; Conway, 1981). The 1% excess NaCl in Na-DNA gives a correction of ~ 0.6 water molecules per nucleotide pair. If the hydration number of Li-DNA is about the same as the Na-DNA, we should have about 3 more water molecules per nucleotide pair in the primary hydration shell of Li-DNA, simply due to the hydration of the excess ions. Indeed, Table 9.1 indicates that this is the case. Our data also imply that the ions (both excess ions and counterions) are somewhat dehydrated at 0% r.h.

Water structure in DNA and other biopolymers is often regarded as ice-like. Weiss *et al.* (1986) simulated water structure near a lipid surface using percolation theory (Blumberg *et al.*, 1984). They found that the connectivity of hydrogen bonds of water near the molecular surface is higher, and therefore more ice-like. It was suggested that this result may apply to DNA (Nimtz, 1986). Using Raman spectroscopy, we have demonstrated the disruption of the hydrogen-bond network. None of our results support ice-like structure in the DNA hydration shell. A critical assumption of Weiss *et al.* (1986) is that all water molecules are perfectly fixed on the lipid surface, which implies that the period of the surface structure is commensurate with the period of the tetrahedral packing of water at the interface. This may be correct for lipid bilayers, but DNA could be very different. It has been found (Grimm *et al.*, 1987) that the structure of the DNA skeleton does not match that of water in the primary hydration shell. The competition between the two structures stabilizes the DNA double helical structure and also disrupts the structure of water.

A relation has been established between the frequency of the O–H stretching mode in crystal and amorphous ice phases and O–O bond distance based on the results of high pressure Raman and X-ray studies (Klug and Whally, 1984; Klug *et al.*, 1987). If this relation is valid also for the water molecules in the DNA hydration shells, we can conclude that the maximum 'red shift' for the water molecules in the DNA primary hydration shell corresponds to a O–O bond distance change of 0.2 Å.

Summary

(1) The decrease in the collective O−H stretching band intensity indicates that the tetrahedral hydrogen bond network of bulk water is disrupted by the presence of DNA; therefore, the primary hydration shell of DNA cannot have an ice-like structure.

(2) The intensity of the collective band gives us a direct measurement of the primary hydration number of DNA of ~ 30 water molecules per nucleotide pair (the 6 H_2O at 0% r.h. are included). This number is constant after the primary hydration shell is completed, which means that interhelical interactions do not influence that primary hydration number significantly.

(3) There are about 6 water molecules per nucleotide pair that are very difficult to remove from DNA.

(4) The ~ 3400 cm^{-1} O−H stretching frequency of the water molecules in the primary hydration shell is shifted to a lower value than that in free water, an indication of structure change of the individual water molecules in the hydration shell.

(5) Water content as a function of r.h. measured from the intensities of the local O−H stretching mode and C−H stretching mode is in agreement with gravimetric measurements.

4 Dynamics of DNA Hydration Shell Studied by Brillouin Spectroscopy

Origin of Brillouin Spectrum in Hydrated DNA

A typical Brillouin spectrum of ordinary solid or liquid consists of two symmetric peaks, the so-called Brillouin doublet. The positions of the peaks gives the vibration frequency, ω_s, of the sound wave in the system (from which the speed of sound can be easily calculated), and the width of the peaks gives the attenuation of the sound wave. In fact, the attenuation coefficient, $\alpha\lambda_s$, is related to the half-width at half maximum intensity of the doublets, Γ, by $\alpha\lambda_s = 2\pi\Gamma/\omega_s$. A typical Brillouin spectrum of hydrated DNA (Na-DNA film at 75% r.h. and room temperature) is shown in Figure 9.6. In contrast to the typical spectrum of solid or liquid, the linewidth is very broad and, in addition, there is a central mode. In general, contributions to a sound wave attenuation are two: (1) structure

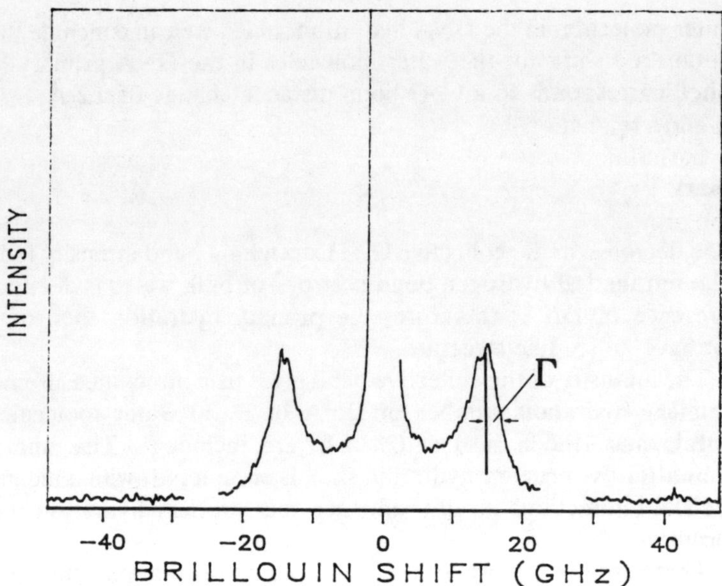

Figure 9.6 A typical Brillouin spectrum of the hydrated DNA films

inhomogeneities which scatter the sound wave and cause a distribution in frequency of the sound wave; (2) the sound wave coupling with other modes dynamically, e.g. a relaxation mode.

We ruled out the attenuation due to structure inhomogeneities in the hydrated DNA films because: (1) the linewidths of the Brillouin doublets in the DNA single crystals (structure is homogeneous) (Tao, 1988) are about the same as the DNA films; (2) the electron microscopic images showed that the microcrystallites are much greater than the wavelength of the sound wave; and (3) the damping is a strong function of temperature. Therefore, we must conclude that the attenuation of the sound wave is dynamic in origin.

The central mode in the spectrum usually indicates a relaxation mode (or overdamped vibration mode). The width of the central mode gives the relaxation time of the relaxation mode. We conclude that the relaxation mode is due to the water molecules in the hydration shells. Relaxations of DNA would occur at a much slower rate, and the diffusion of ions (much less than that of water molecules, in any event) also corresponds to slower relaxations. Furthermore, our conclusion is supported by the fact that the D_2O-hydrated sample has systematically different linewidths.

Mode-coupling Model

The problem of the coupling between the sound wave (Q_d) in the hydrated DNA and a relaxation mode (Q_w) of the hydration shell can be represented by a harmonic oscillator and an overdamped oscillator (Barker and Hopfield, 1964). Keeping only the lowest-order coupling, the equations of motion are:

$$(1/q^2)\rho(\omega_0{}^2 - \omega^2 - 2i\omega\gamma)Q_d + k_{12}Q_w = f_d \tag{9.7}$$

$$m'\omega_w{}^2(1 - i\omega\tau)Q_w - k_{12}Q_d = f_w \tag{9.8}$$

where ω_0 and ω_w are the uncoupled frequencies of the sound wave and the water relaxation mode, and f_d and f_w are the driving forces on the sound wave and the relaxation mode, respectively; ρ is the density of the hydrated DNA film, m' is the effective mass of the hydration shell, τ is the relaxation time of the relaxation mode, and the k_{12} measures the coupling between the sound wave and the relaxation of the hydration shell. The response function of the sound wave is given by solving Equations (9.7) and (9.8):

$$\chi_d(q, \omega) = 1/[\omega_0{}^2 - \omega^2 - 2i\omega\gamma - \delta^2/(1 - i\omega\tau)] \tag{9.9}$$

where $\delta^2 = k_{12}{}^2q^2/m'\rho\cdot\ m' = n_w m_0$ and $\rho = n_d m_d$, where m_0 and m_d are the masses of a water molecule and mass of a nucleotide pair, respectively. $n_w = n_d N_w$, where N_w is the number of water molecules of primary hydration shell for one nucleotide pair. δ^2 can then be expressed as

$$\delta^2 = n_d{}^2 N_w a^2 q^2/m_0 m_d \tag{9.10}$$

where a is the average coupling strength between the sound wave and one water molecule in the hydration shell. From Equation (9.9), we can find that the frequency and attenuation of the sound wave are

$$\omega_s^2 \cong \omega_0^2 + \delta^2 \frac{\omega_s^2\tau^2}{1 + \omega_s^2\tau^2} \tag{9.11}$$

and

$$\alpha\lambda_s{}^2 \cong k\frac{\omega_s}{V_s^2} + \frac{\delta^2}{V_s^2 q^2}\frac{\omega_s^2\tau^2}{1 + \omega_s^2\tau^2} \tag{9.12}$$

where $V_s = \omega_s/q$ is the sound speed. For $\omega_s\tau = 1$, $\alpha\lambda_s$ has a maximum (k is normally small), so we can estimate τ from the maximum position of $\alpha\lambda_s$. If we divide Equation (9.11) by q^2, then we have

$$V_s^2 \cong V_0^2 + \frac{\delta^2}{q^2}\frac{\omega_s^2\tau^2}{1 + \omega_s^2\tau^2} \tag{9.13}$$

where $V_0 = \omega_0/q$.

Experimental Results

Brillouin Scattering as a Function of Temperature

Three samples of both Na- and Li-DNA films were studied at 0%, 75% and 92% r.h., corresponding to no hydration shell, a primary hydration shell and both a primary and a secondary hydration shell, respectively. The experiments were performed in near-backscattering in a temperature range of between room temperature and about 130 K. Detailed experimental procedures are given by Tao *et al.* (1987).

Figure 9.7(a) shows typical spectra at 300 K, 260 K and around 200 K for highly hydrated Na-DNA. The pair of the peaks are the longitudinal sound waves. They move out in frequency as the samples are cooled, showing some increase in linewidth initially but then sharpening considerably at the lowest temperatures. The maximum width of the central spectrum is at around 260 K (as well as the asymmetric phonon line shapes). These are typical indications for a vibrational mode coupling with a relaxation mode, as found in many solid state systems (e.g. Brody and Cummins, 1968).

Data for the longitudinal sound frequencies are summarized in Figure 9.8 (a) for Na-DNA and (b) for Li-DNA. The large change in the sound frequency as a function of temperature in the hydrated samples suggests that the water in DNA also influences the interactions among DNAs and subgroups of DNA. Detailed discussions can be found in Tao *et al.* (1987).

The attenuations ($\alpha\lambda_s$) are plotted in Figure 9.9 for (a) Na-DNA and (b) Li-DNA. We can see that there is virtually no change in the attenuation and the frequency for 0% r.h.; however, a dramatic change is seen near 260 K for the hydrated samples, and added water clearly controls the attenuation.

Fitting the attenuations of the primary hydration sample (75% r.h.), using Equation (9.12), the relaxation time for the primary hydration shell, τ_1, was obtained as a function of temperature, as plotted in Figure 9.14. For the highly hydrated sample (92% r.h.), since both a primary and a secondary are present, the analysis is more complicated. We shall return to this.

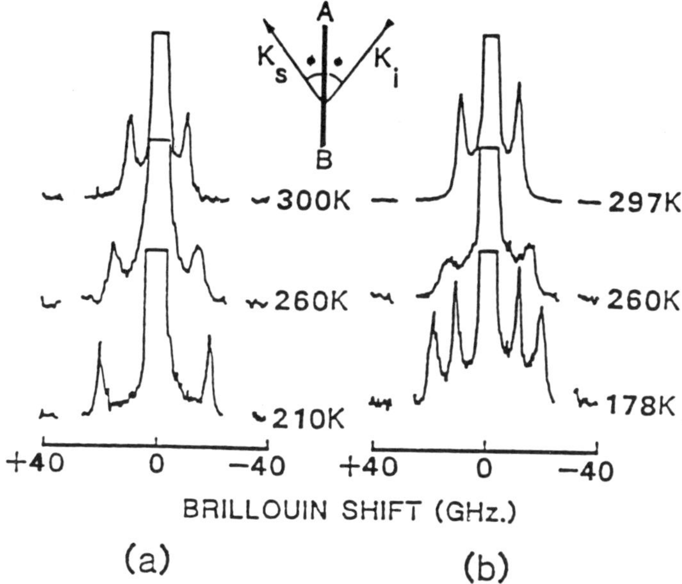

Figure 9.7 Brillouin spectra as a function of temperature for the hydrated Na-DNA (a) and Li-DNA (b). These spectra were obtained in near-backscattering ($\theta \sim 150°$). After Tao *et al.* (1987)

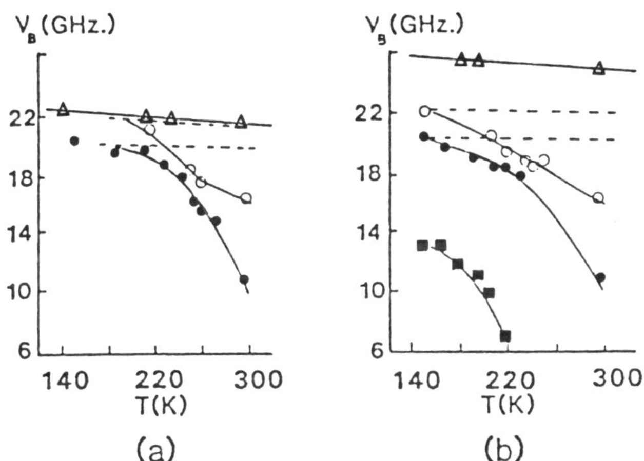

Figure 9.8 Sound frequencies ($\omega_s = 2\pi\nu_B$) as a function of temperature for (a) Na-DNA and (b) Li-DNA. ●, ° and Δ are for the 92% r.h., 75% r.h. and 0% r.h., respectively. The solid lines are to guide the eye. After Tao *et al.* (1987)

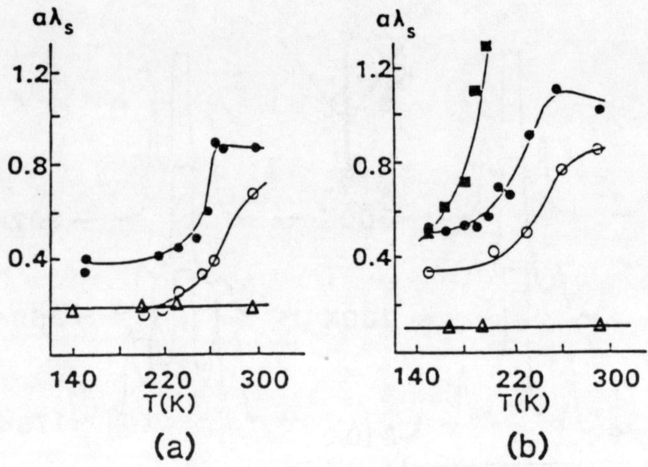

Figure 9.9 Sound attenuations as a function of temperature for (a) Na-DNA and (b) Li-DNA. ●, ° and Δ are for the 92% r.h., 75% r.h. and 0% r.h., respectively. The solid lines are to guide the eye. After Tao *et al.* (1987)

Brillouin Scattering as a Function of Scattering Angle

A typical spectrum of the Na-DNA at 95% r.h. is shown as a function of scattering angle in Figure 9.10 (for sound propagating perpendicular to the helix axis). We can see clearly that there is a maximum at $\theta = 60$, corresponding to a sound wave frequency of 3.5 GHz. According to Equation (9.12), this is another indication that a relaxation time, $2\pi\tau$ is ~ 1/3.5 GHz (~ 40 ps) at room temperature.

The attenuation (for sound propagating perpendicular to the helix axis) is plotted as a function of the corresponding frequency as measured directly from the Brillouin spectrum in Figure 9.11. The data are displaced in increments of 0.4 on the $\alpha\lambda_s$ scale for each humidity, so as to separate the points.

Figure 9.12 shows that the sound frequency increases almost linearly with q (departure from the linear behaviour due to relaxation is very slight). The slopes (sound speeds) fall with increasing hydration, and are, at a given level of hydration, greater for propagation normal to the helix axis, a consequence of the acoustic anisotropy of the DNA samples. This behaviour is mainly due to the hydration-induced potential softening (Lee *et al.*, 1988; Tao *et al.*, 1988).

At low hydration, $\alpha\lambda_s$ increases linearly with frequency (Figure 9.11), reflecting a classical acoustic loss mechanism. As water is added, a maximum develops near 4 GHz. This loss maximum grows in strength, peaking at 93% r.h. at a value of 1.1 for q perpendicular to the helix axis, and about 0.85 for q along the helix axis. At higher humidity, dilution of the sample

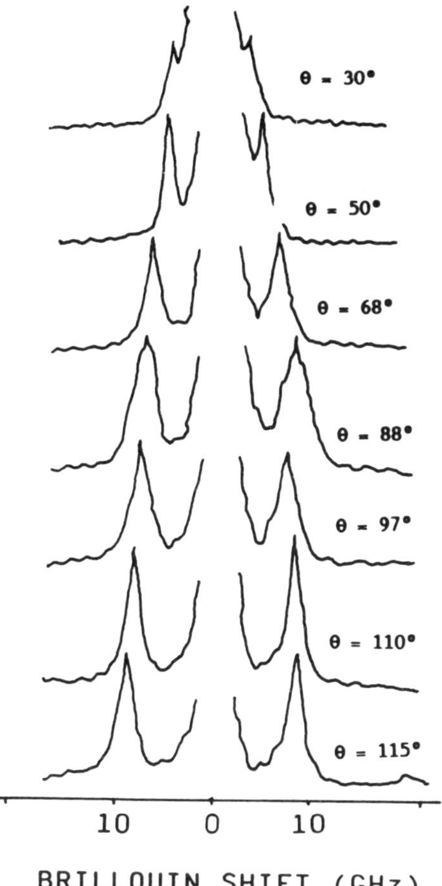

Figure 9.10 Brillouin spectrum of Na-DNA at 95% r.h. as function of scattering angle for sound propagating along the helix. After Lindsay and Tao (1988)

leads to a reduction in the 4 GHz maximum (number of DNA molecules per unit volume decreases), although the build-up of the tail of the secondary hydration loss becomes more obvious. A clear loss maximum is not seen for the relaxation associated with the secondary hydration, as, at room temperatures, its relaxation rate is fast enough to appear here only as a classical $\alpha \sim \omega^2$ tail.

The theoretical curves in Figures 9.11 and 9.12 are calculated using Equations (9.11) and (9.12). These fits determine τ and δ (δ_z and δ_x are for along and perpendicular to the helix axis, respectively) at each degree of hydration (Table 9.2). Knowing τ and δ, the whole spectrum can be calculated using Equations (9.3) and (9.9). Figure 9.13 shows the experimental (dots) and fitted spectra (solid curves). The fitted spectra were obtained using the parameters in Table 9.2 and properly adjusted spectrum

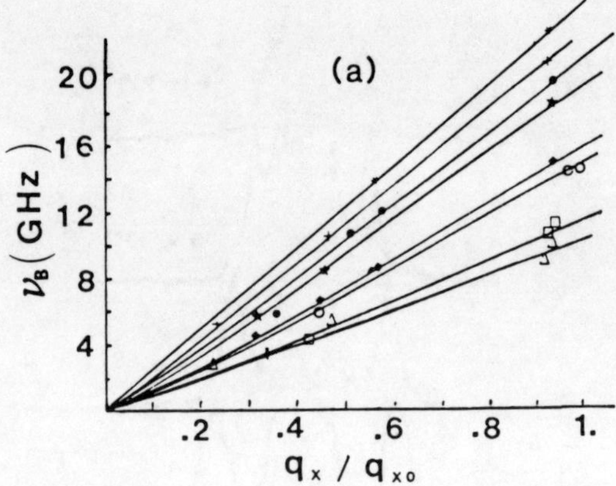

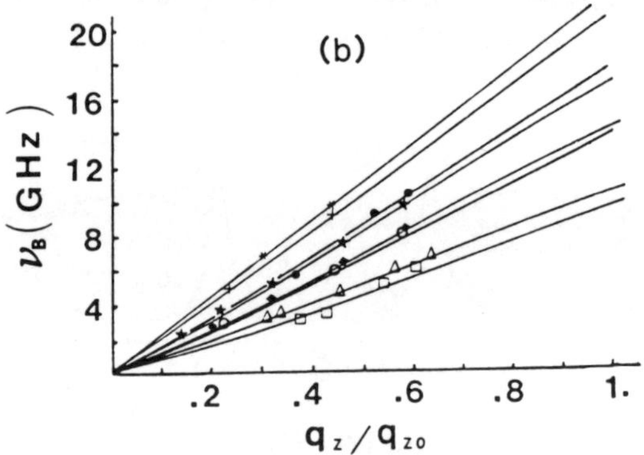

Figure 9.11 Frequencies of the sound waves ($\omega_s = 2\pi\nu_B$) plotted against the reduced wave vector, q/q_0 ($q_0 = 4n\pi/\lambda$). *, +, ●, solid star, solid lozenge, ○, Δ and □ are the experimental results for 0%, 23%, 45%, 59%, 80%, 86%, 93% and 95% r.h., respectively. The solid lines are fittings using Equation (9.11). (a) Is for sound propagating perpendicular to the helix, and (b) is for along the helix. After Tao *et al.* (1988)

intensities and backgrounds. The reasonable agreement between the experimental and the calculated spectra means that our analysis is self-consistent.

Discussion

We note that, for the samples with only the primary hydration shell, the spectra in Li- and Na-DNA are quite similar, but for the samples with both primary and secondary hydration shells, the change of the linewidth is

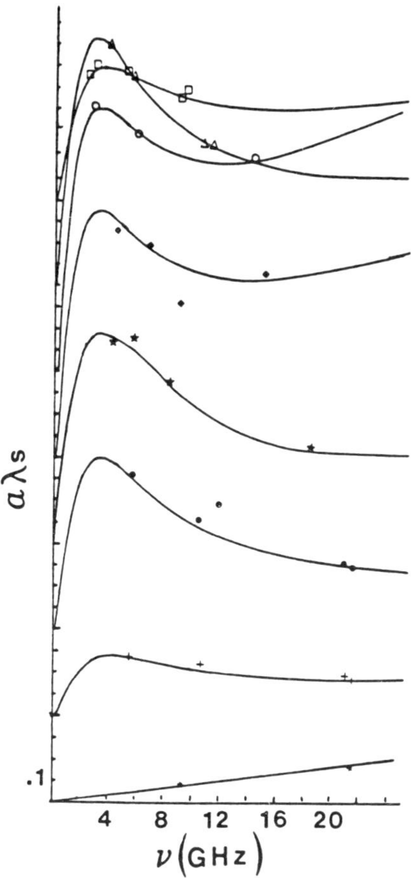

Figure 9.12 Attenuation of the sound waves plotted against the sound frequency. *, +, ●, solid star, solid lozenge, ○, Δ and □ are the experimental results for 0%, 23%, 45%, 59%, 80%, 86%, 93% and 95% r.h., respectively. The solid lines are fittings using Equation (9.12) for sound propagating along the helix. After Tao *et al.* (1988)

sharper for Na-DNA. This may imply that the secondary hydration shell is influenced by the ion species.

Since the parameter *a* measures the average coupling strength between a nucleotide pair and a water molecule in the hydration shell, we have extracted *a* for *q* both along and perpendicular to the helix axis using Equation (9.10). Results are listed in Table 9.3. Within the calculated overall uncertainty (~15%), *a* is remarkably constant, having a value of about 550 × 10^{-52} kg $s^{-1}m^{-2}$ for sound waves in the *xy*-plane and about 415 × 10^{-52} kg $s^{-1}m^{-2}$ for the *z*-direction. These values mean that below about 75% r.h. (as the primary shell is being formed), each added water molecule interacts with the DNA in much the same manner as the molecules pre-

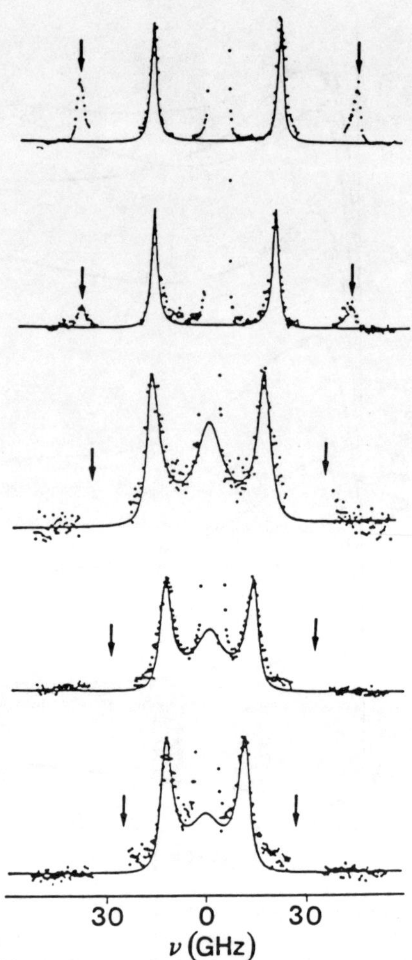

Figure 9.13 From top to bottom, the spectra are 0, 23, 59, 86 and 93% r.h., taken with a 90° scattering angle and q perpendicular to the helix axis. The solid lines are calculated according to Equations (9.3) and (9.9) and the parameters in Table 9.2. Back-scattering peaks due to the reflections from the sample surfaces are indicated by arrows. After Tao *et al.* (1988)

Table 9.2 Parameters used for fitting the dispersion of Brillouin shift, ω_s, with q, and $\alpha\lambda_s$ with ω_s and the spectra

r.h. (%)	δ_x/q (m/s)	δ_z/q (m/s)	V_x (m/s)	V_z (m/s)
23	1400	930	3300	3200
45	2100	1500	3350	2850
59	2100	1700	2380	2230
80	1750	1550	1850	1850
86	1680	1370	1750	1800
93	1400	1000	1400	1350
95	980	710	1500	1700

Table 9.3 Water contents and derived coupling constants for Na-DNA films. The uncertainty in the values of a_x and a_z is about 15%

r.h. (%)	*g* H_2O *per g DNA*	N_w	n_d $10^{27}m^3$	a_x $10^{-52}kg\,s^{-1}\,m^{-2}$	a_z $10^{-52}kg\,s^{-1}\,m^{-2}$
23	0.113	4.1	1.2	576	388
45	0.258	9.5	1.1	600	428
59	0.361	13.2	1.0	578	467
80	0.572	20.0	0.85	460	408
86	0.785	20.0	0.83	452	369
93	–	20.0	0.46	614	439
95	1.563	20.0	0.38	571	417

viously added. Above 75% r.h. (as the secondary shell is formed), the coupling of the primary shell is not disturbed much. This leads to a picture of the primary shell as being bonded most strongly to the DNA itself, and poorly connected with the external water (when compared with the connectivity of 'free' water). This is consistent with our Raman studies of the structure of the hydration shell, which show that the number of reconstructed water molecules surrounding the DNA (primary hydration shell) does not depend on the extent of the secondary shell.

Given the fact that both the Raman studies and the room-temperature Brillouin dispersion results show that the primary hydration is not perturbed by the secondary hydration shell, we shall assume that this is also true for lower temperatures. Using the parameters obtained from the 75% r.h. data for the primary hydration shell, we have extracted the relaxation time, τ_2, for the secondary hydration shell. We summarize our results in Figure 9.14, where we plot ln τ_1 (primary relaxation) and ln τ_2 (secondary relaxation) against $1/T$. It gives a value of about 5 kcal/mol for the activation energy of the primary relaxation mode (we assume the primary shell to remain unaffected by the addition of the secondary shell at all temperatures). The Arrhenius plot of the relaxation time of the secondary hydration shell also follows a simple Arrhenius behaviour over the entire temperature range, with an activation energy of about 7 kcal/mol. Over the same temperature range, dielectric measurements show that the activation energy for the relaxation of supercooled water is about 7.7 kcal/mol (Bertolini *et al.*, 1982) and about 12 kcal/mol (Hasted, 1972) for ice relaxation. If we keep in mind the fact that the relaxation time of the secondary hydration shell at room temperature is also close to the Debye relaxation time of water at room temperature, we might conclude that the behaviour of the secondary hydration shell (below 0 °C) is quite similar to that of supercooled water. The primary hydration shell is very different from pure water. The activation energy of about 5 kcal/mol (less than that of supercooled water) seems to indicate that the cluster size of the hydrogen bond network is smaller than that in supercooled water. This is consistent with the Raman measurements (Section 2), which show that the connectivity of

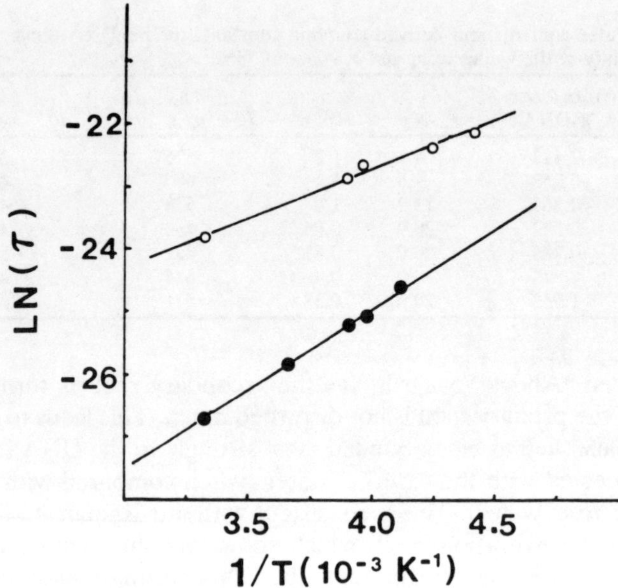

Figure 9.14 Arrhenius plot for the primary relaxation (○) and secondary relaxation (●). The solid lines correspond to activation energies of 5 and 7 kcal/mol. After Tao *et al.* (1988)

the hydrogen-bonded network of the water in the primary hydration shell is reduced relative to free water.

We have compared the relaxations of the DNA hydration shells with the dielectric relaxations of water because the relaxation times of the hydration shells are quite consistent with the dielectric measurements (Wittlin *et al.*, 1986; Cross and Pethig, 1983), and a computer simulation of hydrated DNA which extracts dipole correlation times (Swamy and Clementi, 1987). This leads to a conclusion that the relaxation is due to the orientational motion of the water molecules and the coupling occurs as the acoustic vibration generates an oscillating electric field through motion of the charged groups of the DNA, which, in turn, drives the water molecules in the hydration shells. To confirm this picture, more experiments are necessary.

Summary

(1) The sound wave in hydrated DNA couples strongly to the relaxation modes of the DNA hydration shells.

(2) The relaxation time and coupling strength of the primary hydration shell relaxation do not change significantly over a large range of hydration of the DNA films. This suggests that the primary shell does not change its structure much as hydration proceeds. The relaxational coupling is quite

anisotropic, the coupling constant having a value 550×10^{-52} kg $s^{-1}m^{-2}$ normal to the helix axis and 415×10^{-52} kg $s^{-1}m^{-2}$ along it. Above about 75% r.h., relaxation of the secondary hydration contributes to phonon damping.

(3) The primary shell relaxation time is ~ 39 ps at room temperature, and follows an Arrhenius dependence with an activation energy of about 5 kcal/mol.

(4) The secondary relaxation time is about 2 ps at room temperature. The relaxation time also follows Arrhenius behaviour with an activation energy of 7 kcal/mol, a value similar to that of the Debye relaxation of supercooled water.

5 Conclusions

We have shown that by studying the O–H stretching modes using Raman spectroscopy, the number of water molecules in the primary hydration shell of DNA can be determined and information about the structure of the water molecules in the primary hydration shell can be obtained. In addition, Raman spectroscopy provides a convenient way to determine the water content in a hydrated DNA sample. In principle, similar studies can be conducted in other biopolymer systems. Since water has many vibrational modes, a careful study of other modes should also give us information about the structure of the DNA hydration shells.

Using Brillouin spectroscopy, we have obtained important dynamic information about both the primary and the secondary hydration shells at GHz frequencies. We found that the onset frequencies from elastic to viscous behaviour of the hydration shells are ~ 3.5 GHz and ~ 20 GHz for the primary and the second hydration shells, respectively. Above ~ 20 GHz the water molecules surrounding DNA behave like an elastic medium, while below ~ 3.5 GHz, they act as a viscous medium. Furthermore, we found a strong coupling between the sound wave of DNA and the hydration shells.

So far these methods have been used in hydrated DNA systems only; in principle, they can be applied to other hydrated biological molecular systems.

Acknowledgements

I wish to thank A. Rupprecht for his high-quality samples, and S. A. Lee, T. Weidlich, J. B. Page, M. Soumpasis, A. Garcia, L. Van Zandt, E. Prohofsky, C. A. Angell, J. Green and J. Scott for helpful discussions. Especially I should like to thank S. M. Lindsay for his constant guidance and encouragement throughout this work. This was supported in part by the NSF (BBS 8615653) and ONR (N00014-84-0487).

References

Barker, A. S. and Hopfield, J. J. (1964). Coupled optical phonon mode theory of infrared dispersion in $BaTiO_3$, $SrTiO_3$ and $KTaO_3$. *Phys. Rev.*, **135**, 1732–1737

Berne, B. J. and Percora, R. (1976). *Dynamic Light Scattering*. Wiley, New York

Bertolini, D., Cassettari, M. and Salvetti, G. (1982). The dielectric relaxation time of supercooled water. *J. Chem. Phys.*, **76**, 3285–3290

Blinska, B. and Wilczk, T. (1976). IR hydration studies of DNA salts. *Stud. Biophys.*, **55**, 81–86

Blumberg, R., Stanley, H. E., Geiger, A. and Mausbach, P. (1984). Connectivity of hydrogen bonds in liquid water. *J. Chem. Phys.*, **80**, 5230–5241

Brody, E. M. and Cummins, H. Z. (1968). Brillouin scattering study of ferroelectric transition in KH_2PO_4. *Phys. Rev. Lett.*, **21**, 1263–1266

Brunauer, S., Emmett, P. H. and Teller, E. (1938). Adsorption of gases in multimolecular layers. *J. Am. Chem. Soc.*, **60**, 309–319

Conway, B. E. (1981). *Ionic Hydration in Chemistry and Biophysics*. Elsevier, New York

Corongiu, G. and Clementi, E. (1981). Simulation of the solvent structure for macromolecules: Structure of water solvation Na^+ B-DNA at 300 K and a model for conformational transition induced by solvent variation. *Biopolymers*, **20**, 2427–2483

Cross, T. E. and Pethig, R. (1983). Microwave studies of the interaction of DNA and water in the temperature range of 90–300 K. *Int. J. Quant. Chem. Quant. Biol. Symp.*, **10**, 143–152

DeMarco, C., Lindsay, S. M., Pokorny, M., Powell, J. and Rupprecht, A. (1986). Interhelical effects on the low-frequency modes and phase transitions of Li- and Na-DNA. *Biopolymers*, **24**, 2035–2040

Eyster, J. M. and Prohofsky, E. W. (1974). Lattice vibrational modes of poly(rU) and poly(rA). *Biopolymers*, **13**, 2505–2526

Falk, M. (1966). A gravimetric study of hydration of polynucleotides. *Can. J. Chem.*, **44**, 1107–1111

Falk, M., Hartman K. A., Jr., and Lord, R. C. (1963). Deoxyribonucleic acid, a gravimetric study. *J. Am. Chem. Soc.*, **84**, 3843–3846

Falk, M., Poole, A. G. and Goymour, C. G. (1970). Infrared study of the state of water in the hydration shell of DNA. *Can. J. Chem.*, **48**, 1536–1542

Green, J. L., Lacey, A. R. and Sceats, M. G. (1986). Spectroscopic evidence for spatial correlations of hydrogen bonds in liquid water. *J. Phys. Chem.*, **90**, 3958–3964

Green, J. L., Lacey, A. R. and Sceats, M. G. (1987a). Collective proton motions in H_2O/H_2O_2 mixtures — evidence for defects and network reconstruction. *J. Chem. Phys.*, **86**, 1841–1847

Green, J. L., Lacey, A. R. and Sceats, M. G. (1987b). Determination of the total hydration number of a LiCl cation anion pair via collective proton motions. *J. Chem. Phys. Lett.*, **134**, 385–391

Green, J. L., Lacey, A. R. and Sceats, M. G. (1987c). Collective small amplitude proton motions in hydration shells of aqueous alcohol solutions. *J. Chem. Phys. Lett.*, **137**, 537–542

Grimm, H., Stiller, H., Majkrzak, C. F., Rupprecht, A. and Dahlborg, U. (1987).

Observation of acoustic umklamp phonons in water stabilized DNA by neutron scattering. *Phys. Rev. Lett.*, **59**, 1780–1783

Hakim, M. B., Lindsay, S. M. and Powell, J. W. (1984). The speed of sound in DNA. *Biopolymers*, **23**, 1185–1192

Hasted, J. B. (1972). In Frank, F. (Ed.), *Water: A Comprehensive Treatise*, Vol. 1. Plenum Press, New York

Hearst, J. E. (1965). Determination of dominant factions which influence net hydration of native sodium deoxyribonuclates. *Biopolymers*, **3**, 57–68

Hearst, J. E. and Vinograd, J. (1961). Net hydration deoxyribonucleic acid. *Proc. Natl Acad. Sci. USA*, **47**, 825, 999, 1005–1014

Kennard, O., Cruse, W. B. T., Nachman, J., Prange, T., Shakked, Z. and Rabinovich, D. (1986). Ordered water structure in an A-DNA octamer at 1.7 Å resolution. *J. Biomol. Struct. Dyn.*, **3**, 623–647

Klug, D. D., Mishima, O and Whally, E. (1987). High density amorphous ice, Raman spectrum of the uncoupled O–H and O–D oscillators. *J. Chem. Phys.*, **86**, 5323–5331

Klug, D. D. and Whally, E. (1984). The uncoupled OH stretch in ice, the infrared frequency and integrated intensity up to 189 kbar. *J. Chem. Phys.*, **81**, 1220–1228

Kopka, M. L., Fratini, A. V., Drew, H. R. and Dickerson, R. E. (1983). Ordered water structure around a B-DNA dodecamer, a quantitative study. *J. Mol. Biol.*, **163**, 129–146

Kubo, R. (1966). Fluctuation dissipation theorem. *Rep. Prog. Phys.*, **29**, 255–284

Kuntz, I. D., Brassfield, T. S., Law, G. D. and Purcell, G. V. (1969). Hydration of macromolecules. *Science*, **163**, 1329–1331

Lee, S. A., Lindsay, S. M., Powell, J. W., Weidlich, T., Tao, N. J., Lewen, G. D. and Rupprecht, A. (1988). A Brillouin study of the hydration of Li- and Na-DNA films. *Biopolymers*, **26**, 1637–1665

Lewin, S. J. (1967). Some aspects and stability of native state of DNA. *Theoret. Biol.*, **17**, 181–212

Lindsay, S. M., Anderson, M. W. and Sanderscock, J. R. (1981). Construction and alignment of a high performance multipass vernier tandem Fabry–Perot interferometer. *Rev. Sci. Instrum.*, **52**, 1478–1486

Lindsay, S. M., Lee, S., Powell, J. W., Weidlich, T., DeMarco, C., Lewen, G. D., Tao, N. J. and Rupprecht, A. (1988). The origin of A to B transition in DNA fibers and films. *Biopolymers*, **27**, 1015–1043

Lindsay, S. M., Powell, J., Prohofsky, E. W. and Devi-Prasad, K. V. (1985). Lattice modes and local modes in double helical DNA. In Clementi, E., Corogiu, G., Sarma, M. H. and Sarma, R. H. (Eds), *Structure and Motion in Membranes, Nucleic Acids and Proteins*. Adenine Press, Schenectady, N.Y., pp. 531–551

Lindsay, S. M. and Tao, N. J. (1988). The active role water in DNA. In Sarma, R. H. and Sarma, M. H. (Eds), *Structure and Expression*, Vol. 2: *DNA and Its Drug Complexes*. Adenine Press, Schenectady, N.Y., pp. 217–218

Nimtz, G. (1986). Magic number of water molecules bounds between lipid bilayers. *Phys. Scripta*, **T13**, 172–177

Pilet, J., Blicharski, J. and Brahms, J. (1975). Conformation and structure transition in polydeoxynucleic acids. *Biochemistry*, **14**, 1869

Prescott, B., Steinmetz, W. and Thomas, G. J. (1984). Raman spectral studies of nucleic acids, characterization of DNA structures by laser Raman spectroscopy. *Biopolymers*, **23**, 235–256

Rupprecht, A. and Forslind, B. (1970). Variation of electrolyte content in wet-spun lithium and sodium DNA. *Biochem. Biophys. Acta*, **204**, 304–316

Saenger, W. (1984). *Principles of Nucleic Acid Structure*. Springer-Verlag, Berlin

Sandercock, J. R. (1982). Trends in Brillouin scattering — studies of opaque materials, supported films and central modes. *J. Appl. Phys.*, **51**, 173–206

Sceats, M. G. and Rice, S. A. (1983). In Frank, F. (Ed.), *Water: A Comprehensive Treatise*, Vol. 7. Plenum Press, New York

Schreiner, L. J., Pintar, M. M., Dianoux, A. J., Volino, F. and Rupprecht, A. (1987). Hydration of Na-DNA by neutron quasi-elastic neutron scattering. *Biophys. J.*, **53**, 119–122

Singh, G. P., Parak, F., Hunklinger, S. and Dransfield, K. (1981). Role of adsorbed water in the dynamics of metmyoglobin. *Phys. Rev. Lett.*, **47**, 685–688

Sokoloff, J. B. (1988). *Bull. Am. Phys. Soc.*, **33**, 556–586

Swamy, K. N. and Clementi, E. (1987). In Clementi, E. and Chin, S. (Eds), *Structure and Dynamics of Nucleic Acids, Proteins and Membranes*. Plenum Press, New York, pp. 219–238

Tao, N. J. (1988). *Structure and Dynamics of Hydrated DNA Studied by Laser Spectroscopy*. PhD Dissertation, Arizona State University

Tao, N. J., Lindsay, S. M. and Rupprecht, A. (1987). Dynamics of DNA hydration shells at GHz frequencies studied by Brillouin scattering. *Biopolymers*, **26**, 171–186

Tao, N. J., Lindsay, S. M. and Rupprecht, A. (1988). Dynamic coupling of DNA and its primary hydration shell. *Biopolymers*, **27**, 1655–1671

Tao, N. J., Lindsay, S. M. and Rupprecht, A. (1989). Structure of DNA hydration shell studied by Raman scattering. *Biopolymers*, **28**, 1019–1030

Texter, J. (1978). Nucleic acid–water interaction. *Prog. Biophys. Mol. Biol.*, **33**, 83–97

Tominaga, Y., Shida, M., Kubota, K., Urabe, H., Nishimura, Y. and Tsuoi, M. (1985). Coupled dynamics between DNA double helix and hydration water by low frequency Raman spectroscopy. *J. Chem. Phys.*, **83**, 5972–5975

Tunis, M. J. B. and Hearst, J. E. (1968). On hydration of DNA. 1. Preferential hydration and stability of DNA in concentrated trifluoroacetate solution. *Biopolymers*, **6**, 1325–1344

Van Zandt, L. L. (1986). Resonant microwave absorption by dissolved DNA. *Phys. Rev. Lett.*, **57**, 2085–2088

Van Zandt, L. L. and Davis, M. E. (1986). Theory of the anomalous resonant absorption of DNA at microwave frequencies. *J. Biomol. Struct. Dyn.*, **3**, 1045–1053

Walrafen, E. (1964). Raman spectral studies of water structure. *J. Chem. Phys.*, **40**, 3249–3256

Walrafen, E. (1971). In Frank, F. (Ed.), *Water: A Comprehensive Treatise*, Vol. 1. Plenum Press, New York

Wang, J. H. (1955). The hydration of deoxyribonucleic acid. *J. Am. Chem. Soc.*, **77**, 258–260

Weiss, W., Enders, A. and Nimtz, G. (1986). Connectivity of hydrogen bond in water between lecithin bilayers. *Phys. Rev.*, **A33**, 2137–2139

Whally, E. (1977). Detailed assignment of O−H stretching bands of ice I. *Can. J. Chem.*, **55**, 3429–3441

Wittlin, A., Genzel, L., Kremer, F., Haseler, S. and Poglitsch, A. (1986). Far infrared spectroscopy on oriented films of dry and hydrated DNA. *Phys. Rev.*, **A34**, 493–500

Wong, P. T. T. and Whally, E. (1977). Optical spectra of orientationally disordered crystal, Raman spectrum of ice I in range of 4000–350 cm^{-1}. *J. Chem. Phys.*, **62**, 2418–2425

Part 4
Polysaccharides and Lipids

10
Polysaccharide Interactions with Water

Serge Pérez

1 Introduction

Carbohydrates are among the most abundant organic molecules in living nature. Carbohydrate molecules are distinguished by a long history of involvement in the fundamental process of life (Marchessault, 1984). They are also ubiquitous in Nature, where they play an essential role in promoting structure and texture or establishing storage. Their implications in biological recognition through *carbohydrate-mediated information transfer* have also been established. The primary structures of polysaccharides vary in composition, sequence, molecular weight, anomeric configuration, linkage position and charge density. As a consequence, an almost infinite array of chemical structures and conformations can be generated for polysaccharides. Additional variability may arise from environmental changes such as ionic strength and degree of hydration. More information about polysaccharides can be found in an excellent monograph by Yalpani (1988).

Of the three major components of biological structures, the proteins, nucleic acids and polysaccharides, least is known about the polysaccharides at the secondary and tertiary level of molecular structures. This is because the polysaccharides cannot be obtained in crystals which are large enough for single-crystal X-ray or neutron structure analysis. In their native states, the polysaccharides are either completely amorphous or semicrystalline, as in the case of starch; they may form fibrous structures, as in the celluloses and many of the bacterial polysaccharides.

As with any other carbohydrate molecules, polysaccharides offer an exceptional ratio of hydroxyl groups per saccharide residues. Such a hydrogen-bonding potential has to be statisfied in association with either neighbouring carbohydrate polymers, glycoproteins or surrounding water molecules.

Before beginning a review and analysis of structural features displayed by polysaccharide–water interactions, a brief overview of the experimental methods and aspects of elucidating three-dimensional structures of polysaccharides is presented. This involves X-ray fibre diffraction and electron crystallography, which have to be complemented by extensive molecular modelling. The accuracy to which hydration is characterized is also presented. In Section 3, the water–polysaccharide interactions are analysed according to increasing structural complexities. Examples of water location in crystalline arrangements found in both neutral and anionic polysaccharides are presented first. Then the role of water in crystalline polymorphism of polysaccharides is illustrated. The range of examples covers the structural changes which are found in hydrated and dry polymorphs, as in dextran, as well as those observed between two allomorphs displaying differences in their hydration schemes, as in starch. Moreover, the role of hydration in continuous structural transitions, such as in galactomannans, or as providing intermediate template in irreversible polymorphic transitions is exemplified. The final section is intended to provide illustrations associated with the difficulties and limits of water location in crystalline polysaccharides.

2 Analysis of the Structural Methods

Unlike globular proteins, the length in polysaccharide chains is not fixed, but can vary, often widely, from one molecule to another, and an average molecular weight or molecular weight distribution must be considered. In polysaccharides one end of each chain has the reducing character of monosaccharides and the other end is non-reducing. The determination of the relative orientation of such polar chains in crystalline or semicrystalline structures has been a central theme in the field of polysaccharide crystallography.

X-Ray Fibre Diffraction

The most important method for the structure determination of crystalline polymers has been X-ray fibre diffraction. It has been observed that linear polysaccharides prefer to exist as long helices rather than more complexely folded structures. After dissolution, one usually can produce samples in which such helical macromolecules are aligned with their long axis parallel or antiparallel (by convention, this is the *c*-direction). Further lateral organization may occur, but rarely to the degree of a three-dimensionally ordered single crystal. Fibrous structures typically provide diffraction data of low resolution. Many helical polysaccharides yield no more than 50

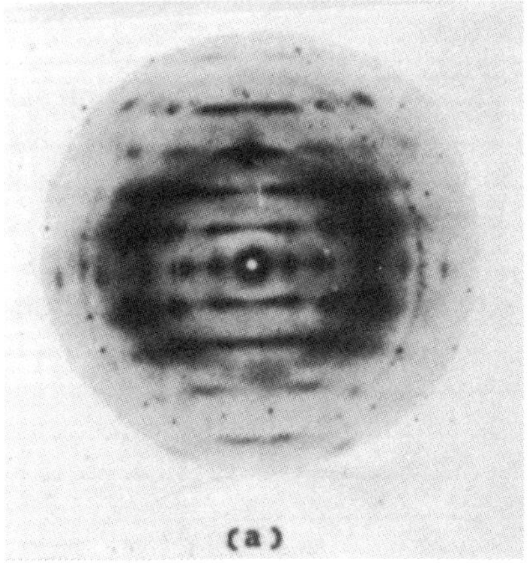

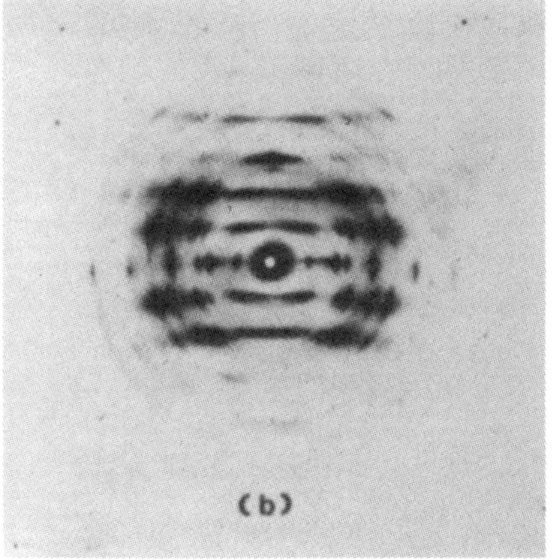

Figure 10.1 Typical example of a fibre diffraction diagram of a polysaccharide: Ca^{2+} and Sr^{2+} salt of iota-carrageenane. From S. Arnott, W. E. Scott, D. A. Rees and C. G. A. McNab, *J. Mol. Biol.*, 1974, 90, 253–267. Reprinted with permission from Academic Press

independent X-ray reflections that can be used to determine the molecular geometry of the crystallographic asymmetric unit (Figure 10.1). Other shortcomings of this approach are the difficulty in assigning the unit-cell parameters and the ambiguities regarding the choice of the space group.

Electron Crystallography

As with many other stereoregular polymers, simple linear polysaccharides, once dissolved and recrystallized, can yield single crystals (see review by Chanzy and Vuong, 1985). Polysaccharides often crystallize with the incorporation of water or solvent molecules. In most cases, a well-defined morphology is obtained, the most popular being plate-like. These thin lamellar surfaces have lateral dimensions of several micrometres for only a few tenths of angstroms in thickness. Most usually, the polymer chain axis lies normal to the lamellar surface. Crystalline domains of such dimensions are well suited to examination by transmission electron microscopy in both imaging and diffraction modes (Figure 10.2). A problem frequently encountered when studying crystalline biopolymers with the electron microscope relates to the vacuum dehydration of the specimen when inserted inside the instrument column. This is particularly critical when water or solvent is part of the crystalline structure. In such instances, total or partial decrystallization takes place, in a matter of minutes, accompanied by drastic distortion of the sample. Several methods have been developed whereby the sample is either viewed inside an hydration chamber (Matricardi *et al.*, 1972; Hui and Parsons, 1974) or quenched in a cryogenic bath prior to insertion into the electron microscope (Chanzy *et al.*, 1971; Taylor

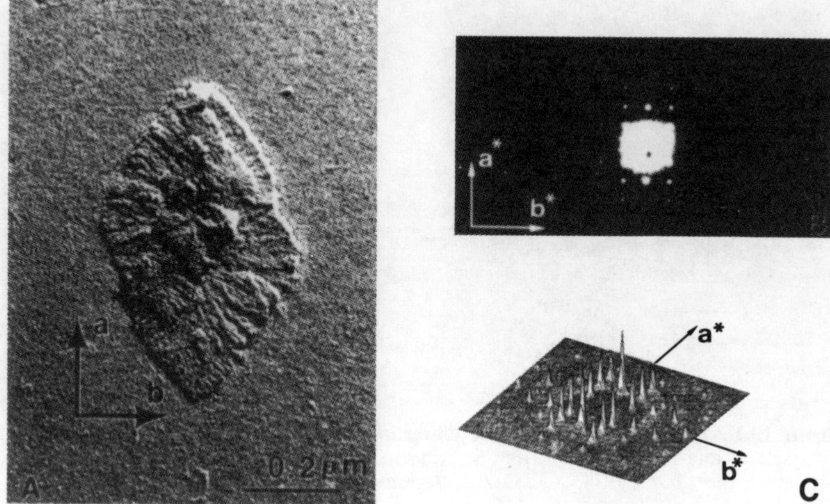

Figure 10.2 Single crystal of a polysaccharide and its corresponding base plane electron diffraction pattern. (A) A mannan I single crystal; (B) its electron diffractogram properly oriented with respect to the crystal as in A; (C) Digitized electron diffractogram as in B after noise removal. From Pérez and Chanzy (1989). Reprinted by permission of Alan R. Liss

and Glaeser, 1974; Taylor *et al.*, 1975). In these instances, the observations are performed at temperature close to that of boiling liquid nitrogen, where the water of crystallization is stable in high vacuum. With such a technique, frozen wet electron diffractograms are readily recorded.

Elucidation of the Crystal Packing

Provided that the data set is of sufficient quality and/or the unit-cell dimensions and space group symmetry are well assigned, the final stage of elucidation involves a complete structural determination of the unit-cell content. In order to do so, a linked-atom description similar to that reported by Smith and Arnott (1978) or Zugenmaier and Sarko (1980) may be used. Optimization procedures seek to fit observed and calculated structure amplitudes with simultaneous optimization of the non-bonded interactions and preservation of helix pitch and symmetry as well as ring closure. Interatomic energy functions, mimicking intermolecular non-bonded interactions (Williams, 1969) are used extensively for this purpose.

Placement of Water Molecules

On the basis of crystalline density, a putative number of water molecules is assumed in agreement with the space group symmetry. Several strategies have been used in the location of these crystalline water molecules (Nishimura and Sarko, 1991). In an initial stage, the water positions are determined in the *ab*-projection using the available *hkO* reflections intensities. The *ab*-plane corresponding to the asymmetric unit is divided into small rectangular regions, and the *x*,*y*-coordinates of a single water molecule are refined in each such region in turn. This produces a weighted *R*-factor map (where *R* has its usual crystallographic meaning) which usually indicates the positions of the water molecules. The *z*-coordinate of each water molecule is then quickly refined with the *hkl* reflection intensities, starting from several trial positions.

In some instances, water molecules do not occupy crystallographically defined positions. In these cases, the water-weighted scattering factors can be used. The use of the latter factors is based on the assumption that the volume of the unit cell is filled with an electron gas whose density is equal to the mean electron density of water.

In the final stages of the refinement, the water molecules are introduced and the structure is minimized. The general scheme for defining and refining polymer structures is given in Figure 10.3.

Molecular Modelling

In contrast to other macromolecules, the diffraction data obtainable from polysaccharides are not sufficient to permit crystal structure determination based on the data alone. A modelling technique must be used which allows the calculation of diffraction intensities from various models for comparison with the observed intensities. It is little more than 20 years since the first, computer-assisted and reasonably detailed structure analysis of a polysaccharide was completed. Now the number of apparently successful structure solutions of polysaccharides exceeds 100, counting all polymorphs, variants, derivatives and complexes. The joint use of molecular modelling and electron diffraction has been invaluable in the quantitative elucidation of crystals and molecular structures (see review by Pérez, 1991). The modelling technique can be combined with the information derived from electron diffraction as a complement to fibre diffraction, to help resolve in a quantitative and unambiguous fashion the three-dimensional polysaccharide structures. The essential molecular features, i.e. chain conformation, chain polarity, packing habits, intermolecular interactions, place and role of water molecules, have been elucidated.

Even when an unequivocal crystal structure is established for a polysaccharide, the hydrogen bonding and the subsequent characterization of the hydration are likely to remain ambiguous in the absence of direct evidence relating the orientation of the hydroxyl groups. Therefore, hydrogen bonds are identified by the O· · ·O separation. It is a reasonable assumption from monosaccharide and oligosaccharide crystal structures that all hydrogen-

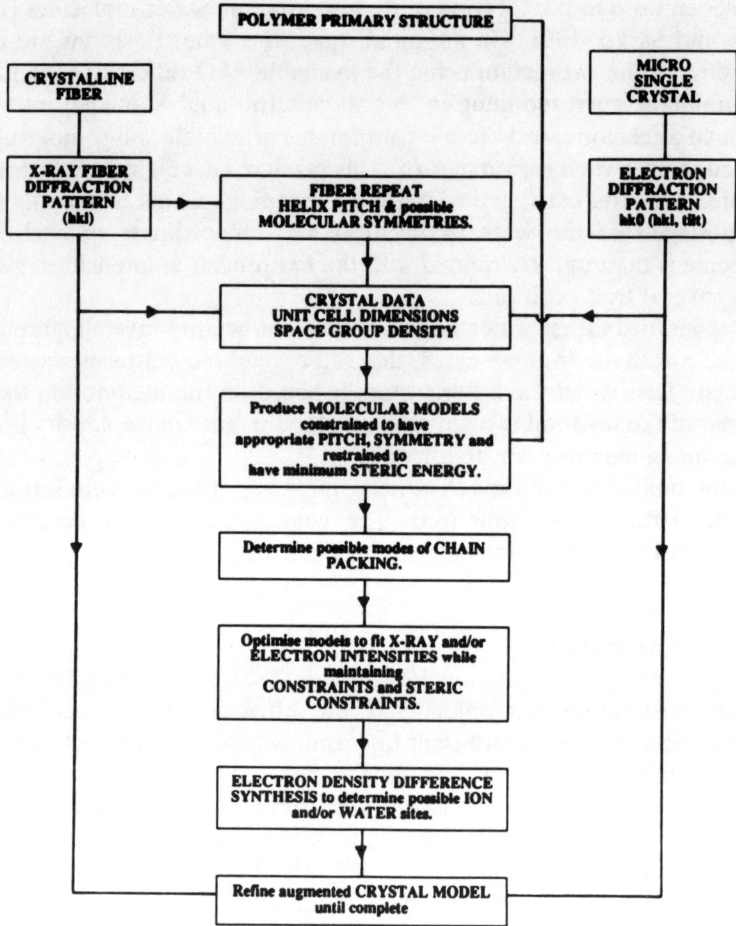

Figure 10.3 Scheme for defining and refining polymer structures. Adapted from Arnott (1980) and Brisse (1991)

bonding donor functional groups will form hydrogen bonds. It is also desirable, albeit not necessary, that all potential hydrogen-bond acceptor atoms be acceptors. Most of these features are derived from small carbohydrate crystal structural studies. However, as the molecules become larger, the hydration number increases, as does the complexity (Jeffrey and Saenger, 1991). It has to be stated that characterizing water–polysaccharide interactions is very much hampered by the absence of information about the hydrogen atoms and is likely to remain speculative.

3 Water Location in Crystalline Polysaccharides

In their excellent review of crystal structure hydration, Bluhm *et al.* (1980) examined six hydrated neutral polysaccharides: (1–4)-β-D-xylan, nigeran, amylose, galactomannan, (1–3)-β-D-glucan and (1–3)-β-D-xylan. They observed that water may occur as columns or as sheets in these structures. Their review was restricted to neutral glycans, in order to avoid hydrating phenomena related to the polyelectrolyte effect.

Example of a Neutral Hydrated Polysaccharide: (1–3)-α-D-Mannan

Mannans are widely distributed in plants and microorganisms, as both homo- and heteropolysaccharides. The (1–3)-α-D-mannans are found as backbone chains of branched heteropolysaccharides in the fruiting bodies of some edible mushrooms. The crystal structure of the hydrated form of (1–3)-α-D-mannan was analysed by combined X-ray diffraction and stereochemical model refinement techniques, from 98 *hkl* intensities (Yui *et al.*, 1992). The polysaccharide crystallizes in a four-chain, monoclinic unit cell (a = 1.133, b = 1.836, c = 0.825 nm and γ = 101.75°), the most probable space group being $P2_1$. There are 16 water molecules in the unit cell. The chains pack with antiparallel polarity and are connected by pairs of intermolecular hydrogen bonds that form an infinite zig-zag sheet. The water molecules, generally found between the sheets (Figure 10.4), provide additional hydrogen bonding that contributes to the intricate three-dimensional hydrogen-bond network in the crystal structure. These water molecules are thought to reside firmly in their crystallographic positions, since it has been observed that a fibre pattern of the hydrate form was still obtained from the sample in vacuum at room temperature.

Example of an Anionic Hydrated Polysaccharide: Gellan

Gels of varying strength are formed by a number of microbial polysaccharides, including curdlan, XM-6, xanthan and gellan, and those produced by *Klebsiella aerogenes* serotype K54 and *Bacillus polymyxa*. Using X-ray fibre diffraction analysis, the molecular structure of gellan has been determined. The structures of the lithium (Chandrasekaran *et al.*, 1988b) and potassium (Chandrasekaran *et al.*, 1988a) complexes have been established. In both, gellan takes up a double-helical structure which is made up of two half-staggered, parallel, three-fold left-handed chains of pitch 5.63 nm. The double helices are packed in an antiparallel fashion in the trigonal unit cell of dimensions a = b = 1.58 nm and c = 2.82 nm, with a lateral separation of 0.9 nm. The gellan molecules are crosslinked by double

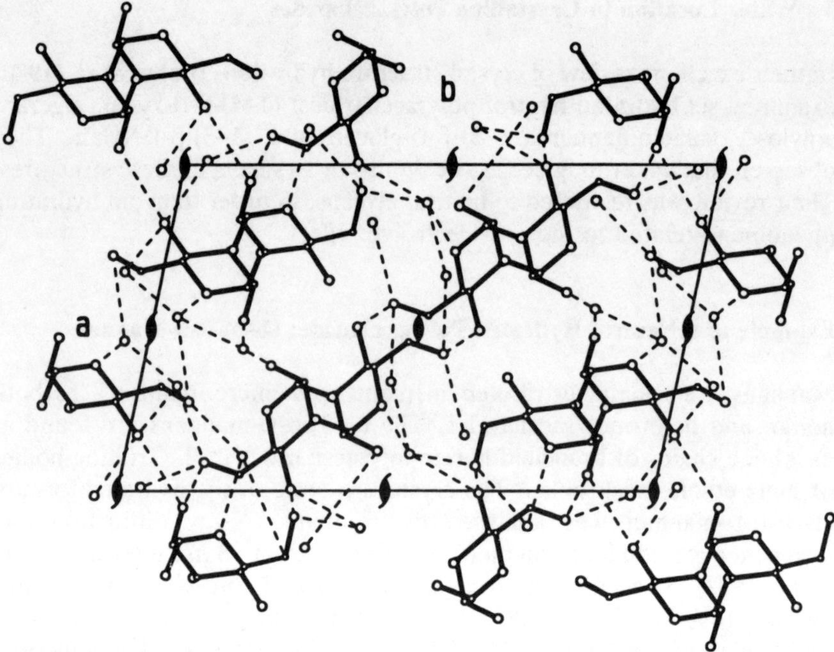

Figure 10.4 Three-dimensional structure of (1–3)-α-D-mannan. Projections of the structure atoms on the *ab*-plane. The 16 water molecules in the unit cell are shown as o. All hydrogen atoms have been omitted and hydrogen bonds are shown as dashed lines. From Yui *et al.* (1992). Reprinted by permission of Elsevier

helix–potassium–water–potassium–double helix interactions (Figure 10.5). These are thought to be responsible for subsequent gelation process.

4 Water in the Crystalline Polymorphism of Polysaccharides

All naturally occurring polysaccharides display an extensive hydroxylated character of the chain. During crystallization, these chains will become interconnected through a network of hydrogen bonds. Quite frequently, there is more than one possibility to achieve such an interconnection. This leads to several three-dimensional arrangements for the crystallizing chains and to the occurrence of polymorphic crystalline forms.

Comparison between Hydrated and Dry Polymorph: Dextran

Dextran is a generic term used to describe D-glucans that contain a substantial number of (1–6)-linked-α-D-glucopyranosyl residues (Sidebotham, 1974). This type of macromolecule is not crystallizable, because of the

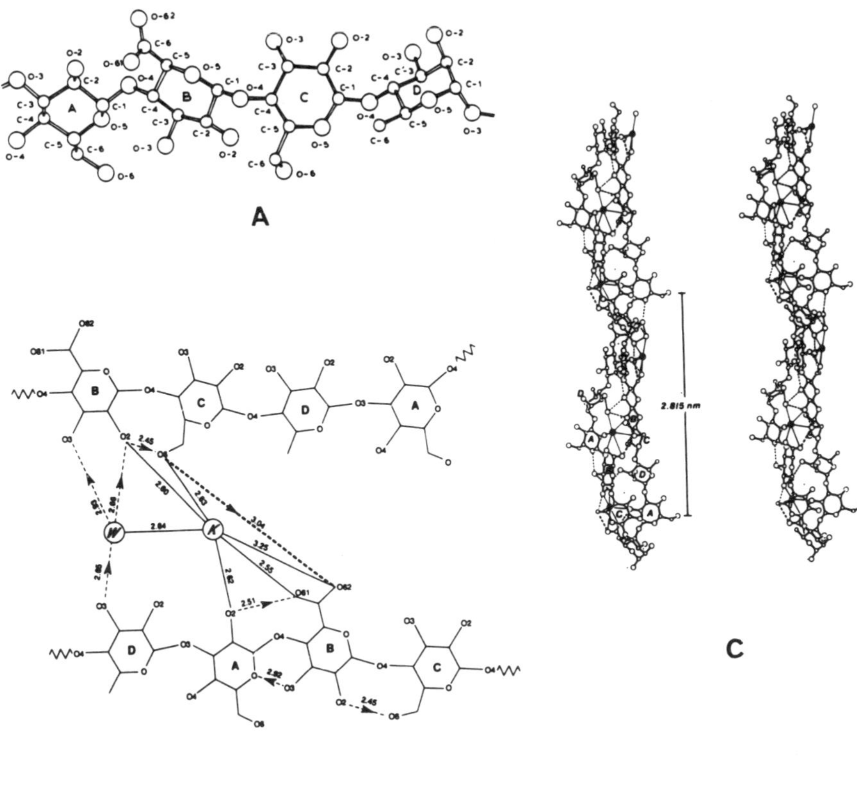

Figure 10.5 Structure and hydration of gellan. (A) Schematic representation of the tetrasaccharide repeating unit of gellan (A, B, C, and D are β-D-glucopyranose, β-D-glucopyranuronate, β-D-glucopyranose and α-L-rhamnopyranose, respectively). From Chandrasekaran *et al.* (1988a). Reprinted by permission of Elsevier. (B) Schematic illustration of the hydrogen-bonding and potassium ion (K) coordination in the gellan double helix, including the first shell water molecules (W). The distances are given in Å units. (C) A stereoscopic representation of the gellan double helix featuring the intrachain hydrogen bonds (thin dashed lines), interchain hydrogen bonds (thick dashed lines), potassium ions (filled circles) and water molecules (open circles) along with the six ligands attached to each potassium ion (thin lines). From Chandrasekaran and Tailambal (1990). Reprinted by permission of the American Chemical Society

presence of branches. In contrast, linear dextran can be chemically synthesized. It has been found (Chanzy *et al.*, 1980, 1981) that such a linear dextran can be crystallized from dilute solution in the form of lamellar single crystals, provided that low molecular weight and monodisperse fractions are used. In addition, linear dextran exhibits two polymorphs; a 'low-temperature' polymorph obtained below 100 °C and a 'high-temperature' polymorph crystallized above 120 °C. The principal difference between these polymorphs is that the high-temperature polymorph is anhydrous, while the low-temperature crystalline form is hydrated. For both polymorphs, the crystal structure was established through a combined electron and X-ray diffraction analysis and stereochemical model refinement.

The asymmetric unit of the anhydrous form is made up of two glucose-independent glucose residues linked α(1–6) (Guizard *et al.*, 1984). The conformation of the chain is relatively extended and ribbon-like, with successive residues in a near twofold screw relationship. Two antiparallel-packed chains pass through the unit cell. The chains of like polarity pack into sheets with extensive intrasheet hydrogen bonding. The sheets, packed antiparallel, are, in turn, extensively bonded together by intersheet hydrogen bonding (Figure 10.6A).

The unit cell of the low-temperature, hydrated dextran polymorph contains six chains and eight water molecules, with three chains of the same polarity and four water molecules constituting the asymmetric unit (Guizard *et al.*, 1985). As in the anhydrous form, the chain is extended, and close to a twofold screw symmetry. However, the chain packing is different in the two structures, in that rotational positions of the chains about the helix axes are considerably different. In the hydrated form, the chain packing is still quite tight (Figure 10.6B), as is evident from intermolecular contacts, numerous hydrogen bonds and the high crystalline density. Water molecules occupy positions between sheets of chains, and play an essential role in hydrogen bonding. The water molecules of the asymmetric unit participate into 18 hydrogen bonds. There is no hydrogen bonding between these water molecules. The fact that these water molecules are tightly held in the crystal lattice explains the experimental observation that crystals of dextran hydrate do not deteriorate rapidly in the electron beam of the microscope. This is also reflected by the unusually high density, 1.68, of the crystalline samples.

Hydration of the dextran structure does not cause significant alterations in the conformation of the chains: it simply separates them. The distances between neighbouring chains of the same polarity along one axis of the unit cell are similar (0.92 nm in the anhydrous, 0.85 nm in the hydrate). However, the setting angles differ by approximately 60° (Figure 10.6C). In both structures, the dextran chains form sheets. In the hydrate, water separates sheets.

Comparison between Two Polymorphs with Different Hydration: Amylose

The starch granule is Nature's chief way of storing energy in green plants over long periods. The granule is well suited to this role, being insoluble in water and densely packed, but still accessible to the plant's metabolic enzymes. Native granules have crystallinity between 15% and 45% and can yield X-ray diffraction patterns, generally of low quality, with cereal starch

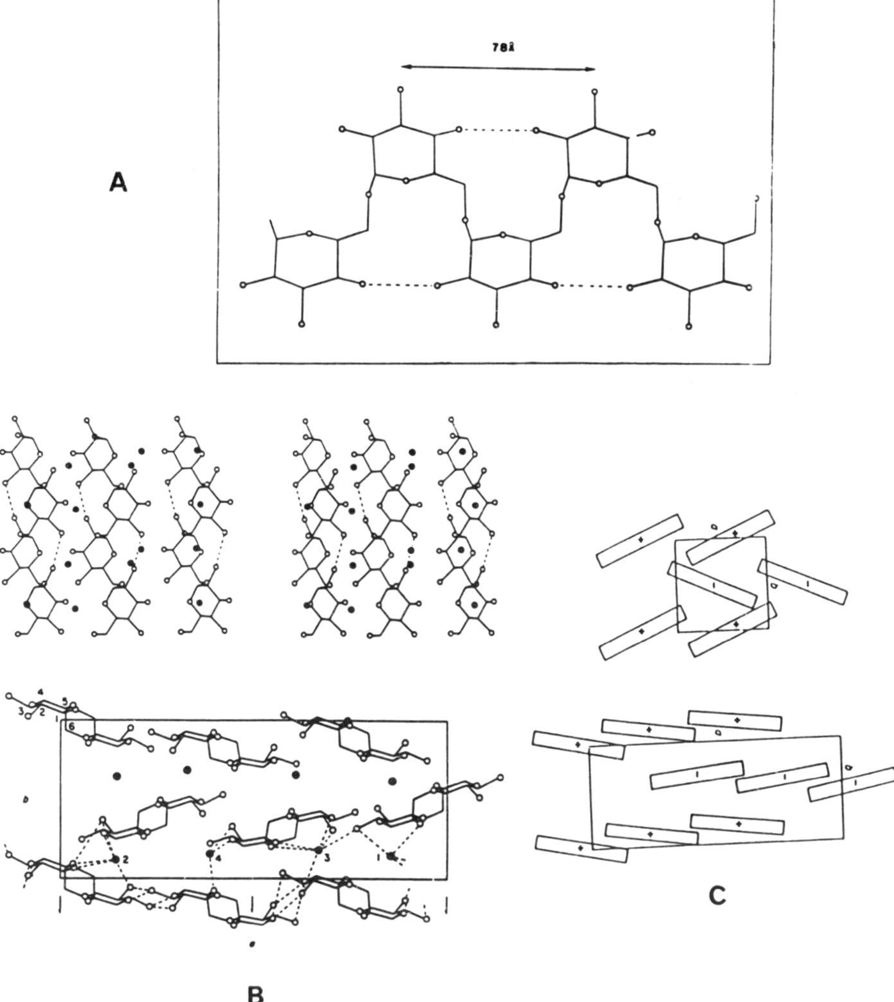

Figure 10.6 Crystalline features of dextran. (A) Molecular drawing of the dextran chain, with its twofold symmetry. Intrachain hydrogen bonding (represented as broken lines) occurs between residue i and $i + 2$. From Guizard (1980). (B) (a) Stereoscopic representation of one chain of dextran in its low-temperature, hydrated polymorph state; (b) projection of the structure of the unit cell down the c-axis. Hydrogen bonds are indicated by broken lines and water molecules by (•). (C) Comparison of chain packing in the anhydrous (left) and the hydrate (right) forms of crystalline dextrans). From Guizard *et al.* (1985). Reprinted by permission of Academic Press

typically giving A-type patterns and tuber starch usually yielding B-diagrams. Partial conversion of B-starch to A-starch can be accomplished by adjusting temperature and humidity conditions; complete conversion has not been achieved. Upon dissolution of starch in hot water, a spontaneous gelation occurs followed by subsequent crystallization; this is the well-known process of retrogradation (Whistler and Smart, 1953). The polymorphic form of retrograded starch is B-type, irrespective of the form of starch initially dissolved.

The two components of starch are amylose, the linear α(1–4)-linked glucan, and amylopectin, a α(1–4)-linked glucan with 1–6 branch points. It is now widely accepted that the amylopectin molecule is the crystalline component in granules, with the short-branched chains forming local organizations.

The structure of the crystalline part of A-starch was established (Imberty *et al.*, 1988) from joint use of electron diffraction on single crystals, X-ray powder diffraction diagrams decomposed into individual peaks (Tran and Buléon, 1987) and previously reported X-ray fibre diffraction (Wu and Sarko, 1978b) after appropriate reindexing. Chains are crystallized in a monoclinic lattice with a = 2.124 nm, b = 1.172 nm, c = 1.069 nm and γ = 123.5°, the c-axis being parallel to the helix axis. Systematic absences are compatible with the space group B2. The unit cell contains 12 glucose residues located in two left-handed, parallel-stranded double helices packed in a parallel fashion; four water molecules are located between these double helices (Figure 10.7a).

The structure of the crystalline part of B-starch was solved (Imberty and Pérez, 1988) using partial information extracted from the base-plane electron diffractogram, i.e. unit cell parameters and two-dimensional symmetry, the diffraction data being taken from the previously reported X-ray fibre diffractogram (Wu and Sarko, 1978a). Chains are crystallized in the hexagonal space group $P6_1$, with lattice parameters a = b = 1.85 nm, c = 1.04 nm. The unit cell contains 12 glucose residues located in two left-handed, parallel-stranded double helices packed in a parallel register; 36 water molecules are located between these helices. With a hydration around 27% w/w, the structure corresponds to a well-ordered crystalline sample, since all the water molecules could be located, with no apparent sign of disorder. This is in agreement with an NMR study (Lechert, 1981) that indicates that 'freezable' water can be observed only when the hydration is above 33%. In the present model, half of the water molecules are tightly bound to the double helices; the remainder forms a complex network centred around the sixfold screw axis of the unit cell (Figure 10.7a). The columnar nature of the hydration is evident.

The double helices in both A-starch and B-starch are both almost perfect left-handed sixfold structures with a crystallographic repeating distance of 1.05 nm. In both forms, the parallel arrangement of all molecules is imposed by the observed space group. When projected on the a,b-plane, the double helices of both forms are seen to be packed in hexagonal or pseudohexagonal arrays (Figure 10.7a, b). The lattice of B-starch has a

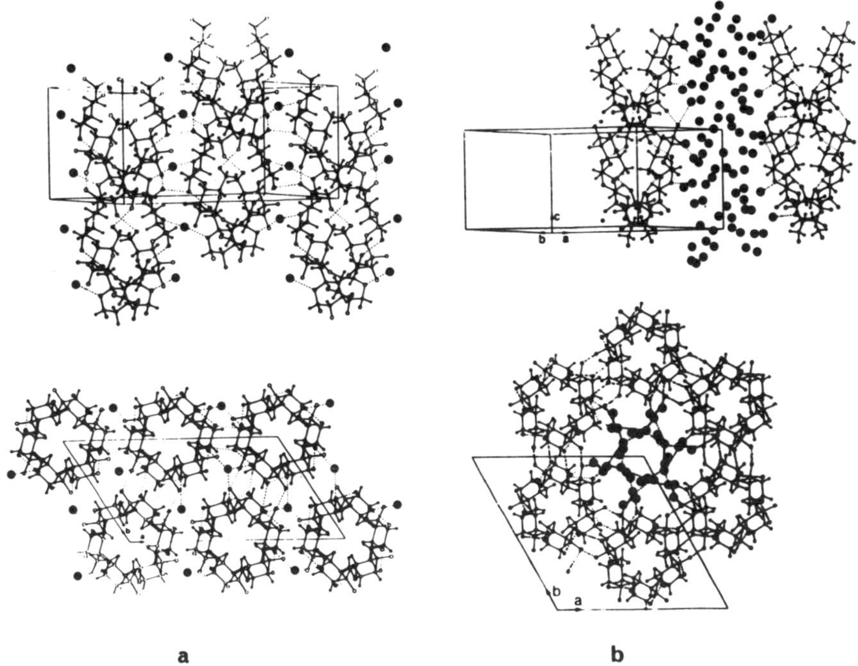

Figure 10.7 (a) A-starch crystalline structure (hydrogen bonds are indicated as broken lines and (•) indicate water molecules). Three-dimensional representation along the fibre axis (*c*-direction) and projection of the structure onto the *a*, *b*-plane. From Imberty *et al.* (1988). Reprinted by permission of Academic Press. (b) B-starch crystalline structure. Three-dimensional representation along the fibre axis (*c*-direction). Hydrogen bonds between the two strands of the double helices and the water molecules are shown as dashed lines. The numerous hydrogen bondings involving only water molecules are not represented in the picture; projection of the structure onto the *a*, *b*-plane. The unit cell content and some neighbouring double helices are represented in order to show the location of the water molecules in a channel. From Imberty and Pérez (1988). Reprinted by permission of John Wiley and Sons

large void in which numerous water molecules can be accommodated. This void is not present in A-starch. In both arrangements, there is a pairing of double helices that corresponds to a 1.1 nm distance between axes of the two double helices. The *c*-axis origins of the two double helices in the pair differ by about *c*/2, allowing a very close nesting of the crests and troughs of the paired double helices (Figure 10.8a). From the foregoing, it appears that the pair of double helices is a common structural element in both extensively hydrated B-starch and the slightly hydrated A-form (Pérez *et al.*, 1990). In a model of the solid-state transition from B-starch to A-starch, these pairs are preserved while the water is removed and the

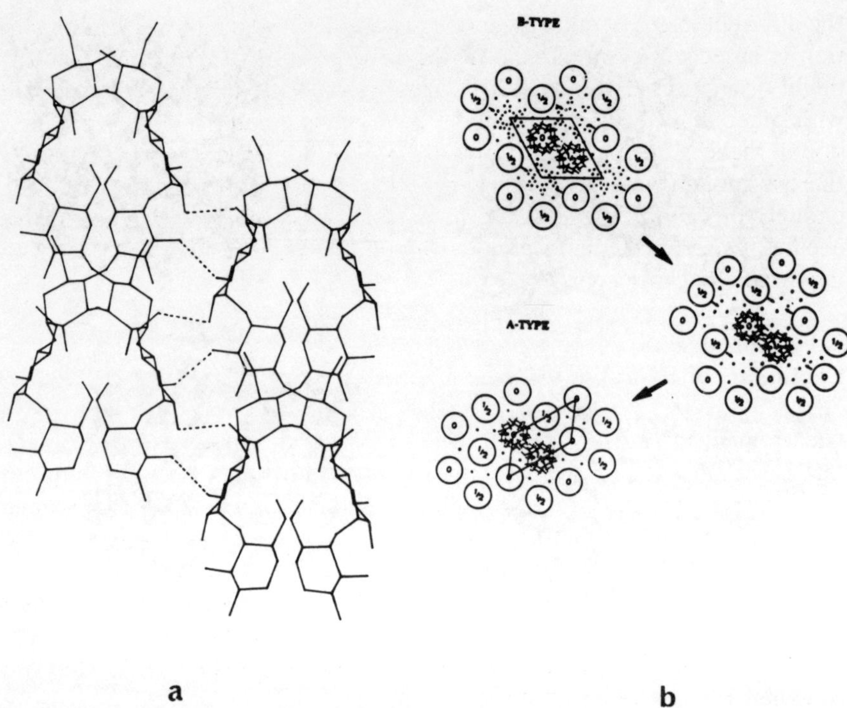

Figure 10.8 (a) Molecular drawing of the best arrangement of two parallel double helices, projected perpendicular to the chain axis; (b) model of the polymorphic transition from the B- to the A-starch in the solid state. The parallel double helices which form the duplex are labelled O and 1/2 (this indicates their relative translation along the *c*-axis). The water molecules are shown as (•). From Pérez *et al.* (1990). Reprinted by permission of the American Chemical Society

helices rearranged to fill the void left by removal of water (Figure 10.8b). On the other hand, it appears that the reverse transition could not take place without a gelatinized intermediate that would disrupt the crystalline architecture of the dense A-form.

In *in vitro* experiments, B-starch crystallizes in pure water and low temperatures, while A-starch requires dehydrating conditions. The cool, wet conditions were thought to be analogous to the *in vivo* conditions of tubers, which make B-starch, and the dryer, warmer conditions were thought to be similar to those in cereals, which make A-starch. More recently, however, the importance of the length of the branched segments has been recognized. Amylosic fragments with DP < 10 do not crystallize, while the A-form results from chains with DP from 10 to 12; chains having DP > 50 result in tangled networks. Thus, it appears that selection of a starch form is controlled by both amylosic chain length and degree of hydration. Gidley (1987) has proposed that chain length effect is a result of the different losses of entropy upon crystallization. An alternative explanation is based on a comparison of the ways in which the A- and B-forms might occur (Imberty *et al.*, 1991). Single strands of amylose can associate with other single strands so long as their DP is greater than 9. Next, two double helices are paired to form the stable duplex described earlier. If there is enough water to fill the central cavity, and if the duplexes are long enough to organize those water molecules in a stable column, the B-form occurs. Otherwise, the pseudohexagonal A-form results, with its less favourable packing energy.

Continuum of Hydration: Galactomannan

Galactomannans are a family of seed endosperm polysaccharides of various Leguminosae. In the seed the polysaccharides protect the embryo from desiccation and provide a source of nutrient. These polymers have linear-core poly(1–4)-β-D-mannan main chains with varying degrees of α-D-galactosyl substituents attached at the C-6 primary hydroxyl group (Figure 10.9a). A wide variety of galactomannans with different mannose/galactose ratios has been studied. Whereas galactomannan polysaccharides are water-soluble materials, the parent linear poly(1–4)-β-D-mannose (mannan I) is water insoluble.

Crystalline, oriented films of these materials can be obtained by evaporation. It has been observed that the best X-ray patterns are obtained when the relative humidity is in the middle range (40–80%). At 98% relative humidity, the amount of water begins to solubilize the chains, thus destroying the crystallinity. Since these early observations, a systematic investigation of unit-cell variations of several members of the series as a function of relative humidity has been conducted (Song *et al.*, 1989) (Table 10.1). Detailed three-dimensional descriptions have been reported from X-ray diffraction, in conjunction with reinvestigations of the structure of the parent mannan I using electron diffraction intensity data derived from single crystals of ivory nut mannan (Chanzy *et al.*, 1987) or X-ray fibre diffraction (Atkins *et al.*, 1988). All galactommannans have essentially the same structure, i.e. an antiparallel sheet stabilized by mannan–mannan hydrogen bonds.

It is apparent from Table 10.1 that two of the unit cell dimensions, i.e. *a* and *c*, remain essentially constant over the entire range of galactose substitution, i.e. from 0% in mannan I to 93% in fenugreek polysaccharides. Conservation of the 1.035 nm periodicity along *c* is explainable on the basis of an energetically preferred main-chain conformation. The constancy of the *a* parameter implies the existence of a regular lateral association, which is independent of the degree of galactose substitution in the galactomannans. A representation of a galactomannan sheet as it occurs in mannan I

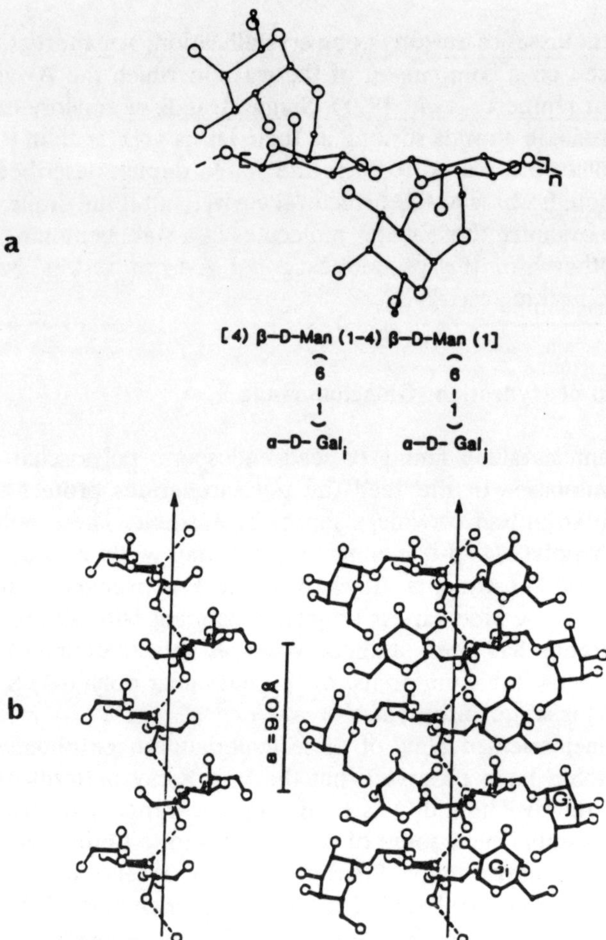

Figure 10.9 (a) Primary structure of the average crystallographic repeating unit in galactomannans such as fenugreek and lucerne gums. The labelling Gal$_i$ and Gal$_j$ denotes that successive galactosyl residues are crystallographically independent and may have different linkage conformations. (b) Comparison of a hydrogen-bonded sheet in mannan I and the same packing arrangement entirely populated with (1–6) linked galactose residues. The different conformations taken by galactosyl residues on either side of a given mannan main backbone are clearly apparent. From Song *et al.* (1989). Reprinted by permission of the American Chemical Society

and the same sheet fully substituted with one galactosyl unit at each mannosyl-6 position, is shown in Figure 10.9b. The decrease in *b* of about 0.3 nm, going from 71% to 0% relative humidity, suggests that a substantial amount of water is bound in the interlamellar space. Any gaps created by the absence of galactose residue at a given site are expected to be filled with water at high humidity. Removal of this water upon drying would induce disorder and subsequent loss of crystallinity. That would be more

Table 10.1 Galactomannan lattice constant

Ref.	*Source*	*r.h. (%)*	*Gal/Man*	a *(nm)*	b *(nm)*	c *(nm)*
1	Fenugreek	71	0.93	0.912	3.327	1.035
1	Fenugreek	0	0.93	0.894	2.950	1.027
1	Lucerne	50	0.92	0.898	3.332	1.033
1	Lucerne	0	0.92	0.899	3.109	1.031
2	Guar	81	0.64	0.913	3.283	1.035
3	Tara	88	0.39	0.891	3.062	1.040
2	Locust bean	88	0.32	0.904	2.061	1.024
4	Mannan I	0	0.00	8.92	7.21	10.27

References: 1, Song *et al.* (1989); 2, Marchessault *et al.* (1979); 3, Chien and Winter (1985); 4, Chanzy *et al.* (1987).

severe at intermediate galactose/mannose ratio. This would also account for the observation that highly substituted galactommanans and therefore more regular polysaccharides, such as those found in fenugreek and lucerne, show less loss of crystallinity upon drying.

Hydration as an Intermediate Form in Industrial Processing: Cellulose

Native cellulose (cellulose I) is found in a highly crystalline fibrillar form suitable for diffraction studies. Like chitin, cellulose is one of the polysaccharides which are not hydrated. Such an intrinsic insolubility is displayed in spite of the plethora of hydroxyl groups, and is readily explained by its three-dimensional crystal structure. The major three-dimensional characteristics of cellulose I include extended chains stabilized by two hydrogen bonds; there is a parallel arrangement of these chains organized in a sheet-like fashion (Gardner and Blackwell, 1974; Sarko and Muggli, 1974; Woodcock and Sarko, 1980). Native cellulose may be converted into the more stable cellulose II polymorph by two methods: regeneration and mercerization. Mercerization refers to the conversion accomplished by swelling native cellulose fibres into concentrated sodium hydroxide solution. Although no dissolution occurs, the swelling allows for reorganization of the chains, and cellulose II results when the swelling agent is removed. Mercerization results in cellulose II, which fibre diffraction pattern is shown to be based on an antiparallel-chain structure (Stipanovic and Sarko, 1976; Kolpak and Blackwell, 1976). The transformation pathways between cellulose and Na-cellulose crystal structures (Okano and Sarko, 1984) are shown in Figure 10.10 (a).

The three-dimensional structures of Na-cellulose I (Nishimura *et al.*, 1991), Na-cellulose IV (Nishimura and Sarko, 1991) and Na-cellulose IIB (Okano and Sarko, 1985) have been proposed. All of the experimental evidence suggests that the conversion step from cellulose I to Na-cellulose I

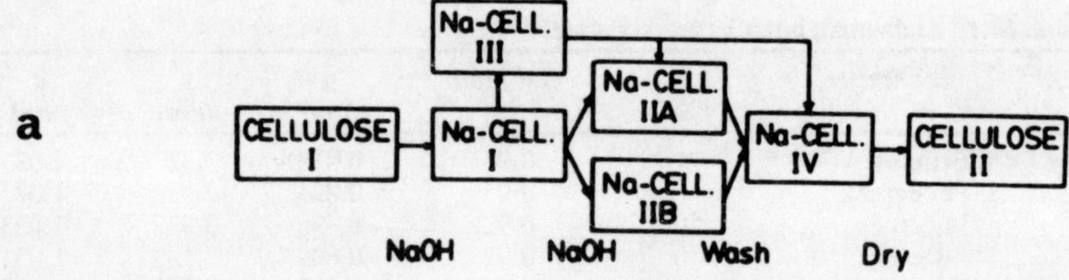

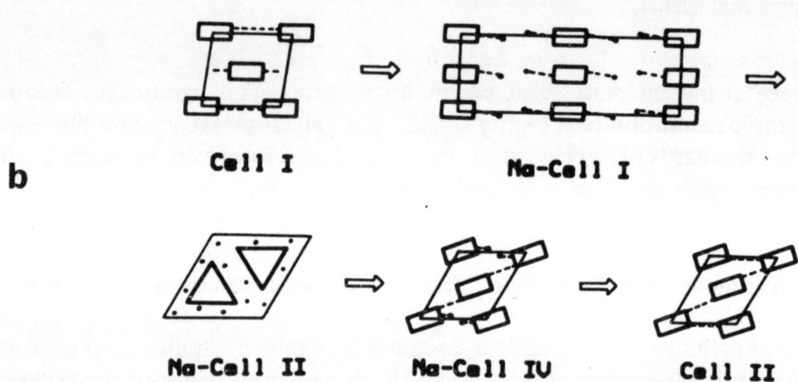

Figure 10.10 (a) Different steps from cell I to cell II; (b) similar structural characteristics of celluloses I and II and Na-celluloses I, II, and IV. Directions of hydrogen bonding are indicated by dashed lines. From Nishimura and Sarko (1991). Reprinted by permission of the American Chemical Society

is the one in which a transformation of the parallel chain polarity to the antiparallel chain polarity occurs. The structure crystallizes in a four-chain unit cell with eight Na^+ ions and an undetermined amount of water. The antiparallel arrangement of the chains has been proposed on the basis of more stable stereochemistry and better symmetry. In the Na-cellulose II structure, the polysaccharide chain departs from its 2_1 symmetry to adopt a threefold helix; the unit cell contains more than 60% of NaOH and water. After all of the alkali has been washed from the Na-cellulose II complex, but prior to its drying, Na-cellulose IV is obtained. The crystal structure of this intermediate is based on a two-chain, monoclinic unit cell which resembles that of cellulose II. The differences arise from the presence of two water molecules in the unit cell; both water molecules take part in four hydrogen bonds each. The removal of these water molecules induces a conformational change of some primary hydroxyl groups and subsequent reformation of the sheet-like structure of the cellulose II type (Figure 10.10b).

5 Difficulties and Limits of Water Location in Crystalline Polysaccharides

Aperiodicity of Water Molecule: Chondroitin 4-sulphate

The mammalian glycosaminoglycans (GAGs) constitute a complex and heterogeneous family of linear sulphated polysaccharides. These proteoglycans play a variety of roles in tissue function. They are extended structures occupying a large hydrodynamic volume, exhibit high viscosity and are reversibly compressible in solution. They provide the cushioning properties of cartilage on bone surfaces. The GAGs are composed of disaccharide repeating units consisting of an amino sugar and a uronic acid or an amino sugar and galactose. Representatives of the GAGs are chondroitin-4-sulphate, chondroitin-6-sulphate, dermatan sulphate, heparan sulphate, heparin, hyaluronate and keratan sulphate. A number of GAG structures have been determined and their features have been reviewed (Arnott and Mitra, 1984).

Although the organization in a fibre specimen is usually artificial, its details can help one to understand the ordered state that can occur in solutions and gels. We have chosen chondroitin 4-sulphate to illustrate what insight can be gained about the interactions between polysaccharides with water molecules. The three-dimensional structure of potassium chondroitin-4-sulphate has been determined through X-ray fibre diffraction (Millane *et al.*, 1983). The polysaccharide chains have a left-handed threefold helical secondary structure stabilized by intra- and intermolecular hydrogen bonds. Two antiparallel chains pass through each trigonal unit cell ($a = b = 1.385$ nm; $c = 2.776$ nm) with space group symmetry $P322_1$. The cations and water molecules in the crystals are not all periodic, and only one K^+ and four water molecules per disaccharide were located by difference Fourier methods. This particular work illustrates how it is difficult to determine helix chirality with certainty in heavily hydrated structures. The presence of large amounts of solvent which are linked to the polyanion in an initially unknown way can mask the handedness of the polyanion when there is a limited amount of diffraction data available. A detailed description of the arrangement of the water molecules with the polysaccharide chain is shown in Figure 10.11. Three of these provide intramolecular stabilization and the other one stabilizes adjacent chains.

Monte Carlo Study of the Possible Hydration Site: Agarose

Among the gel-forming polysaccharides, carrageenans and agarose have been extensively studied. All carrageenans are linear polysaccharides of alternating (1–3)- and (1–4)-linked galactose units. The two principal gell-

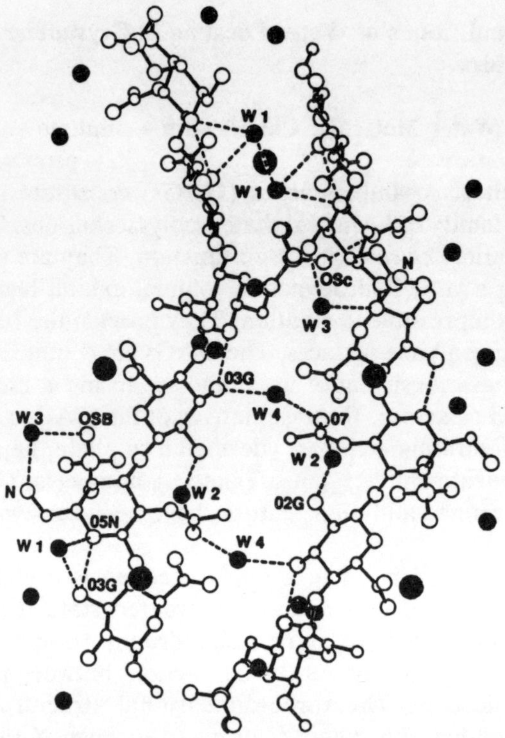

Figure 10.11 Hydration sites of 4-chrondroitin sulphate. From Millane *et al.* (1983). Reprinted by permission of Academic Press

ing fractions, kappa- and iota-carrageenan, contain β-D-galactose and 3,6-anhydro-α-D-galactose units. Galactose units carry a half-ester sulphate group at O4; the difference between them being that 3,6-anhydro-galactose units are sulphated at O2 in iota- but not in kappa-carrageenan. Agarose differs from the carrageenans only by being totally unsulphated and having the anhydrogalactose units replaced by its L-enantiomer.

The structures of iota- and kappa-carrageenan have been determined by X-ray diffraction (Arnott *et al.*, 1974b; Millane *et al.*, 1988). These results indicate that double rather than single helices are formed in the condensed state. In the case of iota-carrageenan, the double-helix structure contains two right-handed parallel chains of pitch 2.66 nm that are translated by exactly half the pitch. It displays a compact carbohydrate core, with sulphate protruding in pairs away from the centre of the helix. By examining the X-ray diffraction patterns from the Ca^{2+} salt of this polymer, the calcium salts were shown to bind to the sulphate groups of adjacent double-helices. Owing to the statistical nature of the crystal structure, details of the locations of water molecules could not be determined. As for kappa-carrageenan, the structure is not a half-staggered double helix.

By analogy with iota-carrageenan, a double-helical structure has been proposed for agarose (Arnott *et al.*, 1974a). It is made up of two parallel left-handed strands with threefold symmetry, a pitch of 1.9 nm and an internal cavity of 0.45 nm. Alternatively, structural models based on the occurrence of an extended single helix have been proposed (Foord and Atkins, 1989; Jimenez-Barbero *et al.*, 1989). A notable feature of the agarose double helix is an interior cavity that extends along the helix axis, and the subsequent suggestion that it is occupied by water molecules that would contribute to the stability of the two strands. The structure of the water around the agarose double helix has been analysed by *ab initio* quantum mechanical calculations and Monte Carlo simulation (Corongiu *et al.*, 1983). The agarose hydration sites have been determined. Under the conditions selected in the Monte Carlo simulation, i.e. a rigid double-helix structure, no hydration site was found inside the double-helix cavity. Connectivity pathways of hydrogen-bonded water molecules around the polysaccharide were found, which resemble those observed in DNA (Clementi and Corongiu, 1981).

6 Discussion and Conclusion

Many polysaccharides have been observed to crystallize in a hydrated form with water molecules either in crystallographic positions or as statistically random components of the crystal structure. All such water molecules contribute to the hydrogen network in the crystalline polysaccharides and are thus important components of the structure. In some cases, hydrated forms are found in the native state. In other cases, hydration has been found to occur in intermediate stages of an industrial process.

As observed with low-molecular carbohydrate molecules, there are examples, such as those provided by cellulose and chitin, where the crystalline native arrangements are devoid of water molecules. It is clear that whenever the macromolecules are relatively simple in shape and pack very efficiently, they do not tend to form hydrates. As the shape of the macromolecules increases in complexity, so does the hydration number. In such particular instances, the water molecules are small enough to be able to stabilize crystal structures by filling voids between polysaccharide chains for more efficient packing. Obviously, they also contribute to the lattice energy by additional hydrogen bonds. The hydrogen-bonding schemes included in the models of the crystal structure of the polysaccharide may still be regarded as speculative. In order to elucidate the type of hydrogen-bonding scheme such as two- and three-centre bonds (Jeffrey and Saenger, 1991), further theoretical calculations should be carried out, starting from the structural models proposed for polysaccharides.

Usually hydration of polysaccharide chains does not cause significant alterations in the conformation of the chains: it simply separates them. Different types of structure and subsequent type of hydration have been observed: these are column or sheet types. They are intimately related to the symmetry of the polysaccharide chains. Obviously, cylinder-type structures, such as those found in multiple helices, induce a columnar type of hydration, and conversely for sheet-type structures. It is also apparent that columnar hydration is more reversible and certainly less morphologically damaging.

Despite the lack of accurate characterization of some of the features displayed between polysaccharides and water, the elucidation of complex structures has evidenced the role played by water molecules during structural transitions. Understanding such transitions at the molecular level may help rationalize complex phenomena such as seed germination, which requires hydration of the polysaccharide in the endosperm. As already pointed out by Bluhm *et al.* (1980), galactomannans could be tailored to adapt to the environmental requirements of plants located in tropical areas where moisture is seasonal. A somewhat more gradual and reversible hydration scheme occurs in starch, a situation more in keeping with temperate climate. In a completely different register, the participating role of water molecules in conjunction with alkali illustrates how cellulose chains can acquire considerable mobility during the solid state transformations.

It has to be recognized that, in some respects, the organization in a fibre specimen is quite artificial. Nevertheless, the structural elucidation of such arrangements helps one to understand the locally ordered state that can occur in solutions and gels. The structure of water plays a critical role in the interaction and stability of polysaccharides in solution. It may be affected by agents that alter the extent of hydrogen-bonded interactions between the components.

The following are some of the future avenues concerned with the water–polysaccharide interactions in the post-crystallography era. It is obvious that further understanding of the role of water in molecular interactions and recognitions remains a challenge. Direct spectroscopic measurements yield average properties over accessible time-scales which are frequently orders of magnitude higher than the correlation times of water molecules. In order to build reasonable kinetic models or to partition water molecules into characteristic subpopulations, one needs to identify specific behaviour capable of describing the fluctuations and propagations of molecular arrangements of water upon the direct influence of conformational changes. Another drastic limitation is due to the present state of molecular modelling. Typically, the time-scale is about nanoseconds for one solute molecule in a periodic molecular dynamics box of 1014 water molecules and solute–solute interactions are hardly considered. Also, the point charges atomic descriptions of systems may be a somewhat limiting feature and polarizable effects are generally ignored for molecular solutes. Moreover, the importance of 'rare events' such as nucleation processes interconnected with density fluctuations is difficult to estimate.

The accumulation of reliable structural data on water–polysaccharide interactions in the condensed phase has provided some major insights into the role played by interactions of this type where hydrogen bonding seems to be a dominating feature. Nevertheless, the features which have been described so far may only partly illustrate the fascinating world of polysaccharide–water interactions in their functional associations.

Acknowledgements

The author would like to express appreciation to Dr Anne Imberty for helpful discussions and careful reading of the manuscript. Appreciation is extended to Mrs Chantal Nicolas for help with the art work.

References

Arnott, S. (1980). 30 years hard labor as a fiber diffractionist. In French, A. D. and Gardner, K. H. (Eds), *Fiber-Diffraction Methods*. American Chemical Society, Washington, D.C., pp. 1–30

Arnott, S., Fulmer, A., Scott, W. E., Dea, I. C. M., Moorhouse, R. and Rees, D. A. (1974a). The agarose double-helix and its function in agarose gel structure. *J. Mol. Biol.*, **90**, 269–284

Arnott, S. and Mitra, A. K. (1984). In Arnott, S., Rees, D. A. and Morris, E. R. (Eds), *Molecular Biophysics of the Extracellular Matrix*. Humana Press, Clifton, N.J., pp. 41–67

Arnott, S., Scott, W. E., Rees, D. A. and McNab, G. C. A. (1974b). *i*-Carrageenan. Molecular structure and packing of polysaccharide double helices in oriented fibers of divalent cation salts. *J. Mol. Biol.*, **90**, 253–267

Atkins, E. D. T., Farnell, S., Mackie W. and Scheldrick, B. (1988). Crystalline structure and packing of mannan I. *Biopolymers*, **27**, 1097–1105

Bluhm, T., Deslandes, Y., Marchessault, R. H. and Sundarajan, P. R. (1980). New insights into the crystal structure hydration of polysaccharides. In *Water in Polymers*. ACS Symposium Series, pp. 253–272

Brisse, F. (1991). Electron diffraction of polymer single crystal. In Fryer, J. R. and Dorset, D. C. (Eds), *Electron Crystallography of Organic Molecules*. Kluwer, Dordrecht, pp. 63–75

Chandrasekaran, R., Millane, R. P., Arnott, S. and Atkins, E. D. T. (1988a). The crystal structure of gellan. *Carbohydr. Res.*, **175**, 1–15

Chandrasekaran, R., Puigjaner, L. C., Joyce, K. L. and Arnott, S. (1988b). Cation interactions in gellan: An X-ray study of the potassium salt. *Carbohydr. Res.*, **181**, 23–40

Chandrasekaran, R. and Tailambal, V. G. (1990). A new generation of gel-forming polysaccharides: an x-ray study. In French, A. D. and Brady, J. W. (Eds), *Computer Modelling of Carbohydrate Molecules*. American Chemical Society, Washington, D.C., pp. 300–314

Chanzy, H., Excoffier, G. and Guizard, C. (1981). Single crystals of dextran: High temperature polymorph. *Carbohydr. Polym.*, **1**, 67–77

Chanzy, H., Guizard, C. and Sarko, A. (1980). Single crystals of dextran: Low temperature polymorph. *Int. J. Biol. Macromol.*, **2**, 149–153

Chanzy, H. D., Pérez, S., Miller, D. P., Paradossi, G. and Winter, W. T. (1987). An electron diffraction study of the mannan I crystal and molecular structure. *Macromolecules*, **20**, 2407–2413

Chanzy, H., Roche, E. and Vuong, R. (1971). Electron diffraction of cellulose triacetate single crystals. *Kolloid Z.Z. Polym.*, **198**, 1034–1035

Chanzy, H. and Vuong, R. (1985). Ultrastructure and morphology of crystalline polysaccharide. In Atkins, E. D. T. (Ed.), *Polysaccharides—Topics in Structure and Morphology*. Macmillan Press, London, pp. 41–71

Chien, Y. Y. and Winter, W. T. (1985). Accurate lattice constants for tara gum. *Macromolecules*, **18**, 1357–1359

Clementi, E. and Corongiu, G. (1981). In Sarma, R. H. (Ed.), *Biomolecular Stereodynamics*. Adenine Press, Schenectady, N. Y., pp. 209–259

Corongiu, G., Fornili, S. L. and Clementi, E. (1983). Hydration of agarose double helix: A Monte Carlo simulation. *Int. J. Quantum Chem.*, **10**, 277–291

Foord, S. A. and Atkins, E. D. T. (1989). New X-ray diffraction results from agarose: Extended single helix structures and implications for gelation mechanism. *Biopolymers*, **28**, 1345–1365

Gardner, K. H. and Blackwell, J. (1974). Structure of native cellulose. *Biopolymers*, **13**, 1975–2001

Gidley, M. (1987). Factors affecting the crystalline type (A–C) of native starches and model compounds: A rationalization of observed effects in terms of polymorphic structures. *Carbohydr. Res.*, **161**, 301–304

Guizard, C. (1980). PhD Dissertation Thesis, University of Grenoble, France

Guizard, C., Chanzy, H. and Sarko, A. (1984). Molecular and crystal structure of dextrans: A combined electron and X-ray diffraction study. I. The anhydrous high-temperature polymorph. *Macromolecules*, **17**, 100–107

Guizard, C., Chanzy, H. and Sarko, A. (1985). Molecular and crystal structure of dextrans: A combined electron and X-ray diffraction study. II. A low temperature, hydrated polymorph. *J. Mol. Biol.*, **183**, 397–408

Hui, S. W. and Parsons, D. F. (1974). Electron diffraction of wet biological membranes. *Science*, **184**, 77–78

Imberty, A., Buléon, A., Tran, V. and Pérez, S. (1991). Recent advances in knowledge of starch structure. *Starch/Stärke*, **43**, 375–384

Imberty, A., Chanzy, H., Pérez, S., Buléon, A. and Tran, V. (1988). The double helical structure of A-starch. *J. Mol. Biol.*, **201**, 365–378

Imberty, A. and Pérez, S. (1988). A revisit to the three-dimensional structure of B-amylose. *Biopolymers*, **27**, 1205–1221

Jeffrey, G. A. and Saenger, W. (1991). *Hydrogen Bonding in Biological Structures*. Springer-Verlag, Berlin, Heidelberg

Jimenez-Barbero, J., Bouffar-Roupe, C., Rochas, C. and Pérez, S. (1989). Modeling studies of solvent effects on the conformational stability of small relatives of agarose. *Int. J. Biol. Macromol.*, **11**, 265–272

Kolpak, F. J. and Blackwell, J. (1976). Determination of the structure of cellulose II. *Macromolecules*, **9**, 273–278

Lechert, H. T. (1981). Water binding in starch: NMR studies on native and gelatinized starch. In Rockland, L. B. and Stewart, G. F. (Eds), *Water Activity: Influences on Food Quality*. Academic Press, London, pp. 223–245

Marchessault, R. H. (1984). Carbohydrate polymers: Nature's high performance materials. In Vadenberg, E. J. (Ed.), *Contemporary Topics in Polymer Sciences*, Vol. 5. Plenum Publishing Corporation, New York, pp. 15–53

Marchessault, R. H., Buléon, A., Deslandes, Y. and Goto, Y. (1979). Comparison of x-ray diffraction data of galactomannans. *Colloid Interface Sci.*, **71**, 375–382

Matricardi, U. R., Moretz, R. C. and Parsons, D. F. (1972). Electron diffraction of wet proteins. Catalase. *Science*, **177**, 278–280

Millane, R. P., Chandrasekaran, R., Arnott, S. and Dea, I. C. M. (1988). The molecular structure of kappa-carrageenan and comparison with iota-carrageenan. *Carbohydr. Res.*, **182**, 1–17

Millane, R. P., Mitra, A. K. and Arnott, S. (1983). Chondroitin 4-sulfate: Comparison of the structures of the K^+ and Na^+ salts. *J. Mol. Biol.*, **169**, 903–920

Nishimura, H., Okano, T. and Sarko, A. (1991). Mercerization of cellulose. 5. Crystal and molecular structure of Na-cellulose I. *Macromolecules*, **24**, 759–770

Nishimura, H. and Sarko, A. (1991). Mercerization of cellulose. 6. Crystal and molecular structure of Na-cellulose IV. *Macromolecules*, **24**, 771–778

Okano, T. and Sarko, A. (1984). Mercerization of cellulose. I. X-ray diffraction evidence for intermediate structures. *J. Appl. Polymer Sci.*, **29**, 4175–4182

Okano, T. and Sarko, A. (1985). Mercerization of cellulose. II. Alkali-cellulose intermediates and a possible mercerization mechanism. *J. Appl. Polymer Sci.*, **30**, 325–332

Pérez, S. (1991). Molecular modelling and electron diffraction of polysaccharides. In Langone, J. (Ed.), *Methods in Enzymology: Molecular Design and Modeling: Concepts and Application*. Academic Press, San Diego

Pérez, S. and Chanzy, H. (1989). Electron crystallography of linear polysaccharides. *J. Electron Microsc. Techn.*, **11**, 280–285

Pérez, S., Imberty, A. and Scaringe, R. P. (1990). Modeling of the interactions of polysaccharide chains: Application to the crystalline polymorphism of starch granule. In French, A. D. and Brady, J. W. (Eds), *Computer Modeling for Carbohydrate Molecules*. ACS Symposium Series, 430, American Chemical Society, Washington, D.C., pp. 281–299

Sarko, A. and Muggli, R. (1974). Packing analysis of carbohydrates and polysaccharides. III. *Valonia* cellulose and cellulose II. *Macromolecules*, **7**, 486–494

Sidebotham, R. L. (1974). Dextrans. In Tipson, R. S. and Horton, D. (Eds), *Advances in Carbohydrate Chemistry and Biochemistry*, Vol. 30. Academic Press, San Francisco, pp. 371–444

Smith, P. J. C. and Arnott, S. (1978). LALS: A linked-atom least-squares reciprocal space refinement system incorporating stereochemical restraints to supplement sparse diffraction data. *Acta Cryst.*, **A34**, 3–11

Song, B. K., Winter, W. T. and Taravel, F. R. (1989). Crystallography of highly substituted galactomannans: Fenugreek and lucerne gums. *Macromolecules*, **22**, 2641–2644

Stipanovic, A. J. and Sarko, A. (1976). Packing analysis of carbohydrates and polysaccharides. 6. Molecular and crystal structure of regenerated cellulose II. *Macromolecules*, **9**, 851–857

Taylor, K. A. and Glaeser, R. M. (1974). Electron diffraction of frozen, hydrated protein crystals. *Science*, **186**, 1036–1037

Taylor, K. J., Chanzy, H. and Marchessault, R. H. (1975). Electron diffraction for hydrated crystalline biopolymers: Nigeran. *J. Mol. Biol.*, **92**, 165–167

Tran, V. and Buléon, A. (1987). Diffraction peak shapes: A profile refinement method for badly resolved powder diagrams. *J. Appl. Crystallogr.*, **20**, 430–436

Whistler, R. L. and Smart, C. L. (1953). *Polysaccharide Chemistry*. Academic Press, New York

Williams, D. E. (1969). A method of calculating molecular crystal structures. *Acta Cryst.*, **A25**, 464–470

Woodcock, C. and Sarko, A. (1980). Packing analysis of carbohydrates and polysaccharides. 11. Molecular and crystal structure of ramie cellulose. *Macromolecules*, **13**, 1183–1187

Wu, H. C. H. and Sarko, A. (1978a). The double-helical molecular structure of crystalline B-amylose. *Carbohydr. Res.*, **61**, 7–25

Wu, H. C. H. and Sarko, A. (1978b). The double-helical molecular structure of crystalline A-amylose. *Carbohydr. Res.*, **61**, 27–40

Yalpani, M. (1988). *Polysaccharides: Syntheses, Modifications and Structure/Property Relations*. Elsevier, Amsterdam

Yui, T., Ogawa, K. and Sarko, A. (1992). Molecular and crystal structure of the regenerated form of (1–3)-α-D-mannan. *Carbohydr. Res.*, **229**, 41–55

Zugenmaier, P. and Sarko, A. (1980). The variable virtual bond modelling techniques for solving polymer crystal structures. In Gardner, K. and French, A. D. (Eds), *Fiber Diffraction Methods*. ACS Symposium Series, 141, American Chemical Society, Washington, D.C., pp. 225–237

11
The Role of Structural Water Molecules in Protein–Saccharide Complexes

Yves Bourne and Christian Cambillau

1 Introduction

Polysaccharides have very diverse biological roles. The best-known among these roles are the less noble, such as those concerning their structural functions in bacteria and plant cell walls, and their storage functions in glycogen and starch. Nevertheless, structure and flexibility of saccharides are recognized as having a deep and specific influence in several biological processes: protein protection, maintenance of conformational integrity, and recognition events involving viruses, enzymes and lectins (Lis and Sharon, 1990; Walker, 1988; Diaz *et al.*, 1989). According to Nathan Sharon (Sharon and Lis, 1986), 'the specificity of many natural polymers is written in terms of sugar residues, not of amino acids or nucleotides'. Since carbohydrates are very hydrophilic molecules, the question addressed in this chapter is whether discrete water molecules, as seen in crystallographic structures of protein–carbohydrate complexes, play a significant role in these interactions. The presence of structural water molecules in proteins and nucleic acids is well documented (Malin *et al.*, 1991; Westhof, 1988). Since a well-refined high-resolution structure contains hundreds of water molecules, the Protein Data Bank (Bernstein *et al.*, 1977) provides a very large structural database of protein-bound water molecules. In nucleic acids, Westhof (1988) has shown that water molecules are an integral part of the structure. The number of high-resolution well-refined crystallographic structures containing at the same time protein, saccharides and water molecules is limited. These structures are good candidates for studying the conformation of the saccharide itself, and the rules dominating its interaction with the protein, although the roles and functions of the proteins which bind saccharides are very diverse. In this chapter, we examine the

Table 11.1 List of the X-ray structures presented in this review

Protein–saccharide complex	*Resolution*	R-*factor*	*Ref.*
Arabinose binding protein/arabinose	1.7 Å	0.139	Quiocho and Vyas (1984)
Arabinose binding protein/galactose	1.9 Å	0.134	Quiocho *et al.* (1989)
Galactose binding protein/glucose	1.8 Å	0.132	Vyas *et al.* (1988)
Maltose binding protein/maltose	2.3 Å	0.25	Spurlino *et al.* (1991)
Lysozyme/tri-NAG	1.75Å	0.23	Johnson *et al.* (1988)
Phosphorylase b/heptenitol			
Phosphorylase b/heptulose-2-P			
Phosphorylase b/maltoheptaose	2.5 Å	0.146	Johnson *et al.* (1983)
Wheat germ agglutinin isolectins 1 and 2		0.172	
N-Acetylneuramyl lactose	2.2 Å	0.153	Wright (1990)
Con A/α-methyl mannoside	2.8 Å	0.19	Derewenda *et al.* (1989)
Lathyrus ochrus isolectin I (LOL I)/			
α-Methyl mannoside	2.0 Å	0.182	Bourne *et al.* (1990b)
LOL I/α-methyl glucoside	2.2 Å	0.179	Bourne *et al.* (1990b)
LOL I/α-Man (1–3) β-Man (1–4)			
GlcNac	2.1 Å	0.185	Bourne *et al.* (1990c)
LOL I/octasaccharide	2.3 Å	0.19	Bourne *et al.* (1992)

role of water molecules in structures of protein–saccharide complexes, determined at high resolution and well-refined (Table 11.1): periplasmic binding proteins, enzymes such as lysozyme and phosphorylase b, and lectins.

2 Periplasmic Binding Proteins: Complexes with Monosaccharides and Disaccharides

Periplasmic sugar binding proteins (PSBP) are located in the periplasmic space of Gram-negative bacteria. They serve as primary receptors of the osmotic shock-sensitive uptake systems for various sugars (Furlong, 1987). Several structures of PSBP have been solved at high resolution by the group of F. Quiocho. The protein–saccharide interactions have been thoroughly described by Quiocho (for reviews, see Quiocho, 1986, 1988, 1989; Vyas *et al.*, 1991), who first established the basic rules of protein–saccharide recognition. Among these rules, hydrogen bonds are mentioned as the main factors in conferring specificity and affinity. All the polar groups of the saccharides are involved in such bonds. Numerous van der Waals contacts are also formed, including one or two aromatic residues stacked on the sugar ring. We shall focus now on the few water molecules involved in the saccharide binding.

In the arabinose binding protein (ABP) complexes with arabinose (Quiocho and Vyas, 1984) and galactose (Quiocho *et al.*, 1989), the sugar molecules are well buried (solvent accessibility of 2%). Two buried water molecules (W309 and W310) are conserved for both structures in the first

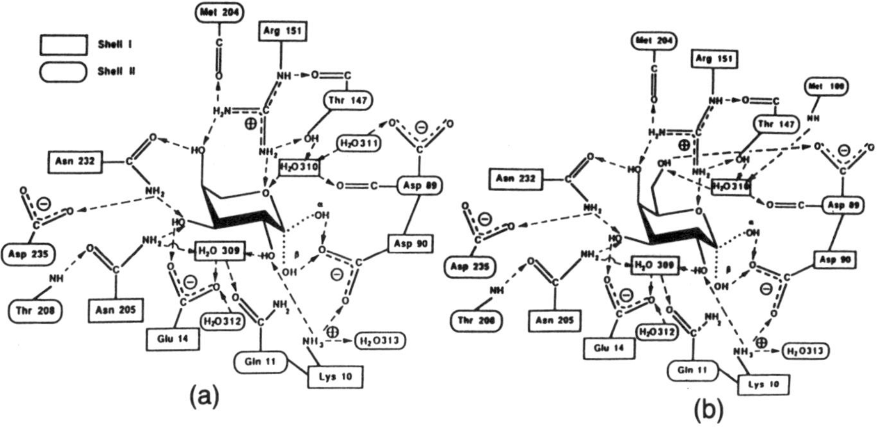

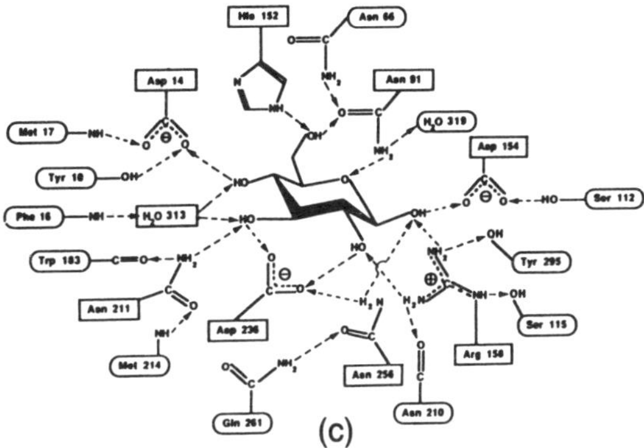

Figure 11.1 Sugar–protein interactions — schematic representation in three different complexes: (a) ABP–arabinose; (b) ABP–galactose; (c) GBP–glucose. From Quiocho (1989) (a) and (c), and Vyas *et al.* (1991) (b)

solvation shell (Figure 11.1a). There is evidence that these molecules, forming hydrogen bonds with sugar atoms O2 and O5, bring the sugar and the two domains of ABP together in the cleft. In the arabinose complex, water molecule W311 replaces the galactose atom O6 in binding Asp 89 (Figure 11.1b). These water molecules have no contact with the bulk solvent.

The galactose binding protein (GBP) complex with D-glucose (Vyas *et al.*, 1988) contains one first-shell water molecule (W313) (Figure 11.1c). This molecule forms bidentate hydrogen bonds with glucose atoms O3 and O4 (on the one hand), and is hydrogen bonded to the amide of Phe 16 (on the other hand).

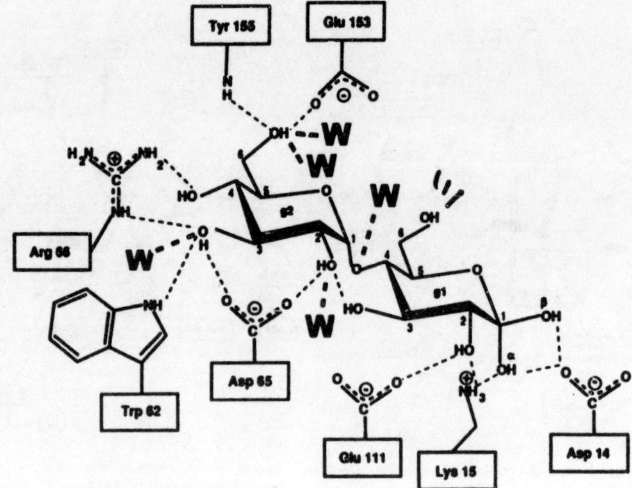

Figure 11.2 Sugar–protein interactions — schematic representation in the MBP–maltose complex. Adapted from Spurlino *et al.* (1991)

The maltose binding protein (MBP) maltose complex was solved to a slightly lower resolution and is not yet fully refined, especially in regard to water molecules (Spurlino *et al.*, 1991). The water accessibility of the maltose molecule is 3.6%, which is slightly higher than in previous cases. There are five water molecules bound to the maltose. In glucose 1, two water molecules are bound to O6, one to O2 and one to O3; in glucose 2, one water molecule is bound to O4 (Figure 11.2). All these water molecules are also hydrogen bonded to protein residues.

3 Enzymes

Lysozyme

Lysozyme catalyses the hydrolysis of the β(1–4) linkage of bacterial cell wall polysaccharides. Many structures of complexes between lysozyme and saccharides (up to a tetrasaccharide) have been determined (Johnson *et al.*, 1988, and references therein), most of these at low resolution. However, even the best-characterized structure, lysozyme complexed with tri-NAG (Blake *et al.*, 1967; Johnson *et al.*, 1988) at 1.75 Å resolution, could be refined to an *R*-factor of only 0.23. Therefore, there is little information about the bound water molecules in this structure. As opposed to the structures of ABP, GBP and MBP, sugar side-chains are accessible to solvent. Three water molecules are bound to the reducing end O1 of NAG 3 (141 OW7 and 8 and 142 OW0) (Figure 11.3). They form part of a

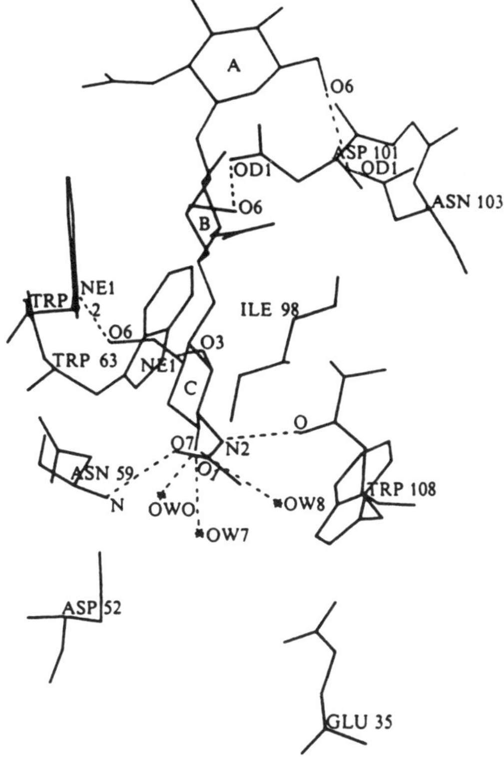

Figure 11.3 $(GlcNac)_3$ bound in the active site cleft of hen egg-white lysozyme. Hydrogen contacts are indicated by dashed lines (cutoff 3.3 Å). From Johnson *et al.* (1988)

network of hydrogen bonds in the middle region of the active site. One of them (OW8) is part of a patch of water molecules found in the native structure.

Glycogen Phosphorylase

Glycogen phosphorylase (Cohen, 1983) catalyses the reversible phosphorylation of α(1–4) linkages of glycogen, leading to the formation of glucose-1-P. This reaction occurs at the phosphorylase catalytic site, distinct from a separate glycogen binding site, called the storage site (Metzger *et al.*, 1967; Weber *et al.*, 1978; Kasvinsky *et al.*, 1978). Regarding the phosphorylase b –saccharide interaction, the most informative studies are those in which heptenitol or heptulose 2-phosphate is bound to the catalytic site (McLaughlin *et al.*, 1984; Hadju *et al.*, 1987), and that in which maltoheptaose is bound to the storage site (Johnson *et al.*, 1983, 1988). As in the case of ABP and GBP, monosaccharides bound to phosphorylase b

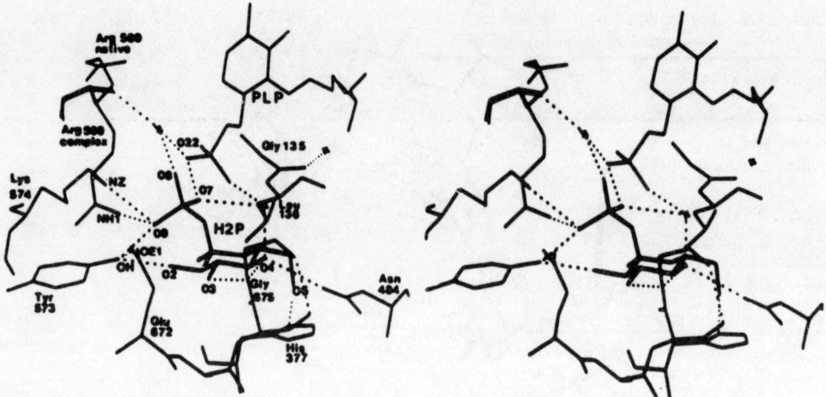

Figure 11.4 Stereo view of heptulose 2-P bound in phosphorylase b catalytic site. From Johnson *et al.* (1988)

active site are buried. The hydrogen bonding potential of every polar group is satisfied. Little or no access is given to the bulk solvent. One water molecule is linked to heptenitol atom O4, and two water molecules link the highly polar phosphorus atoms of heptulose-2-P to the protein (Figure 11.4).

The glycogen storage site is situated on the surface of the phosphorylase. It is composed of two subsites, the major site and the minor site, which are almost contiguous. In the maltoheptaose complex, the major site binds five sugar residues, and the minor site two. Maltoheptaose is water-accessible: 10 water molecules interact with the saccharide (Table 11.2). All these water molecules hydrogen-bond to the saccharide and to protein residues. The minor site requires special attention: among 11 saccharide–protein interactions, only 1 is established directly with the protein, 9 being mediated via water molecules; 1 water molecule binds internally sugar atoms S8 O4 and S9 O5 (Table 11.2; Figure 11.5). A similar result has been found in a *Lathyrus ochrus* lectin–trisaccharide complex, described in Section 4 (Bourne *et al.*, 1990b).

4 Lectins

Lectins are carbohydrate-binding proteins that have been used widely in studying the structure and function of cell surface carbohydrates. Interest in plant lectins was raised because they recognize certain carbohydrates at the surface of malignant cells (Walker, 1988). In the plant kingdom, they are a determining factor in the symbiosis of Rhizobia with legumes (Diaz *et al.*, 1989). Thus far, studies on the structure of lectin–sugar complexes have involved ConA with mannose (Derewenda *et al.*, 1989), favin with

Table 11.2 Hydrogen bonds to saccharide at the glycogen storage site of phosphorylase b. (Adapted from Johnson *et al.*, (1988)

Sugar atom	*Water number*	*Protein or sugar atom*
Major site (water contacts only)		
S5 O5	W901.2	
S6 O6	W908.6	Gln 401 O
S7 O6		
Minor site (all contacts)		
S8 O3		Arg 398 NH1, NH2
O4	W908.8	S9 O5
O6	W861.9	Arg 409 NH2
	W908.7	Arg 409 NH1
S9 O2	W848.8	Ala 213 N
	W848.9	His 208 ND1
O3	W848.8	Ala 213 N
	W849.0	Gln 211 O
	W890.6	Leu 359 O
O5	W908.8	S8 O4
O6	W861.9	Arg 409 NH2

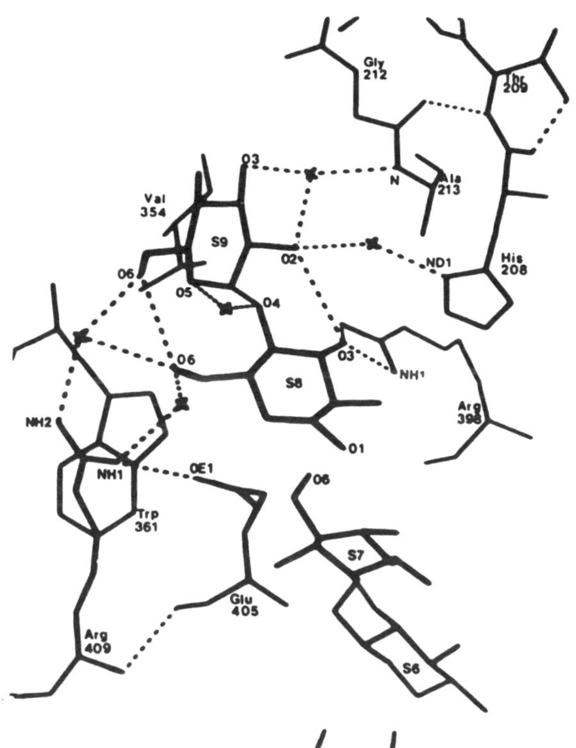

Figure 11.5 Interactions between maltoheptaose and phosphorylase b minor storage site. Water molecules are shown as crosses. From Johnson *et al.* (1988)

glucose (Reeke and Becker, 1986), pea lectin with a mannose trisaccharide (Rini *et al.*, 1986), WGA isolectin with a *N*-acetylneuraminyl-lactose (Wright, 1990; Reeke and Becker, 1988) and *Griffonia simplicifolia* lectin with a blood determinant tetrasaccharide (Delbaere *et al.*, 1990). We shall now discuss some of these structures in regard to their water saccharide and protein interactions.

Wheat Germ Agglutinin

Wheat germ agglutinin (WGA) is a lectin of the Graminae family, which possesses properties distinct from other plant lectins (Leguminosae, etc.). For example, they are specific for two different sugars: *N*-acetyl-D-glucoseamine (GlcNAc) and *N*-acetylneuraminic acid (NeuNAc) (Peters *et al.*, 1979; Monsigny *et al.*, 1980). The crystal structures of complexes of isolectins 1 and 2 of WGA with *N*-acetylneuraminyl lactose have been described recently (Wright, 1990). The NeuNAc residue performs numerous hydrogen bonds with the lectin residues in site 1, whereas the two other residues (Gal and Glc) are much more exposed to the solvent. One water molecule is linked to Tyr 73 in the unoccupied combining site. It repositions slightly upon sugar binding, and exhibits a lower *B*-factor. This water molecule (for example, W180) in one of the sites is linked tetrahedrically between NeuNAc atom O4 and lectin residues Ser 114 (NH), Ser 43 (OH) and Tyr 73 (OH). It further binds to Ser 43 (NH) via water molecule 201 (Figures 11.6, 11.7). Another water molecule (W255) establishes a weak hydrogen bond with NeuNAc atom O8 and lectin residue Glu 115 (OE2). The last one (W277) hydrogen links the saccharide internally, between Glc atom O6 and Gal atoms O1 and O2 (Figures 11.6, 11.7).

Lathyrus ochrus Lectin

Isolectin I (LOL I), isolated from the seeds of *Lathyrus ochrus*, consists of two identical subunits each composed of a light chain (α) with 52 amino-acids, and a heavy chain (β) with 181 residues (Rougé and Sousa-Cavada, 1984). The lectin binds specifically mannose and glucose. Our studies allowed us to describe precisely the monosaccharide-binding site of LOL I (Bourne *et al.*, 1990c) and to emphasize the role of water in binding a trisaccharide to LOL I (Bourne *et al.*, 1990b). More recently, the structure description of a complex between lectin and a biantennary octasaccharide was a big step forward in the understanding of lectin–glycoprotein interactions (Bourne *et al.*, 1991).

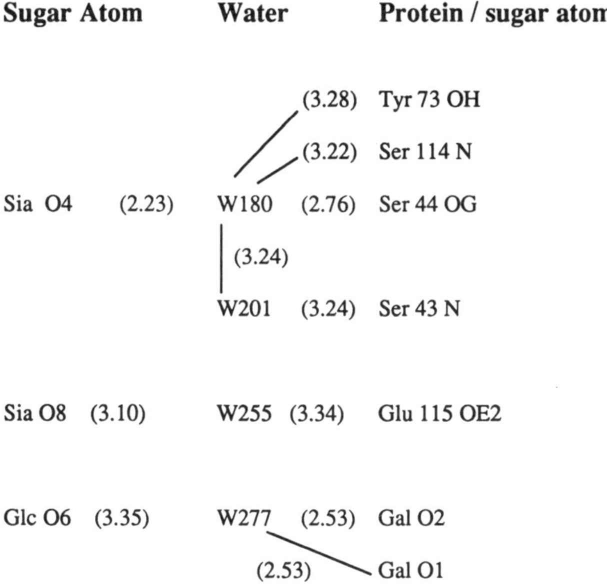

Figure 11.6 Some hydrogen-bond distances (Å) in the WGA–*N*-acetyl neuraminyl structure. From Wright (1990)

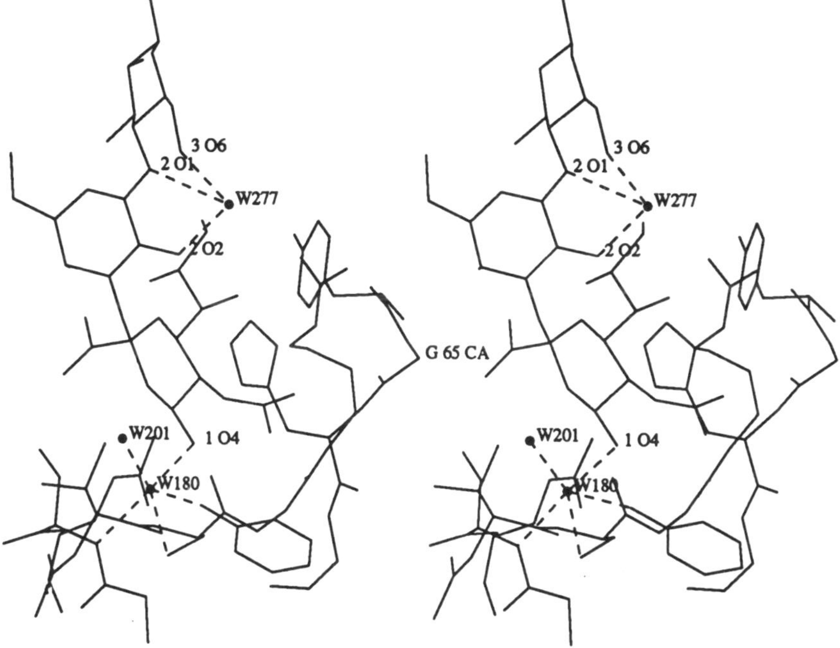

Figure 11.7 Stereoscopic view of the interaction of the trisaccharide *N*-acetylneuraminyl lactose with WGA and water molecules

Monosaccharide Complexes

The structures of isolectin I from *Lathyrus ochrus* (LOL I) complexed to α-methyl-mannoside (LOLM) and α-methyl-glucoside (LOLG) have been solved (Table 11.1). Sugar hydroxyl groups involving oxygens O3, O4, O5 and O6 are hydrogen-bound to the amino acids constituting the monosaccharide binding site of the lectin. Sugar atoms O1 (mannose) and O2 (glucose) are hydrogen-linked to water molecules, which are further linked to second-shell water molecules. The same situation was found in the 2.8 Å resolution structure of ConA with α-methyl-mannoside, where sugar atom O2 binds a water molecule (Derewenda *et al.*, 1989). In the LOL I native crystal structure (LOLF) (Bourne *et al.*, 1990a), four well-ordered water molecules and a disordered one establish hydrogen bonds with residues involved in the binding of monosaccharides (Bourne *et al.*, 1990a). These molecules are close to positions occupied by the monosaccharide hydroxyl groups.

Complex with an α-Man(1–3) β-Man(1–4) GlcNac Trisaccharide

The structure of LOL I complexed to the α-Man(1–3)–β-Man (1–4)–GlcNac trisaccharide (TRI) has been solved at 2.1 Å resolution (LOLT structure) (Bourne *et al.*, 1990b). The trisaccharide is a part of the complex type glycan (Montreuil, 1980; Spik *et al.*, 1982) (Figure 11.8). The TRI conformation in the LOL I/TRI complex is very close to that of unbound TRI (Warin *et al.*, 1979). In contrast to the direct interaction of Man 4 of TRI with the residues of the monosaccharide-binding site, the interaction of Man 3 and GlcNac 2 of TRI with the lectin is completely mediated by sugar–water(s)–lectin linkages (Figure 11.9). Most water molecules are positioned identically in the LOLT and LOLF models. In the neighbourhood of the TRI moiety, only 2 water molecules out of 20 are also present in the native LOLF structure, and none in the LOLM structure. These water molecules are involved in long-chain linkages between TRI and LOL I. An example of such a long chain is the 9 water molecules connecting the atoms of Man 3 and GlcNac 2 to LOL I over a 13 Å distance (Figures 11.9, 11.10).

Complex with a Biantennary Octasaccharide

This complex is formed with an almost complete biantennary *N*-acetyllactosamine-type glycan (Montreuil, 1980; Spik *et al.*, 1982), a structure occurring in many glycoproteins. The two octasaccharide molecules (OCTA) are located in clefts at each end of the long axis of the lectin (Figure 11.11). A large portion of the highly polar saccharide molecule is facing the bulk water. The tightest OCTA–lectin interaction is that of the

antenna A
6
5
4
α-NeuAc-(2-6)-ß-Gal-(1-4)-ß-GlcNAc-(1-2)-α-Man-(1-3)
3
2
1
ß-Man-(1-4)-ß-GlcNAc-(1-4)-ß-GlcNAc-(1-)-Asn
α-NeuAc-(2-6)-ß-Gal-(1-4)-ß-GlcNAc-(1-2)-α-Man-(1-6)
α-Fuc 1, 6
6'
5'
4'
1'
antenna B

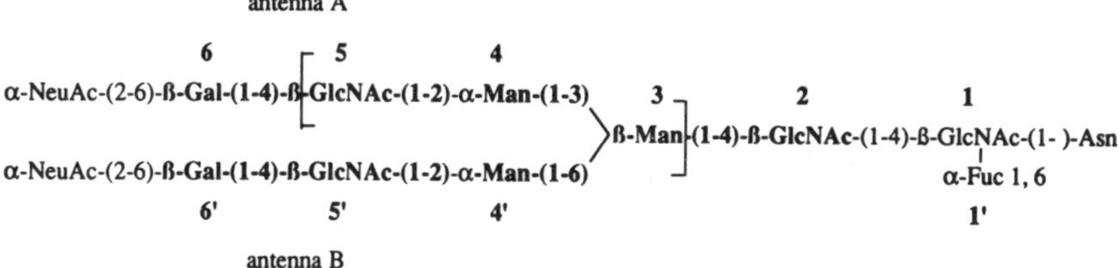

Figure 11.8 Structure of a biantennary glycan of human lactotransferrin; the sugar residues in brackets are those present in the trisaccharide; the sugar residues in bold type are those present in the octasaccharide.

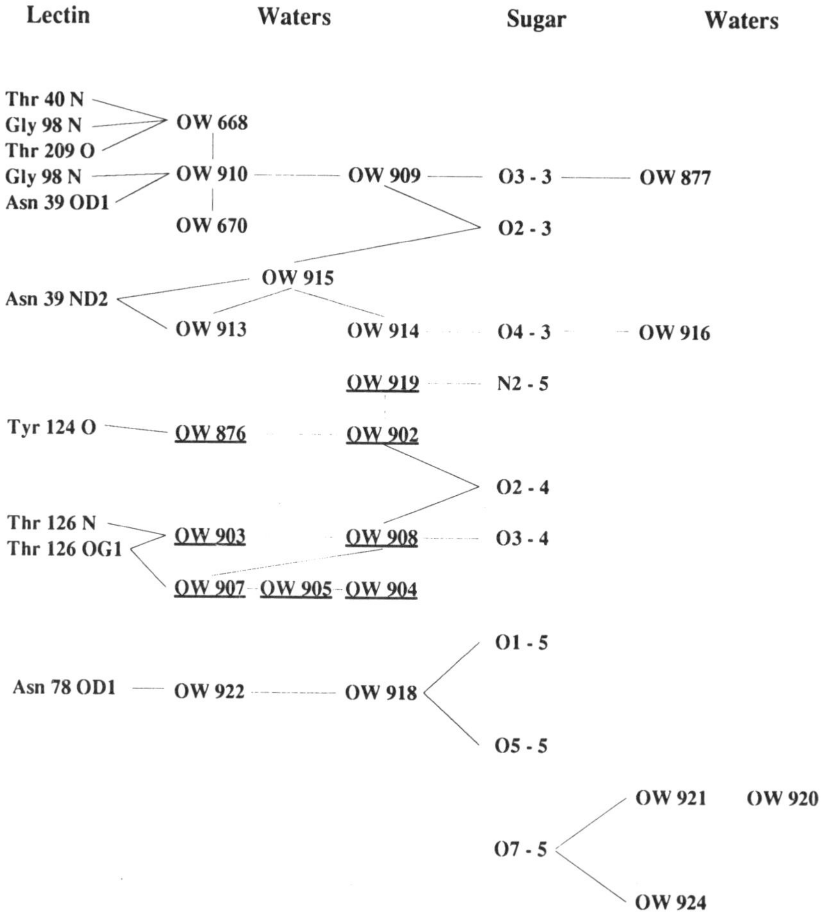

Lectin
Waters
Sugar
Waters
Thr 40 N
Gly 98 N
Thr 209 O
Gly 98 N
Asn 39 OD1
OW 668
OW 910
OW 670
OW 909
O3 - 3
OW 877
O2 - 3
OW 915
Asn 39 ND2
OW 913
OW 914
O4 - 3
OW 916
OW 919
N2 - 5
Tyr 124 O
OW 876
OW 902
O2 - 4
Thr 126 N
Thr 126 OG1
OW 903
OW 908
O3 - 4
OW 907
OW 905
OW 904
O1 - 5
Asn 78 OD1
OW 922
OW 918
O5 - 5
OW 921
OW 920
O7 - 5
OW 924

Figure 11.9 Hydrogen-bonding scheme of the lectin–water–trisaccharide interaction

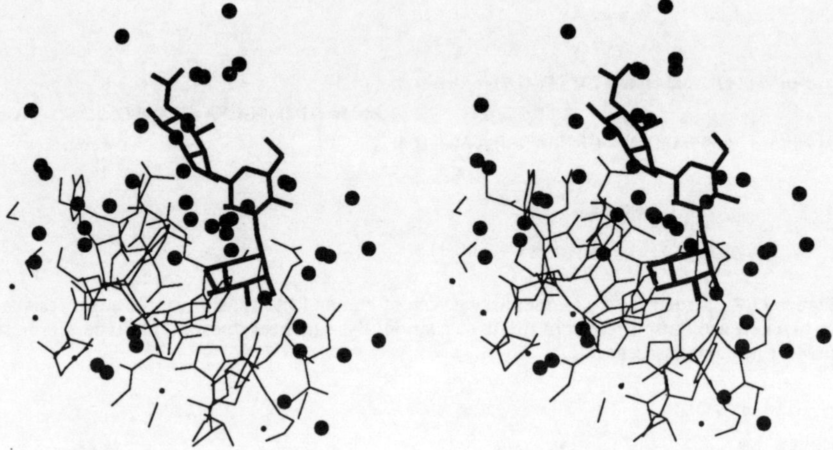

Figure 11.10 Stereoscopic view of the interaction of the trisaccharide Man–Man–GlcNac with LOL I and water molecules

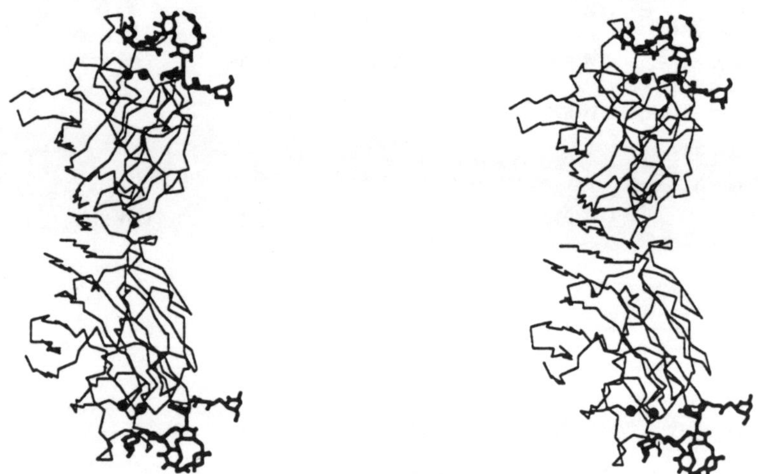

Figure 11.11 Stereo figure showing the Cα plot of LOL I (thin lines) complexed to two octosaccharide molecules (thick lines). The metal ions are also represented

most buried sugar residue, Man 4, in the monosaccharide binding site. The saccharide–lectin complex is stabilized by 23 hydrogen bonds, 14 directly between the protein and the saccharide and 7 by way of 1 water molecule (Figures 11.12, 11.13). In addition, 14 water molecules are involved in longer indirect interactions linking OCTA to the lectin or to itself. As an example of such an interaction, the chain of three water molecules (824, 828, 830) linking the O3 and O4 atoms of Man 3 to the O3 (Man 4) and O7 atoms (GlcNAc 5) stabilizes the conformation of antenna A (Figure

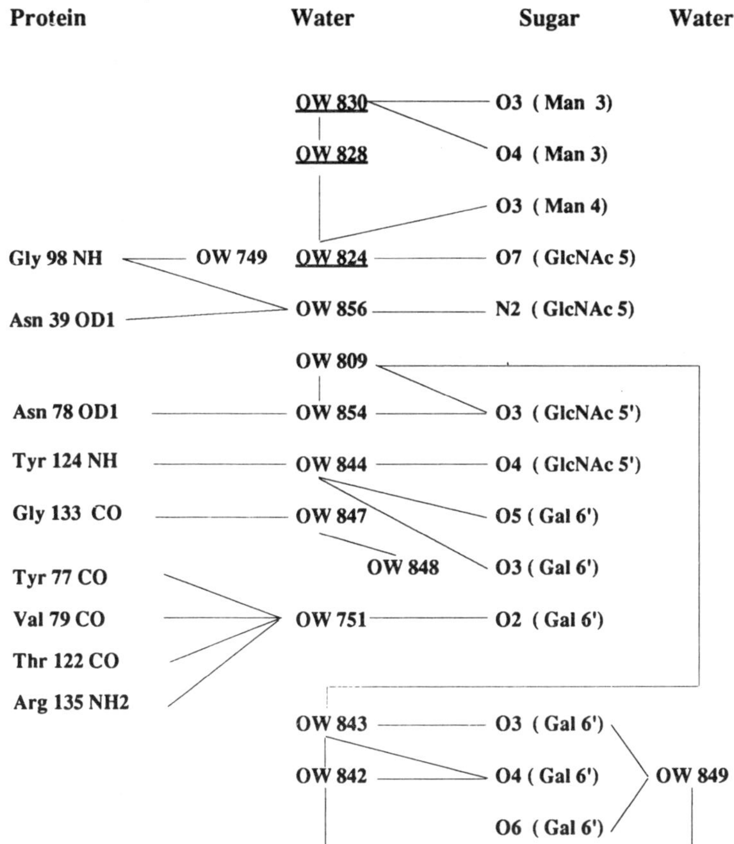

Underlined water molecules stabilize the A antenna.

Figure 11.12 Schematic diagram of the networks of hydrogen bonds in the LOL I–OCTA complex

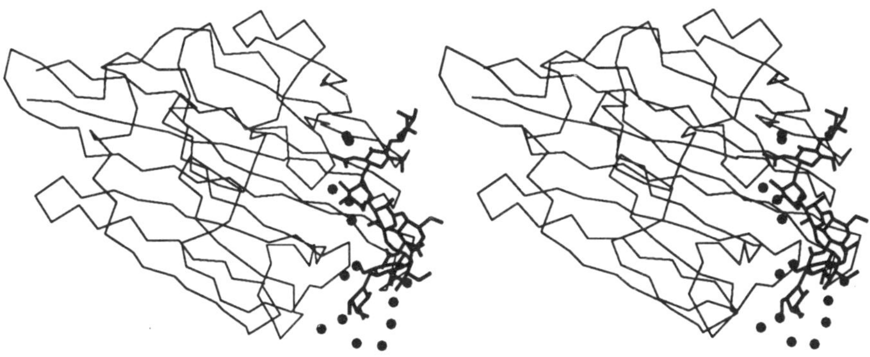

Figure 11.13 Stereoscopic view of the interaction of the biantennary octasaccharide (OCTA) with LOL I and water molecules

11.12). The conformation of the OCTA itself is stabilized by 18 hydrogen bonds, 1 directly and 17 by way of 1 water molecule. This illustrates the important role that water molecules play, in the interactions between saccharide residues within the OCTA molecule itself and between the OCTA molecule and the lectin. A large number of hydration water molecules are also linked to the saccharide.

5 Discussion and Conclusion

The examples presented in this study cover a large range of observations, such as (1) the size of the saccharide (1–8 residues); (2) the type of protein (and its function) the saccharide is complexed with; (3) the nature of the environment of the saccharide. According to Quiocho (1989), owing to the exceptional ratio of hydroxyl groups per saccharide residue, 'hydrogen bonds are the main factors in conferring specificity and affinity to protein–carbohydrate interactions'. As a result, 'all the polar groups of the monopyrannoside substrates are involved in hydrogen bonding'. For the completely buried saccharides described, this means hydrogen bonding to the protein, and to very few (1–3) water molecules. Whereas partially or totally buried sugars tend to satisfy their hydrogen-bonding potential with protein partners, the more exposed sugars form hydrogen bonds to protein-bound water molecules. This statement, partially true for the WGA-trisaccharide complex, is strikingly confirmed in the phosphorylase b glycogen storage site complexes to maltoheptaose and in LOL I–TRI or LOL I–OCTA complexes. Two examples show an extreme and comparable pattern: the glycogen storage minor site and its disaccharide, and the LOL I and the GlcNAc–Gal residues of the TRI saccharide. All the interactions between the protein and the disaccharides are mediated by water molecules. In the latter case, the α-Man(1–3)–β-Man(1–4)–GlcNac motif can achieve two very different conformations, as shown when comparing the lower affinity LOL I–TRI complex with the higher-affinity LOL I–OCTA complex (Bourne *et al.*, 1990c). The isolated α-Man(1–3)–β-Man(1–4)–GlcNac motif, complexed to LOL I as in the LOL I–TRI structure, adopts a very-low-energy conformation, and interacts with the protein with the aid of water molecules.

In conclusion, whereas partially or totally buried sugars tend to satisfy their hydrogen bonding potential with protein partners, the more exposed sugars form hydrogen bonds to protein-bound water molecules. In some cases, these protein–water(s)–saccharide hydrogen bonds can stabilize non-specific sugar–protein interactions: the saccharide adopts a low-energy conformation, and the sugar residues are bound to the protein exclusively through water molecules. The strong trend of saccharides to satisfy their hydrogen-bonding potential, highlighted by Quiocho, is therefore also observed for water molecules squeezed between protein and sugar.

Acknowledgment

The very valuable comments of H. Van Tilbeurgh were highly appreciated.

References

Bernstein, F. C., Koetzle, T. F., Williams, G. J. B., Meyer, E. F., Jr., Brice, M. D., Rodgers, J. R., Kennard, O., Shimanouchi, T. and Tasumi, M. (1977). A computer based archival file for macromolecular structures. *J. Mol. Biol.*, **112**, 535–542

Blake, C. C. F., Johnson, L. N., Mair, G. A., North, A. C. T., Phillips, D. C. and Sarma, V. R. (1967). Crystallographic studies of the activity of hen egg-white lysozyme. *Proc. R. Soc. Lond. (Biol.)*, **B167**, 378–388

Bourne, Y., Abergel, C., Cambillau, C., Frey, M., Rougé, P. and Fontecilla-Camps, J. C. (1990a). X-Ray crystal structure determination and refinement at 1.9 Å resolution of isolectin I from the seeds of *Lathyrus ochrus*. *J. Mol. Biol.*., **214**, 571–584

Bourne, Y., Rougé, P. and Cambillau, C. (1990b). X-Ray structure of a (α-Man(1–3) β-Man (1–4) GlcNac)–lectin complex at 2.1 Å resolution: the role of water in sugar–lectin interaction. *J. Biol. Chem.*, **265**, 18161–18165

Bourne, Y., Rougé, P. and Cambillau, C. (1991). X-Ray structure of a biantennary octasaccharide–lectin complex refined at 2.3 Å resolution. *J. Biol. Chem.* (in press); X-ray structure (2.3 Å) of a complex of *Lathyrus ochrus* isolectin I with a biantennary octasaccharide. *J. Biol. Chem.*, **267**, 197–203

Bourne, Y., Roussel, A., Frey, M., Rougé, P., Fontecilla-Camps, J. C. and Cambillau, C. (1990c). Three-dimensional structures of complexes of *Lathyrus ochrus* isolectin I with glucose and mannose: fine specificity of the monosaccharide binding site. *Proteins*, **8**, 365–376

Cohen, P. (1983). *Control of Enzyme Activity*, 2nd edn. Chapman and Hall, London

Delbaere, L. T. J., Vandonselaar, M., Prasad, L., Quail, J. W., Pearlstone, J. R., Carpenter, M. R., Smillie, L. B., Nikrad, P. V., Spohr, U. and Lemieux, R. U. (1990). Molecular recognition of a human blood group determinant by a plant lectin. *Can. J. Chem.*, **68**, 1116–1121

Derewenda, Z., Yariv, J., Helliwell, J. R., Kalb (Gilboa), A. J., Dodson, E. J., Papiz, M. Z., Wan, T. and Campbell, J. (1989). The structure of the saccharide-binding site of concanavalin A. *EMBO Jl*, **8**, 2189–2193

Diaz, C. L., Melchers, L. S., Hooykaas, P. J. J., Lugtenberg, B. J. J. and Kijne, J. W. (1989). Root lectin as a determinant of host-plant specificity in the Rhizobium–legume symbiosis. *Nature*, **338**, 579–581

Furlong, C. E. (1987). In Neidhardt, F. C. (Ed.), *Escherichia coli and Salmonella typhimurium: Cellular and Molecular Biology*. American Society for Microbiology, Washington, D.C., pp. 768–796

Hadju, J., Acharya, K. R., Stuart, D. I., McLaughlin, P. J., Barford, D., Oikonomakos, N. G., Klein, H. and Johnson, L. N. (1987). Catalysis in the crystal: synchrotron radiation studies with glycogen phosphorylase b. *EMBO Jl*, **6**, 539–546

Johnson, L. N., Cheetham, J., McLaughlin, P. J., Acharya, K. R., Barford, D. and Phillips, D. C. (1988). Protein–oligosaccharide interactions: lysozyme, phosphorylase, amylases. *Current Topics in Microbiology and Immunology*, **139**, 81–134

Johnson, L. N., Stura, E. A., Sansom, M. S. P. and Babu, Y. S. (1983). Oligosaccharide binding to glycogen phosphorylase b. *Biochem. Soc. Trans.*, **II**, 142–144

Kasvinsky, P. J., Madsen, N. B., Fletterick, R. J. and Sygusch, J. (1978). X-ray crystallographic and kinetic studies of oligosaccharide binding to phosphorylase. *J. Biol. Chem.*, **253**, 1290–1296

Lis, H. and Sharon, N. (1986). Lectins as molecules and tools. *Ann. Rev. Biochem.*, **53**, 35–67

McLaughlin, P. J., Stuart, D. I., Klein, H. W., Oikonomakos, N. G. and Johnson, L. N. (1984). Substrate–cofactor interactions for glycogen phosphorylase b: A binding study in the crystal with heptenitol and heptulose 2-phosphate. *Biochemistry*, **23**, 5862–5873

Malin, R., Zielenkiewicz, P. and Saenger, W. (1991). Structurally conserved water molecules in ribonuclease T1. *J. Biol. Chem.*, **266**, 4848–4852

Metzger, B., Helmreich, E. and Glaser, L. (1967). The mechanism of activation of skeletal muscle phosphorylase b by glycogen. *Proc. Natl Acad. Sci. USA*, **57**, 994–1001

Monsigny, M., Roche, A. C., Sene, C., Magnet-Dana, R. and Delmotte, F. (1980). Sugar–lectin interactions: how does wheat-germ agglutinin bind sialoglycoconjugates? *Eur. J. Biochem.*, **104**, 147–153

Montreuil, J. (1980). Structure and conformation of glycoprotein glycans. *Adv. Carbohydr. Chem. Biochem.*, **37**, 157–223

Peters, B. P., Ebisu, S., Goldstein, I. J. and Flashner, M. (1979). Interaction of wheat germ agglutinin with sialic acid. *Biochemistry*, **18**, 5505–5511

Quiocho, F. A. (1986). Carbohydrate-binding proteins: tertiary structures and protein–sugar interactions. *Ann. Rev. Biochem.*, **55**, 287–315

Quiocho, F. A. (1988). Molecular features and basic understanding of protein–carbohydrate interactions: the arabinose-binding protein–sugar complex. *Current Topics in Microbiology and Immunology*, **139**, 135–148

Quiocho, F. A. (1989). Protein–carbohydrate interactions: basic molecular features. *Pure Appl. Chem.*, **61**, 1293–1306

Quiocho, F. A. and Vyas, N. K. (1984). Novel stereospecificity of the L-arabinose-binding protein. *Nature*, **310**, 381–386

Quiocho, F. A., Wilson, D. K. and Vyas, N. K. (1989). Substrate specificity and affinity of a protein modulated by bound water molecules. *Nature*, **340**, 404–407

Reeke, G. N., Jr., and Becker, J. W. (1986). Three-dimensional structure of favin: saccharide binding-cyclic permutation in leguminous lectins. *Science*, **234**, 1108–1111

Reeke, G. N., Jr., and Becker, J. W. (1988). Carbohydrate-binding sites of plant lectins. *Current Topics in Microbiology and Immunology*, **139**, 35–58

Rini, J. M., Carver, J. P. and Hardman, K. D. (1986). Crystallization and preliminary X-ray diffraction studies of a pea lectin-methyl 3,6-di-*O*-(α-D-mannopyranosyl)-α-D-mannopyranoside complex, *J. Mol. Biol.* **189**, 259–260

Rougé, P. and Sousa-Cavada, B. (1984). Isolation and partial characterization of two isolectins from *Lathyrus ochrus* (L.) DC. seeds. *Plant Sci.*, **37**, 21–27

Sharon, N. and Lis, H. (1990). Carbohydrate–protein interactions. *Chem. Brit.*, 679–682.

Spik, G., Strecker, G., Fournet, B., Bouquelet, S., Montreuil, J., Dorland, L., Halkeck, H. V. and Vliegenthart, J. F. G. (1982). Primary structure of the glycans from human lactotransferrin. *Eur. J. Biochem.*, **121**, 413–419

Spurlino, J. C., Lu, G.-Y. and Quiocho, F. A. (1991). The 2.3 Å resolution structure of the maltose- or maltodextrin-binding protein, a primary receptor of bacterial active transport and chemotaxis. *J. Biol. Chem.*, **266**, 5202–5219

Vyas, N. K., Vyas, M. N. and Quiocho, F. A. (1988). Sugar and signal-transducer binding sites of the *Escherichia coli* galactose chemoreceptor protein. *Science*, **242**, 1290–1295

Vyas, N. K., Vyas, M. N. and Quiocho, F. A. (1991). Comparison of the periplasmic receptors for L-arabinose, D-glucose/D-galactose, and D-ribose. *J. Biol. Chem.*, **266**, 5226–5237

Walker, R. A. (1988). In Bog-Hansen, T. C. and Freeds, D. L. J. (Eds), *Lectins—Biology, Biochemistry, Clinical Biochemistry*, Vol 6. Sigma Chemical Co., St. Louis, pp. 591–600

Warin, V., Baert, F., Fouret, R., Strecker, G., Fournet, B. and Montreuil, J. (1979). The crystal and molecular structure of *O*-α-D-mannopyranosyl-(1–3)-*O*-β-D-mannopyranosyl-(1–4)-2-acetamido-2-deoxy-α-D-glucopyranose. *Carbohydr. Res.*, **76**, 11–22

Weber, I. T., Johnson, L. N., Wilson, K. S., Yeates, D. R. G., Wild, D. L. and Jenkins, J. A. (1978). Crystallographic studies on the activity of glycogen phosphorylase b. *Nature*, **274**, 433–437

Westhof, E. (1988). Water: an integral part of nucleic acid structure. *Ann. Rev. Biophys. Biophys. Chem.*, **17**, 125–144

Wright, C. S. (1990). 2.2 Å resolution structure analysis of two refined *N*-acetylneuraminyl–lactose–wheat germ agglutinin isolectin complexes. *J. Mol. Biol.*, **215**, 635–651

12
Lipid Hydration

Gregor Cevc

1 Lipids

The prefix lipo-, or lip-, indicates a fat or a fatty substance. Lipids, therefore, are substances which are not soluble or, at the best, are dispersible in water. In spite of this, and often because of this, lipids are invaluable and ubiquitous in Nature. They are found in minerals and organic tissue, and have even been reported to occur in the interstellar space. In water, lipids spontaneously form boundaries, which are a prerequisite for the creation of individual organisms, and thus constitute about half of the mass of animal cell plasma membrane.

Apolar lipids, such as paraffins and other waxes, fatty alcohols, simple fatty acids, di- and triglycerides, cholesterol, certain steroids, etc., normally do not interact with water at all or hydrate only weakly.

Polar lipids bind water spontaneously at their hydrophilic sites. Most of these sites reside on so-called lipid headgroups, the remaining ones being located on an alcohol, such as glycerol, or a sphingosine backbone and on the upper part of the acyl chains. The corresponding lipids are then called glycerophosphatides and sphingophosphatides, respectively; glycosphingosides contain a sphingosine and an additional sugar moiety.

The most abundant glycerophosphatides are phosphatidylcholines and phosphatidylethanolamines. These represent some 35–70% of the total mass of cell membrane lipids, carry one positive and one negative charge, and thus are zwitterionic. The most widely distributed anionic phospholipids are phosphatidylserine, phosphatidic acid and phosphatidylinositol; these amount together to approximately 10–20% of total membrane lipids. The negatively charged phosphatidylglycerol is abundant especially in microorganisms and plants; the doubly charged cardiolipin is related to phosphatidylglycerol.

Sphingolipids are relatively widespread as well. The zwitterionic sphin-

Headgroup	*Systematic name*	*Generic name*
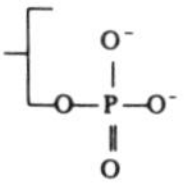 $-O-P(=O)(O^-)-O^-$	1,2-Diacyl-*sn*-glycero-3-phosphoric acid	*sn*-3-Phosphatidic acid
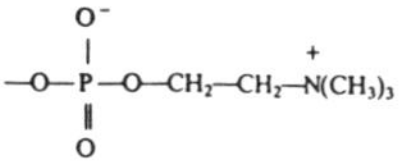 $-O-P(=O)(O^-)-O-CH_2-CH_2-\overset{+}{N}(CH_3)_3$	1,2-Diacyl-*sn*-glycero-3-phosphocholine	*sn*-3-Phosphatidylcholine
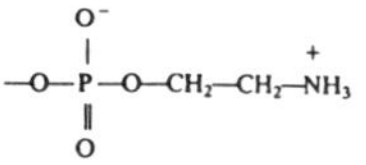 $-O-P(=O)(O^-)-O-CH_2-CH_2-\overset{+}{N}H_3$	1,2-Diacyl-*sn*-glycero-3-phosphoethanolamine	*sn*-3-Phosphatidylethanolamine
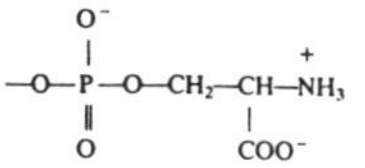 $-O-P(=O)(O^-)-O-CH_2-CH(COO^-)-\overset{+}{N}H_3$	1,2-Diacyl-*sn*-glycero-3-phosphoserine	*sn*-3-Phosphatidylserine
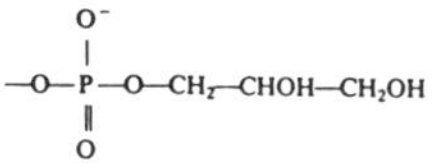 $-O-P(=O)(O^-)-O-CH_2-CHOH-CH_2OH$	1,2-Diacyl-*sn*-glycero-3-phospho-1′-*sn*-glycerol	*sn*-3-Phosphatidylglycerol
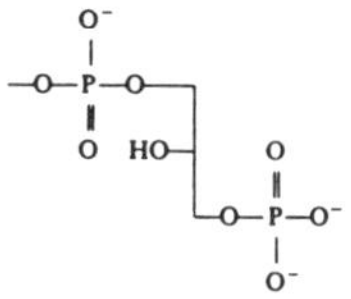 $-O-P(=O)(O^-)-O-CH_2-CH(OH)-CH_2-O-P(=O)(O^-)-O^-$	1,2-Diacyl-*sn*-glycero-3-phospho-1′*sn*-glycero-3′-phosphoric acid	*sn*-3-Phosphatidylglycerol phosphate
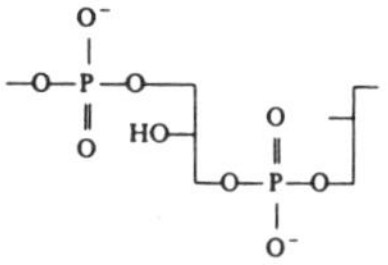 $-O-P(=O)(O^-)-O-CH_2-CH(OH)-CH_2-O-P(=O)(O^-)-O-$	1,2-Diacyl-*sn*-glycero-3-phospho-1′-*sn*-glycero-3′-phospho-1″,2″-diacyl-*sn*-glycerol	*sn*-3,3′-Diphosphatidylglycerol
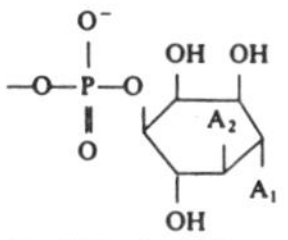 $-O-P(=O)(O^-)-O-$inositol (OH, OH, A_2, A_1, OH); $A_1{=}OH$, $A_2{=}OH$	1,2-Diacyl-*sn*-glycero-3-phospho-1′-*myo*-inositol	*sn*-3-Phosphatidylinositol
$A_1{=}PO_4^{2-}$, $A_2{=}OH$	1,2-Diacyl-*sn*-glycero-3-phospho-1′-*myo*-inositol-4′-phosphate	*sn*-3-Phosphatidylinositol phosphate
$A_1{=}PO_4^{2-}$, $A_2{=}PO_4^{2-}$	1,2-Diacyl-*sn*-glycero-3-phospho-1′-*myo*-inositol-4′, 5′-diphosphate	*sn*-3-Phosphatidylinositol diphosphate

Figure 12.1 Chemical structure of some commonly occurring phospholipids and sphingolipids

gomyelin, for example, accounts for 10–20% of the eukaryotic cell lipids, while sphingoplasmalogenes and sulphatides are less frequent in the biological membranes. The same pertains to the non-charged glycolipids, such as cerebrosides (galacto-, gluco-, lactocerebroside, etc.) and psychosine (1-β-D-galactosylsphingosine), or to gangliosides, the latter containing one or more negatively charged sialic acid residues. All such lipids together do not exceed 5% of the total membrane lipid mass and are largely concentrated in the brain and nervous tissues. A schematic representation of the most common phospholipids is given in Figure 12.1.

2 Sources of Lipid Hydration

The entropy decrease associated with the 'hydration' of apolar lipid chains increases the free energy of the system and makes it a non-negligible positive quantity; this is made possible by the very small internal energy contribution from the apolar lipid part. Lipid chains are thus *hydrophobic*.

The entropy decrease associated with the hydration of the polar lipid heads is much greater than that near the tail part. Notwithstanding this, the internal energy contribution from the lipid heads is sufficiently negative to make the overall free energy of the polar lipid part strongly negative. Lipid heads, therefore, may be called *hydrophilic*.

Polar lipids are thus always amphiphilic (amphiphatic): that is, they are hydrophilic at the headgroup end and hydrophobic at the other, chain, end.

Polar Heads

The first lipid residues to bind water are the well-accessible, strongly polar and, in the majority of cases charged, phosphate, phosphonate, sulphate, carboxylate, etc., groups.

The most strongly bound water molecule associated with an isolated phosphate group bridges the phosphodiester (Peinel *et al.*, 1976). A further four water molecules tend to approach this group with one of their bonds directed towards the anionic oxygen, which bears the largest excess charge, and the other pointing into space (Pullman *et al.*, 1975b; Frischleder *et al.*, 1977). In phospholipids, as well as in many other phosphorus-containing biomolecules, including DNA, the first hydration step thus involves the free oxygen atoms on PO_4^-; crystal structure data shown confirm this (Figure 12.2). Subsequently, the two-bonded oxygen atoms of PO_4^- become hydrated. The total number of water molecules strongly bound to a phosphate group is 7 ± 2, on the average (Finer and Darke, 1974).

Hydroxyl OH groups bind directly fewer than 1–2 water molecules (Harvey *et al.*, 1976). All common cerebrosides thus bind some 12 ± 2 water molecules per hexose. Correspondingly, monoglycerides normally associate strongly with 3–4 water molecules, independent of the lipid chain type (Reed and Shipley, 1989). The preferred direction of water approach to a hydroxyl group or a hydrogen is along the hydrogen bond axis: that is, with the oxygen atom first (Pascher and Sundell, 1977; Pearson and Pascher, 1978).

Water-binding affinity of the methylated ammonium groups is compar-

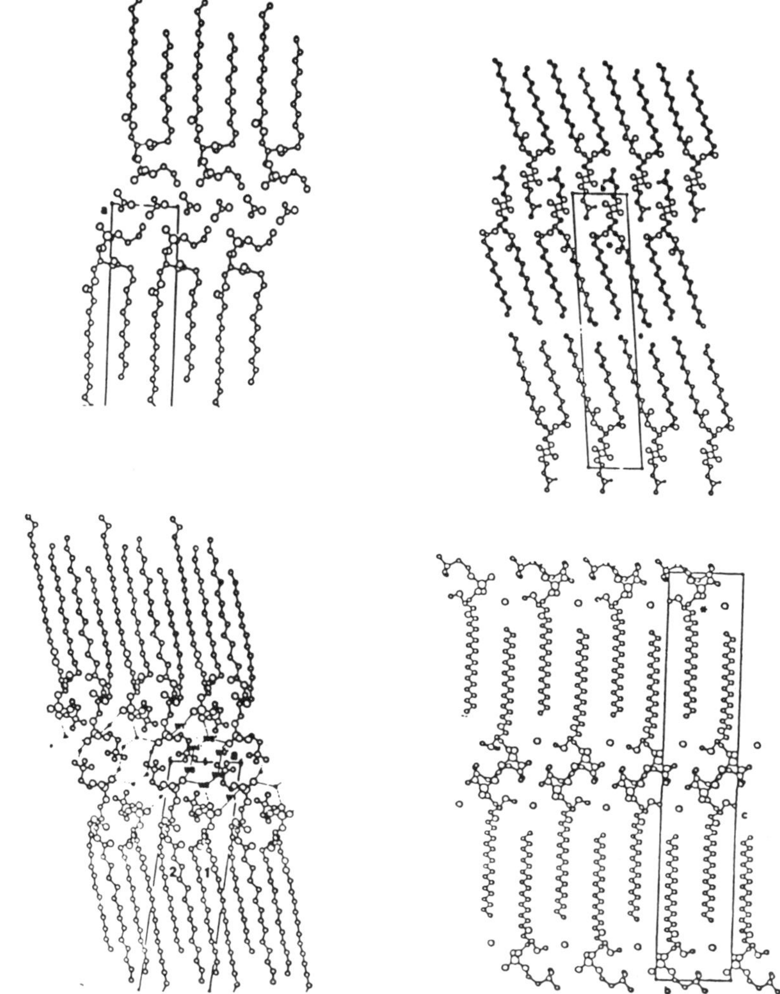

Figure 12.2 Crystal structures of 1,2-dilauroyl-*sn*-3-glycero-phosphoethanolamine acetate (Elder *et al.*, 1977), 1,2-dilauroyl-*sn*-3-glycero-phosphoethanolamine-(*N*, *N*)-dimethyl (Pascher and Sundell, 1986a), 1,2-dimyristoyl-*sn*-3-glycero-phosphocholine dihydrate (Pearson and Pascher, 1979) and 1-octadecyl-2-methyl-*sn*-3-glycero phosphocholine monohydrate (Pascher and Sundell, 1986b), from top left to bottom right, respectively

able to, or perhaps slightly higher than, that of the well-exposed OH groups. On the average, a total of 3 ± 1 water molecules is thus strongly bound to each former group. The optimal hydration pattern of $(CH_3)_4N^+$ involves simultaneous interaction of water with several methyl groups (Port and Pullman, 1973).

Carbonyl groups of the acyl chains or ether-bond oxygens of the alkyl chains are normally less easily accessible to water, owing to the lipid aggregation.* Consequently, these groups are also less capable of binding water. Polar residues on the lipid glycerol or sphingosine backbones or on the lipid chains are thus only hydrated under certain conditions, most notably in the fluid lamellar phase.

Water binding is normally a quantum mechanical phenomenon involving a hydrogen-bond formation, in the majority of cases, as a first step.† It has been reported (Port and Pullman, 1973) that some charge is transferred between the lipid and water molecules even in the absence of direct water–lipid hydrogen bonding. The decrease in water binding energies with distance (i.e. for the higher hydrates) is therefore always very rapid: the first hydration shell of $(CH_3)_2PO_4^-$ accounts for more than 70% of the total binding energy.

Detailed energy results of *ab initio* and semiempirical quantum chemical investigations are available for water binding to dimethylphosphate (Pullman *et al.*, 1975b; Frischleder *et al.*, 1977) and tetraalkylammonium ions (Port and Pullman, 1973) as model fragments for the lipid headgroups. These studies show that seven water molecules bind to an isolated dimethylphosphate anion, $(CH_3)_2PO_4^-$, with energies ranging from −110 to −43 kJ/mol, the energy gain from the association of the first four water molecules with $N^+(CH_3)_4$ being less, from −40 to −30 kJ/mol per water (Figure 12.3). Total hydration energy of the phosphorylcholine 'headgroup' from such a model is concluded to be −660 kJ/mol.

A single sodium cation is strongly attracted to a phospholipid group near the double-bonded phosphate oxygen, with an interaction energy of approximately −67 kJ/mol. In the presence of water molecules the ionic solvation is partly replaced by the phosphate-group oxygens, all within a distance of 0.2 nm. The hydration energies of the residual five ion-bound water molecules at distances of 0.23–0.25 nm is −40 ± 5 kJ/mol (Pullman *et al.*, 1978b).

Apolar Chains

Methylene and other purely hydrocarbon groups normally do not attract water appreciably. Among the apolar groups, the CH_2 groups adjacent to the head group segments with relatively high excess charge densities (e.g.

* Carbonyls on the acyl chains also affect lipid hydration indirectly by affecting the hydrocarbon packing density and, above all, propensity for the formation of interdigitated bilayers, the latter having a much higher limiting hydration than ordinary lipid bilayers (Laggner *et al.*, 1987; Lohner *et al.*, 1987).

† The fact that lipid properties in normal and heavy water are very similar, which has been vindicated in numerous neutron diffraction (Büldt *et al.*, 1978, 1979) and thermodynamic (Chen, 1982) studies, is not in contradiction with this conclusion, as the hydrogen-bond energies in certain range show no isotope effect.

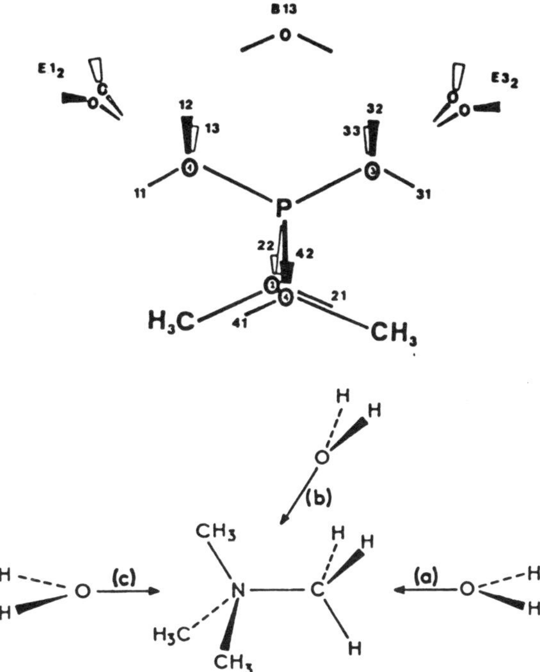

Figure 12.3 Hydration pattern of model lipid headgroups: dimethylphosphate (Frischleder *et al.*, 1977) and tetraalkylammonium ions (Port and Pullman, 1973), as obtained from computer simulations

near a NH_3^+ group) exhibit the highest relative hydrophilicity; but even such apolar groups interact with water without giving rise to significant water binding. Methylene groups adjacent to the non-charged polar groups are really quite non-attractive for water, the unitary free energy of transfer of such groups being 3.7 kJ/mol CH_2. Water has been reported to solvate and to interact directly with alkenes, however (Conrad and Strauss, 1985), presumably at the sites of double bonds. This may explain why the introduction of a single double bond, in terms of hydration energy, is roughly equivalent to a reduction of the hydrocarbon chain length by one methylene group.

The molar energy of hydration of entirely non-polar alkanes is thus quite high: ΔG_{trans} (alkane) $= G_{\text{water}} - G_{\text{bilayer}} = 4.3 + 1.3n_c$, where n_c is the number of carbon atoms per chain. For fatty alcohols this value decreases to ΔG_{trans} (alkanol) $= 0.9 + 1.11n_c$, however (Cevc and Marsh, 1987).[‡]

The first shell water-binding energies and hydrogen-bonding profiles for

[‡] It has been postulated that in addition to an entropic term an enthalpic contribution contributes significantly to the hydrophobic effect. This is not unreasonable, since the en-

the non-polar lipids, such as n-butane, are similar to those found in the bulk water (Jorgensen, 1982), as a result of molecular hydrophobia. Isolated 'solvated' alkanes, therefore, are likely to reside in a cavity formed by the first hydration shell, which loosely resembles a clathrate structure (Geiger *et al.*, 1979). The coordination numbers here are lower by 0.2–0.3, however, in comparison with the corresponding bulk values, owing to enforced ordering (Nakanishi *et al.*, 1984). Water molecules in the first hydration shell of an isolated hydrocarbon molecule thus have hydrogen bonds with each other and with outer water molecules, causing this layer to blend smoothly into the bulk.

Amphiphiles

Monte Carlo simulation for an isolated, rigid amphiphile, 1-butyl-glycero-3-phosphoethanolamine (short-chain lysophosphatidylethanolamine) molecule (Kim and Clementi, 1987) illustrates how more complex lipids can interact with water (Figure 12.4).

Solute molecules vicinal to the hydrophobic chains of such lipid molecule are stabilized by the water–water interactions and not by the phospholipid–water forces, as in the case of simple short-chain hydrocarbons. A large proportion of the directly bound water molecules is associated with the phosphate-group oxygens, as expected, even if the affinity for water of the hydrogen atoms on the ammonium group is relatively higher.* Most of the 'lipid-bound' water molecules reside within 0.16 and 0.22 nm from the headgroup, the most strongly bound water molecules being located approximately 0.3 nm away from the ammonium group.† The total hydration energy of such lysophospholipid molecule is approximately −330 kJ/mol, according to this calculation. It is as yet unclear whether or not this decrease by a factor of 2 relative to the sum of the values obtained for an isolated headgroup, −660 kJ/mol, is theoretically significant. It is nearly certain, however, that this change reflects the correct trend.

---

hanced solvent structuring near a hydrophobic surface may lead to less than optimal interactions between the solute and the first layer of water molecules. But it is widely believed that the entropic term normally prevails and is probably responsible for the observed small solubilities of non-polar species in water. The systematic increase of the entropic term with the size of the non-polar group thus provides the most probable explanation for the decrease in lipid solubility in water with the increasing hydrophobic chain length (Tanford, 1976).

* The water-binding energies for the double-bonded oxygens of the phosphate group in this molecule are −43 and −37, −32 and −22 kJ/mol; for the corresponding ester oxygens the binding energy values are −16, −13, −10, −8 kJ/mol. Water-binding energies for the ammonium group in the same molecule are somewhat higher: −57, −48 and −45 kJ/mol (Kim and Clementi, 1987).

† As one would expect, the positively charged ammonium group attracts chiefly the water

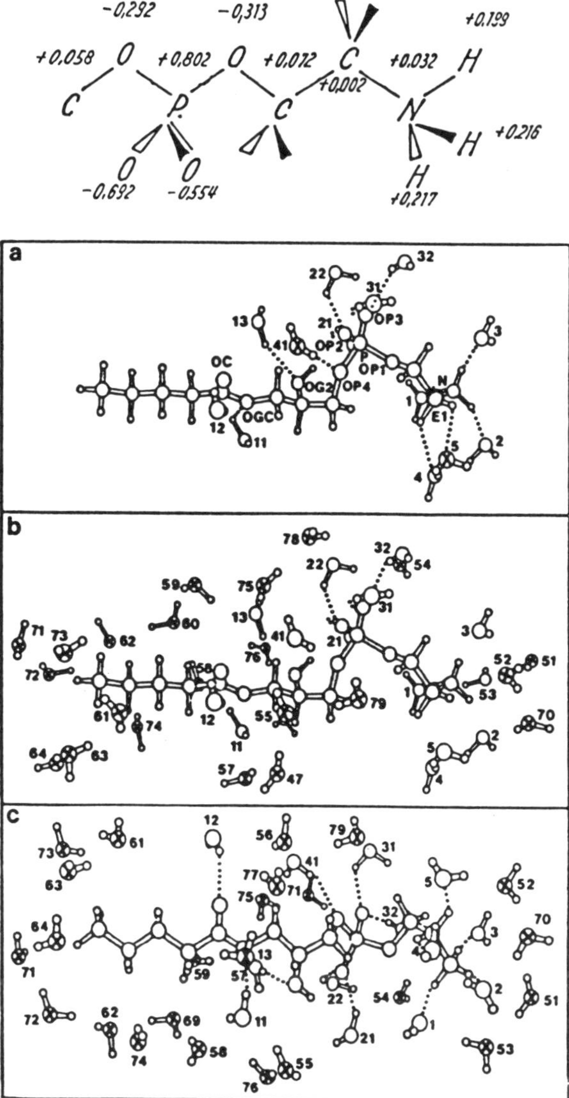

Figure 12.4 Monte Carlo results for the hydration of 1-butyl-glycero-3-phosphoethanolamine (short-chain lysophosphatidylethanolamine, bottom, from Kim and Clementi, 1987) and the distribution and magnitude of the local excess charges on various atoms of the phosphoethanolamine headgroup (top, from Peinel, 1975).

---

oxygens of the proximal solvent molecules. In contrast to this group, the PO_4^--segment tends to repel water oxygens and to attract water hydrogens. The average value of water polarization perpendicular to the 'macroscopic' dipole moment of the lipid headgroup, consequently, is probably quite small.

Lipid (Bi)Layers

Extrapolation of computer simulation results obtained for one lipid molecule to a lipid aggregate must be performed with the utmost caution. The reason for this is that the hydration pattern of a lipid aggregate, such as lipid micelle or bilayer, is prone to differ from the results obtained for the corresponding individual molecules. On the one hand, this may be due to the steric constraints and different molecular and segmental mobilities. Liquid near an extended hydrophobic surface, moreover, tends to maximize the number of hydrogen bonds and molecular packing density (Lee *et al.*, 1984), while in the case of a hydrophilic surface the main trend is to form a hydration layer rather than a hydration shell. The solubility limit for isolated lipid headgroups provides clear evidence for this. Its value is frequently in the range of 1 mol/l and thus typically lower than the lipid headgroup concentrations in a bilayer membrane, ~5–10 mol/l (Schumacher and Sandermann, 1976).

For a model 'lipid membrane' consisting of mobile, but surface-associated, carboxylic groups, a Monte Carlo simulation has revealed the existence of two such distinct layers of (TIP4P) water, with a thickness of approximately 0.5 nm; these are merged with a hydration region in which the structural order decays continuously with separation. The decay length of water order in the lateral direction in such a model is by approximately a factor of 1.5 greater than in the transverse direction, oscillations in the corresponding oxygen–oxygen distribution function being detectable even at separations beyond 1 nm (Nicklas *et al.*, 1990; Schlenkrich *et al.*, 1990; Figure 12.5).

Furthermore, computer simulations of a hydrated lattice of phosphoethanolamine molecules show that going from a single molecule to the lipid 'monolayer', which consists of tens of lipid headgroups, causes a marked decrease of the lipid–water interaction potential minima (Peinel *et al.*, 1983a). Three hydration layers near a phosphoethanolamine lattice, according to theory, should contain 4, 3 and 3 molecules, respectively (Frischleder, 1981), in reasonable agreement with the experimental value of 8 ± 1 (Cevc and Marsh, 1985; McIntosh and Simon, 1986a).

Water structure near a polar surface is always strongly perturbed (cf. Figure 12.5). The water–water interaction energy in the hydration layers of phosphatidylethanolamine lattices is only −23.4 kJ/mol. This is less than one-half of the bulk value, −52.6 kJ/mol. The stereospecific headgroup–water interactions, which amount to −39.7 kJ/mol, can partly compensate for this loss, however. This causes the total interaction energies per hydration layer to be still appreciable, −105.7, −3.3 and −14.2 kJ/mol, for the three hydration layers of the lattice, respectively (Peinel *et al.*, 1983b). The sum of all values, −123 kJ/mol, is smaller than the result for an individual phosphatidylethanolamine molecule but still much greater than the ex-

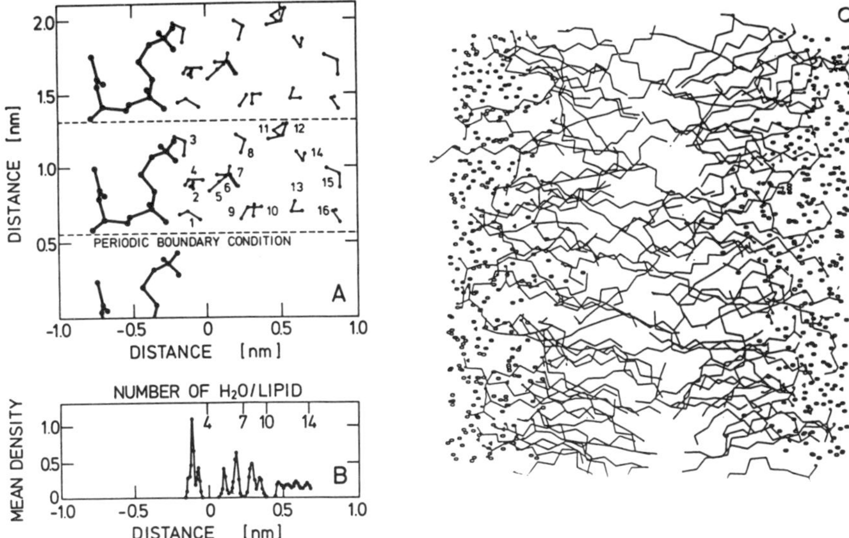

Figure 12.5 Hydration pattern of 'crystalline' (left) and fluid (right) lipid bilayers. (A) Organization of water molecules above a periodic glycerophosphoethanolamine lattice calculated by Monte Carlo simulation. (B) Water density distribution perpendicular to this lipid lattice. The three hydration shells contain a total of 10 water molecules with other molecules already forming the upper boundary of the bulk water phase. The fine structure is due to the small lattice-cell size used in the calculation. Modified from Frischleder (1981). (C) Molecular dynamics simulation of a sodium decanoate–decanol fluid bilayer in water. Water molecules (oxygens denoted as •) penetrate deep in the hydrophobic region but preferentially hydrate the polar carboxylic groups (oxygens denoted as °). The surface excess charges are compensated by the charges on the water molecules (from Egberts and Berendsen, 1988)

perimental value for lipid bilayers, −1.6 kJ/mol. Limiting sample size effects are one possible reason for this discrepancy. Experimental values for other phospholipids are higher, however, for phosphatidylcholines between −17 kJ/mol and −30 kJ/mol (Cevc, 1985).

Owing to the tight lipid–water interactions, the density of interfacially bound water is by a factor of about 2 higher than in the bulk, according to computer simulations (Frischleder, 1981; Nicklas *et al.*, 1990). This compression is restricted to the innermost hydration layers: multiple density oscillations, which are measured with uniform, planar surfaces (Israelachvili and Pashley, 1983; Christenson and Horn, 1985; Israelachvili, 1985; Pashley *et al.*, 1986; Christenson and Blom, 1987) are never observed between lipid bilayer surfaces (Marra and Israelachvili, 1985; Marra, 1986). At best, 2 discrete lipid hydration 'layers' between 2 adjacent lipid bilayers are seen, which contain up to 4 water molecules per lipid.

Several lines of evidence vindicate this conclusion. For example, the chain-melting phase transition temperatures of phosphatidylethanolamines (Cevc and Marsh, 1985), phosphatidylcholines (Kodama *et al.*, 1982a, b;

Jürgens *et al.*, 1983) and other lipids (Madelmont and Perron, 1974) form a sequence of distinct, quantized steps at low water concentrations (see also further discussion). Even more directly, synchrotron X-ray diffraction studies of solid-supported dipalmitoylphosphatidylcholine multibilayers indicate that water binding in the low hydration region ($< 15\%$ or approximately 2 waters/lipid) does not proceed in a continuous fashion, but instead occurs in discrete steps. These correspond to homogeneously hydrated states of the multibilayer stack in which an integral, small number of 1–4 water 'monolayers' is associated with two adjacent lipid bilayer surfaces (Seul and Eisenberger, 1989a). Such states may correspond to the observed, well-defined chain-melting transition temperatures. No signs of the discreteness of water binding beyond the first, or perhaps the second, 'hydration layer' are seen in the calorimetric (Kodama *et al.*, 1982b; Cevc and Marsh, 1985) or dielectric relaxation (Nimtz and Binggeli, 1985) measurements. Lipid-solvation in this region is thus nearly continuous.[‡]

3 Temperature Effects

The propensity for lipid–water association increases with increasing temperature. This is largely due to the better accessibility of water-binding sites at higher temperatures.

At least two arguments support this conclusion. On the one hand, the direct effects of absolute temperature variations on the lipid hydration have been measured to be quite small: the interfacial separation between phospholipid bilayers in the ordered lamellar phase does not change (Smith *et al.*, 1987), even over large temperature intervals (Kirchner and Cevc, to be published). On the other hand, the water-binding strength decreases with increasing temperature, owing to the thermal excitations, weaker dispersion forces and diminished hydration bonding capacity. This effect is disproportionally strong for the lipid-bound water (Nimtz and Weiss, 1987).

Relative temperature variations, $T-T_c$, therefore, as well as absolute temperature, both affect lipid hydration. The chief reason for this are the temperature-induced changes of lipid conformation, which occur especially at the phase transition temperatures, T_c, and above. Increasing temperature in the fluid bilayer phase, for example, lowers the energetic barriers for most intramolecular and some intermolecular motions. Rotation of the individual segments on phospholipid headgroups or orientational head-

[‡] Disordered intermediate states, diagnostic of the coexisting multiple hydration states, may arise if non-integer, small numbers of 'bound waters' — for example, 1.5/lipid — are used. This is probably due to the inhomogeneous water intercalation between adjacent surface layers (Seul and Eisenberger, 1989a, b). These also may give rise to intermediate bilayer phase transition temperatures (Kodama *et al.*, 1982a; Cevc and Marsh, 1985).

group oscillations both speed up with the increasing relative temperature (see also further discussion). Lipid-packing density simultaneously decreases, allowing water molecules a better access to the deeply located water-binding sites, such as acyl-chain carbonyls. All this has also been observed directly, by using ^{31}P nuclear magnetic resonance (Griffin, 1976), dielectric measurements (Enders and Nimtz, 1984), modulated excitation infrared spectroscopy (Fringely and Guenthard, 1976) or deuterium neutron diffraction techniques (Büldt *et al.*, 1979).

4 Ion and Solute Effects

The chief non-specific effect of ions in a solvated lipid system is the screening of the electrostatic intermolecular and interfacial, such as surface double-layer, repulsion.

In an electrolyte containing more than 1–2 mol of salt per litre, the electrostatic Coulombic repulsions between the lipid bilayers are largely screened out. When the bilayer hydrophilicity is too small, this may result in total collapse of the colloidal lipid stability (Marra, 1986). For sufficiently hydrophilic bilayers, however, the residual interfacial repulsion of hydration can keep the suspensions stable. Owing to the latter stabilizing force, the separation between charged lipid bilayers in highly concentrated electrolytes is very similar to that characteristic of the non-charged lipid bilayers in water (Cowley *et al.*, 1978; Loosley-Millman *et al.*, 1982). Hydration force between charged bilayers is not seen at separations greater than 2 nm (Marra, 1986).

The 'specificity' of association between small inorganic ions and lipids is also, at least partly, due to the bilayer hydration effects.

For small ions and uncharged lipid layers the interfacial hydration effects are relatively trivial. Ions are depleted from the close vicinity of hydrated zwitterionic lipid surfaces, owing to the mutual short-range repulsion of hydration. In the case of charged lipid membranes the interfacial ion concentration, conversely, may be anomalously high. This is true at least at low and intermediate electrolyte concentrations, owing to the hydration-dependent increase in the electrostatic surface potential near the membrane surface (Cevc, 1990a).

The situation with organic ions is more complex. Such ionic substances can adsorb, or even bind, to a lipid membrane because of the enthalpic binding, as well as entropic (hydrophobic) effects. The greater the apolar region of an ion, the more likely it is that such an organic ion will bind to a membrane, owing to the hydrophobic effect. To an even larger extent, this holds true for the macromolecules with multiple charges and apolar residues — for example, proteins.

At high ion concentrations the counterion capacity to screen the

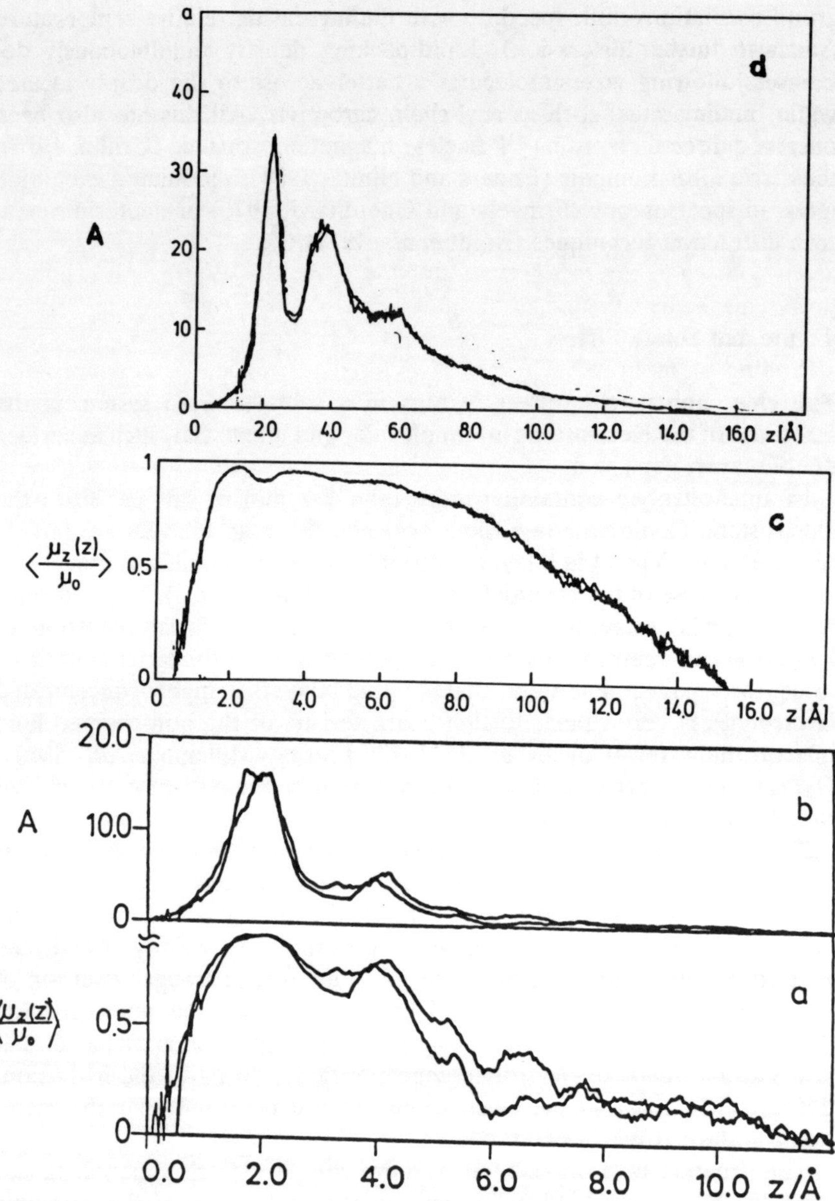

Figure 12.6 Average water polarization perpendicular to the membrane surface as a function of the membrane–oxygen distance for a model bilayer immersed in water (a) or NaCl solution (c). Parameter A in Langevin function, $\mu_z/\mu_0 = \coth(A) - 1/A$, in similar systems (b, d, respectively). Results for two adjacent surfaces are always given. Ion adsorption near the charged membrane surface affects significantly the water orientation profile, which, however, in no case decays exponentially with separation. From Schlenkrich *et al.* (1990) and Nicklas *et al.* (1990)

Coulombic membrane potential is limited by the lipid and ion tendency to remain hydrated. Detailed numerical (Nicklas *et al.*, 1990) and analytical, mean-field (Gruen and Marčelja, 1983, 1984; Cevc and Marsh, 1987; Cevc, 1990a) models corroborate this. For a model membrane consisting of a layer of carboxylic groups, for example, the first adsorbed layer of partly dehydrated ions compensates only some 40% of the net membrane charge. More distant layers of hydrated ions thus mainly screen, rather than neutralize, the membrane charge and form a diffuse layer. Owing to the strong ion–water interactions, the interfacial water polarization profile is always markedly affected by interfacial ion layers (Nicklas *et al.*, 1990) (Figure 12.6).

Binding of monovalent cations to phospholipid headgroups normally occurs 'through water' (Pullman *et al.*, 1978a; Pullman, 1982; Cevc, unpublished data); to the anionic phosphate group, as a rule. Exceptions to this rule are small ions, such as lithium. These can compete with water for the same binding sites and thus lead to at least partial lipid dehydration (Hauser and Shipley, 1981; Cevc *et al.*, 1985). For the same reason, and because of their capacity to form interbilayer ion-bridges, divalent and polyvalent ions not only strongly bind to lipid headgroups but also can dehydrate bilayers completely. Calcium complexes with anionic phosphatidylserine, for example, form water-free, thermodynamically stable lamellar crystals with a transition temperature above 150 °C (Hauser and Shipley, 1984). Complexes of polylysine and charged phospholipids also are essentially devoid of water and highly stable (Hartmann and Galla, 1978). Calcium binding to the zwitterionic lipids is less tight and involves some water; calcium ions have been directly measured (Herbette *et al.*, 1984) to reside approximately 0.2 nm away from the phosphate group.

Anions probably bind to the 'counterionic' ammonium segments on the lipid headgroups together with their water of hydration (Westman and Eriksson, 1979). By all means, common anions increase the overall hydrophilicity of standard phospholipids and mediate additional uptake of water; at sufficiently high concentrations they may even induce chain interdigitation (Cevc, Löbbecke and Nagel, to be published).

Non-ionic solutes affect the properties of lipids indirectly, as a rule, via osmotic effects (Cevc, 1988), although some direct binding (and partial dehydration), after long equilibration times, is also possible (Cevc, 1991b).

5 Modelling Lipid–Water Interactions

To date, no theoretical approach has been developed which would be perfectly suitable for the description of lipid–water association. On the one hand, this is due to lipid complexity. Each lipid molecule contains over 100 atoms, one or more large hydrophobic chains and an extended polar head.

On top of this, some 25–75 atoms reside on the lipid-associated water molecules. The system properties thus depend on multiple water–water, water–lipid and lipid–lipid interactions. For a minimally required cluster of approximately 150 lipid molecules this implies that 3×10^4 atoms should be considered in a detailed system simulation, all transbilayer and interbilayer interactions being still neglected.

One solution to this problem is to retain the detailed description of the interparticle interactions but to introduce simplifications about the hydrated bilayer structure. For example, only one section of a lipid monolayer can be selected; this can be taken to consist of small parts of lipid molecules only and the latter can be confined, for example, in just a few possible conformations.* Bulk solvent can be mimicked by using appropriate periodic boundary conditions.†

Results of such studies, discussed in Section 2, demonstrate clearly the main strengths and weaknesses of such an approach. While a certain feeling for the effects of water and interlipid interactions on the molecular conformation and energetics can be obtained, no cooperative phenomena, such as phase transitions, can be studied and accurate analysis of the structural and thermodynamic system properties as a function of lipid or water concentrations is impossible.

To tackle the latter problem, macroscopic thermodynamics (Khan *et al.*, 1985; Jönsson and Wennerström, 1981; Jönsson, 1987; Jönsson and Persson, 1987; Wolfe and Brockman, 1988; van Oss *et al.*, 1988) or statistical mechanics (Cevc and Marsh, 1987) can be used. Integration of the thermodynamic relation

$$dG_h(c_w) = RT \ln a_w(c_w)dc_w$$

for example, allows direct, model-independent evaluation of the free energy of lipid hydration

$$G_h(c_w) = RT \int_0^{c_w} \ln a_w(c'_w)dc'_w \qquad (12.1)$$

As input information for such an integration, the water adsorption data can be used, the overall water concentration c_w and the number of bound water

* It would be desirable not to restrain the lipid headgroup configuration into a state similar to that observed in an organic lipid solution or dry lipid crystals since water, in both cases, is prone to modify this conformation (Pullman *et al.*, 1975a; Hauser *et al.*, 1980; Cevc and Seddon, 1986; Nicklas *et al.*, 1990).

† It would seem that present water models are as yet insufficiently accurate for a realistic description of the hydrated polar surface. Current disagreement between the results of various simulations of water enclosed between two hard walls (Jönsson, 1982; Barabino and Marchesi, 1984; Marchesi, 1983; Gardner and Valleau, 1987; Parsonage and Nicholson, 1986; Hautman *et al.*, 1989) and the discrepancies between calculated (Scott, 1984) and measured hydration (force) profiles for lipid bilayers (Cevc and Marsh, 1985; Rand and Parsegian, 1990) indicate how far this goal is as yet from being achieved.

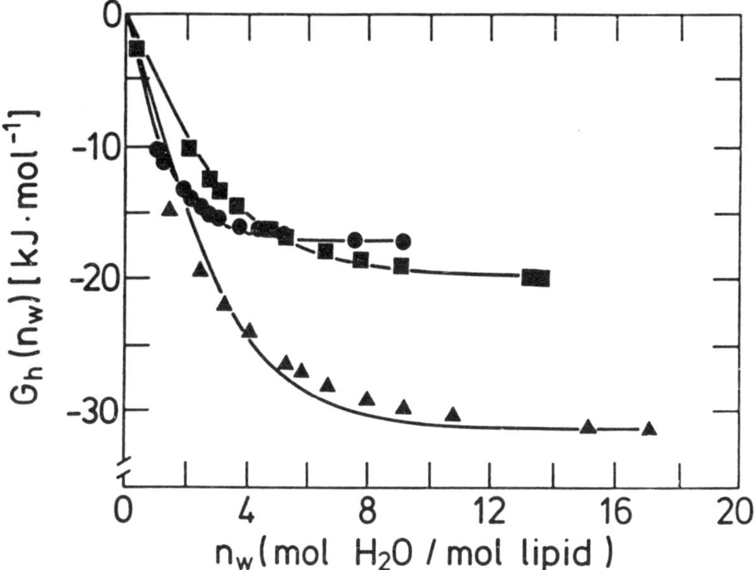

Figure 12.7 Free energy of bilayer hydration as a function of number of bound water molecules, n_w for dioleoyl- (▲), egg-yolk- (■) and dipalmitoyl-phosphatidylcholine (●) at 22 °C. The experimental points were calculated from the sorption data of Jendrasiak and Mendible (1976a) using Equation (12.1), and the curves were evaluated from non-local electrostatic model of hydration. From Cevc and Marsh (1987)

molecules, n_w, being interdependent: $c_w \simeq (1 + M_l/18N_An_w)$ (Cevc and Marsh, 1987; cf. Figure 12.7).

Macroscopic thermodynamics enables reasonably accurate analysis of the water sorption data or descriptions of certain features of the lipid polymorphism, but it provides little insight into the interdependence between such system characteristics and the molecular lipid properties.

As a viable compromise, input knowledge about certain molecular lipid properties in combination with suitable mean-field descriptions of the hydration phenomena can be used. Non-local electrostatics is useful for this purpose. The starting assumption of any non-local electrostatic model is that even a local perturbation influences the system over a finite region; this allows a quasi-electrostatic description of the surface hydration phenomena in a simple, mean-field manner. Normally, the size of the perturbed region is taken to be proportional to the solvent correlation length (Dogonadze *et al.*, 1973; Kornyshev and Vorotyntsev, 1980; Gruen and Marčelja, 1983; Kornyshev, 1983; Cevc, 1985), owing to the intersolvent coupling. This size, in the first approximation, depends on the most pertinent solvent-polarization mode (Dogonadze *et al.*, 1973) and, for water close to a polar surface, is believed to be similar to the intrinsic value of the solvent structure decay length, ξ. Thus, it should be of the order of a tenth of a nm. Two more solvent characteristics must normally also be known: the static and high-frequency dielectric constant, $\varepsilon \leqslant 82$ and $\varepsilon_\infty \simeq 2$, respectively (Cevc, 1990b).[‡]

To describe the effective lipid hydrophilicity in a simple, but self-consistent, logical and useful way, it is convenient to introduce the concept of the hydration bilayer potential, ψ_h. This accounts for all short-range fields stemming from a polar molecule or a polar surface and is therefore a good measure of the overall interfacial hydrophilicity.

The chief mechanism of the lipid–water association is the direct binding of water molecules to the polar residues, as already noted, rather than ordinary electrostatic polarization. It is, therefore, plausible to assume that the main sources of the bilayer hydration potential are the atomic charges in a polar region.*

This can be understood as follows (Cevc, 1985, 1990b). All polar lipids contain at least a few groups, r, with a net local charge, e_r. The main charged regions on a polar lipid molecule are typically separated by an intramolecular barrier. This prevents charge exchange and efficient charge neutralization within one such molecule.[†] Each lipid aggregate and its hydrophilicity, therefore, can be characterized by some spatial distribution, $\rho_p(x)$, and the corresponding total magnitude of the *local-excess* lipid charge.

The spatial distribution of the local-excess charges on lipid polar residues is found by summing up all the corresponding individual contributions from the atoms located at a given distance, x. But the very first step is to weigh individual atom contributions in order to account for their different accessibility to water, $\alpha_r(x)$. Effective individual local-charge density of an atom or a polar residue r can thus be estimated from the relation: $\sigma_{p,\,r}(x) = \sigma_r(x)e_r/A_r$, where A_r is the appropriate surface area.

Finally, the *total surface local-excess charge distribution* can be found by averaging over the entire membrane surface:

‡ Sometimes the magnitude of the virtual 'charge of the water molecules' (Onsager and Dupuis, 1962), indicative of and characteristic for the molecular polarizability and interatomic charge-transfer in the direct surface-water (often hydrogen) bonds, also plays a role. This latter quantity has been estimated to be approximately $e_w \simeq 3.5 \times 10^{-20}$ Cb (Cevc, 1985) but different models yield different values.

* It has long been a matter of debate whether the transition from an initially delocalized state of the excess electron to a localized state, which stands at the beginning of the hydration process, is initiated by attractive long-range, dipolar interactions, or whether localization at pre-existing trapping sites occurs; in the latter case, the initial stability would be due to attractive short-range interactions. Recent experimental results for alcohols and alcohol–alkane mixtures and computer simulation studies (Schnitker *et al.*, 1986) seem to support the latter notion. The solvation process thus appears to begin with an electron localization and with attractive short-range electron–medium interactions, even when no direct hydrogen bonding occurs.

† Owing to the atomic origin of such charges, the precise molecular configuration or organization (monomer or lattice) influences only to a small extent the intramolecular distribution of excess charge (Peinel *et al.*, 1983b), despite the fact that they are of paramount importance for the interbilayer interactions.

$$\rho_p(x) = \sum_r \alpha_r(x)\sigma_{p,\,r} \times \sum_r A_r\, A_{\text{surface}}^{-1} \tag{12.2}$$

The dipolar and multipolar contributions can be accounted for accordingly in terms of the appropriate densities τ_p, ν_p, etc. (Cevc and Marsh, 1985; Belaya *et al.*, 1986, 1987); normally these contributions are quite small, however, since the average size of a lipid headgroup dipole is much greater than the water correlation length, ξ.

Obviously, only such polar groups and their charges must be considered in the sum in Equation (12.2) which are not compensated in the bilayer surface directly. Therefore, only such polar residues should be counted which are not directly coupled to other lipids or groups within the same molecule, as only such groups are really free to bind water. Similar restrictions are customary in computer simulations (Rossky and Karplus, 1979). For lipids they are vindicated by the experimental studies of the phospholipid phase behaviour (Cevc, 1987).‡

The interfacial polarity profile thus corresponds, in the simplest electrostatic approximation, to a distribution throughout the interfacial region of the densities of all available lipid local-excess charges. When the precise polarity distribution is unimportant, the corresponding integral value, $\sigma_p = \int \rho_p(x)\,dx$, i.e. the *surface density of the local-excess charge density*, can be used (Cevc, 1985).

In both cases the bilayer hydration potential can be calculated by an integration throughout the interfacial region

$$\psi_h(x) = \frac{\xi}{\varepsilon_\infty \varepsilon_0}\left(-0.5\int_0^\infty \rho_p(x')\exp[(x - x')/\xi]\mathrm{d}x' + \int_0^x \rho_p(x')\sinh[(x - x')/\xi]\mathrm{d}x'\right)$$

and expressed in terms of known or estimable lipid polarity parameters (Cevc, 1985, 1990b, 1991a). For example, in the low potential approximation for an infinitely narrow interface one gets

$$\psi_h(x) = (\sigma_p\xi/\varepsilon_\infty\varepsilon_0)\exp(-x/\xi) \tag{12.3}$$

where ε_0 is the permittivity of free space.

To calculate the spatial variation of the free energy of bilayer hydration,

‡ When polar parts of the membrane constituents interact in the absence of water, the effective basic 'electroneutrality unit' of each molecular layer consists of the nearest-neighbour charged groups. For lipids, these are typically the phosphate and the ammonium groups. But the same groups can also form intermolecular hydrogen bonds and may partly exchange charges along these bonds via the charge-transfer process. (For one special case this has been already demonstrated in direct experiments: Jean *et al.*, 1984.) This is the reason why in any realistic model of membrane hydration the *intra*molecular as well as the *inter*molecular charge neutralization must be considered.

a virtual charging process can be performed (Verwey and Overbeek, 1948), which is a generalization of the corresponding standard procedure from the theory of surface electrostatics: $G_h(x) = (1/\sigma_p)(1 - \varepsilon_\infty/\varepsilon) \int_0^{\sigma_p} \psi_h(x, \xi, \sigma)d\sigma \int_0^x \rho_p(x')dx'$ (Cevc, 1991a). For the limiting case mentioned (cf. Equation (12.5)), one gets

$$G_h(x) \simeq (1/\varepsilon_\infty - 1/\varepsilon)(\sigma_p^2\xi/\varepsilon_0)\exp(-x/d_p)$$

the corresponding non-linear result being given by Cevc (1985). The difference of inverse permittivities in this expression indicates that, in the present, and any other, electrostatic model of hydration, the free energy is proportional to the work done upon transferring (lipid-associated) atomic charges from the bulk, where $\varepsilon_{ef} = \varepsilon$, to the interface, where, by definition, $\varepsilon_{ef} = \varepsilon_\infty$. This is reminiscent of the Born theory of ion hydration (Bucher and Porter, 1986).

From the profile of hydration free energy it is easy to deduce other thermodynamic variables by means of straightforward textbook procedures (Cevc *et al.*, 1982). The interfacial repulsion of hydration, for example, is given by $p_h(x) = -(1/A)(dG_h(x)/dx)$. The result of the corresponding general derivation shows that hydration force and hydration profiles of lipid bilayers, in general, contain several contributions (Cevc, 1991a). The first depends on the decay length of the solvent structure, ξ; the second depends on the effective distribution of the lipid local-excess charges, this distribution *per se* being a function of all existing positions and motions through space of the lipid polar residues. More accurate representations should also allow for the contributions from the interbilayer correlations (Kjellander, 1984; Podgornik, 1988, 1989; Kornyshev and Leikin, 1989; Leikin and Kornyshev, 1990, 1991).

Lipids with long or highly mobile headgroups or lipids which can penetrate easily out of a membrane are thus expected to form bilayers with a thick interfacial region, $d_p \gg 0$, and extensive interfacial swelling (Cevc *et al.*, 1986).

If the effective surface polarity distribution profile is exponential, $\rho_p(x) = (\sigma_p/d_p)\exp(-x/d_p)$, the interfacial repulsion of hydration attains the following form:

$$p_h(x, d_p \gg \xi) \simeq (1/\varepsilon_\infty - 1/\varepsilon)(\sigma_p^2/\varepsilon_0)\exp(-x/d_p)(\xi/d_p)^2\, x,\ d_p \gg \xi \qquad (12.4)$$

in the case that the interfacial thickness d_p and separation are much greater than the hydration decay length. This expression is similar to the standard non-local electrostatic expression for the evaluation of the hydrational repulsion between two infinitely narrow lipid bilayer surfaces:

$$p_h(x, d_p = 0) = (1/\varepsilon_\infty - 1/\varepsilon)(\sigma_p^2/\varepsilon_0)\exp(-x/\xi) \qquad (12.5)$$

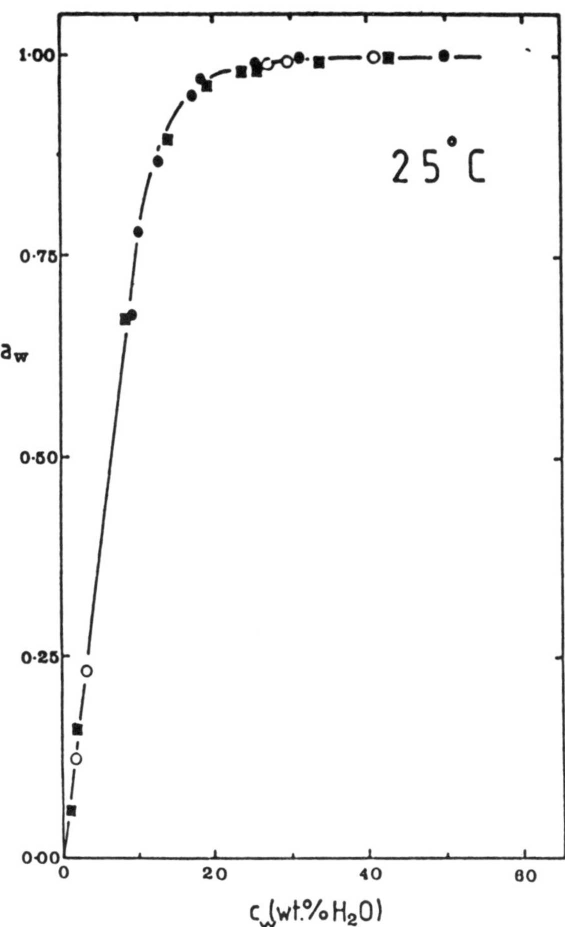

Figure 12.8 Water activity coefficient in a stack of phosphatidylcholine–phosphatidylinositol multibilayers as a function of the bulk water concentration (Hammond *et al.*, 1987)

The essential difference is that in the former case an interfacial thickness parameter appears in place of hydration decay length. Effective decay length of hydration force may also be affected by the inhomogeneous distribution of hydration sources and their relative positions on the interacting surfaces (see, e.g., Leikin and Kornyshev, 1989); these do not appear in models which rely on the use of surface averages.

The interfacial hydration profile from Equation (12.4) is seen qualitatively to parallel the interfacial polarity distribution when the interfacial width is much greater than the hydration decay length. Of course, this pertains only to regions which are accessible to at least some of the surface polar residues: at separations greater than the largest distance accessible to a lipid headgroup standard expression for the hydration decay, Equation (12.5), is valid. The magnitude of d_p for phospholipids being at least 0.6 nm (Helm *et al.*, 1987; Wiener *et al.*, 1989) and sometimes greater than 1 nm (Helm *et al.*, 1987; Pfeiffer *et al.*, 1990; Als-Nielsen and Möhwald, 1991), the significance of finite interfacial thickness is more than obvious, however.

It is remarkable that the interfacial hydration strength in the described model depends quadratically on the relative magnitude of the intersolvent correlation length, $(\xi/d_p)^2$ (cf. Equation (12.4)). This suggests that only a limited number of water molecules can be bound per unit volume of interface. Hydrated lipid headgroups with numerous polar residues thus *must* 'escape in the third dimension'. By doing so they not only occupy more space perpendicular to the bilayer surface but, moreover, also increase the propensity for out-of-plane fluctuations.

6 Consequences of Lipid Hydration

Lipids and their water of hydration are one thermodynamic entity. Lipid–water interactions therefore affect the characteristics of a suspending solvent as well as the structural, dynamic and thermodynamic lipid properties. On the one hand, water activity coefficient and mobility both increase with increasing water concentration (Hammond *et al.*, 1987; see also Equation (12.1)) (Figure 12.8). On the other hand, the lipid packing density simultaneously decreases, while the segmental lipid mobility increases with this concentration (Volke and Gawrisch, 1982). Supramolecular order is also influenced by the bound water content, as is seen from Figure 12.9.

Bound Water Properties

Water Distribution

Water tends to occupy more or less uniformly the interbilayer space between adjacent membranes and also penetrates to some extent into the hydrophobic bilayer region. From the value of the dielectric constant in the 'interfacial' region it was concluded that water may penetrate up to the second methylene group of both chains (Ashcroft *et al.*, 1981). From the positive contribution to the apparent molar heat capacities arising from 'hydrophobic hydration' (Blume, 1983), as well as from the partitioning studies (Griffith *et al.*, 1974), up to 7 methylene groups per fluid chain were concluded to be hydrated. This value is similar to the estimate obtained for lipid micelles (Kresheck, 1975; Kumar and Raghunathan 1986; Casal,

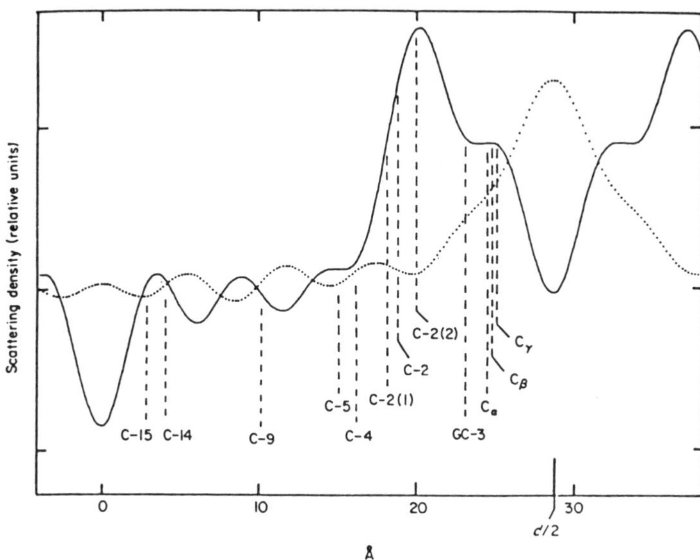

Figure 12.9 Water distribution profile deduced from neutron diffraction data (dotted) for a multilamellar stack of phosphatidylcholine bilayers at 6% water concentration and 20 °C. Labels identify positions of different deuteriated carbons in the chain (numerical indices) and headgroup (alphanumerical indices) regions. From Büldt *et al.* (1979)

1988) and thus should be considered as the upper limit for lipids in excess water. In the low-hydration region, neutron diffraction studies suggest, water penetration involves at least the carbonyl atoms (Büldt *et al.*, 1978, 1979; Smaby *et al.*, 1983)(Figure 12.10).

A substantial proportion of the lipid-bound water molecules in the low-hydration region is thus likely to be confined into the interfacial region. This applies to various lipids and has been shown in some detail for phosphatidylethanolamine (McIntosh and Simon, 1986b) and phosphatidylcholine bilayers (Pfeiffer *et al.*, 1990). For these lipids at least 50% of water (4 and > 10 molecules, respectively) can be inferred to be intercalated between the polar lipid heads.

Water Structure

At the molecular level, water orientation always follows *local electric field* orientation. This conclusion holds both for water molecules as a whole and for single water OH bonds and permits the main trends in the lipid–water interaction energies to be learned from electrostatic calculations. In spite of the resulting local inhomogeneities, the average value of the normal

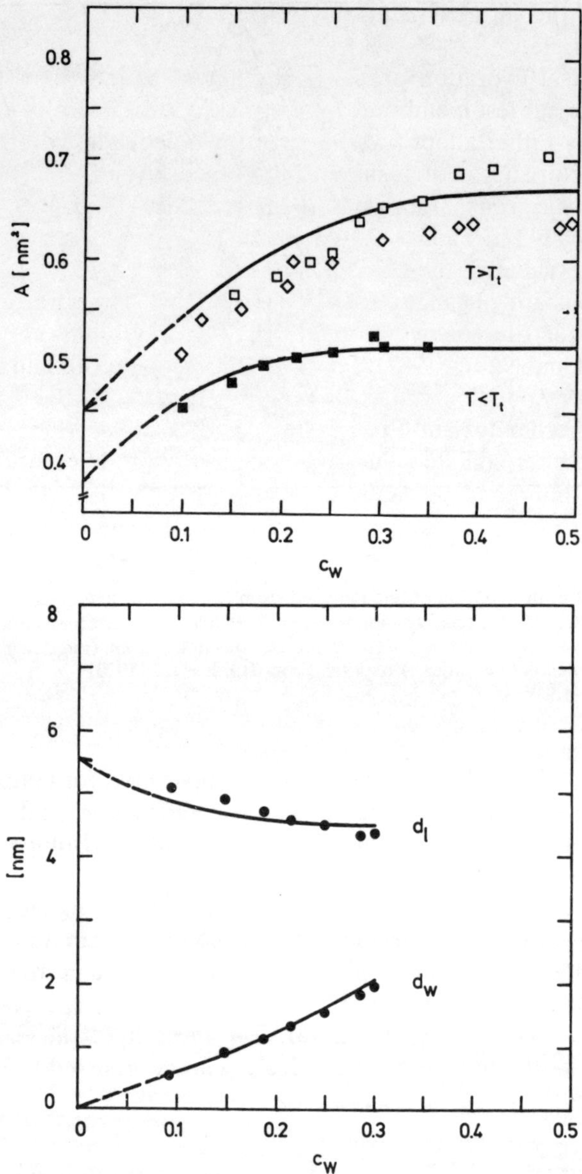

Figure 12.10 Hydration dependence of the lipid packing parameters in phosphatidylcholine multibilayers in systems with increasing water concentration, c_w. A is the area per lipid, d_l bilayer layer thickness and d_w water layer thickness between two polar surfaces. In part from Cevc and Seddon (1986a)

component of the surface-bound water inevitably decreases with distance from a membrane.

The interfacial hydration profile needs not to be exponential, as was assumed in the earliest membrane hydration models (Marčelja and Radić, 1976), but may rather adopt a variety of forms, depending on the precise surface characteristics. For a model membrane consisting of a layer of mobile carboxylic groups, for example, this profile has been found to decay quasiparabolically (cf. Figure 12.6; Schlenkrich *et al.*, 1990).

The average value of the static component of the relative permittivity ('dielectric constant') changes at the chain-melting phase transition from 15.5 to 14.5 for dimyristoylphosphatidylcholine multibilayers in excess water (Enders and Nimtz, 1984). For the membrane interior and interfacial region the corresponding values have been estimated to be 2.5 and 30, respectively (Fernandez and Fromherz, 1978; Cevc *et al.*, 1981; Cevc and Marsh, 1983), intermediate values being measured at different intermediate depths (Kimura and Ikegami, 1985).

Considering the orientation of membrane-associated water molecules, the two possible in-plane librations appear to involve large components of molecular dipole moment oscillations, while the normal component does not. This theoretical implication (Schlenkrich *et al.*, 1990) is in qualitative agreement with the conclusions from the NMR measurements of the orientation dependence of deuterium transverse relaxation time for flat phosphatidylcholine bilayers (Volke, 1984). Surface undulations may introduce additional averaging along the bilayer normal. Ultimately, they may bring the measured water order parameter to zero, even for strongly hydrated lipid bilayers (Füldner, 1981).

On the basis of the dielectric relaxation measurements it has been concluded that the average viscosity of the first 13 and 20 molecules of the lipid-bound water is greater by a factor of approximately 3 or 2, respectively, than that of the bulk water at a comparable temperature. Essentially, all of the interbilayer water is more viscous than the bulk water at 0 °C (Nimtz and Weiss, 1987).

Water Dynamics

A large proportion of the experimental data on the dynamics of lipid-associated water stem from deuterium magnetic resonance measurements (Finer, 1973; Finer and Darke, 1974; Ulmius *et al.*, 1977; Klose and Gawrisch, 1981; Llor and Rigny, 1986). This method has a limited spatial and time resolution, however. Moreover, the quantitative interpretation of its conclusion is complicated by the perturbing effect of interfacial geometry (Füldner, 1981). It has therefore become customary in the magnetic resonance data to distinguish between only a few interlamellar-water classes: the trapped, bound and strongly bound water. Such a distinction is somewhat artificial, however, since the dielectric relaxation measurements indicate that the dynamic properties of bound water change quasi-continuously with increasing lipid hydration (Enders and Nimtz, 1984).

Molecular Relaxation

Theoretical investigations (Geiger *et al.*, 1979; Rossky and Karplus, 1979; Linse, 1981), as well as nuclear magnetic resonance or X-ray and neutron scattering, show that the translational and rotational motions of the 'hydration-shell' water molecules are slower by at least 20%, and sometimes as much as 400%, than those in pure bulk water.

The formation of free electrons in liquid water and solvation are both ultrafast processes; a spectroscopically established upper limit of 0.3 ps has been established for the solvation time (Wiesenfeld and Ippen, 1980). Rotational relaxation times of free water are one order of magnitude longer, 3 ps.

Rotational correlation times of the lipid-associated water are greatly distorted by the strong local fields near the lipid headgroups. These correlation times thus gradually become longer and the exchange rates shorter as the total number of water molecules between adjacent lipid bilayers becomes smaller. Experiments and molecular dynamics simulations of the aqueous core of a reversed ionic micelle (Linse, 1981) or bilayer solution interface (Peinel *et al.*, 1983b) all agree in this (Table 12.1). The rate of effective rotation, for example, increases almost linearly with the water concentration up to a limiting hydration number of 22 ± 2 water molecules per lipid, in the case of phosphatidylcholine (Ulrich *et al.*, 1991). Other lipids achieve similar values at different hydration limits (Llor and Rigny, 1986). The mean amplitude and the activation energy of motion of the deuteriated headgroup segment is hydration-independent, however (Ulrich *et al.*, 1991).

At constant water concentration, the water correlation times increase by a few per cent with chain fluidization, indicative of stronger water–headgroup interactions in the disordered bilayer phase (Enders and Nimtz, 1984).

Diffusion

Since the properties of lipid and water molecules are interdependent, it is not surprising that the self-diffusion constant of the lipid-bound water is strongly affected by the interfacial separation (Table 12.1).

Calculated translational diffusion times for the water molecules near a model membrane surface are by 50–100% longer if motion proceeds along the surface rather than away from the membrane. The corresponding absolute values increase by a factor of approximately 2 for every additional

Table 12.1 Parameters of water between hydrated fluid–lamellar phosphatidylcholine bilayers

Class of water[a]	*Free* ($n_w \geqslant 25$)	*'Trapped'* ($11 < n_w < 25$)	*'Bound'* ($1 < n_w < 11$)	*'Strongly bound'* ($n_w = 1$)	*Ref.*
τ_{rot}	3×10^{-12}s	3×10^{-10}s	8×10^{-10}s	10^{-7}s	Finer (1973)
τ_{exe}	10^{-2}s	10^{-4}s	10^{-4}s		Klose and Gawrisch (1981)
$D_{w,\ lat}$	10^{-5} cm^2/s	(5×10^{-6} cm^2/s)	2×10^{-9} cm^2/s		Klose and Gawrisch (1981)
$D_{w,\ trans}$	3.5×10^{-6} cm^2/s	1.2×10^{-6} cm^2/s	5×10^{-7} cm^2/s		Rigaud *et al.* (1972)
viscosity	1 mP	2 mP	$\geqslant 4$ mP	($\geqslant 5$ mP)	Nimtz and Weiss (1987)
rel. density	1	1	~ 2	$\simeq 2$	various simulations

[a] $D_{w,\ trans}$ from ^{3}HHO diffusion measurements, and all other data from deuterium magnetic resonance experiments. Viscosity data based on dielectric measurements.

0.1 nm of separation, being of the order of 4×10^{-6} cm^2/s and 3.5×10^{-6} cm^2/s, respectively, for the most strongly bound, 'inner layer' water and for the water on the verge of continuum. Here, at approximately 0.7 nm separation, nearly the bulk diffusion constant values are found (Schlenkrich *et al.*, 1990). Concomitantly, the mean water diffusion path along the surface becomes longer and longer as the total number of lipid molecules increases (Klose and Gawrisch, 1981). The diffusion coefficient simultaneously increases from 0.5 to 4×10^{-6} cm^2/s in the case of phosphatidylcholine in 20 – 40% water (Rigaud *et al.*, 1972). The mutual diffusion coefficient for the monoolein–water system, $0.72–1.5 \times 10^{-6}$ cm^2/s, is of similar order of magnitude (Gerritsen and Caffrey, 1990) but by a factor of nearly 10 lower than the self-diffusion values seen in nuclear magnetic resonance experiments (Lindblom *et al.*, 1979).

Some of the electrical bilayer properties reflect similar hydration dependence. The lateral membrane conductivity, for example, increases dramatically upon the binding of the first water molecules to a layer of lipid headgroups (Jendrasiak and Mendible, 1976a, b).

Permeation

The permeability of water across a lipid bilayer depends partly on the lipid headgroup type (Fettiplace, 1978) and, even more importantly, on the physicochemical membrane properties: the more ordered and condensed the lipid bilayer is, the lower is its permeability to water molecules. This holds particularly for cholesterol-free lipid membranes below their chain-melting phase transition. The reason for this is that the rate-limiting step for water permeation through an ordered lipid bilayer seems to be the intracore diffusion. For membranes above the chain-melting phase transition, or for cholesterol-containing vesicles, the hydrocarbon membrane core is a much less effective diffusion barrier, however. This is seen from model calculations (Arakelian and Arakelian, 1983) and even more so from a comparison of the self-diffusion coefficients for water across pure, ordered-phase phosphatidylcholine membranes, which are of the order of 10^{-7} cm^2/s, with the corresponding values for fluid bilayers. The latter are of the order of 2×10^{-5} cm^2/s, i.e. comparable to the self-diffusion coefficient of water in the bulk phase (Lawaczeck, 1979; Engelbert and Lawaczeck, 1985).

In contrast to ion permeation across phosphatidylcholine bilayers, the water permeability, P_w, does not show a maximum at the phase transition. Instead, it reveals an almost discontinuous break (Deamer and Bramhall, 1986). Exceptions are the bilayers at the air–water interface, which show a singularly high permeability for water just at the phase transition (Ginsberg and Gershfeld, 1985).

For diacylphosphatidylcholine membranes, the activation energy for transmembrane water permeation is almost independent of the chain-length, being 104.5–117 kJ/mol, and 46 kJ/mole, below and above the chain melting phase transition temperature, respectively (Böhler *et al.*, 1978). The corresponding permeability values for water across dipalmitoylphosphatidylcholine bilayers are 4.2×10^{-6} cm/s, 3.5×10^{-5} cm/s and 2.4×10^{-3} at 26.5, 37 and 46 °C, respectively. After incorporation of 5–20 mol-% cholesterol the activation energy in gel phase bilayers decreases to 92 kJ/mol.

Effects on Lipids

Dry lipid molecules form crystals in which single chains or their segments are largely parallel and sometimes tilted at an angle relative to the plane of lipid layers (Elder *et al.*, 1977; Hauser *et al.*, 1981). Bulky lipid headgroups tend to interdigitate partly in order to retain close chain packing and to establish networks of the intermolecular hydrogen bonds (Pascher and Sundell, 1986a, b). In general, the crystal structure of nearly anhydrous phospholipids is governed by the steric and van der Waals forces between hydrocarbon chains and by the formation of hydrogen bonds in the hydrophilic headgroup region. If direct interlipid hydrogen bonds are not possible, as in the case of phosphatidylcholines, intermolecular cohesion may be mediated by one or two water molecules (cf. Figure 12.2).

X-Ray crystallography reveals that the bilayers consisting of sugar lipids are organized in very much the same way as phospholipid bilayers. Synthetic cerebrosides, for example, also pack in a bilayer arrangement, with the galactosyl groups aligned almost parallel to the membrane surface (Pascher and Sundell, 1977). Their water binding residues, moreover, are similar to some of the glyceride or phosphatidylglycerol binding sites.

Lipid Structure

Hydration initially does not diminish much the overall lipid packing density and positional correlation lengths within each lipid layer. The latter are typically of the order of 10–20 nm in the gel phase, consisting of dipalmitoylphosphatidylcholines with less than two bound water molecules (Seul and Eisenberger, 1989a); this encompasses a region with some 500–1000 lipid molecules. Extensive hydration may decrease this length, however: the size of cooperativity unit for the phosphatidylcholine chain melting in the fully hydrated state includes just about 150 molecules. Moreover, strong correlations appear to be restricted to the bilayer structures, since the corresponding correlation lengths are dramatically smaller in micellar solutions consisting of single-chain, lysophosphatidylcholines (Kaatze *et al.*, 1979).

Other lipids which can create an extensive intermolecular hydrogen bond network in the headgroup region are more likely to retain strong correlations at maximum hydration. This may be due to the fact that the water-binding capacity of such lipids is not very high. Phosphatidylethanolamines, for example, which are mutually hydrogen-bonded, have a tendency even to expel bound water and then form dry lipid crystals with tilted chains (Seddon *et al.*, 1983b; Tenchov *et al.*, 1984). Upon chain fluidization these crystals rehydrate, the resulting intermembrane water layer thickness being ultimately approximately 9 nm.

Anhydrous natural and synthetic cerebrosides are all metastable, owing to their tendency to form mutual hydrogen bonds. Upon heating, typically between 50 °C and 70 °C for the fully saturated substances, dry cerebroside multilamellae form a set of hydrated two dimensional cerebroside crystals. These are stabilized by both a highly ordered chain-packing mode and a lateral intermolecular hydrogen-bonding network which involves the sphingosine backbone, the galactosyl group and, presumably, 4 ± 1 non-freezable interbilayer water molecules (Ruocco *et al.*, 1981, 1983; Ruocco and Shipley, 1983). These structures thus resemble three-dimensional cerebroside crystals which also contain a few water molecules.

Chain fluidization promotes cerebroside swelling and causes the interlamellar water layer to attain a thickness of approximately 1.4 nm (Ruocco *et al.*, 1981; Reed and Shipley, 1987). This corresponds to some 12 water molecules per cerebroside. If cerebrosides are mixed with glycerophospholipids, even greater interfacial separations, close to 3 nm, can be achieved (Amin *et al.*, 1989). This is probably due to the better accessibility of the individual water binding sites on the glycolipid molecules and to the phospholipid-enforced increase in the interfacial repulsion.

The propensity of maximally hydrated lipids to dehydrate spontaneously, in general, decreases with the increasing temperature, lipid polarity and chain length. The hydrated lamellar gel phase of diacyl phosphatidylethanolamines is stable, for example, when $n_C \geqslant 19$.

Complexes of polar and apolar lipids are normally less hydrophilic than the corresponding pure polar component. Simple addition of inert apolar molecules into lipid bilayers, however, may increase the overall system hydrophilicity, owing to the increased headground accessibility to water.

Cholesterol, for example, promotes swelling of lipid bilayers (McIntosh, 1978; McIntosh *et al.*, 1989b). This occurs in spite of the fact that each cholesterol can adsorb only one water molecule (Jendrasiak and Hasty, 1974). One possible explanation for this is that the additional water uptake results from the increased lipid headgroup mobility (Shepherd and Büldt, 1979), larger interfacial thickness (Cevc, 1991a) and extra sterical repulsion (Simon and McIntosh, 1991) between the mixed phospholipid–cholesterol membranes. Increased probability of transmembrane lipid migration (flip-flop) (Inglefield *et al.*, 1976) supports such explanation.

Interfacially bound hydrophilic or small amphiphilic molecules, on the contrary, can increase the effective bilayer hydration potential considerably. Such molecules, therefore, not only may promote the uptake of water by lipid molecules and increase the interbilayer separation but, moreover, can also induce isothermal phase changes. The best-known examples are the alcohol-induced transitions between an ordinary gel bilayer (L_{β}) and an interdigitated lamellar gel ($L_{\beta,I}$). These have been observed for diacyl- (Rowe, 1983; Simon and McIntosh, 1984) as well as dialkyl-phosphatidylcholines (Veiro *et al.*, 1988).

Lipid Dynamics

Lipid mobility increases with increasing hydration, owing to the diminished intermolecular steric hindrance. This is true for segmental as well as molecular mobilities. An example of increased segmental mobility is the hindered rotation of phosphatidylcholine headgroups, observed in the presence of > 20% of water (Füldner, 1981). Lipids with more than approximately 6 bound water molecules (Volke and Gawrisch, 1982) can exchange between two energetically comparably favourable enantiomeric states of the lipid headgroups. With increasing hydration this rotation speeds up. An effective order parameter of approximately 0.6 is obtained for such fully hydrated phosphatidylcholine headgroups in the lamellar gel (Seelig, 1977), a value of 0.45 being characteristic for the fluid phase (Volke and Gawrisch, 1982). Addition of osmotically active molecules can slow down headgroup motion appreciably (Viti and Minetti, 1981).

The situation with the rotational mobility of lipid chains is quite similar. The frequency at which the hydrocarbon chains are rotating about their long axes increases with the bound water content, essentially unhindered rotation being achieved at approximately 5 H_2O/lipid in the case of phosphatidylcholine. This also pertains to the rotation frequency of chain termini, which is of the order of 2–12 GHz and increases strongly with hydration (Enders and Nimtz, 1984). The chains of other, less strongly hydrated lipids rotate slower than those of phosphatidylcholine at similar relative temperatures, owing to the denser lipid packing.

The lipid lateral diffusion constant, D_l, increases with bilayer hydration (Fisher and James, 1978; McCown *et al.*, 1981). The corresponding functional dependence is similar to that of the lipid molecular area, or its excess value, changing with water concentration (Cevc and Marsh, 1987). Values of $D_l = 10^{-9}$ cm^2/s and 4×10^{-9} cm^2/s have been reported for the phosphatidylcholine di-hydrate and the same lipid in excess water, respectively.

Lipid Polymorphism

The simplest distinction between two lipid phases is in terms of chain order: a lipid membrane is either in a crystalline or ordered (gel-like) or else in a disordered (fluid or liquid-crystalline) phase (Cevc and Marsh,

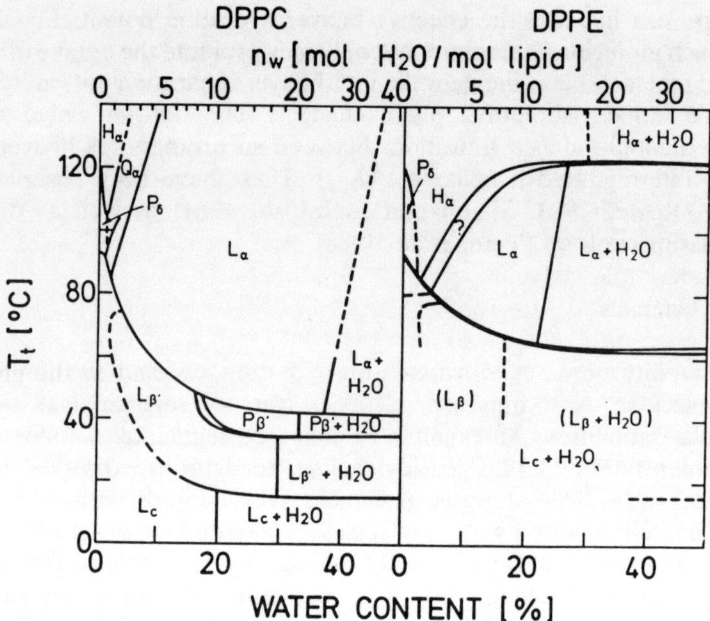

Figure 12.11 Illustrative phase diagrams of dipalmitoylphosphatidylcholine (DPPC) and dipalmitoylphosphatidylethanolamine (DPPE) mixtures with water. Dashed lines indicate tentative phase boundaries. Some of the phases, such as L_β-state of DPPE, are metastable. These revert spontaneously into crystalline phases. Partly based on the work of Jürgens *et al.* (1983), Kodama *et al.* (1982a) and Seddon *et al.* (1983a); modified from Cevc and Marsh (1987)

1987). Transitions between such phases, and also other possible lipid states, have long been known to be strongly influenced by the water content (Small, 1967; Lawrence *et al.*, 1967; Tardieu *et al.*, 1973). Binary phase diagrams of two common phospholipids shown in Figure 12.11 exemplify this.

Subtransition

The structure of hydrated lipids at relatively low temperatures is similar to that of lipid crystals: they form densely packed crystalline lipid lamellae (Mulukutla and Shipley, 1984; Ruocco and Shipley, 1982). With increasing temperature, these become energetically unfavourable owing to the thermal, initially chiefly rotational, chain-excitations. In consequence, at *subtransition* temperature, $T = T_s$, the two-dimensional lipid crystals revert into a more expanded lipid-gel phase of the β- or, more frequently, β′-type, with untilted or tilted chains, respectively (Tardieu *et al.*, 1973). The higher the hydrophilicity of a given lipid, the more likely such a lipid will form β-type structures. Hydrocarbon tilt in lipid crystals is largely a function of the headgroup packing properties, however.

Subtransition temperature changes very strongly with water content in the low-hydration region. Notwithstanding this, headgroup interactions and packing effects are dominant.

Pretransition

Heating lipids in the gel L_{β} - or $L_{\beta'}$ -phase speeds up oscillations of the hydrocarbon chains. This may, but need not, terminate in an essentially unhindered, long-axis chain rotation (Füldner, 1981; Boroske and Trahms, 1983; Trahms *et al.*, 1983) often in a cooperative manner at the *pretransition* temperature, $T = T_p$. Lipid headgroup mobility, most notably the rotation of the lipid headgroups around the P–O-bond to the glycerol backbone (Shepherd and Büldt, 1978), as well as the interfacial area per molecule (Janiak *et al.*, 1979) all increase at $T = T_p$. The area per phosphatidylcholine, for example, increases from 0.525 nm^2 to 0.65 nm^2 at the pretransition temperature (Parsegian, 1983). The concomitant change in the chain area is only a few per cent, however.

Lipid pretransition is observed solely if the polar lipid headgroups are sufficiently hydrated and if the packing between fully saturated chains is sufficiently weak. The amplitude of the bilayer surface ripples (Feuer *et al.*, 1982; Wack and Webb, 1988) and the pretransition enthalpy (Kodama *et al.*, 1982b) decrease from 4 nm and approximately 5 kJ/mol, respectively, to essentially zero with decreasing number of the lipid-bound water molecules. The minimum number of lipid-bound water molecules required for a lipid pretransition is approximately 11 per lipid (Cevc, 1991b)*. This value is essentially chain-length-independent and unaffected by the method of hydration-variation: physical dehydration, the hydrational competition between lipid molecules and substances dissolved in the aqueous subphase, or decreasing the lipid headgroup polarity all affect the pretransition temperature similarly on the appropriate scale (Cevc, 1991b).

Chain-melting Phase Transition

In the low-temperature gel phases the hydrocarbon chains are in an orientationally well-ordered state in which the hydrophobic molecular segments are nearly completely in an all-*trans* configuration. Their extension under such conditions thus is close to the possible maximum (see, e.g., Tardieu *et al.*, 1973; Janiak *et al.*, 1979; Ruocco and Shipley, 1982). At higher temperatures, however, such high chain order is lost, owing to the orientational chain excitations. This results in a cooperative, first-order *chain-melting*

* The minimum lipid hydrophilicity and the critical amount of the lipid-bound water required for the creation of bilayer ripples can be related to the lipid chain-melting transition temperature; the latter must not exceed some maximal, chain-length-dependent value, which is phenomenologically described by equation $T_{p,\,lim}(n) = T_m = 414\ (1 - 2.8/n - 14.5/n^2)$ K, if the lipid pretransition is to exist.

(*order–disorder, gel-to-fluid*) *phase transition* at $T = T_m$ which leads to lateral membrane expansion by 5–15% and to simultaneous thinning of the hydrophobic bilayer core (Janiak *et al.*, 1979).

Chain-melting phase transition temperature and, less so, transition enthalpy inevitably decrease with increasing lipid hydration (Kreissler *et al.*, 1983; Cevc, 1988); the smaller the transition enthalpy, the stronger this effect (see further discussion). Lipids with short or unsaturated chains, as well as asymmetric-chain lipids (Cevc, 1991c), therefore, are most sensitive to the changes in water concentration (Cevc, 1991d; Figure 12.12).

Chain order in the fluid-phase lipid bilayers decreases toward the membrane centre (Gruen, 1980), approximately linearly along the bilayer normal. However, if visualized along the chain axes, this order has a plateau in the upper hydrocarbon region. The width of this plateau, for fully saturated hydrocarbons, decreases with increasing relative temperature and lipid hydration.

Non-lamellar Fluid Phases

When the repulsion between hydrocarbon chains is greater than that of the oppositely directed interfacial repulsion, lamellar lipid bilayers are no longer stable (Cevc, to be published). Strong chain-disorder ($T \geqslant T_m$), chain unsaturation (Lewis *et al.*, 1989), addition of long-chain fat-soluble substances (Kirk and Gruner, 1985; Cevc *et al.*, 1981; Lindblom *et al.*, 1988) and/or a tendency for the tight headgroup packing all lead to such a situation. Consequently, all these factors also may induce the bilayer–non-bilayer phase transitions. Extensive hydration, on the contrary, interferes with the formation of non-bilayer phases (Seddon *et al.*, 1983a; Charvolin, 1985; Yeagle and Sen, 1986).

Hydrocarbon chains in the non-bilayer structures form a continuous lipid matrix in which the water molecules are confined into parallel, hexagonally packed 'cylinders' or one or two water-channel networks for the inverted hexagonal, H_{II} -, or corresponding cubic, Q_α phases, respectively (Seddon, 1990). While the chain-packing density in all non-bilayer phases is lower than that of the corresponding fluid lipid bilayers, the area per lipid headgroup and probably the water-binding strength in the former case correspond to that of the gel phase bilayers at comparable temperature (Seddon *et al.*, 1984; Tate and Gruner, 1989). Nothwithstanding this, the overall water uptake in non-bilayer lipid phases is higher than in the fluid lamellae, for geometric reasons.

Phase Transition Shifts

In fully hydrated systems, the thermodynamic effects of lipid–water association are typically by a factor of 3–10 greater than the effects of Coulom-

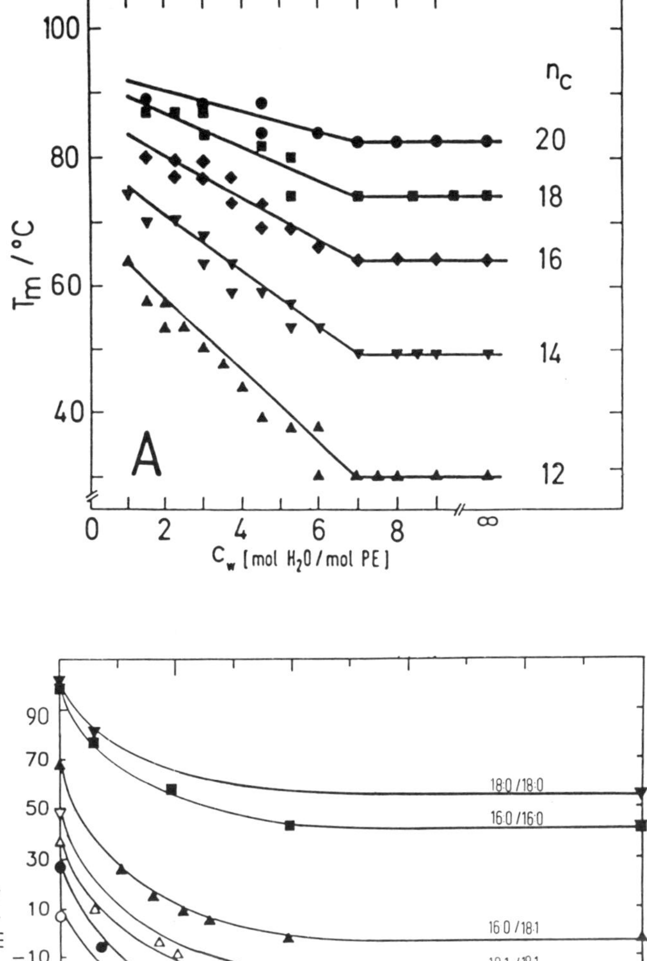

Figure 12.12 Hydration dependence of the chain-melting phase transition temperature: (A) of various diacylphosphatidylethanolamines and (B) of phosphatidylcholines with different degrees of chain-unsaturation. Modified from Cevc and Marsh (1985) and Lynch and Steponkus (1989), respectively. Lines are drawn solely to guide the eye. The sensitivity of lipid-phase transitions to hydration increases with decreasing chain-length and degree of desaturation.

bic membrane potential (Cevc *et al.*, 1986; Cevc, 1989). Variations in lipid ionization which change the net surface charge density, as well as the effects of membrane surface hydrophilicity, are thus prone to cause major shifts of the lipid phase transition temperatures, mainly owing to the latter effect. Chain-melting phase transition temperature therefore decreases with the increasing degree of lipid headgroup deprotonation and thus with pH, irrespective of the resulting net surface charge density.

Different headgroup effects can be rationalized quantitatively (Cevc, 1987) in terms of the various contributions from the polar membrane region to the corresponding total shift of the transition temperature:

$$\begin{aligned}\Delta T_{m,p} &= \sum_i \Delta G_{P_i,m}/\Delta S_{ref,m} \\ &\equiv \sum_i \Delta T_{m,i} = \Delta T_{m,el} + (\Delta T^{PO_4}_{m,h} + \Delta T^{H}_{m,h} + T^{=O}_{m,h}) \\ &\quad + \Delta T_{m,bond} + \Delta T_{m,vdw} + \cdots \end{aligned} \tag{12.6}$$

each individual contribution stemming from one type of the free energy change $\Delta G_{P_i,m}$ at the phase transition. $\Delta S_{ref,m}$ is the reference transition entropy change and in the first approximation identical with the corresponding value measured for the lipid anhydrides:

$$\Delta S_{ref,m}(n_C) \simeq \Delta S_{anh,m}(n_C) \simeq 2(n_C - 5.5) \times (7.5 \pm 1.5)\ \mathrm{J\ K^{-1}\ mol^{-1}} \tag{12.7}$$

For the sake of clarity, the hydration shift in Equation (12.6) is subdivided into three parts (in parentheses). The first one is associated with the phosphate (de)protonation, $\Delta T^{PO_4}_{m,h}$; the second reflects the thermodynamic consequences of a change in lipid hydration arising from changes in the number of available protons on a (charged) headgroup by one, $\Delta T^{H}_{m,h}$; and the third, $\Delta T^{=O}_{m,h}$, pertains to the carbonyl-group hydration.

By means of Equation (12.6) the differences in the transition temperatures of various lipids in excess solution can be analysed and used predictively (Cevc, 1987), once individual shift values have been assigned on the basis of experimental data. For the glycerophospholipids with 18 carbon atoms per chain, for example, one obtains:

$$(\Delta T_{m,bond} \sim 1 \pm 0.75\ \mathrm{K}) \sim (\Delta T^{=O}_{m,h} = 2 \pm 0.5\ \mathrm{K}) < (\Delta T_{m,el} \sim 3.5\ \mathrm{K}) \leqslant (\Delta T^{H}_{m,h} \leqslant 5\ \mathrm{K}) < (\Delta T^{PO_4}_{m,h} \leqslant 9.5\ \mathrm{K})$$

if the shift from van der Waals interactions is neglected. For dimyristoylphospholipids with 14 carbon atoms per chain the corresponding values are greater by $\Delta S_{ref,m}(18)/\Delta S_{ref,m}(14) = 1.5$, giving:

$$(\Delta T_{m,\,bond} \sim 1.5 \pm 1\ K) \leqslant (\Delta T_{m,\,el} \sim 5.5 \pm 0.5)\ K \leqslant$$
$$(\Delta T^{H}_{m,\,h} \sim 7 \pm 1\ K) < (\Delta T^{PO_4}_{m,\,h} \sim 13 \pm 3)\ K$$

The precise values vary somewhat with the detailed lipid type and state. The electrostatic shift arising from the second charge on a lipid molecule is relatively small, $\Delta T_{m,\,el}(++,\,--) - \Delta T_{m,\,el}(+,\,-) \leqslant 3$ K.

Changes in lipid ionization (protonation), consequently, affect the lipid phase behaviour chiefly through the modification in lipid headgroup hydration. This explains why lipid headgroup protonation, which eliminates some of the water-binding sites, always increases the chain-melting phase transition temperature. For the same reason, the (de)protonation-induced phase transition shifts are screenable only incompletely by salts.

While it is clear that the water uptake by lipids increases with the headgroup size, this interdependence is by no means trivial. The chain-melting phase transition temperature of the phosphatidylcholine analogues with an increased phosphate-quaternary ammonium group separation, for example, oscillates between 40 and 45 °C, with an even–odd specificity (Bach *et al.*, 1978). The phase transition temperature of comparably modified phosphatidylethanolamines changes by nearly an order of magnitude more strongly (Seddon *et al.*, 1983a; Cevc, 1989)[†].

7 Lipid Aggregation in Water

Lipids in water normally form micelles (which preferentially arise from substances with a high hydrophilicity/lipophilicity ratio) or lipid bilayers (which are typically found when the energetic contributions from the polar and apolar lipid parts are approximately balanced; Israelachvili *et al.*, 1980). When non-bilayer structures appear, these are diagnostic of dominance of the hydrophilic over the hydrophobic lipid segments.

The propensity for lipid aggregation varies strongly with the *relative* lipid hydrophilicity. The solubility limit for phosphatidylcholine with 10 carbon atoms/chain in water, for example, is approximately 5×10^{-5} mol/l; each additional CH_2 group in the acyl lipid chain decreases this value by a factor of approximately 15 (to reach 5×10^{-10} mol/l for C18-phosphatidylcholines). Headgroup deprotonation has the opposite effect. Charged, deprotonated lipids are typically 3–4 orders of magnitude more water-soluble (Gershfeld, 1989) than are the homologous fully protonated

[†] The average orientation and dynamics of lipid headgroups are always more strongly affected by the lipid headgroup size than is the chain-melting phase transition temperature (Siminovitch *et al.*, 1983). This finding, in combination with the relative constancy of the transition temperature, corroborates the conclusion that nominal dipole moment of the lipid headgroups is not important for the lipid phase behaviour and the strength of direct headgroup–water binding.

substances (De Cuyper *et al.*, 1984; De Cuyper and Joniau, 1985). Chain desaturation also increases lipid solubility in water by approximately one order of magnitude (Daniels *et al.*, 1985).

The free energy of aggregation and the critical lipid micelle (CMC) or bilayer (CBC) concentrations, i.e. the concentrations at which lipid aggregates begin to form, thus interdepend. They increase with the lipid chain-length. This dependence for a single-chain, micelle-forming phospholipid can be expressed as $\ln[\mathrm{CMC}] = 0.2 - 1.1n_{\mathrm{C}}$, where n_{C} is the total number of carbon atoms per chain (excluding the carbonyl group). An identical relation also holds for the transfer of n-alkanols from water to a phospholipid bilayer. For the bilayer forming double-chain lipids one has, however, $\ln[\mathrm{CBC}] = -0.4 - 1.7n_{\mathrm{C}}$, the incremental energy now being appreciably, but by less than a factor of 2, greater.

8 Hydration Force

The repulsion between lipid aggregates in water is dealt with in more detail elsewhere in this book. One possible theoretical approach for its evaluation is introduced in Section 5. Consequently, only a brief interpretation of this important phenomenon, and some of its implications, is next given.

Hydration force arises because of the lipid tendency to bind water. This decreases the local water activity coefficient (cf. Figure 12.8) and creates an excess osmotic pressure between interacting surfaces (Figure 12.13). The resulting inward flow of the solvent molecules then pushes the adjacent polar surfaces apart.

Hydration pressure, consequently, occurs for all hydrophilic surfaces, such as lipid bilayers (Rand and Parsegian, 1990), deoxyribonucleic acids (Rau *et al.*, 1984; Reddy and Berkowitz, 1989), polysaccharides, etc. Within the framework of the model introduced in Section 5, this force is concluded to be proportional to the square of the surface affinity for water binding, the latter being expressed in terms of the surface density of local-excess charge, dipole moment, quadrupole moment, etc. This means that, in principle, $p_{\mathrm{h}} \propto \sigma_{\mathrm{p}}^{2}$ (cf. Equations (12.4) and (12.5)[‡]. Hydration-induced changes of the lipid headgroup conformation (Cevc and Seddon, 1986) or hydration-induced lateral phase separation (Bryant and Wolfe, 1989; Cevc, 1991a; cf. Figure 12.10) therefore affect the water-dependent interfacial repulsion.

Hydration force is believed by many to be just a consequence of the spatial decay of the interfacially induced solvent perturbation, as has been

[‡] A proposition has been put forward that hydration force depends on the macroscopic bilayer dipole potential, which arises from the oriented dipoles in the interfacial region, including lipid headgroups and solvent molecules (Simon and McIntosh, 1989). While it is

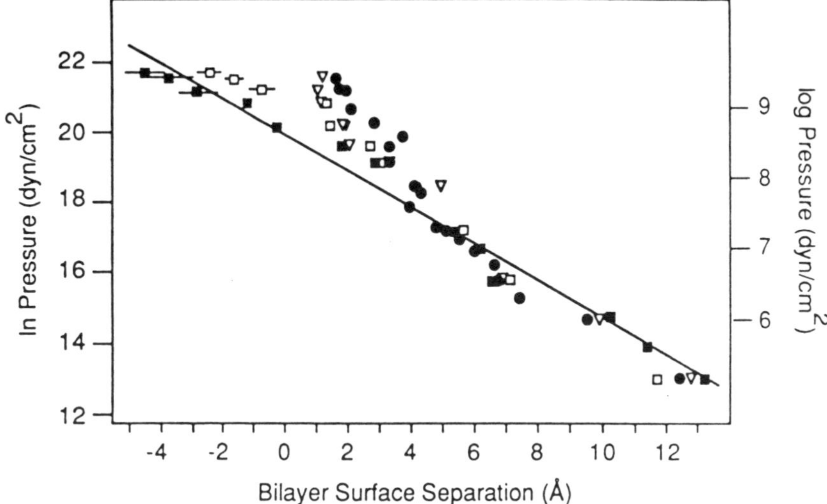

Figure 12.13 Repulsion between hydrated egg-yolk phosphatidylcholine multibilayers containing no (●), 0.2 (▼), 0.33 (□) and 0.5 (■) mole fraction cholesterol. The assignment of bilayer surface separation is a matter of convention. The solid line represents a best fit of the standard, exponential hydration force model to the equimolar phospholipid/cholesterol data, which are clearly non-linear. From McIntosh *et al.* (1989b)

proposed in a seminal, original model of this force (Marčelja and Radić, 1976). This implies that the profile of hydration-force should depend on the solvent characteristics solely and invariably should fall off on the length-scale of mutual correlations between the water molecules. The latter is of the order of either less than 0.1 nm or approximately 0.3 nm, depending on the underlying 'water coupling' mode (Dogonadze *et al.*, 1973).

Recently, however, a more general explanation has been put forward by us (Cevc, 1990b, 1991a) and others (Israelachvili and Wennerström, 1990) which suggests that 'hydration force' and its decay length may also be affected by the lipid size and mobility.

Indeed, the experimental values for the 'hydration-force decay length' vary widely: between 0.13 (McIntosh *et al.*, 1989a) and 0.45 nm (Lyle and Tiddy, 1986) or even more. (For reviews of some of the experimental data see, for example, Rand, 1981, Rand *et al.*, 1985, or Rand and Parsegian, 1990.) In part, this variability may be due to the experimental uncertainty

---

certain that such bilayer surface potential to within 50% stems from the interfacially adsorbed water (Paltauf *et al.*, 1971; Cevc and Gaub, unpublished data), it is doubtful whether the interfacial repulsion of hydration indeed increases with the square of this potential, as has originally been proposed on purely theoretical grounds (Cevc, 1985). If this was the case, it would be difficult to understand why the dialkyl phospholipids, which lack the carbonyl groups and thus have a much lower surface dipole potential, take up amounts of water similar to or greater than the amounts taken up by the corresponding diacyl-phospholipids.

and the discrepancies between various analytical methods (McIntosh and Simon, 1986b; Rand and Parsegian, 1990). But the main variability source could be the dependence of 'hydration force' on the characteristics of and coupling between interacting surfaces.

Correlations between the distributions of the polar lipid headgroups in adjacent bilayers, for example, may modify the effective decay length of hydration force, diminish repulsion, and even give rise to a net intermembrane attraction (Colbow and Jones, 1974; Kjellander, 1984; Podgornik, 1989; Kornyshev and Leikin 1989; Leikin and Kornyshev, 1990, 1991). Short-range coupling between such bilayers, moreover, may reduce hydration force directly (Cevc, 1985). Extensive fluctuations of lipid headgroups may increase the interfacial repulsion (Cevc, 1990b, 1991a; Israelachvili and Wennerström, 1990). And last, but not least, the effects of final interfacial width (Cevc, 1990a, 1991a) and steric headgroup interactions (Simon *et al.*, 1988; McIntosh *et al.*, 1987) may cause hydration force to become a mirror of the detailed structural and surface properties (Cevc, 1991a).

Effective interfacial polarity profile in the latter case is mapped into the spatial profile of the interfacial repulsion. If this profile is temperature-dependent, this is diagnostic of the direct contributions from the changing lipid headgroup conformation and/or location of the individual lipid molecules in lipid bilayers. Strong, sometimes non-linear, increase with temperature of the repulsion between fluid phosphatidylcholine bilayers has, indeed, been observed experimentally (Smith *et al.*, 1987; Kirchner and Cevc, to be published).

Hydration force — and most other hydration phenomena — are not a specialty of lipid–water association. Other associable, normally protic solvents, such as glycerol (Persson and Bergenstahl, 1983; McDaniel *et al.*, 1983) or formamide (Bergenstahl and Stenius, 1987) are similarly efficient. Surface-adsorbed oligosaccharides, which offer numerous OH groups to the lipid headgroups in (partial) replacement for water, thus may preserve the integrity of lipid bilayers upon water removal (Crowe *et al.*, 1987).

9 Relevance of Lipid Hydration

Most of the polar residues encountered on lipids are also found on other biological macromolecules: phospholipids and desoxyribonucleic acid (DNA) or ribonucleic acid (RNA) all carry at least one negatively charged phosphate group; phosphatidylserines or phosphatidylethanolamines, as well as proteins or polypeptides, always possess a terminal amino group and may contain a serine. Phosphatidylinositol, phosphatidylglycerol and cardiolipin offer several OH groups (in the form of cyclic or straight-chain polyalcohols) for the binding of water. Such groups are also found on glycolipids, polysaccharides, DNA or RNA. Lipid hydration, consequently, has many features in common with the hydration of polypeptides, polynucleotides and polysaccharides. Knowledge about lipid hydration thus is enlightening for such systems as well.

Lipid hydration affects the selectivity of surface–solute interactions significantly. Bound water molecules at interfaces between a lipid bilayer and an aqueous solution can mediate the lateral transfer of charge and, consequently, are likely to contribute to the energy transport along membranes. Lipid dehydration — and the resulting lateral phase separation — can also significantly affect the flow of material across a membrane.

Lipid (de)hydration, moreover, appears to be biologically relevant.

Dehydration of lipids in biological membranes caused by different osmotically active polymers, such as polyethylene glycol (Arnold *et al.*, 1983; Tilcock and Fisher, 1979) or oligosaccharides (Sunamoto *et al.*, 1980), can promote the cell–cell fusion. Local water activity gradients can participate in membrane formation and reorganization (Cevc *et al.*, 1990a). Lipid dehydration has been proposed to be the origin of the freezing injury in plants (Lynch and Steponkus, 1989). Lipid hydrophilicity (and/or interbilayer hydration force) affects the sensitivity of various membranes to viral infection (White and Helenius, 1980). Furthermore, lipid hydration has been shown to influence the interactions between lipid bilayers and immunological molecules (Cevc *et al.*, 1990b) or cells (Allen *et al.*, 1989; Blume and Cevc, 1990).

Other examples could be given. They all show that lipid hydration is of paramount importance for the understanding of physicochemical phenomena in biocolloidal systems and for functioning of numerous biological systems.

References

Allen, T. M., Hansen, C. and Rutledge, J. (1989). Liposomes with prolonged circulation times: factors affecting uptake by reticuloendothelial and other tissues. *Biochim. Biophys. Acta*, **981**, 27–34

Amin, N., Collins, J. M., Tamura-Lis, W., Lis, L. J. and Quinn, P. J. (1989). Phase characterization of mixtures of dioleoylphosphatidylcholine, dioleoylphosphatidylethanolamine and cerebrosides. *Colloids and Surfaces*, **36**, 459–467

Arakelian, V. B. and Arakelian, S. B. (1983). Energetic profile of dipole molecules on the border of separation of two phases (in Russian). *Armenian Biol. J.*, **36**, 553–559

Arnold, K., Pratsch, L. and Gawrisch, K. (1983). Effect of poly(ethylene glycol) on phospholipid hydration and polarity of the external phase. *Biochim. Biophys. Acta*, **728**, 121–128

Ashcroft, R. G., Coster, H. G. L. and Smith, J. R. (1981). The molecular organisation of bimolecular lipid membranes. The dielectric structure of the hydrophobic/hydrophilic interface. *Biochim. Biophys. Acta*, **643**, 191–204

Bach, D., Bursuker, I., Eibl, H. and Miller, I. R. (1978). Differential scanning calorimetry of dipalmitoyl phosphatidylcholine analogues and of their interaction products with basic polypeptides *Biochim. Biophys. Acta*, **514**, 310–319

Barabino, G. and Marchesi, M. (1984). Molecular dynamics simulation of water near walls using an improved wall–water interaction potential. *Chem. Phys. Lett.*, **104**, 478–484

Belaya, M. L., Feigel'man, M. V. and Levadny, V. G. (1986). Hydration forces as a result of non-local water polarizability. *Chem. Phys. Lett.*, **126**, 361–364

Belaya, M. L., Feigel'man, M. V. and Levadny, V. G. (1987). Structural forces as a result of nonlocal water polarizability. *Langmuir*, **3**, 648–654

Bergenstahl, B. A. and Stenius, P. (1987). Phase diagrams of dioleoylphosphatidylcholine with formamide, methylformamide and dimethylformamide. *J. Phys. Chem.*, **91**, 5944–5948

Blume, A. (1983). Apparent molar heat capacities of phospholipids in aqueous dispersion. Effects of chain length and head group structure. *Biochemistry*, **22**, 5436–5442

Blume, G. and Cevc, G. (1990). Liposomes for the sustained drug release *in vivo*. *Biochim. Biophys. Acta*, **1029**, 91–97

Böhler, B. A., Gier, J. de and van Deenen, L. L. M. (1978). The effect of gramicidin a on the temperature dependence of water permeation through liposomal membranes prepared from phosphatidylcholines with different chain lengths. *Biochim. Biophys. Acta*, **512**, 480–488

Boroske, E. and Trahms, L. (1983). A^{1}H and ^{13}C NMR study of motional changes of dipalmitoyl lecithin associated with the pretransition. *Biophys. J.*, **42**, 275–283

Bryant, G. and Wolfe, J. (1989). Can hydration forces induce lateral phase separations in lamellar phases?. *Eur. Biophys. J.*, **16**, 369–374

Bucher, M. and Porter, T. L. (1986). Analysis of the Born model for hydration ions. *J. Phys. Chem.*, **90**, 3406–3411

Büldt, G., Gally, H. U., Seelig, A., Seelig, J. and Zaccai, G. (1978). Neutron diffraction studies on selectively deuterated phospholipids. *Nature*, **271**, 182–184

Büldt, G., Gally, H. U., Seelig, J. and Zaccai, G. (1979). Neutron diffraction studies on phosphatidylcholine model membranes. I. Head group conformation. *J. Mol. Biol.*, **134**, 637–691

Casal, H. L. (1988). On the water content of micelles: Infrared spectroscopic studies. *J. Am. Chem. Soc.*, **110**, 5203–5205

Cevc, G., Watts, A. and Marsh, D. (1981). Titration of the phase transition of phosphatidylserine bilayer membranes. Effects of pH, surface electrostatics, ion binding and headgroup hydration. *Biochemistry*, **20**, 4955–4965

Cevc, G., Podgornik, R. and Žekš, B. (1982). The free energy, enthalpy, and entropy of phospholipid bilayer hydration and their dependence on the interfacial separation. *Chem. Phys. Lett.*, **91**, 193–197

Cevc, G. (1985). Molecular force theory of solvation of the polar solutes. The mean-field solvation model, its implications and examples from lipid/water mixtures. *Chem. Scripta*, **25**, 97–107

Cevc, G. (1987). How membrane chain melting properties are controlled by the polar surface of the lipid bilayers. *Biochemistry*, **26**, 6305–6310

Cevc, G. (1988). Effects of lipid headgroups and (nonelectrolyte) solution on the structural and phase properties of bilayer membranes. *Ber. Bunsenges. Phys. Chem.*, **92**, 953–961

Cevc, G. (1989). Colloidal and phase behaviour of biomacromolecules interdepend and are regulated by the supramolecular surface polarity. Examples with lipid bilayer membranes *J. Phys. (Paris)*, **50**, 1117–1134

Cevc, G. (1990a). The molecular mechanism of interaction between monovalent ions and polar surfaces, such as lipid bilayer membranes. *Chem. Phys. Lett.*, **170**, 283–288

Cevc, G. (1990b). Membrane electrostatics. *Biochim. Biophys. Acta. Reviews on Membranes*, **1031–3**, 311–382

Cevc, G. (1991a). How membrane chain-melting phase transition temperature is affected by the lipid chain-asymmetry and degree of unsaturation: Analysis and predictions based on the effective chain-length model. *Biochemistry*, **30**, 7186–7193

Cevc, G. (1991b). Polymorphism of bilayer membranes in the ordered phase and the molecular origin of lipid pretransition and rippled lamellae. *Biochim. Biophys. Acta*, **1062**, 59–69

Cevc, G. (1991c). Hydration force depends on interfacial structure of the polar surface. *J. Chem. Soc. Faraday Trans.*, **II**, **87**, 2733–2739

Cevc, G. (1991d). Isothermal lipid phase transitions. *Chem. Phys. Lipids*, **57**, 293–307

Cevc, G., Fenzl, W. and Sigl, L. (1990a). Surface-induced X-ray reflection visualization of membrane orientation and fusion into multibilayers. *Science*, **249**, 1161–1163

Cevc, G. and Marsh, D. (1983). Properties of the electrical double layer near the interface between a charged bilayer membrane and electrolyte solution. *J. Phys. Chem.*, **87**, 376–379

Cevc, G. and Marsh, D. (1985). Hydration of noncharged lipid bilayer membranes theory and experiments with phosphatidylethanolamines. *Biophys. J.*, **47**, 21–31

Cevc, G. and Marsh, D. (1987a). In *Phospholipid Bilayers. Physical Principles and Models*. Wiley-Interscience, New York

Cevc, G., Seddon, J. M., Hartung, R. and Eggert, W. (1988). Properties of phosphatidylcholine-fatty acid bilayer membranes. I. Effects of protonation, salt, temperature, and chain length on the colloidal and phase behaviour. *Biochim. Biophys. Acta*, **940**, 219–240

Cevc, G. and Seddon, J. M. (1986a). Structural and dynamic consequences of amphiphile hydration. In Mittal, K. L. (Ed.), *Surfactants in Solution*, Vol. 4. Plenum Press, New York, pp. 243–252

Cevc, G., Seddon, J. M. and Marsh, D. (1985). Thermodynamic and structural properties of phosphatidylserine bilayer membranes in the presence of lithium ions and protons. *Biochim. Biophys. Acta*, **814**, 141–150

Cevc, G., Seddon, J. M. and Marsh, D. (1986b). The mechanism of regulation of membrane phase behaviour, structure, and interactions by lipid headgroups, and electrolyte solution. *Faraday Disc.*, **81**, 179–189

Cevc, G., Strohmaier, L., Berkholz, J. and Blume, G. (1990b). Molecular mechanism of protein interactions with the lipid bilayer membrane. *Stud. Biophys.*, **138**, 57–70

Charvolin, J. (1985). Crystals of interfaces: The cubic phases of amphiphile/water systems. *J. Phys. (Paris)*, **3**, 173–190

Chen, C.-H. (1982). Interactions of lipid vesicles with solvent in heavy and light water. *J. Phys. Chem.*, **86**, 3559–3562

Christenson, H. K. and Blom, C. E. (1987). Solvation forces and phase separation of water in a thin film of nonpolar liquid between mica surfaces. *J. Chem. Phys.*, **86**, 419–424

Christenson, H. K. and Horn, R. G. (1985). Solvation forces measured in non-aqueous liquids. *Chem. Scripta*, **25**, 37–41

Colbow, K. and Jones, B. L. (1974). On the stability of the liquid-crystalline lamellar lecithin–water system. *Biochim. Biophys. Acta*, **345**, 91–101

Conrad, M. P. and Strauss, H. L. (1985). The vibrational spectrum of water in liquid alkanes. *Biophys. J.*, **48**, 117–124

Cowley, A. C., Fuller, N. L., Rand, R. P. and Parsegian, V. A. (1978). Measurement of repulsive forces between charged phospholipid bilayers. *Biochemistry*, **17**, 3163–3168

Crowe, J. H., Spargo, B. J. and Crowe, L. M. (1987). Preservation of dry liposomes does not require retention of residual water. *Proc. Natl Acad. Sci. USA*, **84**, 1537–1540

Daniels, C., Noy, N. and Zakim, D. (1985). Rates of hydration of fatty acids bound to unilamellar vesicles of phosphatidylcholine or to albumin. *Biochemistry*, **24**, 3286–3292

Deamer, D. W. and Bramhall, J. (1986). Permeability of lipid bilayers to water and ionic solutes. *Chem. Phys. Lipids*, **40**, 167–188

De Cuyper, M. and Joniau, M. (1985). Spontaneous intervesicular transfer of anionic phospholipids differing in the nature of their polar headgroup. *Biochim. Biophys. Acta*, **814**, 374–380

De Cuyper, M., Joniau, M., Engberts, J. B. F. N. and Sudholter, E., Jr. (1984). Exchangeability of phospholipids between anionic, zwitterionic and cationic membranes. *Colloids and Surfaces*, **10**, 313–319

Dogonadze, R. R., Kornyshev, A. A. and Kuznetsov, A. M. (1973). Phenomenological description of polar media on the basis of an effective Hamiltonian. *Theor. Mat. Fiz.*, **15**, 127–138

Egberts, E. and Berendsen, H. J. C. (1988). Molecular dynamics simulation of a smectic liquid crystal with atomic detail. *J. Chem. Phys.*, **89**, 3718–3731

Elder, M., Hitchcock, P., Mason, R. and Shipley, G. G. (1977). A refinement analysis of the crystallography of the phospholipid, 1,2-dilauroyl-DL-phosphatidylethanolamine, and some remarks on lipid–lipid and lipid–protein interactions. *Proc. Roy. Soc. (London)*, **A 354**, 157–170

Enders, A. and Nimtz, G. (1984). Dielectric relaxation study of dynamic properties of hydrated phospholipid bilayers. *Ber. Bunsenges. Phys. Chem.*, **88**, 512–517

Engelbert, H. and Lawaczeck, R. (1985). The H_2O/D_2O exchange across vesicular lipid bilayers. Lecithins and binary mixtures of lecithins. *Ber. Bunsenges. Phys. Chem.*, **89**, 754–759

Fernandez, M. S. and Fromherz, P. (1978). Lipoid pH indicators as probes of electrical potential and polarity in micelles. *J. Phys. Chem.*, **81**, 1755–1761

Fettiplace, R. (1978). The influence of the lipid on the water permeability of artificial membranes. *Biochim. Biophys. Acta*, **513**, 1–10

Feuer, J., Stamatoff, B., Guggenheim, H. J., Tellez, G. and Yamane, T. (1982). Amplitude of rippling in the P_{β}-phase of dipalmitoylphosphatidylcholine bilayers. *Biophys. J.*, **38**, 217–226

Finer, E. G. (1973). Interpretation of deuteron magnetic resonance spectroscopie studies of the hydration of macromolecules. *J. Chem. Soc. Faraday Trans. II*, **69**, 1590–1600

Finer, E. G. and Darke, A. (1974). Phospholipid hydration studied by deuteron magnetic resonance spectroscopy. *Chem. Phys. Lipids*, **12**, 1–16

Fisher, R. W. and James, T. L. (1978). Lateral diffusion of the phospholipid molecule in dipalmitoylphosphatidylcholine bilayers. An investigation using nuclear spin-lattice relaxation in the rotating frame. *Biochemistry*, **17**, 1177–1183

Fringeli, U. P. and Günthard, H. H. (1976). Hydration sites of egg phosphatidylcholine determined by means of modulated excitation infrared spectroscopy. *Biochim. Biophys. Acta*, **450**, 101–106

Frischleder, H. (1981). Monte Carlo and molecular dynamics simulation of the structure and dynamical properties of molecules in phospholipid/water lamellar systems. In *Sixth School on Biophysics of Membrane Transport*, Poland

Frischleder, H., Gleichmann, S. and Krahl, R. (1977). Quantum-chemical and empirical calculations on phospholipids. III. Hydration of the dimethylphosphate anion. *Chem. Phys. Lipids*, **19**, 144–149

Füldner, H. H. (1981). Characterization of a third phase transition in multilamellar dipalmitoyllecithin liposomes. *Biochemistry*, **20**, 5707–5710

Gardner, A. A. and Valleau, J. P. (1987). Water like particles at surfaces. II. In a double layer and at a metallic surface. *J. Chem. Phys.*, **86**, 4171–4176

Geiger, A., Rahman, A. and Stillinger, F. H. (1979). Molecular dynamics study of the hydration of Lennard-Jones solutes. *J. Chem. Phys.*, **70**, 263–276

Gerritsen, H. C. and Caffrey, M. (1990). Water transport in lyotropic liquid crystals and lipid–water systems: mutual diffusion coefficient determination. *J. Phys. Chem.*, **116**, 332–341

Gershfield, N. L. (1989). Thermodynamics of phospholipid bilayer assembly. *Biochemistry*, **28**, 4229–4232

Ginsberg, L. and Gershfeld, N. L. (1985). Phospholipid surface bilayers at the air–water interface. II. Water permeability of dimyristoylphosphatidylcholine surface bilayers. *Biophys. J.*, **47**, 211–215

Griffin, R. G. (1976). Observation of the effect of water on the ^{31}P nuclear magnetic resonance spectra of dipalmitoyllecithin. *J. Am. Chem. Soc.*, **98**, 851–853

Griffith, O. H., Dehlinger, P. J. and Van, S. P. (1974). Shape of the hydrophobic barrier of phospholipid bilayers (evidence for water penetration in biological membranes). *J. Membrane Biol.*, **15**, 159–192

Gruen, D. W. R. (1980). A statistical mechanical model of the lipid bilayer above its phase transition. *Biochim. Biophys. Acta*, **595**, 161–183

Gruen, D. W. and Marčelja, S. (1983). Spatially varying polarization in water. *J. Chem. Soc. Faraday Trans. II*, **79**, 225–242

Gruen, D. W. and Marčelja, S. (1984). Water structure near the membrane surface. In Perelson, A. S., Delisi, C. and Wiegel, F. W. (Eds), *Cell Surface Dynamics. Concepts and Models*, Dekker, New York, pp. 59–91.

Hammond, K., Lyle, I. G. and Jones, M. N. (1987). Vesicle–vesicle interaction and forces between bilayers in phospholipid systems incorporating phosphatidylinositol. *Colloids and Surfaces*, **12**, 241–257

Hartmann, W. and Galla, H.-J. (1978). Binding of polylysine to charged bilayer membranes. Molecular organization of a lipid–peptide complex. *Biochim. Biophys. Acta*, **509**, 474–490

Harvey, J. M., Symons, M. C. R. and Naftalin, R. J. (1976). Proton magnetic resonance study of the hydration of glucose. *Nature*, **261**, 435–436

Hauser, H., Guyer, W., Pascher, I., Skrabal, P. and Sundell, S. (1980). Polar group conformation of phosphatidylcholine. Effect of solvent and aggregation. *Biochemistry*, **19**, 366–373

Hauser, H., Pascher, I., Pearson, R. H. and Sundell, S. (1981). Preferred conformation and molecular packing of phosphatidylethanolamine and phosphatidylcholine. *Biochim. Biophys. Acta*, **650**, 21–51

Hauser, H. and Shipley, G. G. (1981). Crystallization of phosphatidylserine bilayers induced by lithium. *J. Biol. Chem.*, **256**, 11377–11380

Hauser, H. and Shipley, G. G. (1984). Interactions of divalent cations with phosphatidylserine bilayer membranes. *Biochemistry*, **23**, 34–41

Hautman, J., Halley, J. W. and Rhee, Y.-J. (1989). Molecular dynamics simulation of water between two ideal classical metal walls. *J. Chem. Phys.*, **91**, 467–472

Helm, C. A., Moehwald, H., Kjaer, K. and Als-Nielsen, J. (1987). Phospholipid monolayer density distribution perpendicular to the water surface. A synchroton x-ray reflectivity study. *Europhys. Lett.*, **4**, 697–703

Helm, C. A., Tippman-Krayer, P., Möhwald, H., Als-Nielsen, J. and Kjaer, C. (1991). Phases of phosphatidylethanolamine monolayers studied by synchroton x-ray scattering. *Biophys. J.*, **60**, 1457–1476

Herbette, L., Napolitano, C. A. and McDaniel, R. V. (1984). Direct determination of the calcium profile structure for dipalmitoyllecithin multilayers using diffraction. *Biophys. J.*, **46**, 677–685

Hoekstra, D. (1982). Role of lipid phase separations and membrane hydration in phospholipid vesicle fusion. *Biochemistry*, **21**, 2833–2840

Inglefield, P. T., Lindblom, K. A. and Gottlieb, A. M. (1976). Water binding and mobility in the phosphatidylcholine/cholesterol/water lamellar phase. *Biochim. Biophys. Acta*, **419**, 196–205

Israelachvili, J. N. (1985). Measurements of hydration forces between macroscopic surfaces. *Chem. Scripta*, **25**, 7–4

Israelachvili, J. N., Marčelja, S. and Horn, R. G. (1980). Physical principles of membrane organization. *Q. Rev. Biophys.*, **13**, 121–200

Israelachvili, J. N. and Pashley, R. M. (1983). Molecular layering of water at surfaces and origin of repulsive hydration forces. *Nature*, **306**, 249–250

Israelachvili, J. N. and Wennerström, H. (1990). Hydration or steric forces between amphiphilic surfaces? *Langmuir*, **6**, 873–877

Janiak, M. J., Small, D. M. and Shipley, G. G. (1979). Temperature and compositional dependence of the structure of hydrated dimyristoyl lecithin. *J. Biol. Chem.*, **254**, 6068–6078

Jean, Y. C., Wang, Y. Y. Y. and Yeh, Y. Y. (1984). Temperature dependence of positronium reactivities with charge transfer molecules in bilayer membranes. *J. Chem. Phys.*, **80**, 1671–1676

Jendrasiak, G. L. and Hasty, J. H. (1974). The hydration of phospholipids. *Biochim. Biophys. Acta*, **337**, 79–91

Jendrasiak, G. L. and Mendible, J. C. (1976a). The phospholipid head-group orientation: Effect on hydration and electrical conductivity. *Biochim. Biophys. Acta*, **424**, 149–158

Jendrasiak, G. L. and Mendible, J. C. (1976b). The effect of the phase transition on the hydration and electrical conductivity of phospholipids. *Biochim. Biophys. Acta*, **424**, 133–148

Jönsson, B. (1982). Monte Carlo simulations of liquid water between two rigid walls. *Chem. Phys. Lett.*, **82**, 520–525

Jönsson, B. (1987). Phase equilibria in a three component water–soap–alcohol system. A thermodynamic model. *J. Phys. Chem.*, **91**, 338–352

Jönsson, B. and Persson, P. K. (1987). Effect of charged amphiphiles on the stability of lamellar mesophases. *J. Colloid Interface Sci.*, **115**, 507–512

Jönsson, B. and Wennerström, H. (1981). Thermodynamics of ionic amphiphile–water systems. *J. Colloid Interface Sci.*, **80**, 482–496

Jorgensen, W. L. (1982). Monte Carlo simulation of n-butane in water. Conformational evidence for the hydrophobic effect. *J. Chem. Phys.*, **77**, 5757–5765

Jürgens, E., Höhne, G. and Sackmann, E. (1983). Calorimetric study of the dipalmitoylphosphatidylcholine/water phase diagram. *Ber. Bunsenges. Phys. Chem.*, **87**, 95–104

Kaatze, U., Henze, R. and Eibl, H. (1979). Motion of the lengthened zwitterionic head groups of C16-lecithin analogues in aqueous solutions as studied by dielectric relaxation measurements. *Biophys. Chem.*, **10**, 351–362

Khan, A., Jönsson, B. and Wennerström, H. (1985). Phase equilibria in the mixed sodium and calcium di-2-ethylhexylsulfosuccinate aqueous system. An illustration of repulsive and attractive double layer forces. *J. Phys. Chem.*, **89**, 5180–5184

Kim, K. S. and Clementi, E. (1987). Hydration structures and energetics of phospholipid. *J. Computational Chem.*, **8**, 57–66

Kimura, Y. and Ikegami, A. (1985). Local dielectric properties around polar region of lipid bilayer membranes. *J. Membrane Biol.*, **85**, 225–231

Kirk, G. L. and Gruner, S. M. (1985). Lyotropic effects of alkanes and headgroup composition on the L_α–H_{II} lipid liquid crystal phase transition: hydrocarbon packing versus intrinsic curvature. *J. Phys. (Paris)*, **46**, 761–769

Kjellander, R. (1984). On the image charge model for the hydration force. *J. Chem. Soc. Faraday Trans. II*, **80**, 1323–1348

Klose, G. and Gawrisch, K. (1981). Lipid–water interaction in model membranes. *Stud. Biophys.*, **84**, 21–22

Kodama, M., Kuwabara, M. and Seki, S. (1982a). Successive phase-transition phenomena and phase diagram of the phosphatidylcholine-water system as revealed by differential scanning calorimetry. *Biochim. Biophys. Acta*, **689**, 567–570

Kodama, M., Ogawa, Y., Kuwabara, M. and Seki, S. (1982b). Interactions of water and lecithin molecules as revealed by differential scanning calorimetry. *Studies in Physical and Theoretical Chemistry*, **27**, 449–456

Kornyshev, A. A. (1983). Non-local dielectric response of a polar solvent and Debye screening in ionic solution. *J. Chem. Soc. Faraday Trans. II*, **79**, 651–661

Kornyshev, A. A. and Leikin, S. (1989). Fluctuation theory of hydration forces. The dramatic effects of inhomogeneous boundary conditions. *Phys. Rev. A.*, **40**, 6431–6437

Kornyshev, A. A. and Vorotyntsev, M. A. (1980). Nonlocal electrostatic approach to the double layer and adsorption at the electrode–electrolyte interface. *Surf. Sci.*, **101**, 23–48

Kreissler, M., Lemaire, B. and Bothorel, P. (1983). Theoretical conformational analysis of phospholipids. II. Role of hydration in the gel to liquid crystal transition of phospholipids. *Biochim. Biophys. Acta*, **735**, 23–34

Kresheck, G. C. (1975). In Franks, F. (Ed.), *Water: A Comprehensive Treatise*. Plenum Press, New York, pp. 95ff.

Kumar, V. V. and Raghunathan, P. (1986). Spectroscopic investigations of the water pool in lecithin reverse micelles. *Lipids*, **21**, 764–768

Laggner, P., Degovics, G. and Schuster, A. (1987). Structure and thermodynamics of the dihexadecylphosphatidylcholine–water system. *Chem. Phys. Lipids*, **44**, 31–60

Lawaczeck, R. (1979). On the permeability of water molecules across vesicular lipid bilayers. *J. Membrane Biol.*, **51**, 229–261

Lawrence, A. S., Al-Mamun, M. A. and McDonald, M. P. (1967). Investigation of lipid–water system. Part 2. Effect of water on the polymorphism of long chain alcohols and acids. *Trans. Faraday Soc.*, **63**, 2789–2795

Lee, C. Y., McCammon, J. A. and Rossky, P. J. (1984). The structure of liquid water at an extended hydrophobic surface. *J. Chem. Phys.*, **80**, 4448–4455

Leikin, S. and Kornyshev, A. A. (1990). Theory of hydration forces. Nonlocal electrostatic interaction of neutral surfaces. *J. Chem. Phys.*, **92**, 6890–6898

Leikin, S. and Kornyshev, A. A. (1991). Mean-field theory of dehydration transition. *Phys. Rev. A*, **44**, 1156–1168

Lewis, R. N. A. H., Mammock, D. A., McElhaney, R. N., Turner, D. C. and Gruner, S. M. (1989). Effect of fatty acyl chain length and structure on the lamellar gel to liquid-crystalline and lamellar to reversed hexagonal phase transitions of aqueous phosphatidylethanolamine dispersions. *Biochemistry*, **28**, 541–548

Lindblom, G., Larsson, K., Johansson, L., Fontell, K. and Forsen, S. (1979). The cubic phase of monoglyceride–water systems. Arguments for a structure based upon lamellar bilayer units. *J. Am. Chem. Soc.*, **101**, 5465–5470

Lindblom, G., Sjoelund, M. and Rilfors, L. (1988). Effect of n-alkanes and peptides on the phase equilibria in phosphatidylcholine–water systems. *Liquid Crystals*, **3**, 783–790

Linse, P. (1989). Molecular dynamics study of the aqueous core of a reversed ionic micelle. *J. Chem. Phys.*, **90**, 4992–4504

Llor, A. and Rigny, P. (1986). Some tentative models of molecular motion applied to water in small reversed micelles. *J. Am. Chem. Soc.*, **108**, 7533–7541

Lohner, K., Degovics, G. and Laggner, P. (1987). Thermal phase behaviour and structure of hydrated mixtures between dipalmitoyl- and dihexadecylphosphatidylcholine. *Chem. Phys. Lipids*, **44**, 61–70

Loosley-Millman, M. E., Rand, R. P. and Parsegian, V. A. (1982). Effects of monovalent ion binding and screening on measured electrostatic forces between charged phospholipid bilayers. *Biophys. J.*, **40**, 221–232

Lyle, I. G. and Tiddy, G. J. (1986). Hydration forces between surfactant bilayers: An equilibrium binding description. *Chem. Phys. Lett.*, **124**, 432–436

Lynch, D. V. and Steponkus, P. L. (1989). Lyotrophic phase behaviour of unsaturated phosphatidylcholine species: relevance to the mechanism of plasma membrane destabilization and freezing injury. *Biochim. Biophys. Acta*, **984**, 267–272

McCown, J. T., Evans, E., Diehl, S. and Wiles, H. C. (1981). Degree of hydration and lateral diffusion in phospholipid multibilayers. *Biochemistry*, **20**, 3134–3138

McDaniel, R. V., McIntosh, T. J. and Simon, S. A. (1983). Nonelectrolyte substitution for water in phosphatidylcholine bilayers. *Biochim. Biophys. Acta*, **731**, 97–108

McIntosh, T. J. (1978). The effect of cholesterol on the structure of phosphatidylcholine bilayers. *Biochim. Biophys. Acta*, **513**, 43–58

McIntosh, T. J., Magid, A. D. and Simon, S. A. (1987). Steric repulsion between phosphatidylcholine bilayers. *Biochemistry*, **26**, 7325–7332

McIntosh, T. J., Magid, A. D. and Simon, S. A. (1989a). Repulsive interactions between uncharged bilayers. Hydration and fluctuation pressures for monoglycerides. *Biophys. J.*, **55**, 897–904

McIntosh, T. J., Magid, A. D. and Simon, S. A. (1989b). Cholesterol modifies the short range repulsive interactions between phosphatidylcholine membranes. *Biochemistry*, **28**, 17–25

McIntosh, T. J. and Simon, S. A. (1986a). Hydration force and bilayer deformation. A reevaluation. *Biochemistry*, **25**, 4058–4066

McIntosh, T. J. and Simon, S. A. (1986b). Area per molecule and distribution of water in fully hydrated dilauroylphosphatidylethanolamine bilayers. *Biochemistry*, **25**, 4948–4952

Madelmont, C. and Perron, R. (1974). Etude du systeme myristate de sodium–eau par analyse thermique differentielle. I. Le savon anhydre et la courbe Ti. *Bull. Soc. Chim. France*, **9–10**, 1796–1799

Marčelja, S. and Radić, N. (1976). Repulsion of interfaces due to boundary water. *Chem. Phys. Lett.*, **42**, 129–130

Marchesi, M. (1983). Molecular dynamics simulation of liquid water between two walls. *Chem. Phys. Lett.*, **97**, 224–230

Marra, J. (1986). Direct measurement of the interaction between phosphatidylglycerol bilayers in the electrolyte solutions. *Biophys. J.*, **50**, 815–825

Marra, J. and Israelchvili, J. (1985). Direct measurements of forces between phosphatidylcholine and phosphatidylethanolamine bilayers in aqueous electrolyte solutions. *Biochemistry*, **24**, 4608–4618

Mulukutla, S. and Shipley, G. G. (1984). Structure and thermotropic properties of phosphatidylethanolamine and its *N*-methyl derivatives. *Biochemistry*, **23**, 2514–2519

Nakanishi, K., Ikari, K., Okazaki, S. and Touhara, H. (1984). Computer experiments on aqueous solutions. III. Monte carlo calculation on the hydration of tertiary butyl alcohol in an infinitely dilute aqueous solution with a new water-butanol pair potential. *J. Chem. Phys.*, **80**, 1656–1665

Nicklas, K., Böcker, J., Schlenkrich, M., Bopp, P. and Brickmann, J. (1990). Molecular dynamics simulation of the interface aqueous ionic solution/lipid membrane. In Gassteiger, J. (Ed.), *Software Development in Chemistry*: Proceedings of 4th Workshop 'Computers in Chemistry', Hochfilzen, 1989, Vol. 4. Springer, Berlin, pp. 311–320

Nimtz, G. and Binggeli, B. (1985). Hydration dependence of the head group mobility in phospholipid (DMPC) membranes. *Ber. Bunsenges. Phys. Chem.*, **89**, 842–845

Nimtz, G. and Weiss, W. (1987). Relaxation time and viscosity of water near hydrophilic surfaces. *Z. Phys. B—Condensed Matter*, **67**, 483–487

Onsager, L. and Dupuis, M. (1962). In Pesce, B. (Ed.), *Electrolytes*. Pergamon Press, New York, pp. 27–46

van Oss, C. J., Chaudhury, M. K. and Good, R. J. (1988). Polar interfacial interactions, hydration pressure and membrane fusion. In Ohki, S., Doyle, D., Flanega, T. D., Hui, S. W. and Mayhew, E. (Eds), *Membrane Fusion*. Plenum Press, New York, pp. 113–122

Paltauf, F., Hauser, H. and Phillips, M. C. (1971). Monolayer characteristics of some 1,2-diacyl, 1-alkyl-2-acyl and 1, 2-dialkyl phospholipids at the air–water interface. *Biochim. Biophys. Acta*, **249**, 539–547

Parsegian, V. A. (1983). Dimensions of the intermediate phase of dipalmitoylphosphatidylcholine. *Biophys. J.*, **44**, 413–415

Parsonage, N. G. and Nicholson, D. (1986). Computer simulation of water between metal walls. *J. Chem. Soc. Faraday Trans. II*, **82**, 1521–1535

Pascher, I. and Sundell, S. (1977). Molecular arrangements in sphingo-lipids: the crystal structure of cerebroside. *Chem. Phys. Lipids*, **20**, 175-191.

Pascher, I. and Sundell, S. (1986a). Membrane lipids' preferred conformational states and their interplay. The crystal structure of dilauroylphosphatidyl-*N*, *N*-dimethylethanolamine. *Biochim. Biophys. Acta*, **855**, 68–78

Pascher, I. and Sundell, S. (1986b). The single crystal structure of octadecyl-2-methyl-glycero phosphocholine monohydrate. A multilamellar structure with interdigitated headgroups and hydrocarbon chains. *Chem. Phys. Lipids*, 327–333

Pashley, R. M. (1985). The effects of hydrated cation adsorption on surface forces between mica crystals and its relevance to colloidal systems. *Chem. Scripta*, **25**, 22–27

Pashley, R. M., McGuiggan, P. M., Ninham, B. W., Brady, J. and Evans, F. (1986). Direct measurements of surface forces between bilayers of double-chained quaternary ammonium and bromide surfactants. *J. Phys. Chem.*, **90**, 1637–1642

Pearson, R. H. and Pascher, I. (1979). The molecular structure of lecithin dihydrate. *Nature*, **281**, 499–501

Peinel, G. (1975). Quantum-chemical and empirical calculations on phospholipids. I. The charge distribution of model headgroups of phospholipids obtained by a CNDO–APSG procedure. *Chem. Phys. Lipids*, **14**, 268–273

Peinel, G., Binder, H. and Werge, C. (1983a). The phospholipid–water interface. A theoretical study. *Stud. Biophys.*, **93**, 211–214

Peinel, G., Frischleder, H. and Binder, H. (1983b). Quantum-chemical and empirical calculations on phospholipids. VIII. The electrostatic potential from isolated molecules up to layer systems. *Chem. Phys. Lipids*, **33**, 195–205

Peinel, G., Gründer, W., Kabisch, G. and Arnold, K. (1976). Modelluntersuchungen zur Hydration der Phosphatidylcholin-Kopfgruppe. *Stud. Biophys.*, **59**, 37–46

Persson, P. K. T. and Bergenstahl, B. A. (1985). Repulsive forces in lecithin glycol lamellar phases. *Biophys. J.*, **47**, 743–746

Pfleiffer, W., Henkel, Th., Sackmann, E., Knoll, W. and Richter, D. (1990). *Europhys. Lett.*, **8**, 201–206

Podgornik, R. (1988). Forces between surfaces with surface-specific interactions in a dilute electrolyte. *Chem. Phys. Lett.*, **144**, 503–509

Podgornik, R. (1989). Solvent structure effects in dipole correlation forces. *Chem. Phys. Lett.*, **156**, 71–75

Port, G. N. and Pullman, A. (1973). An *ab initio* study of the hydration of alkylammonium groups. *Theoret. Chim. Acta*, **31**, 231–237

Pullman, A. (1982). Dehydration processes and specificity in biological systems. *Studies in Physical and Theoretical Chemistry*, **27**, 373–382

Pullman, A., Pullman, B. and Berthod, H. (1978a). An SCF *ab initio* investigation of the through water interaction of the phosphate anion with the Na^+ cation. *Theoret. Chim. Acta*, **47**, 175–192

Pullman, B., Berthod, H. and Gresh, N. (1975a). Quantum-mechanical studies on the conformation of phospholipids. The effect of water on the conformational properties of the polar head. *FEBS Lett.*, **53**, 199–269

Pullman, B., Pullman, A., Berthod, H. and Gresh, N. (1975b). Quantum-mechanical studies of environmental effects on biomolecules. VI. *Ab initio* studies on the hydration scheme of the phosphate group. *Theoret. Chim. Acta*, **40**, 93–111

Pullman, B., Pullman, A. and Berthod, H. (1978b). *Ab initio* study of the through water versus direct binding of the Na^+ and Mg^{2+} cations to the phosphate anion. *Int. J. Quantum Chem.: Quantum Biological Symposium*, **5**, 79–90

Rand, R. P. (1981). Interacting phospholipid bilayers: Measured forces and induced structural changes. *Ann. Rev. Biophys.*, **10**, 277–314

Rand, R. P., Das, S. and Parsegian, V. A. (1985). The hydration force, its character, universality and application: some current issues. *Chem. Scripta*, **25**, 15–21

Rand, R. P. and Parsegian, V. A. (1990). Lipid hydration forces between lipid bilayers. *Biochim. Biophys. Acta*, **988**, 351–377

Rau, D. C., Lee, B. and Parsegian, V. A. (1984). Measurement of the repulsive force between polyelectrolyte molecules in ionic solution: Hydration forces between parallel DNA double helics. *Proc. Natl Acad. Sci. USA*, **81**, 2621–2625.

Reddy, R. M. and Berkowitz, M. (1989). Hydration forces between parallel DNA double helices. Computer simulations. *Proc. Natl Acad. Sci. USA*, **86**, 3165–3168

Reed, R. A. and Shipley, G. G. (1987). Structure and metastability of *N*-lignocerylgalactosphingosine (cerebroside) bilayers. *Biochim. Biophys. Acta*, **896**, 153–164

Rigaud, J.-L., Gary-Bobo, C. M. and Lange, Y. (1972). Diffusion processes in lipid–water lamellar phases. *Biochim. Biophys. Acta*, **266**, 72–84
Rossky, P. J. and Karplus, M. (1979). A model for the simulation of an aqueous dipeptide solution. *Biopolymers*, **18**, 825–854
Rowe, E. S. (1983). Lipid chain length and temperature dependence of ethanol–phosphatidylcholine interactions. *Biochemistry*, **22**, 3299–3305
Ruocco, M. J., Atkinson, D., Small, D. M., Skarjune, R. P., Oldfield, E. and Shipley, G. G. (1981). X-Ray diffraction and calorimetric study of anhydrous and hydrated *N*-palmitoylgalactosylsphingoside (cerebroside). *Biochemistry*, **20**, 5957–5966
Ruocco, M. J. and Shipley, G. G. (1982). Characterization of the sub-transition of hydrated dipalmitoylphosphatidylcholine bilayers. Kinetic, hydration and structural study. *Biochim. Biophys. Acta*, **691**, 309–320
Ruocco, M. J. and Shipely, G. G. (1983). Hydration of *N*-palmitoylgalactosylsphingosine compared to monosaccharide hydration. *Biochim. Biophys. Acta*, **735**, 305–308
Ruocco, M. J., Shipley, G. G. and Oldfield, E. (1983). Galactocerebroside-phospholipid interactions in bilayer membranes. *Biophys. J.*, **43**, 91–101
Schlenkrich, M., Nicklas, K., Brickmann, J. and Bopp, P. (1990). A molecular dynamics study of the interface between a membrane and water. *Ber. Bunsengese. Phys. Chem.*, **94**, 133–145
Schnitker, J. and Rossky, P. J. (1986). Electron localization in liquid water: A computer simulation study of microscopic trapping sites. *J. Chem. Phys.*, **85**, 2986–2998
Schumacher, G., Jr., and Sandermann H. (1976). Solubility of phospholipid polar group model compounds in water. *Biochim. Biophys. Acta*, **448**, 642–644
Scott, H. L. (1984). Phosphatidylcholine head groups can induce short-range order in interfacial water in vesicles. *Chem. Phys. Lett.*, **109**, 570–573
Seddon, J. M. (1990). Hexagonal phases. *Biochim. Biophys. Acta*, **1031**, 1–69
Seddon, J. M., Cevc, G., Kaye, R. D. and Marsh, D. (1984). X-Ray diffraction study of the polymorphism of hydrated diacyl- and dialkylphosphatidylethanolamines. *Biochemistry*, **23**, 2634–2644
Seddon, J. M., Cevc, G. and Marsh, D. (1983a). Calorimetric investigation of the gel-to-fluid and lamellar-to-inverted-hexagonal phase transitions in dialkyl and diacyl phosphatidylethanolamines. *Biochemistry*, **22**, 1280–1289
Seddon, J. M., Harlos, K. and Marsh, D. (1983b). Metastability and polymorphism in the gel and fluid bilayer phases of dilauroylphosphatidylethanolamine. Two crystalline forms in excess water. *J. Biol. Chem.*, **258**, 3850–3854
Seelig, J. (1977). Deuterium magnetic resonance: Theory and application to lipid membranes. *Q. Rev. Biophys.*, **10**, 353–418
Seul, M. and Eisenberger, P. (1989a). X-Ray scattering study of water intercalation into thin lyotropic multilayers. I. Molecular ordering and distinct states of hydration. *Phys. Rev. A*, **39**, 4230–4241
Seul, M. and Eisenberger, P. (1989b). X-Ray scattering study of water intercalation into thin lyotropic multilayers. II. Transitions between pure states of hydration via disordered intermediate states. *Phys. Rev. A*, **39**, 4242–4253
Shepherd, J. C. W. and Büldt, G. (1978). Zwitterionic dipoles as a dielectric probe for investigating head group mobility in phospholipid membranes. *Biochim. Biophys. Acta*, **514**, 83–94
Shepherd, J. C. W. and Büldt, G. (1979). The influence of cholesterol on head group mobility in phospholipid membranes. *Biochim. Biophys. Acta*, **558**, 41–47
Siminovitch, D. J., Jeffrey, K. R. and Eibl, H. (1983). A comparison of the head

group conformation and dynamics in synthetic analogs of dipalmitoylphosphatidylcholine. *Biochim. Biophys. Acta*, **727**, 122–134

Simon, A. and McIntosh, T. J. (1984). Interdigitated hydrocarbon chain packing causes the biphasic transition behavior in lipid/alcohol suspensions. *Biochim. Biophys. Acta*, **773**, 169–172

Simon, S. A. and McIntosh, T. J. (1989). Magnitude of the solvation pressure depends on dipole potential. *Proc. Natl Acad. Sci USA*, **86**, 9263–9267

Simon, S. A. and McIntosh, T. J. (1991). Surface ripples cause the large fluid spaces between gel phase bilayer containing small amounts of cholesterol. *Biochim. Phys. Acta*, **1064**, 69–74

Simon, S. A., McIntosh, T. J. and Magid, A. D. (1988). Magnitude and range of the hydration pressure between lecithin bilayers as a function of headgroup density. *J. Colloid Interface Sci.*, **126**, 74–83

Smaby, J. M., Hermetter, A., Schmidt, P. C., Paltauf, F. and Brockman, H. L. (1983). Packing of ether and ester phospholipids in monolayers. Evidence for hydrogen bonded water at the sn-1 acyl group of phosphatidylcholine. *Biochemistry*, **22**, 5808–5813

Small, D. M. (1967). Phase equilibria and structure of dry and hydrated egg lecithin. *J. Lipid Res.*, **8**, 551–557

Smith, G. S., Safinya, C. R., Roux, D. and Clark, N. A. (1987). X-Ray study of freely suspended films of a multilamellar lipid system. *Mol. Cryst. Liq. Cryst.*, **144**, 235–255

Sunamoto, J., Iwamoto, K. and Kondo, H. (1980). Liposomal membranes. VII. Fusion and aggregation of egg lecithin liposomes as promoted by polysaccharides. *Biochem. Biophys. Res. Commun.*, **94**, 1367–1373

Tardieu, A., Luzzati, V. and Reman, F. C. (1973). Structure and polymorphism of the hydrocarbon chains of lipids: A study of lecithin–water phases. *J. Mol. Biol.*, **75**, 711–733

Tate, M. W. and Gruner, S. M. (1989). Temperature dependence of the structural dimensions of the inverted hexagonal (H_{II}) phase of phosphatidylethanolamine containing membranes. *Biochemistry*, **28**, 4245–4253

Tenchov, B. G., Boyanov, A. I. and Koynova, R. D. (1984). Lyotropic polymorphism of racemic dipalmitoylphosphatidylethanolamine. A differential scanning calorimetry study. *Biochemistry*, **23**, 3553–3558

Tenchov, B. G., Boyanov, A. I., Koynova, R. D. and Koumanov, K. S. (1983). Absence of subtransition in racemic dipalmitoylphosphatidylcholine vesicles. *Biochim. Biophys. Acta*, **732**, 711–713

Tilcock, C. P. and Fisher, D. (1979). Interaction of phospholipid membranes with poly(ethylene glycol)s. *Biochim. Biophys. Acta*, **557**, 53–61

Trahms, L., Boroske, W. D. and Klabe, E. (1983). H-NMR study of the three low temperature phases of DPPC–water systems. *Biophys. J.*, **42**, 285–293

Ulmius, J., Wennerström, H., Lindblom, G. and Arvidson, G. (1977). Deuteron nuclear magnetic resonance studies of phase equilibria in a lecithin–water system. *Biochemistry*, **16**, 5742–5745

Ulrich, A. S., Volke, F. and Watts, A. (1991). The dependence of phospholipid headgroup mobility on hydration as studied by deuterium NMR spin–lattice relaxation time measurements. *Biochim. Biophys. Acta* (in press)

Veiro, J. A., Nambi, P. and Rowe, E. S. (1988). Effect of alcohols on the phase transitions of dihexadecylphosphatidylcholine. *Biochim. Biophys. Acta*, **943**, 108–111

Verwey, E. J. W. and Overbeek, J. T. G. (1948). *Theory of the Stability of Lyophobic Colloids*. Elsevier, New York

Viti, V. and Minetti, M. (1981). ^{31}P NMR study of head group behaviour in sonicated phosphatidylcholine liposomes in the gel and in the liquid state. *Chem. Phys. Lipids*, **28**, 215–225

Volke, F. (1984). The orientation dependence of deuterium transverse relaxation rate obtained from unoriented model membrane systems. *Chem. Phys. Lett.*, **112**, 551–553

Volke, F. and Gawrisch, K. (1982). The effect of hydration on the mobility of phospholipids on the gel state. A proton nuclear magnetic resonance spin echo study. *Chem. Phys. Lipids*, **31**, 179–189

Wack, D. C. and Webb, W. W. (1988). Measurements of modulated lamellar P-beta phases of interacting lipid membranes. *Phys. Rev. Lett.*, **61**, 1210–1213

Westman, J. and Eriksson, L. E. (1979). The interactions of various lanthanide ions and some anions with phosphatidylcholine vesicle membranes. A ^{31}P NMR study of the surface potential effects. *Biochim. Biophys. Acta*, **557**, 62–78

White, J. and Helenius, A. (1980). pH-Dependent fusion between semliki forest virus membrane and liposomes. *Proc. Natl Acad. Sci. USA*, **77**, 3273–3277

Wiener, M. C., Suter, R. M. and Nagle, J. F. (1989). Structure of the fully hydrated gel phase of dipalmitoylphosphatidylcholine. *Biophys. J.*, **55**, 315–325

Wolfe, D. H. and Brockman, H. L. (1988). Regulation of the surface pressure of lipid monolayers and bilayers by the activity of water: Derivation and application of an equation of state. *Proc. Natl Acad. Sci. USA*, **85**, 4285–4289

Yeagle, P. L. and Sen, A. (1986). Hydration and the lamellar to hexagonal II phase transition of phosphatidylethanolamine. *Biochemistry*, **25**, 7518–7522

Part 5
Thermodynamics

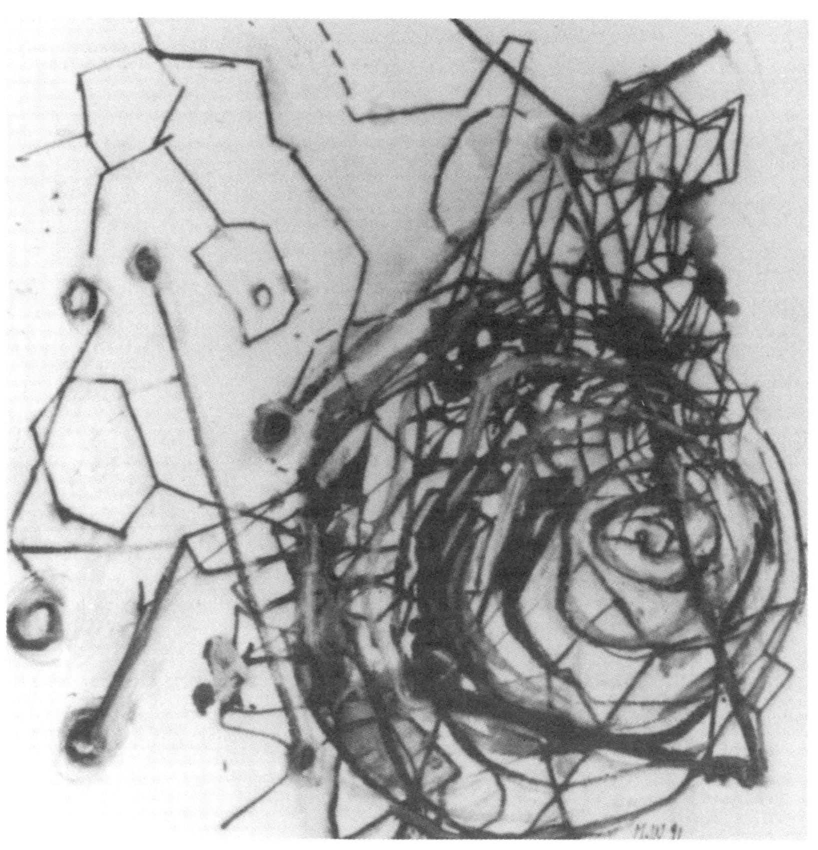

13
Hydration Forces

C. J. van Oss

1 Introduction

The non-covalent physical-chemical interaction forces between polar polymers, biopolymers and/or various inert polar surfaces, particles or cells are of three fundamentally different classes: (1) apolar, electrodynamic, or Lifshitz–van der Waals (LW) interactions (Chaudhury, 1984); (2) polar, or Lewis acid–base (AB) interactions, which in aqueous media are mainly hydrogen-bonding interactions (van Oss *et al.*, 1988a); (3) electrostatic (EL), or Coulombic interactions (van Oss *et al.*, 1988a). In addition, thermal or Brownian movement (BR) interactions may play a role which, while quantitatively fairly unimportant in interactions between large molecules and particles, may not always be negligible when smaller molecules are involved. The first three interactions can be either attractive or repulsive, independent of the sign of the other two varieties, in any particular case (van Oss *et al.*, 1988a); BR interactions are always considered repulsive. In liquids, LW interactions can only be repulsive when acting between two *different* solutes or particles; AB and EL interactions can be repulsive even when acting between identical solutes or particles.

Hydration forces are not in a different class from the non-covalent interactions mentioned above: they are simply the outcome of, mainly, the electron-acceptor/electron-donor, or Lewis acid–base (AB), and to a much smaller extent, Lifshitz–van der Waals (LW) forces. Hydration forces have the important characteristic that they propagate the polar AB forces an appreciable distance away from the initial (usually) electron-donor surface, into the aqueous medium, by means of a preferential orientation of the water molecules of hydration in the adjoining and even, albeit to an exponentially diminishing degree, in subsequent layers of water. Thus, it is understandable that hydration forces have been alluded to as 'structural forces' (Israelachvili, 1985). They have also been called 'disjoining pressures' (see, e. g., Derjaguin, 1989). None of these epithets pertains to a primary force (such as the LW, AB, EL and BR interactions alluded to above), so that their use is best avoided, in order not to lapse into counting the same forces twice (as was done by, e. g., Fowkes, 1972, and by Janczuk and Bialopiotrowicz, 1989), when engaging in energy component summations (van Oss, 1991a).

Hydration interactions have been pioneered by Parsegian and his collaborators (Le Neveu *et al.*, 1977; Parsegian *et al.*, 1979, 1986; Lis *et al.*, 1982), who first measured the non-electrostatic repulsive (hydration) forces between multiple phospholipid lamellae, through osmotic interaction force determinations (Cowley *et al.*, 1978; Parsegian *et al.*, 1986, 1987). The long-range nature of these repulsive forces was confirmed by force-balance measurements: see, e.g., Israelachvili (1985), Claesson (1986), Christenson (1988). Van Oss *et al.* (1987a) showed that the (usually electron-donor) monopolarity of the underlying solute or solid surface is the underlying cause of hydration interactions. Parsegian *et al.* (1985) hypothesized that hydration interactions are propagated through the orientation of water molecules of hydration, and van Oss and Good (1988) experimentally demonstrated the existence and the magnitude of that hydration orientation, as well as its decay with distance, in the first few layers of hydration of serum albumin.

In the following pages the mechanism is shown by which hydration forces stem from the net repulsive hydrogen-bonding interactions between similar as well as dissimilar hydrophilic molecules or surfaces (all of these usually are preponderantly hydrogen-acceptors), and the water molecules of the aqueous medium in which they are dissolved or immersed. With less hydrophilic or with frankly 'hydrophobic' molecules or surfaces, these net hydrogen-bonding interactions tend to be *attractive* and are then known as 'hydrophobic' interactions. Also discussed are the approaches by which hydrophilicity (and its converse, 'hydrophobicity') of solutes, particles or solids can be expressed quantitatively, and finally a number of physicochemical phenomena and applications are treated which are direct consequences of hydration forces.

2 Theory

The derivation or proof of all the following equations can be found elsewhere (see, e. g., van Oss *et al.*, 1988a), or are an established part of the physical chemistry of interfaces (see, e.g., Adamson, 1982) and need not be repeated here. Only the equations necessary for the different applications, and the equations leading up to them, are given below. Here follow the equations which are valid for apolar, Lifshitz–van der Waals (LW), as well as for polar, Lewis acid–base (AB) (electron-acceptor/electron-donor) interactions.

The free energy of cohesion of a (condensed) material 1 is:

$$\Delta G_{11}^{\text{coh}} = -2\gamma_1 \tag{13.1}$$

where γ_1 is the surface tension of material 1. The energy of adhesion between materials 1 and 2 in air or *in vacuo* is, according to the Dupré equation:

$$\Delta G_{12}^{\text{adh}} = \gamma_{12} - \gamma_1 - \gamma_2 \tag{13.2}$$

where γ_1 and γ_2 are the surface tensions of materials 1 and 2, and γ_{12} is their interfacial tension. (From here on the superscripts, 'coh' for cohesion and 'adh' for adhesion, accompanying the symbol for the free energies, will be omitted.) The energy of interaction between materials 1 and 2 immersed in liquid 3 is, according to the Dupré equation in condensed media:

$$\Delta G_{132} = \gamma_{12} - \gamma_{13} - \gamma_{23} \tag{13.3}$$

and the energy of interaction between (macro) molecules or particles of material 1, immersed in liquid 3 then is:

$$\Delta G_{131} = -2\gamma_{13} \tag{13.4}$$

Apolar or Lifshitz–van der Waals Interactions

The existence of a general attraction between neutral atoms was proposed by van der Waals (1873), to account for departures from the gas laws observed in non-ideal gases and liquids. It is now known that 'van der Waals' forces comprise: the randomly orienting dipole–dipole (or orientation) Keesom–van der Waals interactions, the randomly orienting dipole-induced dipole (or induction) Debye–van der Waals interactions, and the fluctuating dipole-induced dipole (or dispersion) London–van der Waals interactions (see, e. g., Overbeek, 1952b). Keesom and Debye interactions only occur among molecules with permanent dipole moments. London interactions are universal and also occur in atom–atom interactions. In *microscopic systems*, the three interaction energies between small molecules or atoms decay very steeply with distance (l), as l^{-6}. For a number of reasons, the London–van der Waals interactions have been considered to be the principal component of van der Waals interactions between *macroscopic entities*, in condensed materials (Overbeek, 1952b; Visser, 1972;

Fowkes, 1983). Hamaker (1937) developed the equation of London–van der Waals interactions in *macroscopic systems*, showing that the additivity of these interactions put them into the realm of long-range interactions: the London–van der Waals energy of attraction between two semi-infinite parallel flat bodies decays with distance, l, as l^{-2} for relatively small values of l ($l < 100$ Å). However, at $l > 100$ Å, retardation sets in, causing the dispersion energy to decay as l^{-3} (see Casimir and Polder, 1946, who first clarified the retardation phenomenon, upon the suggestion of J. Th. G. Overbeek; see also Mahanty and Ninham, 1976 and Israelachvili, 1985).

It was shown by Chaudhury (1984), using the Lifshitz approach, that all three electrodynamic interactions, i.e. the London–van der Waals (dispersion) as well as the 'polar' (in the electrodynamic sense) Keesom–van der Waals (orientation) and Debye–van der Waals (induction) forces, decay with distance at the same rate, and should, *in macroscopic systems*, be treated with the same equations. Together they represent the total apolar, or Lifshitz–van der Waals (LW) surface tension of condensed phase substance, 1 (van Oss *et al.*, 1986, 1988a):

$$\gamma_1^{LW} = \gamma_1^{Keesom} + \gamma_1^{Debye} + \gamma_1^{London} \tag{13.5}$$

Apolar or LW *interfacial* tensions between substances 1 and 2 are expressed as:

$$\gamma_{12}^{LW} = \left(\sqrt{\gamma_1^{LW}} - \sqrt{\gamma_2^{LW}}\right)^2 \tag{13.6}$$

Via Equations (13.1)–(13.6) the various values for ΔG^{LW} can be obtained from the γ_1^{LW} and γ_{12}^{LW} values. The decay with distance of LW interfacial interactions for semi-infinite parallel flat slabs may be expressed (in mJ/m^2) as:

$$\Delta G_{(l)}^{LW} = \Delta G_{(l_0)}^{LW}\left(\frac{\ell_0}{\ell}\right)^2 \tag{13.7}$$

where ℓ_0 is the minimum equilibrium distance, taken to be 0.157 nm (van Oss *et al.*, 1988a), and ℓ may be any distance from 0.157 nm to $\approx$10 nm, beyond which distance retardation effects prevail and Equation (13.7) becomes (at least as far as the London forces are concerned):

$$\Delta G_{(l)}^{LW}\ (\ell > 10\ \text{nm}) = \Delta G_{(l_0)}^{LW}\left(\frac{\ell_0}{\ell}\right)^3 \tag{13.7a}$$

For two equal spheres with radius R, the decay with distance ℓ, between the surfaces of the shells of the spheres, is:

$$\Delta G_{(\ell)}^{LW}\ (\ell < 10\ \text{nm}) = \Delta G_{(\ell_0)}^{LW}\frac{\pi R\ell_0^2}{\ell} \tag{13.7b}$$

(For the interaction between a sphere and a flat plate, $\Delta G_{(\ell)}$ is twice the value shown in Equation 13.7b). In Equations (13.7), (13.7a) and (13.7b), the values for the applicable $\Delta G_{(\ell_0)}^{LW}$ are found via Equations (13.1), (13.2), (13.3) or (13.4); i.e. $\Delta G_{(\ell_0)}$ is the value for the interaction between parallel flat slabs at closest approach which is also the value found from surface or interfacial tension measurements.

Polar or (Lewis) Acid–Base Interactions

Hydrogen bonding or, more generally speaking, electron-donor/electron-acceptor, and thus (Lewis) acid–base (AB), interactions decay with distance at a different rate from LW interactions, and also, in general, obey an entirely different set of rules and equations, due in part to the intrinsic asymmetry of these polar interactions. We first must define γ_1^{AB} of a material 1 (in the condensed state), in terms of its electron-acceptor ($\gamma_1^{\oplus}$) and electron-donor ($\gamma_1^{\ominus}$) parameters:

$$\gamma_1^{AB} = 2\sqrt{(\gamma_1^{\oplus}\gamma_1^{\ominus}} \tag{13.8}$$

where, in general, $\gamma_1^{\oplus} \neq \gamma_1^{\ominus}$. The polar component of the energy of adhesion between materials 1 and 2 is, by definition:

$$\Delta G_{12}^{AB} \equiv -2\,(\sqrt{\gamma_1^{\oplus}\gamma_2^{\ominus}} + \sqrt{\gamma_1^{\ominus}\gamma_2^{\oplus}}) \tag{13.9}$$

and the polar component of the interfacial tension between materials 1 and 2 is:

$$\gamma_{12}^{AB} = 2\,(\sqrt{\gamma_1^{\oplus}\gamma_1^{\ominus}} + \sqrt{\gamma_2^{\oplus}\gamma_2^{\ominus}} - \sqrt{\gamma_1^{\oplus}\gamma_2^{\ominus}} - \sqrt{\gamma_1^{\ominus}\gamma_2^{\oplus}}) \tag{13.10}$$

The polar component of the energy of interaction between materials 1 and 2 immersed in liquid 3 is then:

$$\begin{aligned}\Delta G_{132}^{AB} = {} & 2[\sqrt{\gamma_3^{\oplus}}\,(\sqrt{\gamma_1^{\ominus}} + \sqrt{\gamma_2^{\ominus}} - \sqrt{\gamma_3^{\ominus}}) \\ & + \sqrt{\gamma_3^{\ominus}}\,(\sqrt{\gamma_1^{\oplus}} + \sqrt{\gamma_2^{\oplus}} - \sqrt{\gamma_3^{\oplus}}] \\ & - \sqrt{\gamma_1^{\oplus}\gamma_2^{\ominus}} - \sqrt{\gamma_1^{\ominus}\gamma_2^{\oplus}}\,]\end{aligned} \tag{13.11}$$

and the polar component of the energy of interaction between (macro) molecules or particles of material 1, immersed in liquid 3 is:

$$\Delta G_{131}^{AB} = -4\left(\sqrt{\gamma_1^{\oplus}\gamma_1^{\ominus}} + \sqrt{\gamma_3^{\oplus}\gamma_3^{\ominus}} - \sqrt{\gamma_1^{\oplus}\gamma_3^{\ominus}} - \sqrt{\gamma_1^{\ominus}\gamma_3^{\oplus}}\right) \qquad (13.12)$$

The decay with distance ℓ of AB interfacial interactions for semi-infinite parallel flat slabs may be approximated for $\ell > \lambda$ as:

$$\Delta G_{(\ell)}^{AB} = \Delta G_{(\ell_0)}^{AB} \exp\frac{(\ell_0 - \ell)}{\lambda} \qquad (13.13)$$

where λ is the characteristic correlation length of the liquid, and a measure of the dimension of the molecules of the liquid. For water the value for λ may be approximated as about 0.6 nm (see van Oss, 1990a); however, values of the order of 0.9–1.0 nm have also been proposed (Israelachvili and Pashley, 1984; Marčelja, 1990; Zhou *et al.*, 1990). For two equal spheres with radius r, the decay with distance ℓ is:

$$\Delta G_{(\ell)}^{AB} = \Delta G_{(\ell_0)}^{AB}\, \pi R\lambda \exp\left(\frac{\ell_0 - \ell}{\lambda}\right) \qquad (13.14)$$

where, again, $\Delta G_{(\ell_0)}$ is the value for the parallel flat slab conformation at closest approach.

The Total Interfacial Interaction

The total interfacial (IF) free energy of interaction may be expressed as:

$$\Delta G^{IF} = \Delta G^{LW} + \Delta G^{AB} \qquad (13.15)$$

However, when both LW and AB components have been measured at the minimum equilibrium distance ℓ_0 ($\ell_0 \approx 0.157$ nm for LW and $\ell_0 \approx 0.1$–0.2 nm for AB interactions), which is the case when that measurement has been done by contact angle determination (see below), the values for ΔG^{LW} and ΔG^{AB} at distances $\ell > \ell_0$ must be separately computed via Equations (13.7) and (13.13), respectively, before arriving at $\Delta G_{(\ell)}^{TOT}$ by means of Equation (13.15).

From Equations (13.1) and (13.4), and (13.15), it follows that:

$$\gamma = \gamma^{LW} + \gamma^{AB} \qquad (13.15a)$$

which is valid for surface tensions γ_i, as well as for interfacial tensions γ_{12}.

When $\Delta G < 0$, an interfacial ('hydrophobic') attraction is favoured; when $\Delta G > 0$, a repulsion occurs which has been alluded to as 'hydration pressure'.

Young's Equation

Both LW and AB interactions are incorporated in Young's equation (van Oss *et al.*, 1988a) as follows:

$$(1 + \cos\theta)\,\gamma_L = 2\left(\sqrt{\gamma_S^{LW}\gamma_L^{LW}} + \sqrt{\gamma_S^{\oplus}\gamma_L^{\ominus}} + \sqrt{\gamma_S^{\ominus}\gamma_L^{\oplus}}\right) \quad (13.16)$$

where θ is the contact angle of liquid L on solid S. It thus is possible to characterize a solid completely with respect to γ_S^{LW}, $\gamma_S^{\oplus}$ and $\gamma_S^{\ominus}$ when (advancing) contact angles are measured with three different liquids with known γ_L^{LW}, $\gamma_L^{\oplus}$ and $\gamma_L^{\ominus}$ values.

Contact Angle Determination

Via contact angle determination (see, e.g., Adamson, 1982), employing at least one apolar and two polar liquids, and using Young's equation in the form of Equation (13.16) three times, the three simultaneous equations (13.16) can be readily solved for the γ-values of any solid material, S, yielding γ_S^{LW}, $\gamma_S^{\oplus}$ and $\gamma_S^{\ominus}$. The high-energy apolar liquids best used are: diiodomethane ($\gamma \approx \gamma^{LW} = 50.8$ mJ/m^2) and α-bromonaphthalene ($\gamma \approx \gamma^{LW} = 44.4$ mJ/m^2). The polar liquid pairs that have been sufficiently characterized for this purpose are: water ($\gamma = 72.8$, $\gamma^{LW} = 21.8$, $\gamma^{\oplus} = \gamma^{\ominus} = 25.5$ mJ/m^2), and glycerol ($\gamma = 64.0$, $\gamma^{LW} = 34.0$, $\gamma^{\oplus} = 3.92$ and $\gamma^{\ominus} = 57.4$ mJ/m^2), or formamide ($\gamma = 58.0$, $\gamma^{LW} = 39.0$, $\gamma^{\oplus} = 2.28$ and $\gamma^{\ominus} = 39.6$ mJ/m^2) (see van Oss *et al.*, 1990c).

Thin-layer Wicking with Small Particles

In those cases where the solid materials are in granular form, and cannot be formed into smooth layers or membranes (van Oss *et al.*, 1990b), small particles can be deposited on a glass plate, and the contact angle θ can be determined by thin-layer wicking (Costanzo *et al.*, 1990), using the Washburn equation (Washburn, 1921):

$$h^2 = (tR\gamma_L \cos\theta) / 2\eta \quad (13.17)$$

where h is the height of capillary rise in time t, R the average interstitial radius of the pores between particles, and η the viscosity of the liquid L. R is determined by wicking with a spreading liquid which has a $\gamma_L = \gamma_L^{LW}$ equal to γ_S^{LW} (see Equation 13.17), as $\cos\theta$ then equals 1 (van Oss *et al.*, 1991). Due to the high viscosity of glycerol, it is preferable to use the polar liquid pair, water and formamide, rather than water and glycerol, in thin layer wicking (Costanzo *et al.*, 1990).

Electrostatic Interactions

The electrostatic interaction between two parallel flat plates can be expressed (e.g. in mJ/m^2) as (Overbeek, 1952b):

$$\Delta G_{\ell}^{EL} = 1/\kappa \; 64nkT\gamma_0^2 \exp\,(-\kappa\ell) \qquad (13.18)$$

where ℓ is the distance, $1/\kappa$ is the thickness of the diffuse double layer (Debye length) (see, e.g., Overbeek, 1952a):

$$1/\kappa = \sqrt{(\varepsilon kT) \,/\, (4\pi e^2 \,\Sigma\, v_i^2\, n_i)} \qquad (13.19)$$

n is the number of counterions per unit volume, ε is the dielectric constant of the medium, k is Boltzmann's constant, T is the absolute temperature, v_i is the valency of each ionic species dissolved in the medium, n_i the number of ions of each species per unit volume, e the charge of the electron and

$$\gamma_0 = [\exp(ve\psi_0/2kT) - 1] \,/\, [\exp(ve\psi_0/2kT) + 1] \qquad (13.20)$$

where ψ_0 is the surface potential at the boundary of the particle or cell. For two equal spheres of radius R:

$$\Delta G^{EL} = 0.5\ \varepsilon R\psi_0^2 \ln[1 + \exp(-\kappa\ell)] \qquad (13.21)$$

(expressed, e.g., in ergs, or units of 10^{-7}J), where ε is the dielectric constant of the medium.

The surface potential, ψ_0, of macromolecules or particles at the precise solute–liquid or particle–liquid interface cannot be determined directly by electrokinetic measurements. However, the potential, ζ, at the slipping plane can be readily measured electrokinetically (see, e.g., Overbeek and Wiersema, 1967). Once the value of the ζ-potential is found, ψ_0 can be approximated via (Overbeek, 1952a):

$$\psi_0 = \zeta(1 + z/R)\exp(\kappa z) \qquad (13.22)$$

where z is the distance from the macromolecule's or the particle's surface to the slipping plane (one usually may take z to be $\approx$ 0.3–0.5 nm) and R is the Stokes radius of the macromolecules or particles.

In the case of biopolymers with ζ-potentials of 10 mV or less, i.e. in the case of most biopolymers of an electrophoretic mobility of β-globulins or slower, the electrostatic (EL) interactions usually are negligible, compared with the polar (AB) interactions, whether the latter be repulsive *or* attractive.

The Total Interaction Energy

The total interaction energy is expressed as:

$$\Delta G^{TOT} = \Delta G^{IF} + \Delta G^{EL} \tag{13.23}$$

or, according to Equation (13.15) (van Oss *et al.*, 1988a):

$$\Delta G^{TOT} = \Delta G^{LW} + \Delta G^{AB} + \Delta G^{EL} \tag{13.24}$$

3 Negative Interfacial Tensions and Polar Repulsion

Negative Interfacial Tensions

When we rewrite the expression for the polar component of the interfacial tension between materials 1 and 2 (Equation 13.10) in the form of:

$$\gamma_{12}^{AB} = 2(\sqrt{\gamma_1^{\oplus}} - \sqrt{\gamma_2^{\oplus}})(\sqrt{\gamma_1^{\ominus}} - \sqrt{\gamma_2^{\ominus}}) \tag{13.10a}$$

it becomes clear that γ_{12}^{AB} can assume a negative value when:

$$\gamma_1^{\oplus} > \gamma_2^{\oplus} \textit{ and } \gamma_1^{\ominus} < \gamma_2^{\ominus} \tag{13.25}$$

or when

$$\gamma_1^{\oplus} < \gamma_2^{\oplus} \textit{ and } \gamma_1^{\ominus} > \gamma_2^{\ominus} \tag{13.25a}$$

and when $|\gamma_{12}^{AB}| > \gamma_{12}^{LW}$, *and* $\gamma_{12}^{AB} < 0$, the total interfacial tension (see also Equation (13.6)):

$$\gamma_{12} = (\sqrt{\gamma_1^{LW}} - \sqrt{\gamma_2^{LW}})^2 + 2(\sqrt{\gamma_1^{\oplus}\gamma_1^{\ominus}} + \sqrt{\gamma_2^{\oplus}\gamma_2^{\ominus}} - \sqrt{\gamma_1^{\oplus}\gamma_2^{\ominus}} - \sqrt{\gamma_1^{\ominus}\gamma_2^{\oplus}}) \tag{13.26}$$

will have a negative value. In practice this occurs quite readily in biological and other aqueous systems.

Rewriting Equation (13.4) for the special case of a solute or solid (i) immersed in water (w):

$$\Delta G_{iwi} = -2\gamma_{iw} \tag{13.4a}$$

ΔG_{iwi} is the free energy change in bringing molecules of compound (i), immersed in water from infinity to the minimum equilibrium distance, in which a very thin sliver of water separates two macroscopic bodies of (i) (van Oss and Good, 1984). During the process of bringing the two materials together at the equilibrium separation distance, either the work may be done *by* the system (i.e. when the operating forces are attractive) or work needs to be done *on* the system (i.e. when the operating forces are repulsive). It can therefore be seen that attractive forces will contribute to a positive interfacial tension (free energy per unit area), whereas repulsive forces will give rise to a negative interfacial tension. Since the net interfacial tension (γ_{12} or, in aqueous systems, γ_{iw}) also comprises the apolar (LW) component, which is always positive, the total interfacial tension consists of contributions by both attractive and repulsive forces simultaneously, so that the resultant γ_{12} can be either positive or negative.

A net negative interfacial tension normally gives rise to a thermodynamically unstable interface, so that the system will spontaneously relax by either abolishing the interface or reverting to a more stable geometric configuration, ultimately resulting in $\gamma_{12} \rightarrow 0$. An example of such a process may be found in the formation of microemulsions. Prince (1967), while discussing the formation of micro- and macroemulsions, suggested that a small but positive interfacial tension at an oil–water interface with adsorbed surfactant could favour the formation of macroemulsions, whereas a negative interfacial tension would cause the onset of microemulsion formation (see also van Oss *et al.*, 1987a). Other examples of the effects of the decay of negative interfacial tensions are: 'anomalous' high osmotic pressure of solutions of monopolar compounds in water; phase separation in aqueous polymer solutions; pronounced solubility of biopolymers and other polar compounds in water or in other hydrogen-bonded polar liquids; 'steric' stabilization of hydrophobic particles in aqueous media; and 'depletion' interactions (see below).

There are, however, cases where a net negative interfacial tension does not find scope to relax and tend to $\gamma_{12} \rightarrow 0$. This occurs when polar materials, due to crystallization, or strong covalent cross-linking, notwithstanding considerable negative values of γ_{12}, cannot dissipate or dissolve in an aqueous or other polar liquid medium with which they are in contact. For instance, various mineral solids can display considerable negative γ_{iw} values, which, however, do not normally give rise to a significant degree of solubilization in water (Giese *et al.*, 1989).

Monopolar Repulsion

Inspection of Equation (13.10a) and the conditions for $\gamma_{iw} < 0$, given in Equations (13.25a), (13.25b), shows that when, e.g., $\gamma_i^{\oplus} \approx 0$ and $\gamma_i^{\ominus} > \gamma_w^{\ominus}$, a negative polar interfacial tension can ensue. Upon contact angle measurement, using apolar, as well as a number of well-characterized polar liquids (i.e. water, glycerol, formamide), it became clear that, especially among biopolymers, the occurrence of surfaces manifesting a sizeable value for $\gamma_i^{\ominus}$ and a zero or negligible value for $\gamma_i^{\oplus}$ is very common (van Oss *et al.*, 1987a). Such surfaces may be designated as 'monopolar surfaces', and the compounds manifesting these as 'monopolar compounds'. Compounds in which cohesive hydrogen bonds (or, in more general terms, cohesive Lewis acid–base interactions) occur then are best designated as 'bipolar compounds' (to obviate confusion with '*di*polar' interactions or phenomena, such as electrodynamic dipole interactions). While monopolar compounds with a high $\gamma^{\ominus}$ value and a zero or negligible value for $\gamma^{\oplus}$ are quite prevalent in Nature, monopolar compounds with a sizeable $\gamma^{\oplus}$ and zero $\gamma^{\ominus}$ are rare. (Chloroform is an example of the latter variety.)

From Equation (13.8) it is clear that, for a true monopolar compound, $\gamma_i^{AB} = 0$, so that $\gamma_i^{TOT} = \gamma_i^{LW}$. Thus, once it has been determined for a polar compound that $\gamma_i^{\oplus} = 0$, then its interfacial tension with (e.g.) water can be determined directly from its contact angle with water, using the following version of Young's equation (cf. Equations 13.10 and 13.17):

$$\gamma_{iw} = \gamma_i^{LW} - \gamma_w \cos\theta_w \tag{13.27}$$

It is easily shown that for $\gamma^{\ominus}$ monopolar compounds, with a γ^{LW} value of about 40 mJ/m^2 (which is a rather commonly occurring value for such compounds), $\gamma_i^{\ominus}$ must be of the order of 28.3 mJ/m^2 or higher for γ_{iw} to be negative and $\Delta G_{iwi} > 0$ (cf. using Equations 13.4a and 13.10). Thus, the molecules of such monopolar compounds, immersed in water, *will repel each other*. Monopolar compounds of this class are, e.g., dextran, polyvinyl alcohol, polyethylene glycol (van Oss *et al.*, 1987a), human serum albumin (hydrated), human immunoglobulin G (hydrated) (van Oss, 1989b), fibrinogen (van Oss, 1990b) and many other biopolymers.

4 Quantitative Expression of Hydrophilicity and 'Hydrophobicity'

In the condensed state all compounds attract water molecules to a considerable degree, i.e. with a free energy of adhesion varying from a value of -40 mJ/m^2 for the most apolar compounds to -140 mJ/m^2 for the most strongly monopolar hydrophilic materials; see Table 13.1. Thus, the designation of apolar compounds as 'hydrophobic' is misleading. However, as the use of that adjective in this context has become too widespread to hope for its speedy eradication, we shall continue to use it for the sake of clarity, but in quotation marks.

Table 13.1 Parameters in mJ/m^2 relating to the hydrophilicity or 'hydrophobicity' of various compounds, in the order of increasing ΔG_{iwi} (from van Oss, 1992)

	Compound	*Free energy of interfacial interaction in water,* ΔG_{iwi}	*Free energy of hydration,* ΔG_{iw}	*Electron-donor parameter,* γ_i^{θ}
Molecules which attract each other in water, i.e. *'Hydrophobic'*	Hexane	−102.4	−40.0	0
	CCl_4	−90.0	−54.8	≈0
	Dibromoethane	−75.2	−75.5	?
	Benzene	−69.9	−66.7	2.7
	Ethylhexanoate	−43.5	−76.9	8.5
	Polymethylmethacrylate	−37.8	−94.5	12.0
	Fibrin	−34.2	−99.7	12.0
	Heptaldehyde	−27.4	−82.8	13.9
	Gelatin	−19.4	−100.7	18.5
	Zein	−19.4	−104.8	19.4
	Octanol	−15.8	−91.2	18.4
	Agarose	−3.3	−112.2	26.9
Molecules which repel each other in water, i.e. *Hydrophilic*	Polyvinylpyrrolidone	+1.0	−114.8	29.7
	Dextran	+41.2	−135.4	55
	Fibrinogen	+41.9	−134.3	54.9
	PEO	+52.2	−142.0	64

Table 13.1 shows a list of compounds, varying from totally apolar to very polar, or, if one wishes, from quite 'hydrophobic' to extremely hydrophilic, with, from left to right: (1) their free energies of interfacial interaction; (2) their free energies of hydration; and (3) their electron-donor surface tension parameters.

(1) *The free energy of interfacial interaction* of a compound (i) immersed in water (ΔG_{iwi}) is the most appropriate measure of hydrophilicity with respect to the *solubility of compound (i) in water* (van Oss and Good, 1989). ΔG_{iwi} also is a thermodynamically appropriate way of expressing the hydrophilicity (or otherwise) of compound (i). This is because it is a quantitative expression (in the cases of a negligible LW interaction) of the degree to which the polar attraction of compound (i) to water is greater (hydrophilicity) or smaller ('hydrophobicity') that the sum of the attractions which water molecules have for each other and those which molecules of compound (i) have for each other (cf. Equations 13.4a and 13.26). In other words, when the net free energy of interaction between molecules of compound (i) immersed in water is attractive (i.e. ΔG_{iwi} has a negative value), the molecules of compound (i) have less affinity for water than for themselves; they thus are 'hydrophobic', i.e. they are more autophilic than hydrophilic. When the net free energy of interaction between molecules of compound (i) immersed in water is repulsive (i.e. ΔG_{iwi} has a positive

value), the molecules of compound (i) have more affinity for water than for themselves, in which case they are genuinely hydrophilic. The more negative ΔG_{iwi}, the more 'hydrophobic' compound (i) is; the more positive ΔG_{iwi}, the more hydrophilic compound (i); see Table 13.1. In general (and in contrast with ΔG_{iw}), ΔG_{iwi} is for 95% or more due to polar (AB) forces (van Oss and Good, 1991).

For purposes of comparison of the degree of hydrophilicity between different surfaces, it is appropriate to express ΔG_{iwi} in terms of free energy per unit surface area (i.e. in S.I. units of, e.g., mJ/m^2). However, in special cases, one may wish to express hydrophilicity on a molecular scale. ΔG_{iwi} then is best expressed in units of kT, so that $\Delta G_{iwi} = -\chi_{iw}$ (where χ is the Flory–Huggins solubility parameter; see van Oss *et al.*, 1990a), in which case the molecular degree of hydrophilicity relates directly to the material's solubility (van Oss and Good, 1989). Thus, dextran ($\Delta G_{iwi} = +41.2$ mJ/m^2) is somewhat less hydrophilic than polyethylene oxide (PEO) ($\Delta G_{iwi} = +52.2$ mJ/m^2) (Table 13.1), if one compares overall surface properties. However, the contactable surface area (S_c) between dextran (MW $\approx$ 180 000) chains $S_c \approx 34$ Å^2, while for PEO (MW $\approx$ 6000), $S_c \approx$ 21.2 Å^2, (see van Oss *et al.*, 1990a). Then, expressed in units of kT, ΔG_{iwi} for dextran is $+3.5$ kT, while ΔG_{iwi} for PEO is $+2.77$ kT. Thus, on a molecular level, in aqueous solution, dextran is somewhat more hydrophilic than PEO, which agrees well with the behaviour of these polymers observed in aqueous phase separation, by Albertsson (1986).

(2) *The free energy of hydration* (ΔG_{iw}) also is, by its very meaning, a thermodynamically correct way to characterize hydrophilicity (or its opposite). It should be recalled that, according to the Dupré equation (cf. Equation 13.2):

$$\Delta G_{iw} = \gamma_{iw} + \gamma_i + \gamma_w \tag{13.2b}$$

For protein adsorption, and cell adhesion onto low-energy surfaces, the correlation is quite good, albeit not perfect; see Table 13.2. And, while ΔG_{iw} follows the series shown in Table 13.1 fairly closely, there also are some discrepancies (cf. benzene and octanol). It should be kept in mind, however, that the free energy of hydration (in which one might identify a cut-off of $\Delta G_{iw} \approx -113$ mJ/m^2, where more negative values would indicate genuine hydrophilicity), is largely of an apolar nature, especially for the more 'hydrophobic' compounds (van Oss and Good, 1991).

(3) *The electron-donor surface tension parameter* ($\gamma^{\ominus}$) is a rather good semiquantitative indicator of the degree of hydrophilicity of a compound, but hydrophilicity also remains linked to the values of the γ^{LW} component and the $\gamma^{\oplus}$ parameter (if any) of the surface tension. For instance, when $\gamma_i^{LW} \approx 40$ mJ/m^2 and $\gamma_i^{\oplus} = 0$, it has been shown above that with a $\gamma^{\ominus} \gtrsim 28.3$

Table 13.2 Free energies of interfacial interaction (in mJ/m^2) in water and free energies of hydration of a number of hydrophilic entities, classified in the order (from top to bottom, in each category) of their increasing adsorbability onto low-energy surfaces (data from van Oss, 1991b).

Cells[a]	ΔG_{iw}	ΔG_{iwi}
S. epidermidis (caps)	−141.6	+18.5
HELA cells	−138.3	+6.5
S. epidermidis (non-caps)	−126.7	+37.7
↓Fibroblasts	−119.7	−6.6
Proteins[a]		
HSA (hydrated)	−145.7	+20.4
Fibronectin	−144.4	+24.2
IgG (hydrated)	−138.0	+27.7
Fibrinogen	−134.3	+41.7
HSA (dry)	−105.4	−17.6
Fibrin	−99.7	−34.1
↓IgG (dry)	−105.3	−44.4

[a] The arrows point in the direction of the highest degree of adsorption onto low-energy surfaces.

mJ/m^2, a compound (i) is genuinely hydrophilic, i.e. its molecules, when immersed in water, will tend to repel one another, a condition which normally leads to solubility in water. However, when there is a finite $\gamma_i^{\oplus}$, or when $\gamma_i^{LW} > 40$ mJ/m^2, the cutoff for hydrophilicity occurs for $\gamma_i^{\ominus}$-values that can be different from 28.3 mJ/m^2. Thus, in some cases, judging hydrophilicity of a compound solely by its $\gamma^{\ominus}$-value can be somewhat imprecise.

Absolute Measure of Hydrophilicity or 'Hydrophobicity'

It thus becomes clear that there is *no absolute measure* of hydrophilicity or 'hydrophobicity'. This is due, in part, to the very different roles played by Lifshitz–van der Waals (LW) interactions, on the one hand, and hydrogen-bonding (AB) interactions on the other, and, in part, to the sometimes conflicting roles one ascribes to the terms 'hydrophilicity' and 'hydrophobicity'.

The value of $\gamma_i^{\ominus}$ is only roughly proportional to the hydrophilicity of a compound (i) (when $\gamma_i^{\ominus} \gtrsim 28.3$ mJ/m^2), because the hydrophilicity also depends to some degree on the value of its $\gamma_i^{\oplus}$, as well as of its γ_i^{LW}-value, both of which can, and do, vary independently.

The value of ΔG_{iw} (where $\approx G_{iw} \gtrsim -113$ mJ/m^2 on the more 'hydrophobic' side, and $\Delta G_{iw} \lesssim -113$ mJ/m^2 for the more hydrophilic compounds) would, by the very definition of ΔG_{iw}, be the scale of choice. It is usually germane when predictions are needed of adsorption or adhesion of compound (i), onto a low-energy surface j, immersed in water (see Table 13.2), i.e. when one needs to do a correlation with ΔG_{iwj} (Equations 13.3, 13.6, 13.11):

$$\begin{aligned}\Delta G_{\text{iwj}} &= (\sqrt{\gamma_{\text{i}}^{\text{LW}}} - \sqrt{\gamma_{\text{j}}^{\text{LW}}})^2 - (\sqrt{\gamma_{\text{i}}^{\text{LW}}} - \sqrt{\gamma_{\text{w}}^{\text{LW}}})^2 - (\sqrt{\gamma_{\text{j}}^{\text{LW}}} - \sqrt{\gamma_{\text{w}}^{\text{LW}}})^2 \\ &\quad + 2[\sqrt{\gamma_{\text{w}}^{\oplus}}(\sqrt{\gamma_{\text{i}}^{\ominus}} + \sqrt{\gamma_{\text{j}}^{\ominus}} - \sqrt{\gamma_{\text{w}}^{\ominus}}) + \sqrt{\gamma_{\text{w}}^{\ominus}}(\sqrt{\gamma_{\text{i}}^{\oplus}} + \sqrt{\gamma_{\text{j}}^{\oplus}} - \sqrt{\gamma_{\text{w}}^{\oplus}}) \\ &\quad - \sqrt{\gamma_{\text{i}}^{\oplus}\gamma_{\text{j}}^{\ominus}} - \sqrt{\gamma_{\text{i}}^{\ominus}\gamma_{\text{j}}^{\oplus}}] \end{aligned} \qquad (13.28)$$

However, ΔG_{iw} does not always correlate consistently with solubility or stability in water; see Tables 13.1 and 13.2.

The value of ΔG_{iwi} (where $\Delta G_{\text{iwi}} > 0$ on the hydrophilic side and $\Delta G_{\text{iwi}} < 0$ for 'hydrophobic' compounds), on the other hand, is ideal for predictions connected with the solubility or stability of compound or particle (i), immersed in water, but significantly less than accurate for correlations with adsorption or adhesion onto low-energy surfaces; see Table 13.2.

5 Hydration Orientation

In the section on 'Polar or (Lewis) Acid–Base Interactions', the rate of decay of polar (AB) forces was discussed, as a function of distance, ℓ. The fundamental reason for the possibility of propagating an at first sight short-range interaction, such as the hydrogen-bonding interaction, across a significant distance between the interacting surfaces, is a local change in the organization of the water molecules in the vicinity and under the influence of these surfaces. The existence of a considerable degree of orientation of the water molecules of hydration of proteins and other macromolecules has long been surmised (reviewed by Ling, 1972). The manner of orientation of the water molecules adjoining polar surfaces was first illustrated schematically by Parsegian *et al.* (1985), without, however, indicating (in that illustration) the gradual decay in orientation as a function of distance, ℓ. It is precisely this (exponential) decay in the orientation of water molecules, as a function of ℓ, which is described in Equations (13.13) and (13.14). This orientation of the water molecules near monopolar surfaces, and the decay of that orientation, are shown schematically in Figure 13.1. 'Hydration forces' (Marcelja and Radic, 1976; Cowley *et al.*, 1978; Parsegian *et al.*, 1979, 1985, 1986, 1987; Israelachvili, 1985) thus are the direct manifestation of the polar (AB) interaction forces, as propagated through the orientation of neighbouring water molecules, but mitigated by the decay of that orientation as a function of distance. Therefore, as has been most appropriately stated by Parsegian *et al.* (1987), 'There is

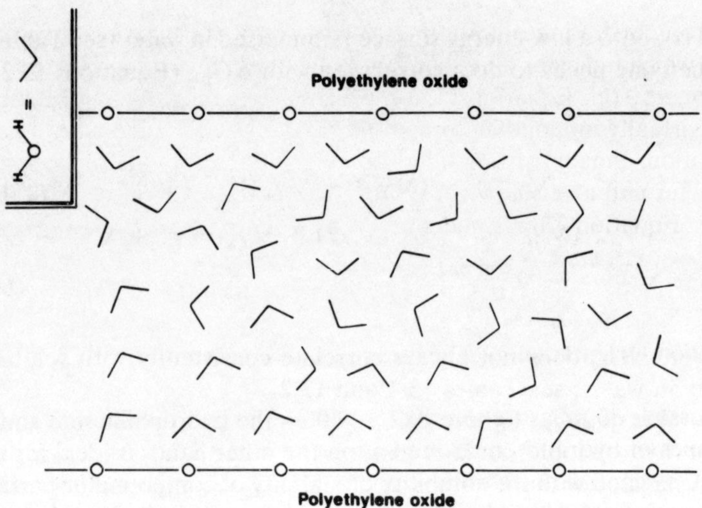

Figure 13.1 Schematic presentation of two opposing monopolar polymer chains with strong electron-donor parameters (e.g. polyethylene oxide: the circles represent the oxygen atoms in the polymer chains), immersed in water. The water molecules (see insert) are represented only as lines which denote the H−O bonds, at an angle of 104.5° (Pauling and Hayward, 1964). In the first layer of hydration most of the water dipoles are strongly oriented, with the oxygen atoms pointing away from the polymer strands. The water molecules residing in layers which are farther distant from the polar polymer strands are less oriented, and at even longer distances they resume a random orientation. However, when two such polymer chains are brought relatively closely together, e.g. within $\ell \approx$ 2–3 nm, as illustrated here, there still is a measurable repulsion between them, brought about by the residual orientation of the water dipoles. In this model the constantly occurring fluctuating complex formation between neighbouring water molecules through hydrogen bonding is not depicted, but it is an additional factor that should be taken into account, as it does give rise to a characteristic decay length (λ) for water, which is somewhat larger than the radius of gyration of single water molecules, i.e. λ is somewhat larger than 0.2 nm (see text)

thus a neat separation between the respective behaviors of surface and medium. The surface determines the coefficient of an exponentially decaying force, while the medium determines the rate of decay.' In other words, in Equations (13.13) and (13.14), $\Delta G^{AB}_{\ell_0}$ describes the (attractive or repulsive) interaction energy of the polar surface itself, while the characteristic decay length, λ, of the liquid medium governs the rate of decay as a function of the distance, ℓ, from the polar surface.

Hydrophilicity and Hydration Orientation

It is easily shown that although hydration orientation is qualitatively different from hydration, the two still are quantitatively linked. To begin with, a high $\gamma_i^{\ominus}$ parameter usually is accompanied by a γ_i^{AB} value which is zero or close to zero. For instance, in practically all cases listed in Table 13.2, monopolarity obtains, which causes both a high degree of hydrophilicity and strong hydration orientation. Then a really strongly negative value for ΔG_{iw} (which indicates a high degree of hydrophilicity) can only be attained when $\gamma_{iw} < 0$ (cf. Equation 13.4a), which, again, only occurs with monopolar or virtually monopolar compounds and thus also is linked to hydration orientation. Finally, strong hydration orientation, originating in a high $\gamma_i^{\ominus}$ parameter and a very low (or zero) $\gamma_i^{\oplus}$ parameter, gives rise to a negative γ_{iw} (cf. Equation 13.26) and thus to a strongly positive ΔG_{iwi} (Equation 13.4a).

Hydration Orientation near Proteins

It is possible to measure the degree of hydration orientation of the water molecules of hydration near, e.g. a protein such as human serum albumin (HSA), by contact angle measurement (van Oss and Good, 1988) on layers of concentrated hydrated HSA. HSA can be concentrated on an ultrafilter membrane, and the amount of residual water of hydration can be determined by weighing the membrane before use and after accumulation of the (known) amount of protein, plus residual water. The contact angles obtained with apolar and polar liquids on layers of HSA with one and two layers of water of hydration per molecule are given in Table 13.3. Using Equation (13.17), the contact angle data thus obtained yield the values of the surface tension parameters given in Table 13.4, as well as the interfacial free energies (ΔG_{1w1}) between dry or hydrated HSA, and water. Water molecules in the bulk liquid are taken to be in a state of random orientation. The degree of orientation (in %) of water molecules in a given hydration layer can be characterized as the decrease in γ_1^{AB} of the hydrated layer, relative to γ_w^{AB} (= 51 mJ/m^2). Ultrafiltration of 25 ml 5% (w/v) HSA yielded a layer of hydrated protein containing 33% (w/w) water, and ultrafiltration of 25 ml 2% (w/v) HSA resulted in a layer of hydrated protein containing 72% (w/w) water (van Oss and Good, 1988).

When the water of hydration is completely oriented, the γ^{AB} of such a hydrated surface becomes zero, as such a surface is then effectively monopolar (see above); i.e. it has a large $\gamma^{\ominus}$ but effectively no $\gamma^{\oplus}$. A 73.5% orientation for 33% hydrated HSA means that the water and glycerol in the drops with which the contact angles are measured, virtually 'see' only the oxygen atoms of the water molecules of hydration. Given the molecular dimensions of the HSA molecule (calculated from Sober, 1968) of a prolate spheroid of $\approx$ 40 $\times$ 140 Å, 33% hydration would correspond to about one molecular layer of hydration, assuming that the hydration is largely peripheral (see Ling, 1972). Similarly, 73.5% hydration corresponds to approximately two layers of water molecules of hydration.

At the isoelectric pH of 4.9 HSA is also still to some extent 'hydrated', but a drastic change in its configuration has taken place, causing it to expose only its apolar moieties at the air (or, initially, the nitrogen)

Table 13.3 Contact angles determined on layers of dry and hydrated Human Serum Albumin

	Water	*Glycerol*	*Hexadecane*	*α-Bromonaphthalene*	*Diiodomethane*
Dry HSA[a]	63.5°	59.5°	—	23.2°	37°
HSA, 33% hydrated[a]	12°	48.5°	15°	—	—
HSA, 72% hydrated[a]	0°	14.5°	13°	—	—

[a] From van Oss and Good (1988).

Table 13.4 Surface-tension components and parameters of dry and hydrated Human Serum Albumin and of water and interfacial interaction energy (ΔG_{1w1}) between HSA molecules, in water[a], and the degree of orientation of the water molecules of hydration[b]. From the data given in Table 13.3. From van Oss (1993)

	γ_1^{LWa}	$\gamma_1^{\oplus a}$	$\gamma_1^{\ominus a}$	γ_1^{ABa}	ΔG_{1w1}^{a}	*Orientation*[b]
Dry HSA	41	0.126	17.2	2.94	−23.0	NA[c]
HSA, 33% hydrated	26.6	0.60	75.9	13.5	+62.2	73.5%
HSA, 72% hydrated	26.8	6.0	51.5	35.2	+21.6	31.0%
H_2O (bulk)	21.8	25.5	25.5	51.0	NA[c]	0%

[a] Expressed in mJ/m^2.
[b] Expressed as % decrease in γ^{AB}, compared with the γ_w^{AB} of bulk water.
[c] Not applicable.

interface, when highly concentrated by ultrafiltration (van Oss and Good, 1988).

It thus appears that there exists a fundamental difference between ordinary attraction of water molecules, a type of 'hydration' which is even exhibited by alkanes (van Oss and Good, 1991), and hydration accompanied by orientation of the water molecules, as in the case of HSA at neutral pH (treated above) and other biopolymers. For instance, with HSA, its monopolarity is not strong enough to cause a repulsion when immersed in water in the dry state, as deduced from the strong negative value of ΔG_{1w1} under these conditions (van Oss *et al.*, 1986). At a contactable surface area between two HSA molecules of 1 nm^2, for dry HSA, $\Delta G_{1w1} = -5.75\ kT$; this would mean virtually total insolubility in water, in that state. However, once hydrated (with strongly positive ΔG_{1w1} values), HSA becomes readily soluble in water, owing to the repulsive effect of the first layer(s) of oriented water of hydration (yielding a value for ΔG_{1w1} of $+\ 15.6\ kT$). But this is possible only *either* because of an original relatively modest monopolarity of the protein in the dry state, which is strongly amplified by the orientation of the closest layers of water molecules, *or* because of a refolding of the protein, which then reassumes its native hydrophilic conformation upon rehydration, thus recovering from the reversible denaturation which had been caused by drying and reassuming a conformation which may be likened to that of a monomolecular micelle.

From hydrodynamic measurements, it would follow that the first layer of hydration is the only layer sufficiently strongly bound to the HSA molecule to migrate with it (Ling, 1972): i.e. the slipping plane is situated at, or close to, the outer boundary of the first layer of hydration. As can be seen from the decrease in orientation in the second layer (Table 13.4), the decay in orientation with distance appears to be quite steep; it correlates well with the exponential decrease proposed by Marcelja and Radic (1976), Parsegian *et al.* (1985, 1986, 1987) (cf. Equation 13.13), although the correlation is better when a value of $\lambda > 0.2$ nm is adopted for water (e.g. $\lambda \approx 0.6$ nm; see van Oss 1990a, 1992; see Section 2, above; see also Zhou *et al.*, 1990).

Linkage between AB and EL Forces

The influence of AB forces on EL forces, and vice versa, is mutual:

(1) The surface of amphoteric molecules tends to become strongly '*hydrophobic*' under conditions of pH close to the isoelectric point.
(2) Very hydrophilic surfaces (i.e. monopolar surfaces with a high $\gamma_i^\ominus$ value) give rise to a high degree of hydration orientation, which causes a decrease in the dielectric constant of the liquid medium near the liquid–solid interface.
(3) The electrophoretic mobility of polar molecules or particles in polar organic media correlates with their degree of electron donicity or accepticity.

(1) Holmes-Farley *et al.* (1985) showed that contact angles obtained with buffered waters of different pHs on amphoteric surfaces were quite high close to the isoelectric point of the amphotere, and low at pHs fairly far removed from the isoelectric point. As indicated above, serum albumin, even though still somewhat hydrated at the isoelectric pH of 4.9, becomes quite hydrophobic (van Oss and Good, 1988). It is well known that proteins in general are least soluble in water at their isoelectric point. It has been shown that the decreased solubility of, e.g., albumin at pH 4.9 is not directly caused by the electroneutrality of the protein at its isoelectric pH (as the phenomenon is not dependent on the dielectric constant of the medium), but rather by the increased 'hydrophobicity' of the isoelectric protein (van Oss, 1989b). Thus, in aqueous media, on a macroscopic level, a net equality of positive and negative charges on an amphoteric surface (i) does not give rise to a finite and equal degree of electron donicity and accepticity at the surface, but rather to the disappearance of both $\gamma_i^\ominus$ and $\gamma_i^\oplus$ (perhaps by mutual elimination).
(2) As shown in Table 13.4 (van Oss and Good, 1988), and as illustrated in Figure 13.1, strongly hydrophilic surfaces (i.e. monopolar surfaces with a high $\gamma_i^\ominus$-value) give rise to a strong orientation of the water molecules in

the first layer adjoining the interface and, to a decreasing extent, in the second and further layers. This orientation of water molecules not only causes 'hydration pressure' which is measurable a fair distance away from the interface, but also locally decreases the permittivity (ε) of water to ε'. For a given electrophoretic mobility, this effect would cause an increase in the real surface potential (ψ_0) by, roughly, $\sqrt{\varepsilon/\varepsilon'}$, as compared with ψ_0 calculated with the assumption that $\varepsilon = 80$. It is reasonable to estimate that, with a degree of orientation of 73.5% in the first hydration layer (see Table 13.4), ε' would be of the order of ≈ 20 in that layer. However, the influence of this phenomenon on the ζ-potential, which is, after all, the potential (at the slipping plane) that is actually obtained from the electrophoretic mobility, the error would be less drastic. For an orientation of the order of 31% (Table 13.4), a rough estimate of ε' would be about 55, which rapidly grows to 80 with increasing distance, l. Then the ζ-potential, derived from the electrophoretic mobility, would be about 17% higher than was at first supposed. At the same time, however, the thickness of the diffuse ionic double layer, $1/\kappa$ (Equation 13.19), would be considerably less than at first assumed, so that the decay of ΔG^{EL} as a function of l will be significantly steeper at higher ionic strengths.

In conclusion, the permittivity gradient in the immediate vicinity of hydrophilic monopolar surfaces gives rise to a considerably higher ψ_0-value than the one obtained when adhering to a constant value of $\varepsilon = 80$ for water. The ζ-potential, taken at the slipping plane, which is one to two layers of water molecules farther out, fares somewhat better and usually will only be underestimated by about 17%. Thus, estimation of an *operative* ψ_0 value, from ζ, via Equation (13.22) will not give rise to large errors. The Debye length ($1/\kappa$) will be considerably smaller than is usually calculated for high ionic strengths, but, at ionic strengths of 0.01 or less, $1/\kappa$ is not significantly affected. (For considerations on the dielectric profile near biological systems, due to ion–ion interactions, see Vaidhyanathan, 1982, 1985, 1986.)

(3) There is a semiquantitative correlation between electrophoretic mobility and electron donicity and accepticity. The degree of electron donicity of a macromolecule or particle can be estimated by its electrophoretic mobility in organic polar liquids which are Lewis acids, and the degree of electron accepticity can be obtained via electrophoresis in Lewis base organic liquids (Fowkes *et al.*, 1982; Labib and Williams, 1984; Labib, 1988). It is not surprising, therefore, that amphoteric entities which are suspended or dissolved in water at a pH corresponding to their isoelectric point, and consequently have no electrophoretic mobility, also have zero electron donicity as well as zero electron accepticity. They must, therefore, manifest maximum apolarity, or 'hydrophobicity' under these conditions (see paragraph 1, above). For observations on a putative correlation between the donicity scale of various 'wet' and 'dry' inorganic oxides and

their 'ζ-potentials' in water, as a function of pH, see Labib and Williams (1986).

6 Consequences and Examples of Hydration Forces

Osmotic Pressure of Water-soluble Polymers

As shown by van Oss *et al.* (1990a), the strong monopolar repulsion between polar polymer molecules in aqueous solution at high concentrations gives rise to a strongly negative Flory–Huggins χ_{12} parameter, which can be expressed as

$$\chi_{12} = -S_c \Delta G_{121}/kT, \text{ or } \chi_{12} = 2\, S_c \gamma_{12}/kT \qquad (13.29)$$

where S_c is the contactable surface area between two polymer molecules (van Oss and Good, 1989). The effect of this is that the second virial coefficient in the expression for the osmotic pressure, Π, becomes strongly positive:

$$\Pi = RTc_1[1/M_1 + B(c_1/d_1^2)\,(0.5 - \chi_{12}) + \cdots] \qquad (13.30)$$

where

$$B = V_w\,(n_1/M_1)^2 \qquad (13.31)$$

and where V_w is the volume of 1 mol of the solvent, water, n_1 the number of monomer units per polymer molecule and M_1 the molecular weight of the polymer (see van Oss *et al.*, 1990a). A strongly positive second virial coefficient term in Equation (13.30) results in unusually high osmotic pressures (of the order of 30 MPa), especially at high concentrations of strongly monopolar polymers dissolved in water. In addition, at high concentrations, these osmotic pressures are virtually independent of the molecular weight of the polymer (at MW $\geqslant$ 1000) (see Figure 13.2 and Table 13.5). These considerations result from the measurements done on the osmotic pressures of aqueous solutions of polyethylene glycols, or polyethylene oxides (PEO) of various molecular weights (see van Oss *et al.*, 1990a), where Π/c_1 *increases* with the concentration, c_1. This is a phenomenon which has been recognized as peculiar to polymers dissolved in 'good' solvents (Napper, 1983). Thus, in the case of PEO, its activity increases with concentration. This is largely due to the fact that, with increasing concentration, PEO becomes less hydrated (because of the lack of a surplus of water molecules), which leads to an increase in the positive value of ΔG_{121}^{AB}, i.e. to an increase in mutual repulsion between PEO mol-

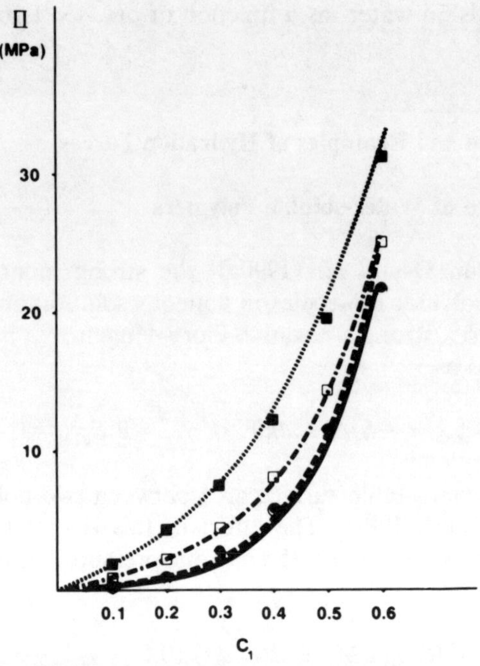

Figure 13.2 Plot of the osmotic pressure, II, against the PEO concentration, c_1, for four different molecular weights of PEO. The *curves* are based upon the experimental results obtained by Arnold *et al.* (1988): . . . M = 150; -.- M = 400; ---- M = 6000; ——— M = 20 000. The *points* are calculated via Equation (13.29): M = 150; M = 400; M = 6000; M = 20 000. From van Oss *et al.* (1990a) and van Oss (1992)

Table 13.5 Osmotic pressures, in MPa, of aqueous solutions of PEG of various molecular weights calculated according to Equation (13.29), broken down into the first three virial coefficient terms (from van Oss *et al.*, 1990a, and van Oss, 1992)

Concentrations, c_1	*MW*	$\left(\frac{1}{2} - \chi\right)$	$\Pi 1 + \Pi 2 + \Pi 3 = \Pi$ *total*
0.6	150	3.28	9.76[a] + 20.25[a] + 1.07[a] = 31.09
	400		3.66 + 20.25 + 1.07 = 24.99
	6 000		0.34 + 20.25 + 1.07 = 21.57
	20 000		0.07 + 20.25 + 1.07 = 21.40
0.5	150	2.36	8.13 + 10.59 + 0.66 = 19.37
	400		3.05 + 10.59 + 0.66 = 14.30
	6 000		0.20 + 10.59 + 0.66 = 11.44
	20 000		0.06 + 10.59 + 0.66 = 11.31
0.4	150	1.74	6.51 + 5.25 + 0.37 = 12.13
	400		2.44 + 5.25 + 0.37 = 8.05
	6 000		0.17 + 5.25 + 0.37 = 5.78
	20 000		0.05 + 5.25 + 0.37 = 5.66
0.3	150	1.33	4.88 + 2.37 + 0.17 = 7.42
	400		1.83 + 2.37 + 0.17 = 4.37
	6 000		0.12 + 2.37 + 0.17 = 2.66

Table 13.5 continued

Concentrations, c_1	MW	$\left(\frac{1}{2} - \chi\right)$	$\Pi 1 + \Pi 2 + \Pi 3 = \Pi$ total
	20 000		0.04 + 2.37 + 0.17 = 2.57
0.2	150	1.05	3.25 + 0.87 + 0.05 = 4.16
	400		1.22 + 0.87 + 0.05 = 2.14
	6 000		0.08 + 0.87 + 0.05 = 1.00
	20 000		0.02 + 0.87 + 0.05 = 0.94
0.1	150	0.87	1.63 + 0.19 + 0.006 = 1.83
	400		0.60 + 0.19 + 0.006 = 0.80
	6 000		0.04 + 0.19 + 0.006 = 0.24
	20 000		0.01 + 0.19 + 0.006 = 0.21

[a] All figures are rounded off to the nearest 10 kPa.

ecules, and a tendency to draw more water into the polymer — in other words, high osmotic pressure. While oriented hydration is, of course, the mechanism by which polar repulsion is propagated some appreciable distance beyond the very close range of direct hydrogen bonds (see above), the hydration orientation decreases exponentially with distance, and the result is the decay of polar repulsion (Figure 13.1).

Thus, the greater the dilution, the weaker the polar repulsion. The strong positive value of ΔG_{121} for PEO in water, and thus its strongly negative χ_{12} value, therefore not only causes the unusually high (and largely molecular weight-independent) osmotic pressure, but it also is the real cause of the strong stabilizing power of PEO for aqueous suspensions of 'hydrophobic' particles (see, e.g., Napper, 1983), as well as of its flocculating power for various proteins (van Oss, 1988) and of its capacity to cause and/or to facilitate cell or liposome fusion (Arnold *et al.*, 1988; van Oss *et al.*, 1988b).

As can be seen clearly from Table 13.5, owing to the decay of χ_{12} with greater dilution, at great dilution, Equation (13.30) simply reverts to the van't Hoff part of the equation, i.e. to:

$$\Pi = RTc_1/M_1 \tag{13.32}$$

where χ_{12} veers toward the value of +0.5. Owing to the widely held belief that χ_{12} should be determined at maximum dilution, the published values for χ_{12} of virtually all polymer–solvent systems vary only between 0.4 and 0.5 (see, e.g., Barton, 1983), which is unhelpful in predicting the real osmotic pressure or solubility behaviour of polymers at higher concentrations. It is only by divesting the molecules of a given polymer of most, if not all, of their coats of solvation, i.e. by operating at *the highest possible concentration*, that one can determine the χ_{12}-value which is most germane to its actual osmotic pressure and solubility (see below).

Stability of Particles and Cells in Aqueous Media

In biological systems, and even more so in observations on suspensions of clay particles, *in aqueous media*, it has come to be realized that the classical DLVO theory (after Derjaguin, Landau, Verwey and Overbeek; see, e.g., Verwey and Overbeek, 1948) usually does *not* furnish an adequate quantitative depiction of the interplay of the various forces which play a role in determining their stability. This is because the DLVO theory only considers van der Waals–London attractions and electrostatic repulsions, as a function of distance, and does not account for the quantitatively usually exceedingly important hydrogen-bonding interactions occurring in water. Only in the rare case where a biological entity (i) immersed in water is 'neutral' from a polar point of view (e.g. when $\gamma_i^{\oplus} = 0$ and $\gamma_i^{\ominus}$ is equal to that of water, i.e. $\gamma_i^{\ominus} = 25.5$ mJ/m^2), will the simple DLVO approach be precisely valid. In the stability of red blood cells, in the aqueous medium of blood, at closest approach, the van der Waals attraction can be expressed as $\Delta G^{LW} \approx -0.6$ mJ/m^2 and the electrostatic repulsion, $\Delta G^{EL} \approx +0.5$ mJ/m^2, while the polar (hydrophilic) repulsion (neglected in the DLVO approach) is as high as $\Delta G^{AB} \approx +25$ mJ/m^2 (van Oss, 1989a). When these interaction energies are projected as a function of distance between red cells, according to the DLVO approach, it should be entirely feasible to squeeze cells closely enough together to become attached to each other. It is known, however, that even upon ultracentrifugation at $\approx$ 300 000 *g*, red cells still will not adhere together (van Oss, 1985). By considering the polar repulsion forces, and by extending these to some distance away from the cell surface by taking the hydration forces into account, it can easily be understood why it is impossible to exert enough outside pressure to make two cells adhere together by purely physical means (van Oss, 1989a). It should be stressed that the surface properties of red cells (and of other cells) are essentially those of these cells' glycocalices, i.e. the layer of fuzzy glycoprotein strands which occupy the cell's outer surfaces.

It can be equally important not to neglect attractive hydrogen-bonding ('hydrophobic') interactions, e.g. in the case of suspensions of clay of relatively low hydrophilicity. For instance, it can be shown with the clay hectorite that while DLVO diagrams predict aqueous suspension stability at all NaCl concentrations from 1 to 10^{-6} molar, when also taking the (attractive) polar forces into account, instability at 0.1M and 1M NaCl must occur and stability can only prevail at ionic strengths lower than 0.01. Only the latter predictions conform to the experimental observations (van Oss *et al.*, 1990b).

'Steric' Stabilization

In considering the stability of suspensions of cells, it is, of course, imperative to take into account the repulsive hydration forces between the most distal parts of the glycocalix molecules surrounding the cells: it is their mutual repulsion due to hydration forces which is mainly responsible for the cells' stability (van Oss, 1989a, 1990a). In the same manner, stabilization of hydrophobic particles occurs through the repulsive hydration forces which hydrophilic polymers (adsorbed or otherwise attached to these particles) exert on each other. This is all that 'steric' stabilization is about. The complicated theories that have been developed to account for the phenomenon of stabilization of particles by means of electrokinetically neutral polymers (see, e.g., Napper, 1983) can be vastly simplified by just taking the polar repulsion between such polymer molecules into account (van Oss, 1991a). The general rule is that 'steric' stabilization is only possible with polymers which are soluble in the liquid in which the stabilization is to occur. There are exceptions to this rule. Thus, it is fruitful to determine what makes polymers soluble in liquids, and especially in water; see the following section.

Hydration Forces and Polymer Solubility

The mechanism by which solute molecules are solubilized in a solvent is precisely the same as the mechanism by which particles form a stable suspension in the liquid, i.e. it occurs only when there is a net repulsion between solute molecules, or between suspended particles.

The best indicator for the solubility of any solute (1), in a given liquid (2), is furnished by the value of the free energy of interaction between two solute molecules, when immersed in the liquid: ΔG_{121}, or for solute (i), immersed in water (w): ΔG_{iwi}. When ΔG_{121} has a negative value, two molecules (1) will attract each other when immersed in liquid (2), but if that attraction energy is less than the 'repulsive' energy of the Brownian, or thermal movement ($\approx 1\ kT$), then solubility still prevails. Thus, in non-polar systems, solubility is favoured when $0 > \Delta G_{121} \gtrsim -1\ kT$ (or, if one assumes that the movement of solute molecules in the liquid has three degrees of freedom, when $0 > \Delta G_{121} \gtrsim -1.5\ kT$) (van Oss and Good, 1989).

In polar solvents, and especially in water, as shown above, ΔG_{iwi} can readily attain a positive value. Thus, solubility, in polar systems, prevails when $-1.5\ kT \lesssim \Delta G_{iwi}$. Here, of course, ΔG_{121} or ΔG_{iwi} is expressed in kT units, i.e. in energy per solute molecule pair. For this it is essential to know the contactable surface area (S_c) between two solute molecules, so that, if ΔG_{121} is expressed in energy units per unit area,

$$\Delta G_{121\,(\text{in }kT)} = S_c\,\Delta G_{121} \tag{13.33}$$

In terms of the Flory–Huggins χ_{12} parameter, we return to Equation (13.29): $\chi_{12} = -S_c\,\Delta G_{121}$, showing that χ_{12} is nothing else but the work of adhesion (W_{121}) of solute molecules (1), when immersed in liquid (2) because, by convention,

$$W_{121} = -\Delta G_{121} \tag{13.34}$$

And therefore (cf. Equation 13.4a):

$$W_{121} = \chi_{12} = 2\gamma_{12}S_c/kT \tag{13.29a}$$

or

$$W_{iwi} = \chi_{iw} = 2\gamma_{iw}S_c/kT \tag{13.29b}$$

Thus, solubility prevails when $\chi_{12} \leqslant -1$–$1.5\,kT$, and especially when $\chi_{12} < 0$, happens most readily in aqueous media, i.e. when $\chi_{iw} < 0$. This occurs only when $\gamma_{iw} < 0$ (Equation 13.29b). It so happens that $\gamma_{iw} < 0$ for a great many polar solutes (van Oss *et al.*, 1987a), because polar solutes all tend to have an appreciable $\gamma_i^\ominus$, and a very low $\gamma_i^\oplus$ parameter, while water has a high $\gamma_w^\oplus$ parameter. Thus, in these cases (cf. Equation 13.10):

$$\gamma_{iw}^{AB} = 2(\sqrt{\gamma_i^\oplus\gamma_i^\ominus} + \sqrt{\gamma_w^\oplus\gamma_w^\ominus} - \sqrt{\gamma_i^\oplus\gamma_w^\ominus} - \sqrt{\gamma_i^\ominus\gamma_w^\oplus}) < 0 \tag{13.10a}$$

when the sum of the terms pertaining to the polar *adhesion* between solute and water $(\sqrt{\gamma_i^\oplus\gamma_w^\ominus} + \sqrt{\gamma_i^\ominus\gamma_w^\oplus})$ has a higher value than the sum of the terms describing the polar *cohesion* of solute and water $(\sqrt{\gamma_i^\oplus\gamma_i^\ominus} + \sqrt{\gamma_w^\oplus\gamma_w^\ominus})$. This, in a nutshell, is why so many polar solutes (especially those which are of biological origin) are soluble in water.

As shown above, it is not uncommon for a protein (e.g. serum albumin or immunoglobulins) to be *insoluble* in water in the dried state, but to become *soluble* after contact with water and rehydration (van Oss *et al.*, 1986; van Oss and Good, 1988). This is probably due to a reversible denaturation of these proteins upon drying. Not all proteins display this behaviour, however: see, e.g., fibrinogen (van Oss, 1990b).

The ζ-potential of polymers (e.g. proteins) plays a direct role in contributing to solubility at $|\zeta| \geqslant 10$ mV. Below that value, ΔG^{EL} becomes rapidly negligible. However, indirectly, through the lowering of ζ (for instance, by changing the pH of the liquid medium until the isoelectric point of the protein is approached) the decrease in ζ-potential (which causes a decrease in ΔG^{EL}) is also accompanied by a decrease in ΔG^{AB}, i.e. the protein surface becomes much more 'hydrophobic' (see Section 6). Thus, the pronounced decrease in the solubility of proteins at or near their isoelectric point, from a quantitative point of view, is principally due to the accompanying change from a hydration repulsion to a 'hydrophobic' attraction.

Interactions between Two Different Particles or Molecules in Aqueous Media

In the interfacial interaction between two dissimilar 'hydrophobic' entities immersed in water, or between one 'hydrophobic' entity and one hydrophilic entity in water, ΔG_{iwj} (see Equation 13.29) usually has a negative value. However, a positive value for ΔG_{1w2} is also possible. In the latter case a repulsion ensues, i.e. generally when the entities (i) and (j) both are hydrophilic, i.e. if they are electron-donor monopolar compounds, and the average value of their $\gamma^{\ominus}$ parameters is greater than 28.3 mJ/m².

There are some important differences between the behaviour of similar and that of dissimilar entities, in the outcome of the repulsive interactions which they undergo, when immersed or dissolved in water. A positive value of ΔG_{iwi} gives rise to a pronounced solubility of compound (i) in water (see above). A positive value of ΔG_{iwj} (where both compounds (i) and (j) are dissolved in water) tends to result in phase separation (van Oss *et al.*, 1987b, 1989). In aqueous phase separation one can obtain as many phases as the number of different dissolved polymers (Albertsson, 1986).

When ΔG_{iwj} has a negative value, and when, e.g., entity (i) is moderately 'hydrophobic' and entity (j) hydrophilic (so that, in *water*, an attraction occurs), the attraction between (i) and (j) can be changed into a repulsion by the admixture of a water-miscible *organic solvent*; see below.

7 Solvation Forces in Non-Aqueous Media

'Hydrophobic' and 'Solvophobic' Interactions

Owing to the much smaller polar energy of cohesion of non-aqueous polar liquids than of water, the net mutual interfacial ('solvophobic') attraction of low-energy solute molecules in such non-aqueous polar liquids is much smaller than the 'hydrophobic' attraction between these molecules in water. And often, in the case of an attraction between low-energy surfaces and hydrophilic compounds dissolved and/or immersed in water ($\Delta G_{132} < 0$, where the subscript 3 denotes the liquid; see Equation 13.29), by admixture of a water-miscible polar solvent to the water, the attraction between (1) and (2) can be turned into a repulsion ($\Delta G_{132} > 0$). This phenomenon is the basis of 'reversed phase liquid chromatography' (RPLC), where hydrophilic compounds (e.g. proteins, peptides, etc.), dissolved in water, are adsorbed onto chromatographic carriers with a moderately low energy ('hydrophobic') surface. The adsorbed hydrophilic compound can then be desorbed, or 'eluted' from the chromatographic column by the admixture of increasing amounts of a water-miscible organic solvent — e.g., ethylene glycol (van Oss, 1990d), dimethylsulphoxide, acetonitrile, etc. (see Table 13.6). In principle the same phenomenon also underlies the mechanism of removing low energy material (e.g., 'dirt') from hydrophilic tissues by means of an aqueous surfactant solution, that is, 'washing', or by 'dry-cleaning' with an appropriate medium to low-energy solvent.

Solubility in Polar Organic Solvents

Non-aqueous polar solvents (e.g. formamide, ethylene glycol) can dissolve many biopolymers. The solvation forces occurring in such solvents are not as strong as those in water, but liquids such as formamide and ethylene glycol do have a polar component of their energy of cohesion (which causes the insolubility in these solvents of apolar compounds such as polystyrene, which are soluble in totally apolar solvents of similar γ^{LW}, such as benzene or α-bromonaphthalene); see Table 13.6. The best organic solvents for (electron-donor) monopolar solutes are those with a pronounced electron-acceptor parameter, such as chloroform. Thus, chloroform dissolves PEG. It should also dissolve biopolymers such as dextran, but it does not. This is most likely due to the residual hydration of dried dextran powder, which is exceedingly difficult to eliminate.

Comparison between Water and Polar Organic Liquids

For both hydration (or solvation) force repulsions, and for 'hydrophobic' (solvophobic) attractions, water is by far the most effective liquid. This is due to two factors: (1) its polar energy of cohesion is higher than that of any other liquid, and (2) the $\gamma^{\oplus}$-parameter of water is higher than that of any other compound measured to date. The latter property is of great importance in view of the fact that it is through its electron-acceptor capacity that water interacts so strongly with most polar organic molecules and especially with those of biological origin, which tend to have an elevated $\gamma^{\ominus}$ (electron-donor) parameter. For this reason, a vastly greater variety of compounds are soluble in water than in any other liquid.

Table 13.6 Tabulation of the influence of hydration* or solvation** forces, 'hydrophobic'[†] or 'Solvophobic'[††], or Lifshitz–van der Waals[§] Interactions on: (a) *solubility* and (b) *attachment to moderately 'hydrophobic' carrier* of various types of solutes in different solvents

	Stronger self-hydrogen bonding liquid (water)	Moderately self-hydrogen-bonding liquids (e.g. glycerol, formamide, ethylene glycol, dimethylsulphoxide; mixtures of water with acetonitrile, lower alcohols, etc.	Monopolar electron-acceptor liquids (e.g. chloroform)	Non-polar liquids and monopolar electron-donor liquids (e.g. (a) alkanes, (b) benzene, methylethyl ketone, tetrahydrofuran)
(A) Solubility (depends on sign and value of ΔG_{131})				
Hydrophilic solutes[a]	Very soluble*	Soluble**	Only soluble if totally non-hydrated**	Insoluble[†§]
Somewhat polar 'hydrophobic' solutes[b]	Often insoluble[†]	Soluble**	Soluble**	Often soluble[§]
Apolar solutes[c]	Very insoluble[†]	Insoluble[††]	Often soluble[§]	Often soluble[§]
(B) Attachment to a moderately 'hydrophobic' carrier (depends on sign and value of ΔG_{132})				
Hydrophilic solutes[a]	Weak attachment[d], or attachment only under dehydrating conditions (e.g. high salt)[e†]	Detachment[f**]	Detachment only if solute is totally non-hydrated**	Usually attachment[†§]
Somewhat polar 'hydrophobic' solutes[b]	Attachment[†]	Detachment[f**]	Detachment[g**]	Usually detachment[g§]
Apolar solutes[c]	Strong attachment[†]	Attachment[††]	Detachment[g§]	Detachment[g§]

[a] For example, proteins, polysaccharides.
[b] For example, lipoproteins, lipopolysaccharides, denatured proteins.
[c] For example, alkanes, lipids.
[d] As in RPLC.
[e] As in 'hydrophobic interaction' liquid chromatography.
[f] Elution step in RPLC.
[g] As in dry-cleaning.

* Due to hydration forces.
**Due to solvation forces.
[†] Due to 'hydrophobic' interactions.
[††] Due to 'solvophobic' interactions.
[§] Partly or mainly due to LW interactions.

Phase Separation in Polar Organic Solvents

Even phase separation between PEO and polymethylmethacrylate, as well as between PEO and polystyrene, can occur in chloroform, which is also due to polar interactions, i.e. to solvation forces alone (van Oss *et al.*, 1989); see Table 13.6. In solutions of two different apolar polymers in apolar solvents, phase separation may also take place (van Oss *et al.*, 1979, 1989). This, however, is solely due to a net Lifshitz–van der Waals repulsion, which can, under certain conditions, occur in ternary apolar systems. Solvation forces of an orientational nature play no role in this phenomenon.

8 How Specific Interactions Overcome Hydration Forces

Dichotomy between Specific and Aspecific Interactions

Specific interactions in aqueous media, i.e. the non-covalent interactions between various ligands and their receptors, such as antigen–antibody, substrate–enzyme and carbohydrate–lectin interactions, are affected by *the same forces* as those keeping cells and/or biopolymers apart (van Oss, 1990c). There lies an apparent paradox in the fact that, using the same interaction forces, specific interactions between biological entities are necessarily *attractive*, whereas the aspecific interactions between the same biological entities, which serve to prevent contact between these entities, are *repulsive*. The explanation of the paradox is to be found in the fact that while the repulsive (aspecific, stabilizing) interactions operate between relatively large surface areas (≈ 10–100 nm^2) or between entities with a fairly large radius of curvature ($R \approx$ 3–5000 nm), the attractive (specific) interactions act between protruding processes or prominent parts of molecules with rather small contactable surface areas (0.5–5.0 nm^2), or between specific sites with a small radius of curvature ($R \approx$ 0.5–2.5 nm) (van Oss, 1990c) (see Table 13.7).

Thus, while normally cells and biopolymers keep their distance under the influence of, principally, repulsive hydration forces, on a macroscopic level, in exceptional circumstances that distance can, on a microscopic level, be bridged by a combination of two physicochemical features:

(1) The *ligand* (e.g. the epitope of an antigen) consists of a prominently placed small patch with a special and characteristic chemical composition, which can specifically bind (through AB, EL and LW forces, usually in that order of importance) to its *receptor* (e.g. the paratope of its corresponding antibody), which is of the same size as the ligand and has an analogous but

Table 13.7 Comparison between aspecific and specific interactions (between cells and cells, cells and biopolymers, biopolymers and biopolymers) (van Oss, 1990c)

	Aspecific interactions	*Specific interactions*
Nature of contact sites	Usually rather homogeneous	Often heterogeneous
Surface area of contact	Large (10–100 nm^2)	Small (0.5–5.0 nm^2)
Total size or radius molecule or particle	May be quite large ($R \approx$ 3–6 μm)	Usually small ($R \approx$ 1–10 nm or with processes with a very small radius of curvature)
Influence of Brownian movement	Small	Large
Binding energy	Usually *repulsive* (+25 mJ/m^2[a], e.g. for erythrocytes)	*Attractive* (–5 to –25 kT per particle, or –10 to –50 mJ/m^2)

[a] Strong repulsive forces of this magnitude can only be overcome by moieties with a very small radius of curvature; hence the smallness or thinness of most molecules comprising a ligand or receptor.

complementary chemical composition, provided, of course, that ligand and receptor be enabled to make contact.

(2) The ligand can only reach the receptor, and establish contact with it, if it can bridge the distance (mainly caused by hydration forces) between the cells and/or biopolymers of which they are an integral part. This is achieved by having the ligand placed on a fairly small biopolymer, or at the end of a part of a cell or a biopolymer with a small radius of curvature (R). Equations (13.7b), (13.14) and (13.21) indicate that the LW, AB and EL repulsion energies are proportional to R, so that processes or molecules with a small enough R can easily pierce a strong repulsion field between larger entities (e.g. through Brownian motion), upon which the specific attraction forces between ligand and receptor can give rise to permanent attachment.

The difference between aspecific and specific interaction forces operating being biological entities, in water, are summarized in Table 13.7. In practice there tends to be a certain degree of asymmetry between the configurations of ligands and receptors (e.g. epitopes and paratopes), for a number of reasons. For instance, carbohydrate biopolymers (polysaccharides) are almost invariably hydrophilic and thus can manifest repulsive hydration forces among themselves, but cannot mutually engage in 'hydrophobic' attractions, because, to do so, at least one 'hydrophobic' moiety is required. Proteins, however, can have both hydrophilic and 'hydrophobic' moieties. Therefore only proteins, and not polysaccharides, can generate enough versatility to function as *antibodies*, which frequently comprise 'hydrophobic' patches in their paratope structure, although both proteins and polysaccharides can be *antigens*. In addition, the 'hydrophobic' patches in the paratopes tend to be situated in concavities (with a negative radius of curvature) in the variable sites of immunoglobulins (see, e.g., Amit *et al.*, 1986), which obviates the occurrence of accidental aspecific attractive 'hydrophobic' interactions with irrelevant biopolymers or cells (van Oss, 1990c).

Specific Interactions and Loss of Water of Hydration

Especially in antigen(Ag)–antibody(Ab) interactions, thermodynamic measurements have shown (see e.g., reviews by van Oss and Absolom, 1984; Absolom and van Oss, 1986) that while the formation of Ag–Ab complexes implies a certain degree of increase in order (and thus decrease in entropy), usually a *decrease* in entropy of the total system has been noted. This decrease in entropy has generally been ascribed to a randomization of water molecules formerly bound as (oriented) water of hydration to Ag and Ab molecules. The increase in entropy accompanying the expulsion of water of hydration, typical for many Ag–Ab interactions, tends to be larger than the decrease in entropy due to the relatively regular structure formation in Ag–Ab complexes. Mukkur (1984) clearly distinguishes between the two types of entropy. The compensating interplay between enthalpy and entropy in Ag–Ab interactions as a function of temperature has also been discussed extensively by Mukkur (1980, 1984). Enthalpy–entropy compensation is a generally observed phenomenon in interactions occurring in water, among small molecules (Lumry and Rajender, 1970), as well as with larger molecules such as proteins (Sturtevant, 1977) or nucleic acids (Breslauer *et al.*, 1987).

The fact that a decrease in hydration so often accompanies Ag–Ab reactions has prompted Steane and Greenwalt (1974, 1977) to propose that the decrease in hydration is a driving force for Ag–Ab reactions. It appears certain that dehydrating conditions favour at least the possibility of better isolation of Ag–Ab complexes (Albini *et al.*, 1984), but *not* of their formation or permanency, when high ionic strengths are used as dehydrating measures (Smeenk *et al.*, 1982). But dehydrating conditions do tend to strengthen the secondary bonds formed between Ag and Ab, when complexes are allowed to 'ripen', as a function of elapsed time (Absolom and van Oss, 1986).

References

Absolom, D. R. and van Oss, C. J. (1986). The nature of the antigen–antibody bond and the factors affecting its association and dissociation. *Crit. Rev. Immunol.*, **6**, 1–46

Adamson, A. W. (1982). *Physical Chemistry of Surfaces*. Wiley-Interscience, New York

Albertsson, P. Å. (1986). *Partition of Cell Particles and Macromolecules*. Wiley-Interscience, New York, p. 17

Albini, B., Fagundus, A. M. and Vladutiu, A. O. (1984). Circulating immune complexes. In Atassi, M. Z., van Oss, C. J. and Absolom, D. R. (Eds), *Molecular Immunology*. Marcel Dekker, New York, pp. 381–401

Amit, A. G., Mariussa, R. A., Phillips, S. E. V. and Poljak, R. J. (1986). Three-dimensional structure of an antigen/antibody complex at 2.8 Å resolution. *Science*, **233**, 747–753

Arnold, K., Herrmann, A., Gawrisch, K. and Pratsch, L. (1988). Water-mediated effects of PEG on membrane properties and fusion. In Ohki, S., Doyle, D., Flanagan, T. D., Hui, S. W. and Mayhew, E. (Eds), *Membrane Fusion*. Plenum Press, New York, pp. 255–272

Barton, A. F. M. (1983). *Handbook of Solubility and Other Cohesion Parameters*. CRC Press, Boca Raton, Florida

Breslauer, K. J., Remeta, D. P., Chou, W. Y., Ferracote, R., Curry, J., Zaunczkowski, D., Snyder, J. G. and Marky, L. A. (1987). *Proc. Natl Acad. Sci. USA*, **84**, 8922–8926

Casimir, H. B. and Polder, D. (1946). Influence of the retardation on the London–van der Waals forces. *Nature*, **158**, 787–788

Chaudhury, M. K. (1984). *Short-range and Long-range Forces in Colloidal and Macroscopic Systems*. Dissertation, SUNY at Buffalo

Christenson, H. K. (1988). Non-DLVO forces between surfaces — Solvation, hydration and capillary effects. *J. Disp. Sci. Tech.*, **9**, 171–206

Claesson, P. M. (1986). *Forces between Surfaces Immersed in Aqueous Solutions*. Dissertation, Royal Institute of Technology, Stockholm

Costanzo, P. M., Giese, R. F. and van Oss, C. J. (1990). Determination of the acid–base characteristics of clay mineral surfaces by contact angle measurements — implications for the adsorption of organic solutes from aqueous media. *Adhesion Sci. Technol.*, **4**, 267–275

Cowley, A. C., Fuller, N. L., Rand, R. P. and Parsegian, V. A. (1978). Measurement of repulsive forces between charged phospholipid bilayers. *Biochemistry*, **17**, 3163–3168

Derjaguin, B. V. (1989). *Theory of Stability of Colloids and Thin Films*. Consultants Bureau/Plenum Press, New York

Fowkes, F. M. (1972). Donor–acceptor interactions at interfaces. *J. Adhesion*, **4**, 155–159

Fowkes, F. M. (1983). Acid–base interaction in polymer adhesion. In Mittal, K. L. (Ed.), *Physical-chemical Aspects of Polymer Surfaces*, Vol. 2. Plenum Press, New York, pp. 583–603

Fowkes, F. M., Jinnai, H., Mostafa, M. A., Anderson, F. W. and Moore, R. J. (1982). Mechanism of electric charging of particles in nonaqueous liquids. In Hair, M. and Groucher, M. D. (Eds), *Colloids and Surfaces in Reprographic Technology*. ACS Symposium Series, **200**, 307–324

Giese, R. F., Van Oss, C. J., Norris, J. and Costanzo, P. M. (1989). Surface energies of some smectite clay minerals. In *Proceedings IXth International Clay Conference*, Strasbourg, France

Hamaker, H. C. (1937). The London–van der Waals attraction between spherical particles. *Physica*, **4**, 1058–1072

Holmes-Farley, S. R., Reamey, R. H., McCarthy, T. J., Deutch, J. and Whitesides, G. M. (1985). Acid–base behaviour of carboxylic acid groups covalently attached at the surface of polyethylene. The usefulness of contact angle in following the ionization of surface functionality. *Langmuir*, **1**, 725–740

Israelachvili, J. N. (1985). *Intermolecular and Surface Forces*. Academic Press, London

Israelachvili, J. N. and Pashley, R. (1984). Measurement of the hydrophobic interaction between two hydrophobic surfaces in aqueous electrolyte solutions. *J. Colloid Interface Sci.*, **98**, 500–514

Janczuk, B. and Bialopiotrowicz, T. (1989). Surface free-energy components of liquids and low energy solids and contact angles. *J. Colloid Interface Sci.*, **127**, 189–204

Labib, M. (1988). The origin of the surface charge on particles suspended in organic liquids. *Colloids and Surfaces*, **29**, 293–304

Labib, M. and Williams, R. (1984). The use of zeta-potential measurements in organic solvents to determine the donor–acceptor properties of solid surfaces. *J. Colloid Interface Sci.*, **97**, 356–366

Labib, M. and Williams, R. (1986). An experimental comparison between the aqueous pH scale and the electron donicity scale. *Colloid Polymer Sci.*, **264**, 533–541

Le Neveu, D., Rand, R. P., Gingell, D. and Parsegian, V. A. (1977). Measurement and modification of forces between lecithin bilayers. *Biophys. J.*, **18**, 209–230

Ling, G. N. (1972). Hydration of macromolecules. In Horne, R. A. (Ed.), *Water and Aqueous Solutions*. Wiley-Interscience, New York, pp. 663–700

Lis, L. J., McAlister, M., Fuller, N., Rand, R. P. and Parsegian, V. A. (1982). Interactions between neutral phospholipid bilayer membranes. *Biophys. J.*, **37**, 657–666

Lumry, R. and Rajender, S. (1970). Enthalpy–entropy compensation phenomena in water solutions and small molecules: a ubiquitous property of water. *Biopolymers*, **9**, 1125–1227

Mahanty, J. and Ninham, B. W. (1976). *Dispersion Forces*. Academic Press, New York

Marcelja, S. (1990). Interactions between interfaces in liquids. In Charvolin, J., Joanny, J. F. and Linn-Justin, J. (Eds), *Liquids at Interfaces*. North-Holland/Elsevier, Amsterdam, New York, pp. 99–51

Marcelja, S. and Radic, N. (1976). Repulsion of interfaces due to boundary water. *Chem. Phys. Lett.*, **42**, 129–130

Mukkur, T. K. S. (1980). Thermodynamics of hapten–antibody interaction(s). *Trends Biochem. Sci.*, **5**, 72–74

Mukkur, T. K. S. (1984). Thermodynamics of hapten–antibody interactions. *Crit. Rev. Biochem.*, **16**, 133–167

Napper, D. H. (1983). *Polymeric Stabilization of Colloidal Dispersions*. Academic Press, New York

Overbeek, J. Th. G. (1952a). Electrochemistry of the double layer. In Kruyt, H. R. (Ed.), *Colloid Science*, Vol. 1. Elsevier, Amsterdam, pp. 115–193

Overbeek, J. Th. G. (1952b). The interaction between colloidal particles. In Kruyt, H. R. (Ed.), *Colloid Science*, Vol. 1. Elsevier, Amsterdam, pp. 245–277

Overbeek, J. Th. G. and Wiersema, P. H. (1967). The interpretation of electrophoretic mobilities. In Bier, M. (Ed.), *Electrophoresis*, Vol. 2. Academic Press, New York, pp. 1–52

Parsegian, V. A., Fuller, N. and Rand, R. P. (1979). Measured work of deformation and repulsion of lecithin bilayers. *Proc. Natl Acad. Sci. USA*, **76**, 2750–2754

Parsegian, V. A., Rand, R. P., Fuller, N. L. and Rau, D. C. (1986). Osmotic stress

for the direct measurement of intermolecular forces. *Meth. Enzymol.*, **127**, 400–416
Parsegian, V. A., Rand, R. P. and Rau, D. C. (1985). Hydration forces: What next? *Chem. Scripta*, **25**, 28–31
Parsegian, V. A., Rand, R. P. and Rau, D. C. (1987). Lessons from the direct measurement of forces between biomolecules. In Satran, S. and Clark, N. (Eds), *Proceedings of the Symposium on Complex and Supermolecular Fluids*. Wiley, New York, pp. 121–142
Pauling, L. and Hayward, R. (1964). *The Architecture of Molecules*. Freeman, San Francisco, p. 7
Prince, L. M. (1967). A theory of aqueous emulsions. I. Negative interfacial tension at the oil/water interface. *J. Colloid Interface Sci.*, **23**, 165–173
Smeenk, R. J. T., van der Lehy, G. and Aarden, L. A. (1982). Avidity of antibodies to dsDNA, Farr assay and PEG assay. *J. Immunol.*, **128**, 73–78
Sober, H. (1968). *Handbook of Biochemistry*. CRC Press, Cleveland, pp. C10, C36
Steane, E. A. and Greenwalt, T. J. (1974). Water of hydration and the mechanism of antigen–antibody interaction. *Transfusion*, **14**, 501
Steane, E. A. and Greenwalt, T. J. (1977). Red cell agglutination. In Mohn, J. F., Plunkett, R. W., Cunningham, R. K. and Lambert, R. M. (Eds), *Human Blood Groups*: Proc. 5th Intl. Convoc. Immunol., Buffalo. Karger, Basel, New York, pp. 36–43
Sturtevant, J. M. (1977). Heat capacity and entropy changes in processes involving proteins. *Proc. Natl Acad. Sci. USA*, **74**, 2236–2240
Vaidhyanathan, V. S. (1982). Inhomogeneous interfacial regions in biological systems. I. Basic differential equations and their implications. *J. Biol. Phys.*, **10**, 153–166
Vaidhyanathan, V. S. (1985). The electric potential profile in inhomogenous interfacial regions of biological systems. *Stud. Biophys.*, **110**, 29–42
Vaidhyanathan, V. S. (1986). A fundamental question about the electrical potential profile in the interfacial region of biological membrane systems. In Blank, M. (Ed.), *Bioelectrochemistry*. Plenum Press, New York, pp. 30–51
van der Waals, J. D. (1873). *Over de continuiteit van den gas — en vloeistoftoestand*. Dissertation, Leiden. (*Concerning the Continuity of the Gas and Liquid States*)
van Oss, C. J. (1985). Stability of human red cell suspensions at 300 000 × *g*. *J. Dispersion Sci. Tech.*, **6**, 139–146
van Oss, C. J. (1988). Coacervation, complex-coacervation and flocculation. *J. Dispersion Sci. Tech.*, **9**, 561–573
van Oss, C. J. (1989a). Energetics of cell–cell and cell–biopolymer interactions. *Cell Biophys.*, **14**, 1–16
van Oss, C. J. (1989b). On the mechanism of the cold ethanol precipitation method of plasma protein fractionation. *J. Protein Chem.*, **8**, 661–668
van Oss, C. J. (1990a). Surface free energy contribution to cell interactions. In Glaser, R. and Gingell, D. (Eds), *Biophysics of the Cell Surface*. Springer Verlag, Berlin, New York, pp. 131–152
van Oss, C. J. (1990b). Surface properties of fibrinogen and fibrin. *J. Protein Chem.*, **9**, 487–491
van Oss, C. J. (1990c). Aspecific and specific intermolecular interactions in aqueous media. *J. Molec. Recognition*, **3**, 128–136
van Oss, C. J. (1990d). Polar interfacial interactions in RPLC. *Israel J. Chem.*, **30**, 251–255
van Oss, C. J. (1991a). Interaction forces between biological and other polar

entities in water. How many different primary forces are there? *J. Dispersion Sci. Tech.*, **12**, 201–219

van Oss, C. J. (1991b). The forces involved in bioadhesion to flat surfaces and particles — Their determination and relative roles. *Biofouling*, **4**, 25–35

van Oss, C. J. (1993). *Interfacial Forces in Aqueous Media*. Marcel Dekker, New York (in preparation)

van Oss, C. J. and Absolom, D. R. (1984). Nature and thermodynamics of antigen–antibody interactions. In Atassi, M. Z., van Oss, C. J. and Absolom, D. R. (Eds), *Molecular Immunology*. Marcel Dekker, New York, pp. 337–360

van Oss, C. J., Arnold, K., Good, R. J., Gawrisch, K. and Ohki, S. (1990a). Interfacial tension and the osmotic pressure of solutions of polar polymers. *J. Macromolec. Sci.-Chem.*, **A27**, 563–580

van Oss, C. J., Chaudhury, M. K. and Good, R. J. (1987a). Monopolar surfaces. *Adv. Colloid Interface Sci.*, **28**, 35–65

van Oss, C. J., Chaudhury, M. K. and Good, R. J. (1987b). The mechanism of partition in aqueous media. *Separ. Sci. Technol.*, **22**, 1515–1526

van Oss, C. J., Chaudhury, M. K. and Good, R. J. (1988a). Interfacial Lifshitz–van der Waals and polar interactions in macroscopic systems. *Chem. Rev.*, **88**, 927–941

van Oss, C. J., Chaudhury, M. K. and Good, R. J. (1988b). Polar interfacial interactions, hydration pressure and membrane fusion. In Ohki, S., Doyle, D., Flanagan, T. D., Hui, S. W. and Mayhew, E. (Eds), *Membrane Fusion*. Plenum Press, New York, pp. 113–122

van Oss, C. J., Chaudhury, M. K. and Good, R. J. (1989). The mechanism of phase separation of polymers in organic media — Apolar and polar systems. *Separ. Sci. Technol.*, **24**, 15–30

van Oss, C. J., Giese, R. F. and Costanzo, P. M. (1990b). DLVO and non-DLVO interactions in hectorite. *Clay Clay Minerals*, **38**, 151–159

van Oss, C. J., Giese, R. F., Li, Z., Murphy, K., Chaudhury, M. K. and Good, R. J. (1991). Contact angles and spreading coefficients of liquids on particulate solids, measured by wicking. *J. Adhesion Sci. Tech.*, to be published

van Oss, C. J. and Good, R. J. (1984). The 'equilibrium distance' between two bodies immersed in a liquid. *Colloids and Surfaces*, **8**, 373–381

van Oss, C. J. and Good, R. J. (1988). Orientation of the water molecules of hydration of serum albumin. *J. Protein Chem.*, **7**, 179–183

van Oss, C. J. and Good, R. J. (1989). Surface tension and the solubility of polymers and biopolymers: The role of polar and apolar interfacial free energies. *J. Macromolec. Sci. Chem.*, **A26**, 1183–1203

van Oss, C. J. and Good, R. J. (1991). Surface enthalpy and entropy and the physico-chemical nature of hydrophobic and hydrophilic interactions. *J. Dispersion Sci. Tech.*, **12**, 273–287

van Oss, C. J., Good, R. J. and Busscher, H. J. (1990c). Estimation of the polar surface tension parameters of glycerol and formamide, for use in contact angle measurements on polar solids. *J. Dispersion Sci. Technol.*, **11**, 75–81

van Oss, C. J., Good, R. J. and Chaudhury, M. K. (1986). The role of van der Waals forces and hydrogen bonds in 'hydrophobic interactions' between biopolymers and low energy surfaces. *J. Colloid Interface Sci.*, **111**, 378–390

van Oss, C. J., Omenyi, S. N. and Neumann, A. W. (1979). Negative Hamaker coefficients. II. Phase separation of polymer solutions. *Colloid Polymer Sci.*, **257**, 737–744

Verwey, E. J. W. and Overbeek, J. Th. G. (1948). *Theory of the Stability of Lyophobic Colloids*. Elsevier, Amsterdam

Visser, J. (1972). On Hamaker constants: A comparison between Hamaker constants and Lifshitz–van der Waals constants. *Adv. Colloid Interface Sci.*, **3**, 331–363

Washburn, E. W. (1921). The dynamics of capillary flow. *Phys. Rev.*, **18**, 273–283

Zhou, Z., Wu, P. and Ma, C. (1990). Hydrophobic interaction and stability of colloidal silica. *Colloids and Surfaces*, **50**, 177–188

14

Solvation Thermodynamics of Biopolymers

A. Ben-Naim

1 Introduction

For any biochemical process taking place in aqueous medium one can view the Gibbs energy change (or any other thermodynamic quantity) as comprising two contributions: one originating from the solutes involved in the process, and the second including all possible solvent effects. Thus, we write

$$\Delta G^{\ell} = \Delta G^{g} + \delta G \tag{14.1}$$

where ΔG^{ℓ} is the Gibbs energy change for a well-defined process (say conformational change or association of biopolymers at some specified conditions) in the liquid state, ΔG^{g} is the Gibbs energy change for the same process carried out in vacuum, and δG, defined in Equation (14.1), includes all the solvent contributions to the Gibbs energy change.

Each of the two quantities ΔG^{g} and δG is too complicated to be calculated exactly for any biochemical process. Yet there is a fundamental difference between these two quantities. ΔG^{g} depends only on the properties of the solutes participating in the process. For instance, for the process of protein folding we need to know the partition functions associated with all the degrees of freedom of the protein molecule. This is an extremely complicated quantity and at present may be approximately estimated for simple cases only. The quantity δG, on the other hand, is far more complicated. It involves all possible interactions between the solutes and the solvent, as well as between all solvent molecules. This quantity may be expressed in terms of averages over all possible configurations of the solvent molecules. At present only a very rough estimate of this quantity may be obtained by simulation methods.

Although it is not possible to make a definite statement regarding the relative magnitude of the two contributions to ΔG^{ℓ} for any biochemical process, it is generally believed that the solvent contribution to ΔG^{ℓ}, is very important, or perhaps even the dominant one of the two terms in Equation (14.1). In this chapter we focus on δG only. It is easy to show that δG may be expressed as a linear combination of solvation Gibbs energies of the various solutes involved in the process (Ben-Naim, 1992). For instance, for protein folding, or any conformational change of a biopolymer, we can write

$$\delta G(\mathrm{U} \rightarrow \mathrm{F}) = \Delta G^{*}_{\mathrm{F}} - \Delta G^{*}_{\mathrm{u}} \tag{14.2}$$

where $\Delta G^{*}_{\mathrm{F}}$ and $\Delta G^{*}_{\mathrm{u}}$ are the solvation Gibbs energies of the folded (F) and of the unfolded (U) forms of the protein, respectively. Similarly, for the association between a protein (P) and a ligand (L), (or for any association between two biopolymers) we can write

$$\delta G(\mathrm{P} + \mathrm{L} \rightarrow \mathrm{PL}) = \Delta G^{*}_{\mathrm{PL}} - \Delta G^{*}_{\mathrm{P}} - \Delta G^{*}_{\mathrm{L}} \tag{14.3}$$

where again ΔG^{*}_{α} is the solvation Gibbs energy of the solute α.

We see that in order to study the solvent contribution to the Gibbs energy change of a biochemical process (e.g. Equations 14.2 or 14.3) we need to know the solvation Gibbs energies of all the solutes involved in that process. For this reason we shall focus, in the rest of this chapter, on the solvation thermodynamics of a single macromolecule.

We begin in the next section by introducing the definitions of the solvation thermodynamic quantities. We next proceed to analyse the various ingredients of ΔG^{*}_{α}. Some order-of-magnitude numerical examples are given in Section 4. In Section 5, a critical examination of the group-additivity assumption is presented and compared with the exact procedure outlined earlier. In Section 6 we present a brief survey of various biochemical processes, in which the solvation Gibbs energies might contribute significantly to the overall driving force of that process. We conclude with a discussion on the relevance of structural changes in the solvent to the solvation process.

Other aspects of solvation of biopolymers have been reviewed by Nemethy *et al.* (1981), Edsall and McKenzie (1983), Finney and Savage (1988).

2 Definitions

The *solvation process* is defined as the process of transferring of a solute α, being at a specific conformation, from a fixed position in vacuum to a fixed position in the liquid, the process being carried out at a given temperature T, pressure P and solvent composition.

The solvation Gibbs energy of α is defined as the change in the Gibbs energy for the solvation process. Similarly, one can define the solvation entropy, enthalpy, volume, etc., for the same process. Clearly, the most fundamental quantity is the solvation Gibbs energy, denoted by ΔG^*_α. In principle having ΔG^*_α as a function of T and P, one may derive all other thermodynamic quantities by standard relationships, e.g.

$$\Delta S^* = -\frac{\partial \Delta G^*}{\partial T} \tag{14.4}$$

$$\Delta V^* = \frac{\partial \Delta G^*}{\partial P} \tag{14.5}$$

$$\Delta H^* = \Delta G^* + T\Delta S^* \tag{14.6}$$

It should be noted that the solvation thermodynamic quantities, as defined above, differ from the more conventional standard thermodynamic quantities defined in terms of different processes of transferring of a solute from the gaseous into the liquid phase. We shall not need the latter in this chapter. The relations between the two sets of quantities have been given elsewhere (Ben-Naim, 1987).

The solvation quantities, as defined above, as well as the corresponding standard quantities, are measurable only for solutes which have measurable vapour pressure. This is clearly not the case for macromolecules. Therefore, other means of studying the solvation thermodynamics of biomolecules are pursued.

In the following two sections we shall analyse the ingredients that combine to build up the solvation Gibbs energy of a macromolecule. Our starting point is the statistical mechanical expression for the solvation Gibbs energy. Before we write down this expression, we define two more fundamental quantities.

(1) The *solute–solvent interaction* is defined as the work required to bring *one* solute α and *one* solvent molecule w from fixed positions and orientations and at infinite separation, to the final configuration denoted by $(\mathbf{X}_\alpha, \mathbf{X}_w)$, where $\mathbf{X}$ stands for the six-dimensional vector comprising the three locational and the three orientational coordinates of the molecule. The corresponding work is referred to as the solute–solvent pair interaction and is denoted $U(\mathbf{X}_\alpha, \mathbf{X}_w)$.

(2) The *total binding energy* of a solute α to a system of N solvent molecules at a specific configuration $\mathbf{X}_\alpha, \mathbf{X}_1 \ldots \mathbf{X}_N$ is defined as the work

required to add a solute at a configuration $\mathbf{X}_\alpha$ given a system of N solvent molecules at a specific configuration $\mathbf{X}_1 \ldots \mathbf{X}_w$, denoted by $\mathbf{X}^N$. This work is referred to as the binding energy of α, and is given by the difference in the energies of the system before and after adding the solute α:

$$B_\alpha(\mathbf{X}_\alpha, \mathbf{X}^N) = U(\mathbf{X}_\alpha, \mathbf{X}_1 \ldots \mathbf{X}_N) - U(\mathbf{X}_1 \ldots \mathbf{X}_N)$$
$$= \sum_{i=1}^{N} U(\mathbf{X}_\alpha, \mathbf{X}_i) \quad (14.7)$$

The second equality on the right-hand side of Equation (14.7) is valid only when the total binding energy can be written as a sum of pairwise interactions between the solute and each of the solvent molecules. We shall assume that this equally holds for our systems.

The *solvation Gibbs energy* of the solute α is now given by (Ben-Naim, 1987, 1992)

$$\Delta G^*_\alpha = -kT\ell n \langle \exp[-\beta B_\alpha] \rangle_o \quad (14.8)$$

where k is the Boltzmann constant, T the temperature, $\beta = (kT)^{-1}$, and the average quantity denoted by $\langle \ \rangle_o$ is over all possible configurations of the solvent molecules in the TPN ensemble, i.e.

$$\langle \exp[-\beta B_\alpha] \rangle_o = \int \cdots \int dV d\mathbf{X}^N exp[-\beta PV - \beta U(\mathbf{X}^N)$$
$$- \beta B_\alpha(\mathbf{X}_\alpha, \mathbf{X}^N)]$$
$$= \int \cdots \int dV d\mathbf{X}^N P_o(V, \mathbf{X}^N) \exp[-\beta B_\alpha(\mathbf{X}_\alpha \mathbf{X}^N)] \quad (14.9)$$

where $P_o(V, \mathbf{X}^N)$ is the probability density of finding the pure solvent (i.e. in the absence of the solute) at a given volume V and at a configuration $\mathbf{X}^N$. Note that the integrations over all $\mathbf{X}^N$ gives an average quantity which is still a function of $\mathbf{X}_\alpha$. However, for a macroscopic, homogeneous and isotropic system, the work required to introduced α into the system at a fixed configuration $\mathbf{X}_\alpha$ is independent of the specific configuration selected. Therefore, we have denoted the solvation Gibbs energy by ΔG^*_α, rather than by $\Delta G^*_\alpha(\mathbf{X}_\alpha)$. Note also that ΔG^*_α is defined for the *solvation process*, i.e. α must be placed at an arbitrarily fixed configuration in the solvent. The work involved in adding α to the system without the constraint of a fixed configuration is simply the chemical potential of α which is different from ΔG^*_α. The two quantities are related by

$$\mu_\alpha = \Delta G^*_\alpha + kT\ell n \rho_\alpha \Lambda^3_\alpha q^{-1}_\alpha \quad (14.10)$$

where ϱ_α is the number density of α and Λ_α^3 is its momentum partition function, and q_α is the internal partition function of α.

Another important quantity is the *average* binding energy of α, defined by

$$\langle B_\alpha \rangle_o = \int \cdots \int \mathrm{d}V \mathrm{d}\mathbf{X}^N P_o(V, \mathbf{X}^N)\, B_\alpha(\mathbf{X}_\alpha, \mathbf{X}^N) \tag{14.11}$$

where the average in Equation (14.11) employs the same distribution function $P_o(V, \mathbf{X}^N)$ as in Equation (14.9). Note that, in general,

$$\langle \exp[-\beta B_\alpha] \rangle_o \neq \exp[-\beta \langle B_\alpha \rangle_o] \tag{14.12}$$

i.e. one cannot interchange the average and the exponential operations.

A quantity related to $\langle B_\alpha \rangle_o$ that appears in the expression of the solvation energy and entropy of α (see below) is the *conditional average binding energy*. This is defined by

$$\langle B_\alpha \rangle_\alpha = \int \cdots \int \mathrm{d}V \mathrm{d}\mathbf{X}^N P(V, \mathbf{X}^N/\mathbf{X}_\alpha)\, B_\alpha(\mathbf{X}_\alpha, \mathbf{X}^N) \tag{14.13}$$

The difference between Equations (14.11) and (14.13) is in the distribution function used in calculating the average quantity. In (14.11) we use $P_o(V, \mathbf{X}^N)$, the distribution function for the *pure* solvent (hence the subscript 'o'), whereas in (14.13) we use the *conditional* distribution function, *given* that a solute has already been placed at a fixed configuration $\mathbf{X}_\alpha$.

In Section 1 we introduced the solvation process at P, T constants. Similarly, one can perform the solvation process at V, T constants; the corresponding work is the Helmholtz energy, given by

$$\Delta A_\alpha^* = -kT\ell\mathrm{n} \langle \exp[-\beta B_\alpha] \rangle_o \tag{14.14}$$

which is formally the same relation as in (14.8), but now the average is in the T, V, N ensemble rather than in the T, P, N ensemble: Although the two quantities ΔG_α^* and ΔA_α^* are defined for *different* processes, they will have the same magnitude if the average volume $\langle V \rangle$ in the T, P, N system is equal to the exact volume V in the T, V, N system. For some theoretical consideration ΔA_α^* is simpler to analyse than ΔG_α^* (Ben-Naim, 1987).

The other thermodynamic quantities of solvation may be calculated by standard relationships, e.g.

$$\Delta S_\alpha^* = -\left(\frac{\partial \Delta G_\alpha^*}{\partial T}\right)_P = -\left(\frac{\partial \Delta A_\alpha^*}{\partial T}\right)_V \tag{14.15}$$

$$\Delta H^*_\alpha = \Delta G^*_\alpha + T\Delta S^*_\alpha \tag{14.16}$$

$$\Delta E^*_\alpha = \Delta H^*_\alpha - P\Delta V^*_\alpha \tag{14.17}$$

$$\Delta V^*_\alpha = \left(\frac{\partial \Delta G^*_\alpha}{\partial P}\right)_T \tag{14.18}$$

For most of the processes carried out in aqueous solutions at T, P constants, the term $P\Delta V^*_\alpha$ is negligible compared with either ΔE^*_α or ΔG^*_α. Therefore, we can neglect the difference between ΔE^*_α and ΔH^*_α, or between ΔA^*_α and ΔG^*_α.

The statistical mechanical expressions for the entropy and the enthalpy of solvation are (neglecting $P\Delta V^*_\alpha$ terms)

$$T\Delta S^*_\alpha = kT\ell n \langle \exp[-\beta B_\alpha]\rangle_o + \langle B_\alpha \rangle_\alpha + \langle U_N \rangle_\alpha - \langle U_N \rangle_o \tag{14.19}$$

$$\Delta E^*_\alpha = \langle B_\alpha \rangle_\alpha + \langle U_N \rangle_\alpha - \langle U_N \rangle_o \tag{14.20}$$

Note that both $T\Delta S^*_\alpha$ and ΔE^*_α include the conditional average binding energy, defined in (14.13). Note also that the solvation energy ΔE^*_α *differs* from the average binding energy. The former also includes the term $\langle U_N \rangle_\alpha - \langle U_N \rangle_o$, which is the change in the total average interaction energy among all solvent molecules caused by adding α to the system. This term may be reinterpreted as arising from structural changes induced by the solute on the solvent (Ben-Naim, 1987, 1992).

An important conclusion follows from the two expressions (14.19) and (14.20) for the entropy and the enthalpy (or energy) of solvation. Both of these quantities include contribution from structural changes induced in the solvent by the solvation process. These contributions could be large or even dominate the values of ΔS^*_α and ΔH^*_α. However, once the Gibbs or the Helmholtz energies of solvation are constructed (i.e. $\Delta H^*_\alpha - T\Delta S^*_\alpha$ or $\Delta E^*_\alpha - T\Delta S^*_\alpha$, respectively), the contribution due to structural changes in the solvent cancels out. Hence, whatever the structural changes induced in the solvent, these cannot affect the solvation Gibbs energy. This conclusion invalidates the common 'explanation' of the large and positive Gibbs energy of solvation of simple non-polar solutes in water. The traditional explanation is based on the following two observations: (1) The values of ΔG^*_α for simple solutes α in water is large and positive. (2) ΔS^*_α is large and negative. If one assumes that the large negative value of ΔS^*_α is due to structural changes induced in water (see Equation 14.19), then it follows that these structural changes also are responsible for the large positive value of ΔG^*_α. The fault in this argument is immediately clear from (14.19) and (14.20). Once we form the combination $\Delta H^*_\alpha - T\Delta S^*_\alpha$

(or $\Delta E^*_\alpha - T\Delta S^*_\alpha$), the contribution due to structural changes in the liquid cancels out and does not appear in the value of ΔG^*_α (see also Section 7 below).

To summarize, we have defined the following quantities: the total binding energy of a solute α to the solvent at a specific configuration $\mathbf{X}^N$, the average binding energy of α, the solvation energy of α and the solvation Gibbs energy of α.

All of these quantities are, in general, different. The differences could be large in some cases. Unfortunately, a considerable confusion still exists in the biochemical literature, where some of the different quantities are treated as if they were equivalent.

Note also that the solvation quantities are, in principle, measurable quantities (whenever α has a measurable vapour pressure). However, the quantities B_α and $\langle B_\alpha \rangle_\alpha$ are not measurable, and can only be estimated theoretically.

3 The Ingredients of ΔG^*_α

As we noted in Section 2, ΔG^*_α is a measurable quantity for solutes for which the ratio of the number densities of α in the gaseous and in the liquid phases can be measured. For large biopolymers ΔG^*_α is not directly measurable. However, one can estimate the values of ΔG^*_α by decomposing ΔG^*_α into its components, which, in turn, may be estimated by either theoretical or experimental means. The analysis of the ingredients of ΔG^*_α is also important for the elucidation of the driving forces for some biochemical process (see Section 6).

To analyse the ingredients that constitute the solvation Gibbs energy, we start by writing the total binding energy, defined in (14.7) in one of the two forms

$$B_\alpha = B^{\mathrm{BB}} + \sum_k B^k \tag{14.21}$$

$$B_\alpha = B^{\mathrm{H}} + B^{\mathrm{S}} + B^{\mathrm{HB}} + \cdots \tag{14.22}$$

In the first form (14.21) we divide the total binding energy of α into contributions originating from various *groups* in the molecule.

For biopolymers, we may distinguish between the backbone (BB) polymer and all the side-chains. The specific division into groups is a matter of convenience and depends also on the availability of relevant experimental data (see Section 4 below). As an example, we divide the total solute–solvent interaction (and, hence, the binding energy) of ethanol

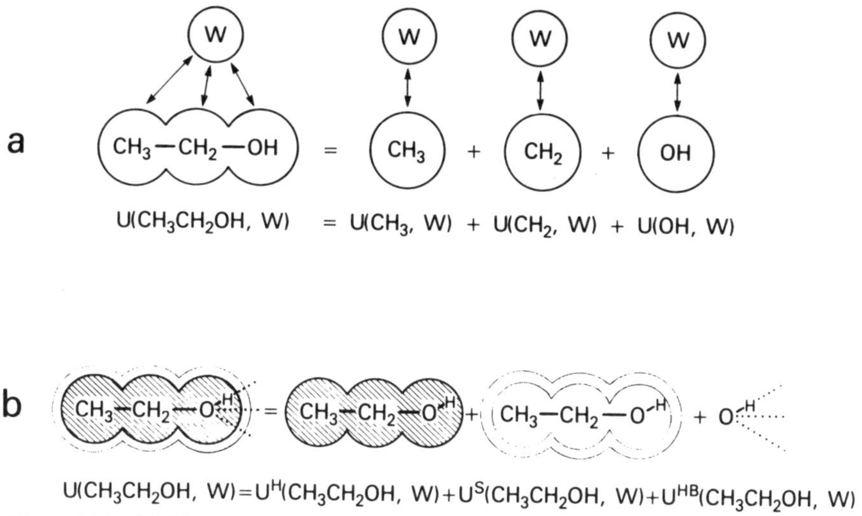
a
W
CH3—CH2—OH = CH3 + CH2 + OH
U(CH3CH2OH, W) = U(CH3, W) + U(CH2, W) + U(OH, W)
b
CH3—CH2—O H = CH3—CH2—O H + CH3—CH2—O H + O H
U(CH3CH2OH, W)=U^H(CH3CH2OH, W)+U^S(CH3CH2OH, W)+U^HB(CH3CH2OH, W)

Figure 14.1 (a) The total ethanol–water interaction is decomposed into three parts corresponding to the methyl, methylene and hydroxyl groups.
(b) The total ethanol–water interaction is decomposed into its three types of contributions: the hard, the soft and the hydrogen-bond interactions

into the interaction due to the methyl, methylene and the hydroxyl group. This is illustrated schematically in Figure 14.1(a).

In the second form, Equation (14.22), we list all the *types* of interactions contributing to the binding energy of α. For example, the 'hard' (H) repulsive interaction, the 'soft' (S), or the van der Waals, and the hydrogen bond (HB) interaction. This is illustrated for ethanol in Figure 14.1(b).

Of course, one can choose a combination of the two divisions shown in (14.21) and (14.22). For instance, B^{BB} may be further viewed as consisting of hard, soft and HB parts. Similarly, B^{H} in (14.22) may be further split according to the various *groups* of the entire molecule α.

It should be noted that the division (14.21) is an approximation based on the assumption that each of the groups or parts of the molecule α interacts with the solvent independently of other groups in the molecule. On the other hand, the division (14.22) lists all the types of interactions, and in principle can serve as a definition of the solute–solvent interaction, and, hence, of the total binding energy. We shall henceforth assume that either (14.21) or (14.22) is given and proceed to examine the corresponding contribution of each of the terms on the r.h.s. of (14.21) or (14.22) to the Gibbs energy of solvation.

To demonstrate the relation between the *additivity* of the binding energy, as expressed in (14.21) or (14.22), and the corresponding *non-additivity* of the Gibbs energy of solvation, we consider a simple example. Let α be a molecule having two parts, denoted by a and b, such that

$$B_\alpha = B_a + B_b \tag{14.23}$$

a and *b* could be, for instance, the two methyl groups of ethane. Accepting (14.23) as a valid approximation for the *binding energy* of α for each configuration of the solvent molecules, it is clear that the additivity property is maintained when we average over all configurations of the solvent molecules, e.g. in the T, V, N ensemble we have

$$\begin{aligned}\langle B_\alpha\rangle_o &= \int\cdots\int d\mathbf{X}^N P_o(\mathbf{X}^N)B_\alpha(\mathbf{X}^N)\\ &= \int\cdots\int d\mathbf{X}^N P_o(\mathbf{X}^N)\,[B_a(\mathbf{X}^N) + B_b(\mathbf{X}^N)]\\ &= \langle B_a\rangle_o + \langle B_b\rangle_o\end{aligned} \tag{14.24}$$

Thus, the group additivity presumed to hold for B_α leads to a corresponding group additivity for $\langle B_\alpha\rangle$. This is, in general, not the case for the solvation *Gibbs* or *Helmholtz* energy. To see this we use the definition of the Helmholtz energy of solvation for α (14.14) and using (14.23) to obtain

$$\begin{aligned}\Delta A_\alpha^* &= -kT\ell n\,\langle \exp[-\beta B_a - \beta B_b]\rangle_o\\ &= -kT\ell n\,[\int\cdots\int d\mathbf{X}^N P_o(\mathbf{X}^N)\exp[-\beta B_a - \beta B_b]\\ &\neq \Delta A_a^* + \Delta A_b^*\end{aligned} \tag{14.25}$$

We see that, in general, ΔA_α^* is not additive with respect to the two contributions of *a* and *b*. Although B_α is additive for *any* specific configuration of the solvent molecules, the average of the product of the two quantities $\exp[-\beta B_a]\exp[-\beta B_b]$ is, in general, not equal to the product of the two average quantities $\langle \exp[-\beta B_a]\rangle$ and $\langle \exp[-\beta B_b]\rangle$. It is only in the rare events when the two 'random variables' $\exp[-\beta B_a]$ and $\exp[-\beta B_b]$ are *independent* that such an additivity of the Helmholtz or the Gibbs energy can be realized. We shall return to some further comments on the group additivity, often assumed for protein, in Section 5. We now show that the average quantity in (14.25) may be split into two terms, corresponding to the successive introduction of the two groups into the liquid. This procedure will then be generalized for biopolymers of any size.

We use the following exact relation (Ben-Naim, 1987, 1992):

$$\begin{aligned}\Delta A_\alpha^* &= -kT\ell n\,[\langle \exp[-\beta B_a - \beta B_b]\rangle_o]\\ &= -kT\ell n\,[\langle \exp[-\beta B_a]\rangle_o\,\langle \exp[-\beta B_b]\rangle_a]\end{aligned}$$

$$= \Delta A_a^* + \Delta A_{b/a}^* \tag{14.26}$$

where ΔA_a^* is the solvation Helmholtz energy of the group a and $\Delta A_{b/a}^*$ is the *conditional* solvation Helmholtz energy of b *given* a. Note that the latter is different from ΔA_b^* appearing on the right-hand side of (14.25). In (14.26), the process of solvation of the entire solute α is carried out in two steps. First, we introduce the group a into the liquid. The corresponding Helmholtz energy is ΔA_a^*. In the second step we introduce the group b, not at *any* point in the liquid, but at a point adjacent to a, which has already been inserted at some fixed point in the first step. The corresponding work is the *conditional* solvation Helmholtz energy of b *given* a denoted by $\Delta A_{b/a}^*$.

Clearly we could change the order of introducing the two groups a and b to obtain, instead of (14.26), the equivalent relation

$$\Delta A_\alpha^* = \Delta A_b^* + \Delta A_{a/b}^* \tag{14.27}$$

Both (14.26) and (14.27) are exact relations, provided that the additivity of the binding energy (14.23) is presumed. These are clearly different from the additivity assumption often used for the solvation Gibbs or Helmholtz energy (i.e. equality sign in 14.25; see also Section 5).

We now apply the same procedure as in (14.27) to either (14.21) or (14.22). The resulting expressions for the Helmholtz energy of a globular protein are

$$\Delta A_\alpha^* = \Delta A_{\mathrm{BB}}^* + \Delta A_{1/\mathrm{BB}}^* + \Delta A_{2/1,\,\mathrm{BB}}^* + \Delta A_{3/1,\,2,\,\mathrm{BB}}^* + \cdots \tag{14.28}$$

$$\Delta A_\alpha^* = \Delta A_{\mathrm{H}}^* + \Delta A_{\mathrm{S/H}}^* + \Delta A_{\mathrm{HB/H,\,S}}^* + \cdots \tag{14.29}$$

In (14.28) we first introduce the backbone (BB), i.e. the molecule α from which all side-chains are cut-off. We next introduce, one at a time, all the side-chains. The *condition* in each step is the presence of the backbone as well as all the side-chains introduced in the previous steps. Similarly in (14.29) we first 'turn on' the hard (H) part of the binding energy of α. This is equivalent to the work required to create a cavity in the liquid. In the next step we 'turn on' the soft (S) interaction, given that the hard interaction has already been turned on. In the third step we turn on the hydrogen bond (HB) interactions, given H and S, and so on.

Both representations of ΔA_α^* are exact within the assumptions (14.21) or (14.22). Similar expressions hold true for the solvation Gibbs energy.

We now return to biopolymers and use the same methodology of splitting the solvation Gibbs or Helmholtz energy into different components. We can use either (14.21) or (14.27) to express the total binding energy of the solute to the solvent. The actual choice we make depends on the

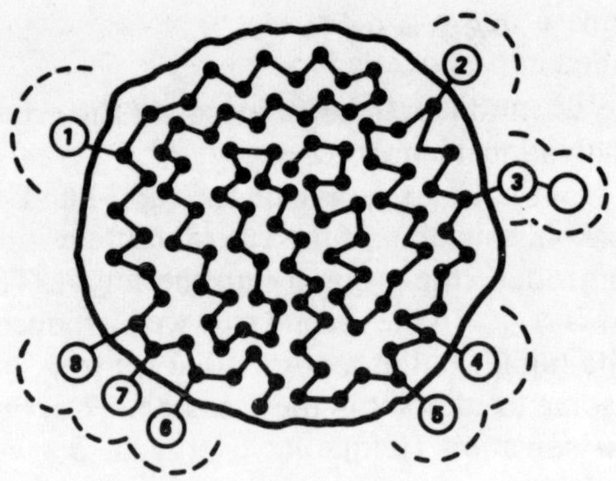

Figure 14.2 A schematic presentation of the 'backbone' of a globular protein and a few side-chains protruding from the 'backbone'. The term 'backbone' refers here to that part of the protein which is included in the closed bold line. The protruding side-chains are indicated by numbers 1–8

availability of the relevant experimental data (see below and next section).

For concreteness we consider a globular protein, from which we may cut off all side-chains that protrude from its surface into the solvent (Figure 14.2). The total binding energy of the protein is written as

$$B_\alpha = B^{BB} + \sum_{k=1}^{8} B^k \tag{14.30}$$

where B^{BB} is the binding energy of the backbone (BB). The term 'backbone' refers here to that part of the protein that is contained within the bold line in Figure 14.2. The remaining parts of the binding energy originating from the various side-chains are denoted by B^k. Here, 'side-chains' are those side chains that protrude from the bold line in Figure 14.2 but do not include the side-chains within the bold line.

Clearly, there are infinitely many ways of presenting B_α as a sum of the form (14.30), depending on how we choose to define the backbone and the protruding side-chains. Once we made such a choice, the solvation Gibbs energy can be written, following the example in (14.26), as

$$\Delta G^*_\alpha = \Delta G^*_{BB} + \Delta G^*_{SC/BB} \tag{14.31}$$

where ΔG^*_{BB} is the solvation Gibbs energy of the backbone and $\Delta G^*_{SC/BB}$ is the conditional solvation Gibbs energy of all protruding side-chains, *given* the backbone.

To proceed further, we must group all the side-chains into independent groups. By 'independent' we mean here two groups that are far apart on the surface of BB, such that their solvation spheres are nonoverlapping. For instance, side-chains 1 and 2 in Figure 14.2 are independent of all other groups. Side chains 4 and 5 are correlated, i.e. their solvation spheres (indicated by dashed curve) overlap. Similarly 6, 7 and 8 are triply correlated. Thus, for the particular example of Figure 14.2, we write the solvation Gibbs energy as

$$\Delta G^*_\alpha = \Delta G^*_{BB} + \sum_{i=1}^{3} \Delta G^*_{i/BB} + \Delta G^*_{4,5/BB} + \Delta G^*_{6,7,8/BB} \qquad (14.32)$$

And for the general case

$$\Delta G^*_\alpha = \Delta G^*_{BB} + \sum_{i} \Delta G^*_{i/BB} + \sum_{k,l} \Delta G^*_{k,l/BB} + \cdots \qquad (14.33)$$

The general presentation of the Gibbs energy of solvation is as follows. The first term is the solvation Gibbs energy of the backbone (BB). The second term includes the *conditional* solvation Gibbs energies of all the independent side-chains (these are 1, 2, and 3 in the example of Figure 14.2). The third term includes the conditional solvation Gibbs energies of pairwise correlated side-chains (4 and 5 in Figure 14.2). Higher-order terms such as triplet, quadruplets, etc., can be added; we shall see in the next section that, because of lack of relevant experimental data, these terms cannot be studied at present. We shall therefore focus only on the three terms presented in Equation (14.33).

The general presentation given in Equation (14.33) lists all contributions to the solvation Gibbs energy due to the different parts of the macromolecule. A further classification is useful in terms of the types of side-chains, for instance a hydrophobic (HφO) side-chain will have a substantially different contribution from a hydrophilic (HφI) side-chain. Likewise, a pair of HφO side-chains will contribute differently from a pair of HφI side-chains. The presentation of ΔG^*_α in (14.33) also suggests a way of estimating the various contributions to the total Gibbs energy of solvation. Some illustrative examples will be presented in the next section.

Clearly, one could also use the second division of the binding energy, (Equation 14.22) to rewrite the total Gibbs energy of solvation in terms of the types of solute–solvent interaction. The latter focuses on the relative contributions of the *types* of interactions (say, hard, soft, hydrogen-bonding, etc.) rather than on the types of groups occupying the surface of the solute.

4 Some Illustrative Numerical Examples

In this section we present some typical, order-of-magnitude examples of the various contributions to the total Gibbs energy of solution.

The Backbone Contribution

At present there exists no experimental method of estimating the quantity ΔG^*_{BB}. However, assuming that the solute α is approximately spherical, we can write

$$\Delta G^*_{BB} = \Delta G^{*H}_{BB} + \Delta G^{*H/S}_{BB} \qquad (14.34)$$

where ΔG^{*H}_{BB} is the hard and $\Delta G^{*S/H}_{BB}$ is the soft contribution to ΔG^*_{BB}. The hard part is equivalent to the work required to create a cavity at some fixed position in the liquid. The best way of estimating this part is using the scaled-particle theory, SPT (Reiss *et al*, 1959; Reiss, 1966; Ben-Naim, 1987, 1992).

Figure 14.3 shows some typical values for the work of cavity formation W/kT as a function of the radius of the solute in different solvents. (The radius of the cavity is $R + R_w$, where R_w is the radius of the solvent molecule.) Note that this function is a monotonically increasing function of R, and the values of W/kT are somewhat larger in water as compared with benzene and ethanol (the values corresponding to D_2O are given by the dotted curve which falls just below the H_2O curve). One can easily compute the cavity work for any given radius. For a typical globular protein of diameter of about 40–50Å we estimate that the Gibbs energy of solvation of the hard part is in the range of +470 to +700 kcal/mol, i.e. one must invest a considerable amount of energy to create such a cavity.

The second contribution on the right-hand side of (14.34) is due to the soft interaction. We have recently estimated this term to be of the order of about −85 to −110 kcal/mol for a typical size of a protein of diameter between 40 Å and 50 Å (Ben-Naim *et al.*, 1989, 1990). Thus, for this size of the protein the soft part is about one-sixth of the hard part of ΔG^*_α (with different signs). As the diameter of the solute becomes larger, the dominant part will be the Gibbs energy of cavity formation.

Conditional Solvation Gibbs Energies of Independent Side-chains

Ideally, it would be useful to have the conditional solvation Gibbs energy of all possible side-chains of the amino acids, the condition being the backbone of the polypeptide. This information is unavailable at present. Once available, such a list of the conditional Gibbs energies of solvation could serve as an 'ideal' hydrophobicity scale (Ben-Naim, 1990a).

A variety of hydrophobicity scales have been published, based on different experimental sources. For instance, Wolfenden and Lewis (1976) and Wolfenden *et al.* (1981) has published the solvation Gibbs energies of the *side-chains* from the gaseous phase into water. These values do repre-

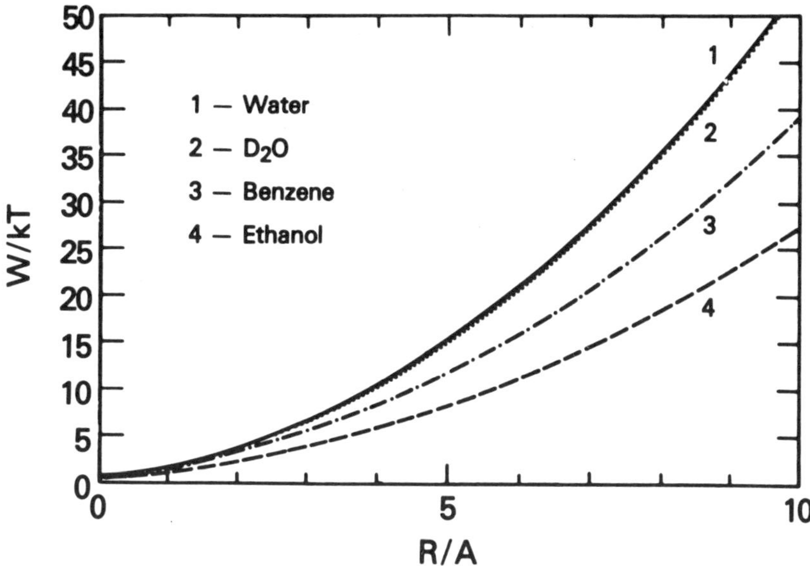

Figure 14.3 The work required to create a cavity as a function of the radius of the solute R that creates the cavity for four liquids at 25°C

sent the intrinsic hydrophobicity of the side-chains in the sense that they measure the relative tendency of the side-chains to dissolve in water. However, this tendency is different from the *conditional* solvation Gibbs energy required above. The difference arises from the effect of the backbone on its immediate vicinity. In other words, the work required to introduce a side-chain into an arbitrary point in water is *different* from the work required to introduce the same side-chains into a point adjacent to the polypeptide backbone. One cannot assume that the backbone will cause a uniform shift in the values of the solvation Gibbs energies of the side-chains. Different side-chains are expected to be affected differently by the presence of the backbone.

A different hydrophobicity scale, that comes closer to the 'ideal' scale, is the one published by Fauchere and Pliska (1983). This scale does take into account the effect of the backbone, but it gives the *difference* in the conditional solvation Gibbs energies of the side-chains between water and octanol. The octanol contribution is not needed for our estimation of the solvation Gibbs energy of a protein.

In Table 14.1 we present some values of the conditional solvation Gibbs energies of some groups next to a hydrocarbon backbone. The methodology of obtaining these values, based on experimental data is explained in great detail elsewhere (Ben-Naim, 1992). It should be pointed out the *conditions* here are not the polypeptide backbone but some hydrocarbon

Table 14.1 Selected values of the conditional solvation Gibbs energies for some hydrophobic and hydrophilic groups

Group	*Backbone*	$\Delta G\alpha$ *(kcal. mol^{-1})*
Methane	None	1.98
Methyl	CH_3–	–0.16
Methyl	CH_3–CH–CH_3	0.31
Methyl	Cyclohexane	0.45
Methyl	Benzene	0.012
Methyl	Naphthalene	0.024
Hydroxyl	CH_3–	–7.11
Hydroxyl	CH_3CH_2–	–6.84
Hydroxyl	CH_3–CH–CH_3–	–6.71
Hydroxyl	Cyclohexane	–6.70
Carbonyl	CH_3–C–CH_3–	–5.8
Carboxyl	$CH_3CH_2CH_2CH_2$–	–5.52
Amine	$CH_3CH_2CH_2CH_2$–	–6.63

molecules. The values do indicate, however, how large the effect of a backbone could be on the conditional solvation Gibbs energy. (The dependence of the conditional Gibbs free energy of various groups on the carrier has been noted earlier (Wolfenden, 1983; Rosenman, 1988; Ben-Naim, 1990a).)

From Table 14.1 we see that an hydroxyl group contributes about −6.7 kcal/mol and a carbonyl group about −5.7 kcal/mol. A hydroxyl group can form at most three hydrogen bonds with water molecules, whereas a carbonyl group can form at most two hydrogen bonds. We have recently estimated that the conditional solvation Gibbs energy 'per one arm' of the polar group (the number of arms is the number of directions along which a polar group can form hydrogen bonds) is about −2.25 kcal/mol (Ben-Naim, 1990a). Taking ribonuclease A as a typical-size protein, we found that a total of about 200 oxygens and nitrogens are exposed to the solvent (Ben-Naim *et al.*, 1989). If each of these can form only one hydrogen bond we have about $-200 \times 2.25 = -450$ kcal/mol. This is clearly an underestimation, since some of the oxygens and nitrogens can form more than one hydrogen bond to the solvent. This value does indicate that a large part of the solvation Gibbs energy is due to functional groups that are exposed to the solvent and can form hydrogen bonds to water molecules.

From Table 14.1, we also see that the contribution of a hydrophobic group is relatively quite small, of the order of about 0.3–0.4 kcal/mol for a methyl group next to a saturated hydrocarbon and much smaller next to an aromatic molecule.

From the data presented above we can conclude that had all functional groups been independent, their contribution to the total solvation Gibbs energy would be of a similar order of magnitude to that of the contribution due to the cavity formation. This conclusion is tentative, pending the

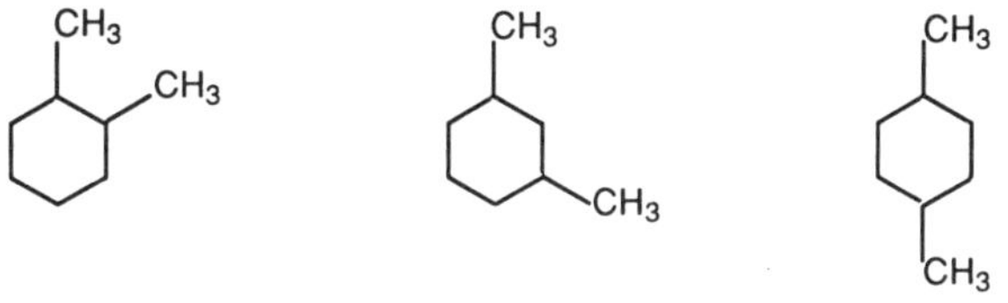

Figure 14.4 Three isomers of dimethylbenzene. The difference in the solvation Gibbs energy is attributed to the intramolecular hydrophobic interaction

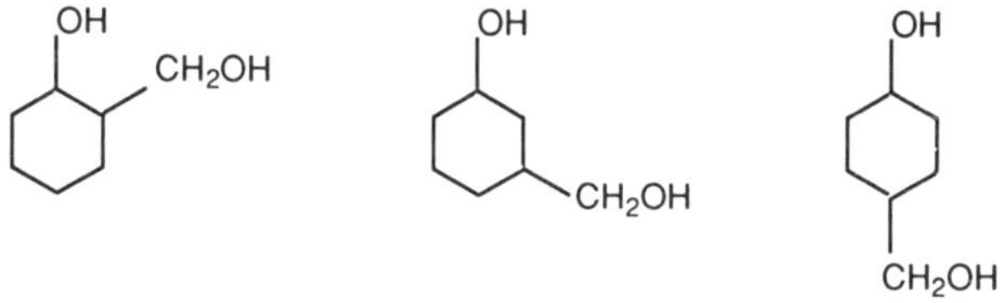

Figure 14.5 Three isomers of a phenol derivative, differing in the extent of hydrophilic interaction

availability of further experimental information on the conditional solvation Gibbs energies of all side-chains, next to the polypeptide backbone. If, however, the functional groups are not independent, then correlations between the functional groups can add significantly to the overall solvation Gibbs energy.

Contributions Due to Pairwise Correlated Groups

In the previous subsection we discussed the contribution of individual and independent groups such as methyl, hydroxyl or carbonyl. When two functional groups are close to each other, such that the solvation spheres of the two groups overlap, there is an additional contribution to the Gibbs energy of solvation.

Consider as an example two methyl groups attached to a benzene ring (the carrier). When these two groups are far from each other (either when they sit on different benzene rings or occupy the 1,4-positions on a single benzene ring — Ben-Naim, 1992), they may be considered as independently solvated. We now bring them to a close separation (say in the 1,2-position of 1,2-dimethylbenzene) (Figure 14.4) and we find that the solvent-induced part of this process involves about -0.3 kcal/mol (Ben-Naim, 1990a). In other words, the solvation Gibbs energy of the 1,2-dimethylbenzene is about -0.3 kcal/mol lower than the corresponding solvation Gibbs energy of the 1,4-dimethylbenzene. Note that although the additional solvation Gibbs energy is not large, it is larger, in absolute magnitude, than the conditional solvation Gibbs energy of a methyl group attached to the benzene ring (Table 14.1).

A more dramatic change in solvation Gibbs energy occurs when two polar groups such as hydroxyl or carbonyl are brought to a close distance, such that a water molecule can bridge between two groups (Figure 14.5). Both experimental data (Ben-Naim, 1990a), theoretical calculation (Ben-Naim, 1989) and simulation results (Mezei and Ben-Naim, 1990) indicate that such pair correlation effects can contribute as much as −3 kcal/mol to the solvation Gibbs energy.

Triplet and Higher-order Correlations

No relevant experimental data are available to estimate the contribution of higher-order correlations to the solvation Gibbs energy. Nevertheless a recent theoretical estimate (Ben-Naim, 1990a) indicates that triplet and quadruplet correlation between HϕI groups could be significant. Nothing is known about the corresponding correlation between HϕO groups.

5 Critique of the Group Additivity Assumption for the Solvation Gibbs Energy

There have been various suggestions of expressing the solvation Gibbs energy, as well as other solvation quantities, in terms of a sum of group's contribution (Eisenberg and McLachlan, 1986; Eisenberg *et al.*, 1987; Kang *et al.*, 1987; Ooi *et al.*, 1987; Cabani *et al.*, 1981; Butler, 1937; Hine and Mookerjee, 1975). For a simple small molecule α having two parts *a* and *b* (where *a* and *b* could be two methyl groups in ethane or a methyl and chloride in methyl chloride). We have seen that, to a good approximation, we may write the total binding energy of the entire molecule α as

$$B_\alpha = B_a + B_b \tag{14.35}$$

However, from (14.35) it does not follow that the solvation Gibbs energy of α can be written as

$$\Delta G^*_\alpha = \Delta G^*_a + \Delta G^*_b \tag{14.36}$$

The reason is that, in general, *a* and *b* are not independently solvated. The exact expression for ΔG^*_α in terms of ΔG^*_a and ΔG^*_b is (Ben-Naim, 1987)

$$\Delta G_\alpha = \Delta G^*_a + \Delta G^*_b + \delta G(\infty \rightarrow a - b) \tag{14.37}$$

where δG is the solvent induced change in the Gibbs energy for the process of bringing a and b from *fixed* positions at infinite separation to the final configuration a–b. The latter quantity could be of the same order of magnitude as ΔG_a^* and ΔG_b^* and cannot be neglected. For example, for CH_3–CH_3, the solvation Gibbs energy of CH_3–, approximated by CH_4 is

$$\Delta G^*_{CH_3} \approx \Delta G_{CH_4} = 1.9 \text{ kcal/mol} \tag{14.38}$$

and we also have

$$\Delta G^*_{CH_3CH_3} = 1.77 \text{ kcal/mol} \tag{14.39}$$

$$\delta G(\infty \rightarrow CH_3CH_3) \approx -2.03 \text{ kcal/mol} \tag{14.40}$$

Clearly, for larger molecules, the assumption of group additivity cannot account for the difference of the solvation Gibbs energies for different isomers. For instance, the three isomers of dimethyl benzene (Figure 14.4), according to the group additivity assumption, would be written as

$$\Delta G^*_{\text{dimethyl benzene}} \approx \Delta G^*_B + 2\Delta G^*_{CH_3} \tag{14.41}$$

where 'B' stands for the benzene ring and CH_3 for the methyl groups.

However, the experimental values of the 1,2-, the 1,3- and the 1,4-dimethylbenzene are different. The differences arise from the extent of correlation between the two methyl groups at the different configurations. This difference could be quite large if the groups are hydrophilic, e.g. the three isomers of the phenol derivative (Figure 14.5) are expected to have large difference in the solvation Gibbs energies (Ben-Naim, 1989, 1990a).

When applying the idea of group of additivity to large protein molecules, one often writes

$$\Delta G^*_\alpha = \sum_i g_i A_i \tag{14.42}$$

where A_i is the solvent accessible area of the group i and g_i is the characteristic contribution, per unit area of the group i. This kind of group additivity is in general unjustified for the following reasons.

(1) In contrast to small molecules discussed above, large macromolecules, such as globular proteins, have many groups that are not exposed to the solvent. These groups contribute significantly to the 'hard' part of the solvation Gibbs energy, see for instance ΔA^*_H in Equation (14.29). However, in the representation (14.42) only those groups that are exposed to the solvent are accounted for and therefore this expression

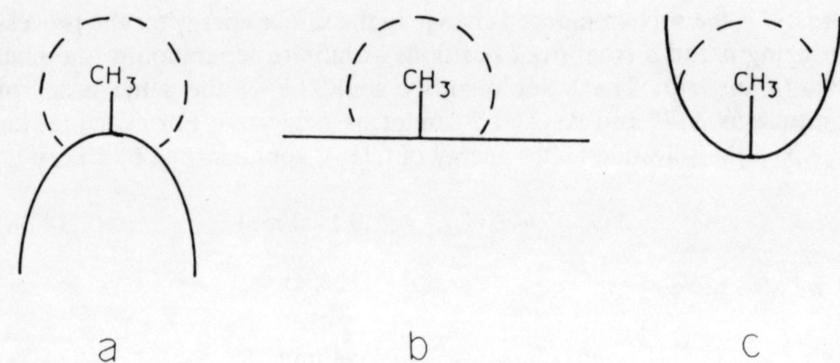

Figure 14.6 Three possible exposures of a methyl group on the surface of a protein. The exposed area decreases from a to b to c

cannot be used to describe the solvation Gibbs energy of a macromolecule having a large hard core which is not directly exposed to the solvent.

(2) As noted above, the group additivity assumption does not take into account pairs and higher-order correlations between groups. In fact, the representation of ΔG^*_α in the form (14.42) implicitly assumes that all groups exposed to the solvent are independently solvated. This, of course, is not the case in general. We have noted above that pair correlations between two HϕO groups could contribute some 0.5–1 kcal/mol to the solvation Gibbs energy, depending on the size and relative location of the groups. Correlation between two HϕI groups could contribute about −3 kcal/mol, depending on the distance and the relative orientation of the two HϕI groups. It is known that molecules containing more than one polar functional groups do not conform to the empirical scheme of group additivity (Cabani *et al.*, 1981; Kang *et al.*, 1987).

Clearly there are many possibilities for pairs and higher-order correlations between groups that are exposed to the solvent. The collective contribution of these correlations could contribute significantly, if not dominantly, to the overall solvation Gibbs energy of the protein. Recently it has been demonstrated that a small change in the conformation, or a mutation in one solvent-exposed side-chain, leaving the right-hand side of (14.42) almost unchanged, could cause a very large change in the solubility of the protein (Ben-Naim, 1992).

(3) Even when focusing on the sum of independently solvated groups, as implied in (14.42), the proportionality of each term to the area exposed to the solvent is in general unjustified. For a simple non-polar side-chain, the assumption of proportionality to the surface area might be reasonable. For instance, in the three situations shown in Figure 14.6, the exposed area of the methyl group decreases from a to b to c.

Therefore, for a *fixed* group, attached to the *same* surface, g_i is constant and $A_a > A_b > A_c$, resulting in a decreasing contribution of the methyl

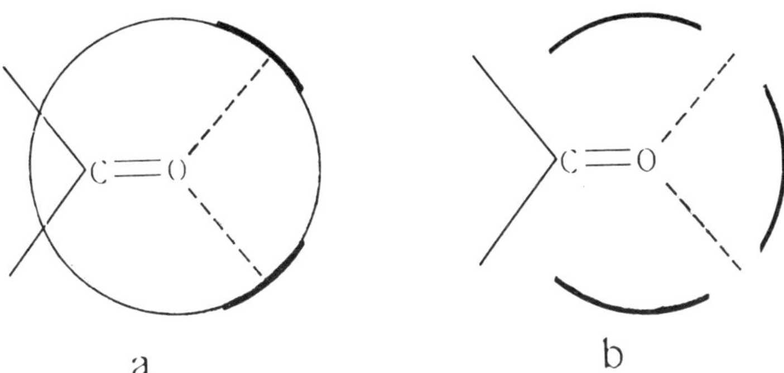

Figure 14.7 A schematic drawing of the surroundings of an exposed carbonyl group to the solvent. (*a*) The two 'arms' along which HBs can be formed are blocked. Hence, no HBs to the solvent may be formed.
(*b*) The two 'arms' of the carbonyl group are exposed to the solvent. Hence, one or two HBs may be formed with water molecules

group to the overall solvation Gibbs energy. However, note that the *same* group, say methyl, attached to different surfaces might have different values of the g_i in Equation (14.42).

As an example, a methyl group attached to a benzene ring contributes about 0.09 kcal/mol, whereas the same group attached to hexane contributes about 0.33 kcal/mol to the solvation Gibbs energy. Thus, while the proportionality to the exposed surface might be a reasonable approximation for non-polar groups, one must assign different values for g_i depending on the type of backbone to which this group is attached.

The situation is far more complex for polar groups that can form hydrogen bonds (HB) to the solvent. In this case, the important factor that determined the contribution of such a group to the solvation Gibbs energy is *not* the exposed area, but the number of 'arms' exposed to the solvent. For instance, a hydroxyl group has three 'arms', i.e. three preferential directions along which it can form HBs with solvent molecules (two as acceptor, one as donor), a carbonyl group has two arms, as acceptors. Figure 14.7 shows schematically the surroundings of a carbonyl group. The two dotted lines indicate the two 'arms', i.e. the directions along which the carbonyl group can form HBs. In Figure 14.7(a) the two arms are blocked (indicated by bold bars) and therefore no HBs can be formed with solvent molecules. In Figure 14.7(b), the two arms are exposed to the solvent; hence, two HBs can be formed.

Clearly the solvation Gibbs energy contributed by the C=O group in a is far smaller than is the case b in spite of the fact that the overall surface exposed in case b is smaller than in case a. We also recall that the contribution of a HφO group such as methyl group to the solvation Gibbs energy is about 1/20 of the contribution of a Hφ*I* group such as hydroxyl or carbonyl. The inevitable conclusion is that groups that contribute significantly to ΔG_α^* are the ones for which the proportionality to the exposed surface does not hold.

To summarize, the application of group additivity assumption is in general unjustified for three reasons: (1) the neglect of the contribution of the unexposed groups; (2) the neglect of pairs and higher-order correlations between groups; (3) the inadequate representation of the contribution of each group as being proportional to its exposed surface area.

6 Examples of Biochemical Processes Affected by Solvation

It has long been recognized that hydration, i.e. solvation by water, might have a significant effect on the driving forces for biochemical processes taking place in aqueous media. Both hydrogen bonds and hydrophobic interactions were considered as ingredients of the solvent contribution to the driving force (Kauzmann, 1959). In this section we shall briefly survey some of the key biochemical processes, the solvation effects on which have been studied. We shall first treat the simplest case of a solute being at a single conformational state. The relevant process is the solubility of the solute. The second example is a solute which can be in one or more conformational states. The relevant process is the transformation from one state to another. The third example is the association of two solutes to form a compound solute. Other, more complicated processes, such as the association of a large number of solutes with or without conformational changes, may be viewed as a combination of these elementary processes. We close this section with a discussion of the solvent induced *forces* on biochemical processes.

Solvation and Solubility

Perhaps the simplest quantity that is affected by the solvation is the *solubility* of the solute itself. Normally, the solubility of a macromolecule α is measured at equilibrium with respect to a solid phase (the latter may be a pure solid phase or contain a fixed quantity of the solvent, i.e. water of hydration).

At equilibrium between the two phases we have (for details see Ben-Naim, 1992)

$$\mu_\alpha^s = \mu_\alpha^\ell = \mu_\alpha^{*\ell} + kT\ell n \rho_\alpha^\ell \Lambda_\alpha^3 \qquad (14.43)$$

where μ_α^* is the pseudochemical potential of α, ρ_α^ℓ is the equilibrium density, or the solubility of the solute α with respect to the solid phase s. Λ_α^3 is the momentum partition function, which is assumed to be independent of the presence of the solvent. If the same system is at equilibrium with respect to an ideal gas phase, the analogous equation is

$$\begin{aligned}\mu_\alpha^s = \mu_\alpha^g &= \mu_\alpha^{*g} + kT\ell n\rho_\alpha^g\Lambda_\alpha^3 \\ &= -kT\ell n q_\alpha + kT\ell n\rho_\alpha^g\Lambda_\alpha^3\end{aligned} \tag{14.44}$$

where q_α is the internal partition function of α, and again is assumed to be approximately independent of the presence of the solvent.

From (14.43) and (14.44) we have

$$\Delta G_\alpha^* \equiv \mu_\alpha^{*\ell} - \mu_\alpha^{*g} = kT\ell n(\rho_\alpha^g/\rho_\alpha^\ell) \tag{14.45}$$

where $(\rho_\alpha^g / \rho_\alpha^\ell)$ is the ratio of the number densities of α in the gaseous and in the liquid phase, respectively. Clearly, for large solutes ΔG_α^* cannot be measured directly, since no measurable density of α can be detected in the gaseous phase. However, ΔG_α^* is still the quantity that determines the solubility of α with respect to the solid phase. Thus from (14.43), (14.44) and (14.45) we have

$$kT\ell n\rho_\alpha^\ell\Lambda_\alpha^3 = (\mu_\alpha^s - \mu_\alpha^{*g}) + \Delta G_\alpha^* \tag{14.46}$$

The quantities Λ_α^3, μ_α^s, μ_α^{*g} are independent of the solvent. Hence, any change in the solvent would affect ΔG_α^* and, hence, the solubility. For example, the solubility of α in the liquid phase ℓ, relative to pure water w is given by

$$kT\ell n(\rho_\alpha^\ell/\rho_\alpha^w) = \Delta G_\alpha^{*\ell} - \Delta G_\alpha^{*w} \tag{14.47}$$

Thus, although $\Delta G_\alpha^{*\ell}$ is not a measurable quantity, the relative solvation Gibbs energy on the r.h.s. of (14.47) is measurable. Note that, in general, the solvation Gibbs energy, as defined above may depend on the solute concentration ρ_α. It is only in the limit of very dilute solutions that ΔG_α^* becomes independent of ρ_α^ℓ (but still dependent on the composition of the solvent). In this limit ΔG_α^* coincides with one of the standard Gibbs energy of transfer from the gaseous to the liquid phase. The treatment above can be easily generalized for solutes having more than one conformational states, and to a solvent mixture of any number of components. (For further details see Ben-Naim, 1987, 1992.)

Conformational Changes in Aqueous Solutions

Let F and U be any two conformations of the solute α. These could be the *cis–trans* conformers of a small molecule, the helix-coil transition in polypeptide, the folded (F) and the unfolded (U) states of a protein, or the A and B structures of DNA.

The involvement of the solvation in the stability of the native structure of proteins has been discussed by many authors: for example, Finney and Savage, 1988; Edsall and McKenzie, 1983; Westhof and Sundaralingan, 1986; Thanki *et al.*, 1988; Sundaralingan and Sekharadu, 1989; Ben-Naim *et al.*, 1989.

Likewise, the solvent effects on the stability and conformational transitions of nucleic acid have been studied extensively. Examples are: Saenger *et al.*, 1986; Saenger, 1987; Breslauer *et al.*, 1986; Dickerson *et al.*, 1982; Falk *et al.*, 1962a, b, 1963; Kopka *et al.*, 1983; Kennard and Hunter, 1989; Kennard *et al.*, 1986; Chevrier *et al.*, 1986; Goodfellow, 1984; Texter, 1978; Westhof, 1987a, b, 1988; Westhof *et al.*, 1985, 1988; Subramanian *et al.*, 1988; Lewin, 1966, 1967.

The equilibrium constant for the conversion reaction

$$U \rightleftarrows F \tag{14.48}$$

is determined by the condition

$$\mu_U^{\ell} = \mu_U^{*\ell} + kT\ell n\rho_U\Lambda_\alpha^3 = \mu_F^{\ell} = \mu_F^{*\ell} + kT\ell n\rho_F\Lambda_\alpha^3 \tag{14.49}$$

or equivalently

$$\begin{aligned} K^l = (\rho_F/\rho_U)_{eq} = \left(\frac{X_F}{X_U}\right)_{eq} &= \exp\left[-\beta(\mu_F^* - \mu_U^*)\right] \\ &= \frac{q_F}{q_U}\exp[-\beta(\Delta G_F^* - \Delta G_U^*)] \\ &= K^g \exp[-\beta\delta G\ (U \rightarrow F)] \end{aligned} \tag{14.50}$$

In (14.50) we wrote the equilibrium constant in the liquid phase K^{ℓ}, as a product of the equilibrium constant in the gaseous phase K^g, and the solvent effect on the equilibrium constant. The latter is simply the difference in the solvation Gibbs energies of the two forms F and U.

It is clear that even a small conformational change can cause a large change in the solvation Gibbs energy and, hence, gives rise to a very large change in the equilibrium constant. Furthermore, suppose we examine the same reaction (14.48) in different solvents, say by adding a solute or by changing the composition of the solvent. Since K^g is independent of the solvent, any change in K^{ℓ} is necessarily a result of the change in the relative solvation Gibbs energies of the two conformers F and U.

In many biochemical processes, regulation is achieved by means of conformational changes induced by an effector. In such processes the solvent effect could play a major role in determining the efficiency of the regulation mechanism (Ben-Naim, 1992).

Binding Equilibria

There are many different binding processes in which solvation effects can play a decisive role. These range from binding of small molecules to plasma proteins, binding of a substrate to an enzyme, binding of hormones or drugs to receptors, association of proteins, binding of proteins to nucleic acid, etc.

The simplest binding process can be written as

$$P + L \rightleftarrows PL \tag{14.51}$$

where P and L stand for protein and ligand, respectively, and PL represents the complex of P and L at some preferential binding sites (these could be either on P or on L or on both).

Using the equilibrium condition

$$\mu_P + \mu_L = \mu_{PL} \tag{14.52}$$

we can write in analogy with (14.50) the equilibrium constant in the liquid as

$$\begin{aligned} K^{\ell} = \left(\frac{\rho_{PL}}{\rho_P \rho_L} \right)_{eq} &= K^g \exp[-\beta(\Delta G^*_{PL} - \Delta G^*_P - \Delta G^*_L)] \\ &= K^g \exp[-\beta \delta G(\mathrm{P+L} \to \mathrm{PL})] \end{aligned} \tag{14.53}$$

Here K^g is the equilibrium constant for the same reaction (14.51) taking place in vacuum. The quantity δG (P+L→PL) is the solvent effect on the binding Gibbs energy. As in the case of conformational equilibria, discussed above, the solvent effect is the difference between the solvation Gibbs energies of the product and the reactant, i.e.

$$\delta G(\mathrm{P + L \to PL}) = \Delta G^*_{PL} - \Delta G^*_P - \Delta G^*_L \tag{14.54}$$

Thus, knowledge of the solvation Gibbs energies of the species involved in the binding process (or in any other, more complex reaction) is sufficient for the determination of the solvent effect on the equilibrium constant pertaining to reaction (14.51).

As in the case of conformational changes, in the binding process solvent effects could change dramatically both the extent of the binding equilibrium and the specificity of the binding processes. For instance, the selection of a specific binding site (i.e. the molecular recognition) could be changed by varying the solvent composition (Ben-Naim, 1992).

Solvation Gibbs Energies and Solvent-induced Forces

In the previous subsections we have considered the total Gibbs energy change involved in the process of conformational change or of a binding reaction. These changes in Gibbs energies can be viewed as the *thermodynamic driving force* for the appropriate process. The actual, instantaneous force operating on a molecule or on a particular group of a macromolecule is related to the infinitesimal change in the solvation Gibbs energy.

Consider first the conformational change as represented in Equation (14.48). Starting from any specific conformation of the molecule α, let $\mathbf{R}^M$ represent the configuration of the entire molecule. Here $\mathbf{R}^M = \mathbf{R}_1 \cdots \mathbf{R}_M$ and $\mathbf{R}_i$ is the locational vector of the *i*-th nucleus of α.

For each configuration $\mathbf{R}^M$ one can define the Gibbs energy of the system of *one* solute α and *N* solvent molecules. It may be shown that the total average force operating on a specific group, say *i*, is given by (Ben-Naim, 1992)

$$\mathbf{F}_i(\mathbf{R}^M) = -\nabla_i U(\mathbf{R}^M) - \nabla_i \Delta G^*_\alpha \tag{14.55}$$

where the first term is the *direct* force exerted by the molecule α, being at $\mathbf{R}^M$ on the group *i*, in the absence of the solvent. This force originates from all the groups belonging to the solute α at a specific conformation $\mathbf{R}^M$. The second term is the solvent-induced force. This is an average force exerted by the solvent molecules on the *i*th group. The average is over all configurations of the solvent molecules, given that the solute α is at a fixed conformation $\mathbf{R}^M$. We see that the solvent-induced force is intimately related to changes in the solvation Gibbs energy. The statistical mechanical expression for the solvent induced force is

$$\begin{aligned} F_i^{(SI)}(R^M) &= -\nabla_i \Delta G^*_\alpha \\ &= -\rho_\omega \int \nabla_i U(\mathbf{R}_i, \mathbf{X}_w)\, g(\mathbf{X}_w, \mathbf{R}^M)\, d\mathbf{X}_w \end{aligned} \tag{14.56}$$

where ρ_w is the solvent density, $\nabla_i U(\mathbf{R}_i, \mathbf{X}_w)$ is the direct force exerted by a solvent molecule at $\mathbf{X}_w$, on the *i*th group at $\mathbf{R}_i$. $g(\mathbf{X}_w, \mathbf{R}^M)$ is the solute–solvent pair correlation function at the specified configuration $(\mathbf{X}_w, \mathbf{R}^M)$.

It has recently been argued that a strong solvent-induced force may be expected when two functional groups (e.g. hydroxyl or carbonyl) are at a distance of about 4.5 Å and oriented in such a way that a water-bridge between the two oxygens may be formed (Ben-Naim, 1990a).

In Equation (14.55) we referred to the force exerted on one group (the ith one) of a macromolecule being at a specific conformation $\mathbf{R}^M$. Likewise, the solvent-induced force on the ith group of a solute α, given a second solute, β, at a fixed conformations $\mathbf{R}_\alpha^{M\alpha}$ and $\mathbf{R}_\beta^{M\beta}$, respectively, is given by

$$F_i^{(SI)} = -\nabla_i \, \Delta G_{\alpha,\beta}^* \, (\mathbf{R}_\alpha^{M\alpha}, \mathbf{R}_\beta^{M\beta}) \tag{14.57}$$

In (14.57), in contrast to (14.55), we assume that the conformations of both α and β are fixed; only the relative distance and orientation of the two solutes can change. The simplest example is when α and β are two simple spherical molecules, say two argon atoms in a solvent, in which case (14.57) reduces to

$$\begin{aligned} F_\alpha^{(SI)} &= -\nabla_\alpha \, \Delta G_{\alpha,\beta}^* \, (\mathbf{R}_\alpha, \mathbf{R}_\beta) \\ &= -\rho_\omega \int \nabla_\alpha U(\mathbf{R}_\alpha, \mathbf{X}_w) g(\mathbf{X}_w/\mathbf{R}_\alpha, \mathbf{R}_\beta) \alpha X_w \end{aligned} \tag{14.58}$$

where $U(\mathbf{R}_\alpha, \mathbf{X}_w)$ is the direct interaction between α and a solvent molecule at $(\mathbf{R}_\alpha, \mathbf{X}_w)$ and $g(\mathbf{X}_w/\mathbf{R}_\alpha, \mathbf{R}_\beta)$ is the 'pair' correlation function between a water molecule and the pair of solutes, being at $\mathbf{R}_\alpha$, $\mathbf{R}_\beta$ and viewed as a single entity. The gradient ∇_α effectively changes the relative distance between the two solutes. Clearly, a solvent-induced force between two solutes (at fixed conformations) exists, if and only if a small change in their relative distance and orientations produces a change in the solvation Gibbs energy of the pair of solutes at the specified conformations.

7 Solvation and Structural Changes in the Solvent

We have noted in Section 2 the existence of an exact compensation effect in the entropy and the enthalpy of solvation. This follows from the formal expressions (14.19) and (14.20). This can also be reinterpreted as arising from the contributions to the entropy and energy (or enthalpy) of solvation due to structural changes in the solvent. It follows from this compensation that structural changes in the solvent, induced by the solvation process cannot affect the solvation Gibbs energy. This result was proved in various forms for the solvation as well as for the association process (Ben-Naim, 1975, 1987, 1991). It was later criticized several times (e.g. Marcelja *et al.*, 1977; Nemethy *et al.*, 1981) on the grounds that it contradicts the principles of thermodynamics, i.e. '. . . if the free energy change accompanying a structural change were zero, there would be no driving force for shifting the structural equilibrium of the solvent' (Nemethy *et al.*, 1981). It seems to me that there exists some confusion regarding the free energy change for the solvation, or dimerization process, on one hand, and the free energy change associated with the structural changes in the solvent. It is the process of solvation, or association, that causes the structural changes in the solvent. The later process involves a flow of water molecules between different structural species at equilibrium — hence, at equal chemical potential. Therefore, this flow of water molecules cannot affect the Gibbs energy of the *solvation* or *association* processes. At the same time the same flow of water molecules can affect both the entropy and the enthalpy of the biochemical processes.

Although an exact proof has been given earlier (Ben-Naim, 1978, 1987), we present here a simpler example to demonstrate the main point of the argument.

Consider pure water in a two-phase system, gas (g) and liquid (l) at equilibrium. In the gaseous phase we assume that water molecules exist only as monomers. In the liquid we may classify water molecules according to the number of hydrogen bonds they are engaged in. Since equilibrium exists throughout the system, we have

$$\mu_w^g = \mu_w^\ell = \mu_0 = \mu_1 = \mu_2 = \mu_3 = \mu_4 \tag{14.59}$$

where μ_w^g and μ_w^ℓ are the chemical potential of the water (w) in the gaseous and in the liquid phases, respectively. The initial state of the system is depicted schematically on the left-hand side of Figure 14.8.

Normally, we add a solute α to the *liquid* phase and induce a structural change within the various species of water molecules. Here we use a simpler process. We first separate the two phases by placing a partition between the phases (see first step in Figure 14.8) and add a solute α (dashed square in the figure) to a fixed position in the gaseous phase. The chemical potential of the water in the gaseous phase in the initial state is given by

$$\mu_w^g = kT\ell n\rho_w \Lambda_w^3 \tag{14.60}$$

which is equal to μ_w^l according to (14.59). Now, after adding the solute the chemical potential of the water is changed. Within the second virial coefficient the new chemical potential is

$$\begin{aligned}\bar{\mu}_w^g &= kT\ell n\rho_w \Lambda_w^3 - \frac{kT\rho_w}{8\pi^2}\int[\exp[-\beta U_{w\alpha}]-1]d\mathbf{X}_w \\ &= kT\ell n\rho_w \Lambda_w^3 + B^{EX} - B^{IN}\end{aligned} \tag{14.61}$$

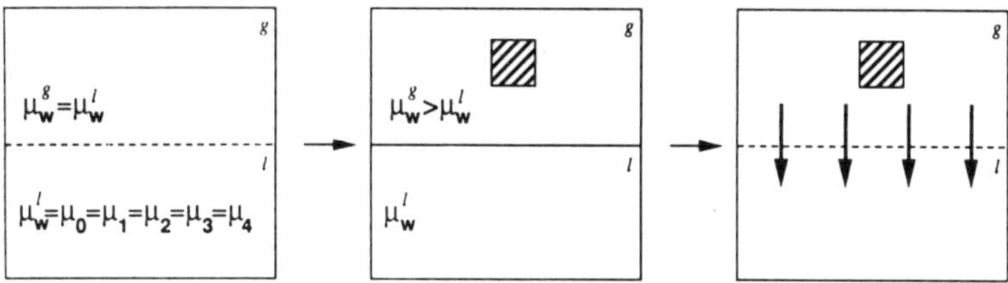

Figure 14.8 Water molecules in two phases at equilibrium. In the first step we place a partition between the two phases (solid line separating g and l), and insert a solute (shaded square) to the gaseous phase. In the second step the partition is removed, causing a flow of water from one phase to another

Note that since the partition separates the two phases, ρ_w is unchanged. In (14.61) we denoted by B^{EX} the effect of the excluded volume on the second virial coefficient and by B^{IN} the contribution due to solute–solvent attractive interaction (e.g. hydrogen-bonds).

Suppose first that α is a hard solute; then B^{EX} is positive; hence,

$$\bar{\mu}_w^g > \mu_w^g = \mu_w^\ell \tag{14.62}$$

i.e. placing α in the gaseous phase will increase the chemical potential of the water in this phase. If, on the other hand, a strong hydrogen bond is formed between α and a water molecule such that $B^{IN} > B^{EX}$, then clearly

$$\bar{\mu}_w^g < \mu_w^g = \mu_w^\ell \tag{14.63}$$

In the second step we remove the partition between the two phases. Clearly in the case (14.62) water molecules will flow from g to ℓ, i.e. the insertion of α will repel water molecules from g into ℓ. In the second case the flow is in the opposite direction, i.e. the strong interaction between α and the water will attract water molecules from the liquid into the gaseous phase. In both cases, the flow of water molecules is a result of the inequality of the chemical potential (14.62) or (14.63). However, in the limit of macroscopic system this difference in $\bar{\mu}_w^g - \mu_w^g$ can be made as small as we wish (note that we have placed one solute α in the macroscopic system). Therefore, the flow of water molecules between the two phases (initially at equilibrium) cannot affect the solvation Gibbs energy of the *solute* itself. At the same time, the same flow may affect both the entropy and the enthalpy of solvation of the solute α.

8 Conclusion

The solvation Gibbs energy of biopolymers is defined as the change in the Gibbs energy for the process of the transferring a single molecule from a fixed position in an ideal gaseous phase into a fixed position in the liquid phase. A general procedure to characterize the solvation Gibbs energy in terms of its various ingredients is outlined. The difference between this procedure and a similar one based on group-additivity assumption is stressed. Some typical biochemical processes, the driving force of which might be significantly affected by solvation, are briefly discussed.

References

Ben-Naim, A. (1975). *Biopolymers*, **14**, 1337
Ben-Naim, A. (1978). *J. Phys. Chem.*, **82**, 874
Ben-Naim, A. (1987). *Solvation Thermodynamics*. Plenum Press, New York
Ben-Naim, A. (1989). *J. Chem. Phys.*, **90**, 7412
Ben-Naim, A. (1990a). *Biopolymers*, **29**, 567
Ben-Naim, A. (1990b). *J. Chem. Phys.*, **93**, 8196
Ben-Naim, A. (1991). *J. Phys. Chem.*, **95**, 1437
Ben-Naim, A. (1992). *Statistical Thermodynamics for Chemists and Biochemists*. Plenum Press, New York
Ben-Naim, A., Ting, K. L. and Jernigan, R. L. (1989). *Biopolymers*, **28**, 1309
Ben-Naim, A., Ting, K. L. and Jernigan, R. L. (1990). *Biopolymers*, **29**, 901
Breslaur, K. J., Frank, R., Blocker, H. and Marky, L. A. (1986). *Proc. Natl Acad. Sci. USA*, **83**, 3346
Butler, J. A. V. (1937). *Trans. Faraday Soc.*, **33**, 229
Cabani, S., Gianni, P., Mollica, V. and Lepori, L. (1981). *J. Sol. Chem.*, **10**, 563–593
Chevrier, B., Dock, A. C., Hartmann, B., Leny, M., Moras, D., Thuong, M. T. and Westhof, E. (1986). *J. Mol. Biol.*, **188**, 707
Dickerson, R. E., Drew, H. R., Conner, B. N., Wing, R. M., Fratini, A. V. and Kopka, M. L. (1982). *Science*, **216**, 475
Edsall, J. T. and McKenzie, H. A. (1983). *Adv. Biophys.*, **16**, 53–183
Eisenberg, D. and McLachlan, A. D. (1986). *Nature*, **319**, 199–203
Eisenberg, D., Wesson, M., Goodsell, D., Wilcox, W., Gribskov, M., Altschuh, D. and Rees, D. C. (1987). *Protein Structure, Folding and Design*, **2**, 203–213
Falk, M., Hartman, K. A. and Lord, R. C. (1962a). *J. Am. Chem. Soc.*, **84**, 3843
Falk, M., Hartman, K. A. and Lord, R. C. (1962b). *J. Am. Chem. Soc.*, **85**, 391
Falk, M., Hartman, K. A. and Lord, R. C. (1963). *J. Am. Chem. Soc.*, **85**, 387
Fauchere, J. L. and Pliska, V. (1983). *Eur. J. Med. Chem.*, **18**, 369
Finney, J. L. and Savage, H. F. J. (1988). In Dogonadze, R. R., Kalman, E., Korhyshev, A. A. and Ulstrup, J. (Eds), *The Chemical Physics of Solvation*. Elsevier, Amsterdam
Goodfellow, J. M. (1984). *J. Theoret. Biol.*, **107**, 261
Hine, J. and Mookerjee, P. K. (1975). *J. Org. Chem.*, **40**, 292–297
Kang, Y. K., Nemethy, G. and Scheraga, H. A. (1987). *J. Phys. Chem.*, **91**, 4109–4117
Kauzmann, W. (1959). *Adv. Protein Chem.*, **14**, 1

Kennard, O., Cruse, W. B. T., Nachman, J., Prange, T., Shakked, Z. and Rabinowich, D. (1986). *J. Biomol. Struct. Dyn.*, **3**, 623
Kennard, O. and Hunter, W. N. (1989). *Q. Rev. Biophys.*, **22**, 327
Kopka, M. L., Fratini, A. V., Drew, H. R. and Dickerson, R. E. (1983). *J. Mol. Biol.*, **163**, 129–146
Lewin, S. (1966). *Arch. Biochem. Biophys.*, **115**, 62
Lewin, S. (1967). *J. Theoret. Biol.*, **17**, 181
Marcelja, S., Mitchell, D. L., Nenham, B. W. and Sculley, M. J. (1977). *J. Chem. Soc. Faraday Trans.*, **73**, 630
Mezei, M. and Ben-Naim, A. (1990). *J. Chem. Phys.*, **92**, 1359
Nemethy, G., Peer, W. J. and Scheraga, H. A. (1981). *Ann. Rev. Biophys. Bioengng*, **10**, 459–97
Ooi, T., Oobatake, M., Nemethy, G. and Scheraga, H. A. (1987). *Proc. Natl Acad. Sci. USA*, **84**, 3086
Reiss, H. (1966). *Adv. Chem. Phys.*, **9**, 1
Reiss, H., Frisch, H. and Lebowitz, J. L. (1959). *J. Chem. Phys.*, **31**, 369
Rosenman, M. A. (1988). *J. Mol. Biol.*, **200**, 513–522
Saenger, E. (1987). *Ann. Rev. Biophys. Biophys. Chem.*, **16**, 93
Saenger, W., Hunter, W. and Kennard, O. (1986). *Nature*, **324**, 385
Subramanian, P. S., Ravishanker, G. and Beveridge, D. L. (1988). *Proc. Natl Acad. Sci. USA*, **85**, 1836
Sundaralingan, M. and Sekharudu, Y. C. (1989). *Science*, **244**, 1333
Texter, J. (1978). *Prog. Biophys. Mol. Biol.*, **33**, 83
Thanki, N., Thornton, J. M., and Goodfellow, J. M. (1988). *J. Mol. Biol.*, **202**, 637–657
Westhof, E. (1987a). *Int. J. Biol. Macromol.*, **9**, 186
Westhof, E. (1987b). *J. Biomol. Struct. Dyn.*, **5**, 581
Westhof, E. (1988). *Ann. Rev. Biophys. Biophys. Chem.*, **17**, 125
Westhof, E., Dumas, P. and Moras, D. (1988). *Biochimie*, **70**, 145
Westhof, E., Prange, T. H., Chevrier, B. and Moras, D. (1985). *Biochimie*, **67**, 811
Westhof, E. and Sundaralingan, M. (1986). *Biochemistry*, **25**, 4868
Wolfenden, R. (1978). *Biochemistry*, **17**, 201
Wolfenden, R. (1983). *Science*, **222**, 1087–1093
Wolfenden, R. (1985). *Proc. Indian Acad. Sci. (Chem. Sci.)*, **94**, 121–138
Wolfenden, R., Andersson, L., Cullis, P. M. and Southgate, C. C. B. (1981). *Biochemistry*, **20**, 849
Wolfenden, R. and Lewis, C. A. (1976). *J. Theoret. Biol.*, **59**, 731
Wolfenden, R. and Williams, R. (1983). *J. Am. Chem. Soc.*, **105**, 1028

Index

Ab initio SCF calculations 245, 246, 259, 260
Acoustic vibration, of DNA 267, 278
Actinidin 114, 116, 120
Activation energy, of DNA hydration shell 287, 289
Agarose 313-315
Agglutinin, wheat germ 328
Ammonium group
  hydration 340, 342, 344, 373
  interactions with phosphates 354
  ion binding 350
Amylose 301, 304, 306, 309
Antibodies 422-423
Arabinose binding protein 124, 126
Azurin 111, 124, 126

Bence-Jones protein (Rhe) 118
Binding
  average 434
  energy 432-433
Brillouin spectroscopy 268-269

Calcium 351
Carbonyl group, hydration 39, 69-81, 341, 348, 359, 372
Carboxylic group 340, 345, 347, 350
  hydration 78-83, 340, 350
Carrageenan 314-315
Catalytic sites 99, 112, 124, 130-132
Cellulose 311-312, 315-316
Cerebroside 340, 365-366
Charge
  effective 354-355
  lipid 338, 343
  local excess 340, 342-343, 350
  surface 340, 349, 372
  transfer 341, 352, 354
Chymotrypsin 101, 125, 127
Cholesterol 338, 364, 366
Chondroitin 313
Clathrate 35, 55
Colloidal stability 349, 376
Compressibility, isothermal (κ_T) 46
Conditional average 434, 439
Counterions 255, 257, 259
Crambin 121
Crystal 98, 101-102
  density 101
  morphology 123
  mosaicity 103
  packing 112, 123
  solvent 98, 101, 104
Crystallization 128, 137
Curdlan 301
Cyclodextrins 34, 175
Cytochrome 101, 124, 125

Dextran 296, 302-304
$d(A)_n.d(T)_n$ 185, 187, 246
$d(AAT)_n.d(TTA)_n$ 248
$d(AC)_n.d(CT)_n$ 248
$d(AG)_n.d(CT)_n$ 248
d(CA) 230
d(CCAACGTTGG) 181
d(CCAAGATTGG) 238
d(CGCGAATTCGCG) 168, 170, 179, 185, 187, 192, 213, 229, 252, 257, 260
d(CG) 177, 258, 259
$d(C)_n.d(G)_n$ 181-184, 191, 194, 246, 248, 257
Dielectric constant
  effects of lipid phase 360
  high-frequency 353
  hydration dependence 360
  interfacial 356-357
  relaxation 346, 360
  static 353, 360
  temperature dependence 360
Diffraction, liquid water 8-11
Dipole moment, gas phase 50
DNA

charge densities 248
drug complexes 177, 204ff, 229, 235
hydration of canonical forms
A-DNA 165, 227, 247-251, 256
B-DNA 165, 179ff, 227, 246, 250, 252, 256, 257, 260
Z-DNA 182, 229, 231, 238, 247-249
hydration of grooves 179ff, 235ff, 244-245, 250, 252-253, 255, 257-258
polymorphism 165, 226
primary hydration shell 266, 277, 288-289
protein complexes 128-129, 207ff
secondary hydration shell 266, 288-289
tertiary structure 226
triplexes 203ff

Dynamics
Brownian 209ff
lipid chain 357
lipid headgroup 369, 373
water 45-57, 360
water binding 360, 362

Electron crystallography 297
Electron carriers 124-125
Endothiapepsin-inhibitor complex 128
Energy calculations 94
Energy minimisation 104, 167, 171, 177, 248, 253, 260
Enthalpy-entropy compensation 455-457
Erabutoxin 122
Ewald summation 173, 182

Flavodoxin 103
Force fields
AMBER 172, 174, 177, 181, 184, 186, 187, 189, 194, 205, 207, 257
CHARMM 172, 175, 177, 185, 187-200, 204, 206
GROMOS 172, 184, 189, 191, 194, 197
Free energy
adhesion, of 395
calculations 253, 255
cohesion 395
hydration 352-353, 430ff
lipid hydration 352-355

Galactomannan 301, 309
Gellan 301
Glucan 301, 306
Glutathione reductase 122
Glycerophospholipids 338
Glycogen phosphorylase 325
Glycosaminoglycans 313
Group additivity 438-439, 446-449

Haemoglobin 133
Heat capacity, molar, Cp 46
Hoogsteen, or non-Watson-Crick pairs 204, 227, 235, 239
Hydration
alkane 342
alkene 342
apolar chain 342
chain fluidity effect 346
chain length effect 342
complex 348
decay length 352, 374
discreteness 346
enthalpy 342
entropy 340, 342
fatty alcohol 342
forces
description of 349, 351, 356, 374, 393ff
for thick interfaces 356
measured 375
spatial decay 356, 374ff
headgroup 340, 351, 355, 359, 369, 372ff
hydrocarbon 342
ion 342, 349
layer 344ff
main chain 69
α-helix 70-71, 73-74, 149
3_{10}-helix 154-155
ß-sheet 72, 76-79, 116, 155
ß-turn 75, 79-80, 154-155
models 342, 351ff
nucleic acid
base 231, 233, 235, 238, 244, 246-248, 250, 256
base pairs 231, 244, 246-248, 250, 256
number of DNA 266, 277

pattern 343, 347
profile 350, 352, 355, 360, 375
relaxation of DNA hydration shell 278-284, 288-289
role in protein folding 113-114, 159-160
secondary structure, distorsion, disruption 151, 155, 157
secondary structures of proteins 149, 154-155
shell 123, 179, 244, 255, 344ff
side chain 75, 84, 86-88
apolar 89, 91
aromatic 90
charged 78, 81-83
polar 81
temperature effects 348, 366, 368
see also Solvation
Hydrocarbon
chain-length effects 371
configuration 370
fluidization 366, 370
interdigitation 341, 367
rotation 367
unsaturation 371, 374
Hydrogen
bonding capacity 66
bonds 98, 247, 252, 256, 257, 259, 341, 352, 355, 359
3-centered 149
criterion 65
disruption by water 152
lifetime, τ_{HB} 50
maximization of 37-38
network 45, 53, 56, 365-366
profile 342
Hydrophilicity 338, 340, 342, 345, 349, 353, 403-406
Hydrophobic
effect 110, 121, 340, 349
hydration 55
interaction 55, 56, 256
scale 442
Hydrophobicity 403-406
Hydroxyl group, hydration 340, 355
Hydroxylase 128, 131

Ice
amorphous structure 18
I hexagonal structure 12
I cubic structure 15
II-IX structures 16-18
polymorphs 11
X 12
Iceberg 55
Insulin 108, 110
Interactions
apolar 395
Brownian movement 393
Debye 395
electrostatic 137, 400-412
interfacial 398
Keesom 395
Lewis acid-base 393, 397, 411
London 395
Lifshitz-van der Waals 393-395
'solvophobic' 419
Interface
geometry 360
hydration 359
thickness 345, 356, 358
Interfacial tensions 401
Interleukin 135
Intermolecular complexes 124-129
Ion
anions and cations 350
binding 349
polyvalent 350
specificity 349
Isotope effect, dynamic 52

Larmor frequency ϖ 52
Lectin, *Lathyrus ochrus* 328ff
Limiting hydration
cerebrosides 340
for pretransition 369
hydroxyl 341
phosphatidylcholine 359, 361
phosphatidylethanolamine 359, 361
Lipid
aggregation 345, 373
apolar 338
area 359
bilayer
activation energy of 365
permeability of 364ff
temperature dependence 365
biodistribution 338ff
charges 339, 370ff
chemical structure 339

complex 348, 366
cooperative unit 366
crystal structure of 341, 365
dehydration 376, 377
diffusion, lateral 357
dipole 354
dry 341, 356
dynamics 348, 355, 373
excess charge 354
flip-flop 367
fluidization of chains 366
headgroups
area 370
charges 343
fluctuations 356-357, 376
ionization 373ff
length 373
mobility 369
polarity 369
protonation 373
rotation 367
structure 339
hexagonal phase, inverted 370
hydrophilicity
effective 353-354
temperature dependence 348, 366, 368
hydrophobicity 340
interdigitation 367
ion binding 349
lattice 347
limiting hydration 359, 361
mobility 367
nomenclature 338
non-charged 339
non-lamellar phase 370
occurrence 338
packing parameters 358, 359
phase transitions 348
polar 338, 355
polar heads 338, 340
polymorphism 368
positional correlations and water 365
pretransition 369
solubility 345, 374
structure 339, 365
subtransition 369
Lysophopholipids 344, 366
hydration 344
Lysozyme 99-100, 110, 112, 132, 324
hydration 99
phage T4 101, 115, 118, 134

Mannan 301, 309
Mercerization 311
Model 351ff
Monte Carlo 344-345, 350
non-local electrostatics 352
perturbative 372
quantum mechanical 342
thermodynamic 352
Modelling of solvent
disordered water structure 7, 256, 260
knowledge-based 92
Molecular dynamics 63, 101, 104, 129-133, 137, 171ff, 183, 199, 245, 256, 258, 260
modelling 104, 296
orbital calculations 245, 246, 259
restrained dynamics 200ff
surface 98
Monte Carlo simulations 174, 179, 245, 248, 253-258, 260
Mutagenesis, site-directed 133-135
Myohemerythrin 107

Neurotoxin 128, 130
Neutron
crystal diffraction 31, 109-110
ices, hydrates 6
scattering
inelastic 52
quasielastic 52
Nigeran 301
Nuclear magnetic resonance 51-57, 101, 135, 169-170, 349, 360-361
Nuclear Overhauser effect 197ff
Nuclease, staphylococcal 135-136

Occupancy, factor 107

Packing efficiency 54
Pair correlation function PCF 8
Pairwise correlation 445
Papain 116
Particle number density 46
Penicillopepsin 120
Pentagonal arrangements 121, 122
Periplasmic binding proteins 322ff
Phosphate group hydration 175, 181, 229-232, 244, 246-250, 255-258, 260, 340, 344, 349-350, 354, 372

Phosphatidylcholine
biological occurrence 338
chain melting 348, 371
discreteness of hydration 346
headgroup 339, 341
hydration of 346, 358-359, 361
hydration energy of 346, 350
limiting hydration 359, 361
packing properties 358-359
phase diagram 371
phase transition 346
pretransition 369
structure 341
subtransition 368
temperature effects 348
Phosphatidylethanolamine
biological occurrence 338
discreteness of hydration 346
headgroup 339, 341
hydration of 346, 358-359
hydrogen bonding 366
limiting hydration of 359, 361
packing 358-359
phase diagram 371
phase transition 346
structure 341
temperature effects 348
Phosphatidylserine 338, 351, 376ff
Phosphodiester backbone hydration 228, 231, 246, 248, 258
Phosphofructokinase 115, 119
Phospholipase 124
Polar
charge distribution 343
group fluctuations 348, 355, 357
heads, lipid 338-339
water binding sites 340
Polarity profile 356, 375
Polarization
orientation 359
profile 350, 352, 355, 360, 375-376
Potential
dipolar 375
electrostatic 349, 372
energy function 245, 247, 248, 254, 257-258, 260
energy surface 54
hydration 349, 354ff
interaction 345
Prealbumin 116
Pressure
high 45, 48
osmotic 413
Protease
A 100, 133
aspartic 102, 115
complexes 125
inhibitors 125, 127
serine 124
Protein
crystals 63
Data Bank 112-113
folding 63
hydration 63, 148
mobility 63
water-sites 66
Pseudorotation 174-175

Raman spectroscopy 268-269
Random network model 45, 52
Rayleigh scattering, depolarized 50
Refinement, of X-ray models 104-109, 137
Resolution
atomic 98
high 101-102, 106, 112, 125, 137
low 102, 110
Relaxation
Debye type 50
dielectric 50, 346, 360
DNA hydration shell (of) 278-284, 288-289
electric quadrupole 51
NMR 50, 51
spin-lattice T_1 55
Retardation effects 396
Ribonuclease 111, 130-131
Ribose hydration 228, 235, 246, 257
Rubredoxin 104, 109, 112-113, 121, 137

Salt
bridges 98-99
ions 112
Scaled particle theory 442
Self-diffusion, coefficient 48, 52
Solubility 450
parameter 413
polymer 417
Solute-solvent interaction 432
Solvation
enthalpy 432
entropy 432, 435

processes 431-432
shells 50
thermodynamics 430-499
time 361
volume 432, 435
Solvent
effects
on association 431, 452
on protein folding 431, 435
induced force 453-455
molecule distribution 67
non-polar 256
Sphingolipids 338, 366
Stacking geometries 226, 239
Starch 295-296, 304-308, 316
Substrate
binding 99, 124, 139
mimicry 99, 100, 128
Subtilisin 134
Sulphate binding protein 117
Synchroton radiation 102

Thermitase 103
Transfer RNA 183, 231, 235, 237
complex with synthetase 231, 233
Trypsin
bovine 106
inhibitor 105, 112, 135
Streptomyces griseus 115
Tryptophan repressor/operator 129

Vitamin B12 coenzyme 31

Water
activity coefficient 353, 357-358
biological relevance 3, 45, 376
bound 98-101, 112, 124, 135
bridges 228, 231ff, 244, 246-250, 256, 258
bulk 99-101, 108-109, 123-124, 131-137
buried 125
clusters 99
complex permittivity 50, 412
concentration 352
constitutive 99, 112, 125, 137
decay length, structure 352
density 46, 346ff
density oscillations 346
diffusion 48-50, 361
insertion into
1—2 hydrogen bond 156
1—3 hydrogen bond 155
1—4 hydrogen bond 150
intermolecular correlations 352
layer 344, 347
model
EMPWI 248
MCY 259
SPC 189
TIP3P 174, 189
TIP4P 257
TIPS2 248, 257
monoalcohol mixtures 55
network 112, 117-123, 128, 130, 135, 138, 257, 260
filament 226, 255, 257
pentagons 177, 248, 258
spine of hydration 179, 226, 227, 228, 253, 255, 257
order decay length 353, 356, 376
ordered 104, 109, 137, 227
orientation 359
permeabiliy in lipid bilayers 364-365
phase transition shift 371
polarization 353, 359
regularities in structure
non-bonded 21
structural 21
relaxation 361
rotation 361-362
rotational
correlation time τ_1, τ_2 50, 52
diffusion model 50, 52
mobility 55
shell 344
sorption 350, 352
structure of
continuum models 45
D- 46, 52, 57
I- 46
mixture models 45
V- 46, 52
temperature, of maximum density, T_{MD} 46
tetrahedral, local arrangement 54
translational
mobility 50, 55
motion 361-362
virtual charge 353
viscosity, η 48, 52, 362

Xanthan 301
X-ray
crystal diffraction 5, 98, 102-103
fibre diffraction 296, 301, 306, 309, 313

Xylan 301

Young's equation 399

Zero point vibrations 46